METHODS IN MOLECULAR BIOLOGY

Series Editor
John M. Walker
School of Life and Medical Sciences
University of Hertfordshire
Hatfield, Hertfordshire, AL10 9AB, UK

For further volumes:
http://www.springer.com/series/7651

Hepatitis C Virus Protocols

Edited by

Mansun Law

Department of Immunology and Microbiology, The Scripps Research Institute, La Jolla, CA, USA

Humana Press

Editor
Mansun Law
Department of Immunology and Microbiology
The Scripps Research Institute
La Jolla, CA, USA

ISSN 1064-3745 ISSN 1940-6029 (electronic)
Methods in Molecular Biology
ISBN 978-1-4939-8975-1 ISBN 978-1-4939-8976-8 (eBook)
https://doi.org/10.1007/978-1-4939-8976-8

Library of Congress Control Number: 2018964717

Cover illustration: The protein complex structure illustrates binding of broadly neutralizing antibody AR3C (ribbon) to the neutralizing face of HCV E2 core domain (blue) surrounded by N-glycans (grey). Background is a super-resolution microscopy image of HCV-infected cells labelled by an anti-E2 antibody (red). Image designed by Joe Grove, Christina Corbaci, Leopold Kong, and Mansun Law.

This Humana Press imprint is published by the registered company Springer Science+Business Media, LLC, part of Springer Nature.
The registered company address is: 233 Spring Street, New York, NY 10013, U.S.A.

Foreword

The history of hepatitis virus discovery has depended greatly on the application of emerging research technologies, which has culminated in the discovery of five major hepatitis viruses within the last 50 years. Such breakthrough methods included electron microscopy in the discovery of HAV, ouchterlony rocket immunodiffusion in the identification of HBV, cDNA expression immunoscreening in the cases of HCV and HEV, and tissue immunofluorescence staining in the discovery of HDV. Not only have emerging methods been instrumental in hepatitis virus discovery but of course they, along with the huge progress in genomics, proteomics, and structural biology, have been instrumental in gaining knowledge of how these agents work and interact in the body. As such, it is vital that new technologies and methods are continually updated and appraised constantly by the research community since such activity is critical to the innovative scientific process. This excellent volume describes in detail a comprehensive and detailed selection of methods now available in HCV research. Not only will the application of these methods teach us more about HCV pathogenesis and help us to develop preventative strategies but they will also be of great relevance to other viruses, including, crucially, ones that are sure to emerge in the future with drastic consequences for mankind. While we are now better placed than ever to combat emerging viral disease, the regularity of their occurrence and the huge challenges that still exist in developing preventative strategies make this volume of HCV methodologies most timely and highly relevant.

Edmonton, AB, Canada *Michael Houghton*

Preface

This third edition of *Hepatitis C Virus Protocols* targets readers interested in basic virology, host-microbe interactions, and antiviral drug and vaccine research. Hepatitis C virus (HCV) is a blood-borne virus responsible for approximately half a million deaths from liver cancer and end-stage liver diseases each year. Until 2011, HCV was the sole member of the *Hepacivirus* genus in the *Flaviviridae* virus family. New hepaciviruses and the genetically related pegiviruses have now been found in nonhuman primates and many other species, revealing HCV-like viruses are much more commonly circulated than previously thought. Although the impact of these findings on human and veterinary health is yet to be determined, the research strategies developed to study HCV are generally applicable to the study of these new viruses.

This volume collects the most updated concepts and experimental protocols developed by leading researchers in the field. The book chapters are organized into five topics:

1. Review of hepatitis C virus and bioinformatic tools
2. Methods for HCV cloning, culture, and purification
3. Methods for the study of HCV life cycle
4. Methods for the study of host immune responses
5. Small animal models

The book focuses on providing an easy guide to readers explaining the essential methods for the study of this interesting virus, and the experimental systems relevant for vaccine development. A broadly effective HCV vaccine is an unmet public health need for the eradication of this human disease. The scientific challenges in designing a vaccine against antigenically variable virus are shared by other viruses including HIV and influenza. Since the publication of the last edition, we have witnessed important conceptual and technological breakthroughs in the study of antibody and T cell responses to HCV to facilitate vaccine development. HCV will continue to serve as an important model for the study of basic virology, virus pathogenesis, human immunology, and vaccinology.

La Jolla, CA, USA *Mansun Law*

Acknowledgments

I sincerely thank all the authors for their advice, support, and patience in the preparation of this book. My effort was supported partly by NIH grants AI079031, AI106005, AI123365, and AI123861.

Contents

Foreword ... *v*
Preface ... *vii*
Acknowledgments ... *ix*
Contributors ... *xv*

PART I REVIEW OF HEPATITIS C VIRUS AND BIOINFORMATIC TOOLS

1 Overview of Direct-Acting Antiviral Drugs and Drug Resistance of Hepatitis C Virus ... 3
Darrick K. Li and Raymond T. Chung

2 Similarities and Differences Between HCV Pseudoparticle (HCVpp) and Cell Culture HCV (HCVcc) in the Study of HCV ... 33
Laura Riva and Jean Dubuisson

3 Hepatitis C Virus Database and Bioinformatics Analysis Tools in the Virus Pathogen Resource (ViPR) ... 47
Yun Zhang, Christian Zmasek, Guangyu Sun, Christopher N. Larsen, and Richard H. Scheuermann

PART II METHODS FOR VIRUS CLONING, CULTURE, AND PURIFICATION

4 Establishment of Replication-Competent HCV Strain with Minimum Modifications ... 73
Takanobu Kato, Asako Murayama, and Takaji Wakita

5 Full-Length Open Reading Frame Amplification of Hepatitis C Virus ... 85
Ulrik Fahnøe and Jens Bukh

6 InFusion Cloning for the Generation of Biologically Relevant HCV Chimeric Molecular Clones ... 93
Barnabas King, Richard Urbanowicz, Alexander W. Tarr, Jonathan K. Ball, and C. Patrick McClure

7 Production and Purification of Cell Culture Hepatitis C Virus ... 105
Cynthia de la Fuente and Maria Teresa Catanese

8 Embryonic or Induced Pluripotent Stem Cell-Derived Hepatocellular Systems for HCV Culture ... 121
Xianfang Wu and Viet Loan Dao Thi

PART III METHODS FOR THE STUDY OF HCV LIFE CYCLE

9 Molecular Identification of Transmitted/Founder Hepatitis C Viruses and Their Progeny by Single Genome Sequencing ... 139
Hui Li, Muhammad N. Zahid, Shuyi Wang, and George M. Shaw

10 A Cell Culture Model for Persistent HCV Infection 157
Victoria Castro, Ginés Ávila-Pérez, Lidia Mingorance, and Pablo Gastaminza

11 Expression Cloning of Host Factors Required for the HCV Replication Cycle 169
Sandra Ciesek and Thomas von Hahn

12 Application of Deep Mutational Scanning in Hepatitis C Virus 183
Nicholas C. Wu and Hangfei Qi

13 Studying the Hepatitis C Virus-Induced Epigenetic Signature After Cure with Direct-Acting Antivirals 191
Shira Perez and Meital Gal-Tanamy

14 Measuring Hepatitis C Virus Envelopment by Using a Proteinase K Protection Assay 209
Allison E. Roder and Stacy M. Horner

15 Membrane Fusion Assays for Studying Entry Hepatitis C Virus into Cells 219
Solène Denolly, François-Loïc Cosset, and Natalia Freitas

16 Functional Characterization of Apolipoproteins in the HCV Life Cycle 235
Luhua Qiao and Guangxiang (George) Luo

17 Investigating Hepatitis C Virus Infection Using Super-Resolution Microscopy 247
Pedro Matos Pereira, Caron Jacobs, and Joe Grove

18 Live Cell Imaging of Hepatitis C Virus Trafficking in Hepatocytes 263
Yasmine Baktash and Glenn Randall

19 Cloning and Analysis of Authentic Patient-Derived HCV E1/E2 Glycoproteins 275
Richard A. Urbanowicz, Jonathan K. Ball, and Alexander W. Tarr

20 Simultaneous Quantification of Hepatitis C Virus Envelope Glycoproteins E1 and E2 by Dual-Color Fluorescence Immunoblot Analysis 295
Fang Chen, Erick Giang, Radhika Gopal, and Mansun Law

21 Overcoming Challenges of Hepatitis C Virus Envelope Glycoprotein Production in Mammalian Cells 305
Samantha A. Yost, Jillian Whidby, Abdul G. Khan, Yuanyuan Wang, and Joseph Marcotrigiano

PART IV METHODS FOR STUDYING IMMUNE RESPONSES

22 Monitoring of Interferon Response Triggered by Cells Infected by Hepatitis C Virus or Other Viruses Upon Cell–Cell Contact 319
Séverin Coléon, Sonia Assil, and Marlène Dreux

23 Complement Regulation and Immune Evasion by Hepatitis C Virus 337
Young-Chan Kwon and Ranjit Ray

24 Direct Ex Vivo Functional Analysis of HCV-Specific T Cells 349
Su-Hyung Park and Eui-Cheol Shin

25 Simultaneous Analysis of HCV-Specific $CD4^+$ and $CD8^+$ T Cells by Multicolor Flow Cytometry 363
David Wolski and Georg M. Lauer

26 Probing the Antigenicity of HCV Envelope Glycoproteins by Phage Display Antibody Technology 381
Erick Giang, Fernando Aleman, and Mansun Law

27 Isolation of HCV Neutralizing Antibodies by Yeast Display 395
Zhen-yong Keck, Yong Wang, Patrick Lau, and Steven K. H. Foung

28 Detection of Antibodies to HCV E1E2 by Lectin-Capture ELISA 421
Marian Major and Mansun Law

29 In Vitro Neutralization Assay Using Cultured Hepatitis C Virus 433
Jannick Prentoe and Jens Bukh

30 Standardized Method for the Study of Antibody Neutralization of HCV Pseudoparticles (HCVpp) 441
Justin R. Bailey, Richard A. Urbanowicz, Jonathan K. Ball, Mansun Law, and Steven K. H. Foung

Part V Small Animal Models

31 Detection of Antigen-Specific $CD8^+$ T Cells in the Livers of HCV Core Transgenic Mice 453
Dustin A. Cobb, Aditya P. Dandekar, and Young S. Hahn

32 Mice with Chimeric Human Livers and Their Applications 459
Donna N. Douglas and Norman M. Kneteman

33 Mouse Models for Studying HCV Vaccines and Therapeutic Antibodies 481
Jenna M. Gaska, Qiang Ding, and Alexander Ploss

Index *505*

Contributors

FERNANDO ALEMAN • *Department of Immunology and Microbiology, The Scripps Research Institute, La Jolla, CA, USA; Navega Therapeutics Inc, La Jolla, CA, USA*

SONIA ASSIL • *CIRI, Inserm, U1111, Université Claude Bernard Lyon 1, CNRS, UMR5308, École Normale Supérieure de Lyon, Univ Lyon, Lyon, France*

GINÉS ÁVILA-PÉREZ • *Department of Molecular and Cellular Biology, Centro Nacional De Biotecnología-Consejo Superior de Investigaciones Científicas (CNB-CSIC), Madrid, Spain*

JUSTIN R. BAILEY • *Division of Infectious Diseases, Department of Medicine, Johns Hopkins University, Baltimore, MD, USA*

YASMINE BAKTASH • *Department of Microbiology, The University of Chicago, Chicago, IL, USA*

JONATHAN K. BALL • *School of Life Sciences, The University of Nottingham, Nottingham, UK; NIHR Biomedical Research Unit in Gastrointestinal and Liver Diseases, Nottingham University Hospitals NHS Trust, The University of Nottingham, Nottingham, UK*

JENS BUKH • *Copenhagen Hepatitis C Program (CO-HEP), Department of Infectious Diseases, Hvidovre Hospital, Hvidovre, Denmark; Department of Immunology and Microbiology, Faculty of Health and Medical Sciences, University of Copenhagen, Copenhagen, Denmark*

VICTORIA CASTRO • *Department of Molecular and Cellular Biology, Centro Nacional De Biotecnología-Consejo Superior de Investigaciones Científicas (CNB-CSIC), Madrid, Spain*

MARIA TERESA CATANESE • *Nouscom SRL, Rome, Italy; Department of Infectious Diseases, King's College London, London, UK*

FANG CHEN • *Department of Immunology and Microbiology, The Scripps Research Institute, La Jolla, CA, USA*

RAYMOND T. CHUNG • *Liver Center and Division of Gastroenterology, Department of Medicine, Massachusetts General Hospital, Boston, MA, USA*

SANDRA CIESEK • *Institute of Virology, University Hospital Essen, University of Duisburg-Essen, Essen, Germany; German Center for Infection Research (DZIF), External Partner Site Essen, Essen, Germany*

DUSTIN A. COBB • *Beirne B. Carter Center for Immunology Research, University of Virginia, Charlottesville, VA, USA*

SÉVERIN COLÉON • *CIRI, Inserm, U1111, Université Claude Bernard Lyon 1, CNRS, UMR5308, École Normale Supérieure de Lyon, Univ Lyon, Lyon, France*

FRANÇOIS-LOÏC COSSET • *CIRI–International Center for Infectiology Research, Team EVIR, Inserm, U1111, Université Claude Bernard Lyon 1, CNRS, UMR5308, Ecole Normale Supérieure de Lyon, Univ Lyon, Lyon, France*

ADITYA P. DANDEKAR • *Beirne B. Carter Center for Immunology Research, University of Virginia, Charlottesville, VA, USA*

CYNTHIA DE LA FUENTE • *National Center for Biodefense and Infectious Diseases, George Mason University, Manassas, VA, USA*

Solène Denolly • *CIRI–International Center for Infectiology Research, Team EVIR, Inserm, U1111, Université Claude Bernard Lyon 1, CNRS, UMR5308, Ecole Normale Supérieure de Lyon, Univ Lyon, Lyon, France*

Qiang Ding • *School of Medicine, Tsinghua University, Beijing, China; Lewis Thomas Laboratory, Department of Molecular Biology, Princeton University, Princeton, NJ, USA*

Donna N. Douglas • *Department of Surgery, Faculty of Medicine and Dentistry, University of Alberta, Edmonton, AB, Canada*

Marlène Dreux • *CIRI, Inserm, U1111, Université Claude Bernard Lyon 1, CNRS, UMR5308, École Normale Supérieure de Lyon, Univ Lyon, Lyon, France*

Jean Dubuisson • *CIIL–Centre d'Infection et d'Immunité de Lille, Institut Pasteur de Lille, U1019-UMR 8204, Univ. Lille, CNRS, Inserm, CHU Lille, Lille, France*

Ulrik Fahnøe • *Copenhagen Hepatitis C Program (CO-HEP), Department of Infectious Diseases, Copenhagen University Hospital, Hvidovre, Denmark; Department of Immunology and Microbiology, Faculty of Health and Medical Sciences, University of Copenhagen, Copenhagen, Denmark*

Steven K. H. Foung • *Department of Pathology, Stanford University School of Medicine, Stanford, CA, USA*

Natalia Freitas • *CIRI–International Center for Infectiology Research, Team EVIR, Inserm, U1111, Université Claude Bernard Lyon 1, CNRS, UMR5308, Ecole Normale Supérieure de Lyon, Univ Lyon, Lyon, France*

Meital Gal-Tanamy • *The Azrieli Faculty of Medicine, Bar-Ilan University, Safed, Israel*

Jenna M. Gaska • *Lewis Thomas Laboratory, Department of Molecular Biology, Princeton University, Princeton, NJ, USA*

Pablo Gastaminza • *Department of Molecular and Cellular Biology, Centro Nacional De Biotecnología-Consejo Superior de Investigaciones Científicas (CNB-CSIC), Madrid, Spain*

Erick Giang • *Department of Immunology and Microbiology, The Scripps Research Institute, La Jolla, CA, USA*

Radhika Gopal • *Department of Immunology and Microbiology, The Scripps Research Institute, La Jolla, CA, USA*

Joe Grove • *Division of Infection and Immunity, Institute of Immunity and Transplantation, University College London, London, UK*

Young S. Hahn • *Beirne B. Carter Center for Immunology Research, University of Virginia, Charlottesville, VA, USA; Department of Microbiology, Immunology, and Cancer Biology, University of Virginia, Charlottesville, VA, USA*

Stacy M. Horner • *Department of Molecular Genetics and Microbiology, Duke University Medical Center, Durham, NC, USA; Department of Medicine, Duke University Medical Center, Durham, NC, USA*

Caron Jacobs • *Quantitative Imaging and Nanobiophysics Group, MRC Laboratory for Molecular Cell Biology, Department of Cell and Developmental Biology, University College London, London, UK*

Takanobu Kato • *Department of Virology II, National Institute of Infectious Diseases, Tokyo, Japan*

Zhen-yong Keck • *Department of Pathology, Stanford University School of Medicine, Stanford, CA, USA*

Abdul G. Khan • *Department of Chemistry and Chemical Biology, Center for Advanced Biotechnology and Medicine, Rutgers University, Piscataway, NJ, USA*

BARNABAS KING • *School of Life Sciences, The University of Nottingham, Nottingham, UK; NIHR Biomedical Research Unit in Gastrointestinal and Liver Diseases, Nottingham University Hospitals NHS Trust, The University of Nottingham, Nottingham, UK*
NORMAN M. KNETEMAN • *Department of Surgery, Faculty of Medicine and Dentistry, University of Alberta, Edmonton, AB, Canada*
YOUNG-CHAN KWON • *Department of Internal Medicine, Saint Louis University, St. Louis, MO, USA; Department of Molecular Microbiology and Immunology, Saint Louis University, St. Louis, MO, USA; Institut Pasteur Korea, Daejeon, Republic of Korea*
CHRISTOPHER N. LARSEN • *Vecna Technologies, Greenbelt, MD, USA*
PATRICK LAU • *Department of Pathology, Stanford University School of Medicine, Stanford, CA, USA*
GEORG M. LAUER • *Gastrointestinal Unit and Liver Center, Massachusetts General Hospital and Harvard Medical School, Boston, MA, USA*
MANSUN LAW • *Department of Immunology and Microbiology, The Scripps Research Institute, La Jolla, CA, USA*
DARRICK K. LI • *Department of Medicine, Massachusetts General Hospital, Boston, MA, USA*
HUI LI • *Perelman School of Medicine, University of Pennsylvania, Philadelphia, PA, USA*
GUANGXIANG (GEORGE) LUO • *Department of Microbiology, University of Alabama at Birmingham School of Medicine, Birmingham, AL, USA*
MARIAN MAJOR • *Division of Viral Products, Center for Biologics Evaluation and Research, Food and Drug Administration, Silver Spring, MD, USA*
JOSEPH MARCOTRIGIANO • *Laboratory of Infectious Diseases, National Institute of Allergy and Infectious Diseases, National Institute of Health, Bethesda, MD, USA*
C. PATRICK MCCLURE • *School of Life Sciences, The University of Nottingham, Nottingham, UK; NIHR Biomedical Research Unit in Gastrointestinal and Liver Diseases, Nottingham University Hospitals NHS Trust, The University of Nottingham, Nottingham, UK*
LIDIA MINGORANCE • *Department of Molecular and Cellular Biology, Centro Nacional De Biotecnología-Consejo Superior de Investigaciones Científicas (CNB-CSIC), Madrid, Spain*
ASAKO MURAYAMA • *Department of Virology II, National Institute of Infectious Diseases, Tokyo, Japan*
SU-HYUNG PARK • *Laboratory of Translational Immunology and Vaccinology, Graduate School of Medical Science and Engineering, Korea Advanced Institute of Science and Technology (KAIST), Daejeon, Republic of Korea*
PEDRO MATOS PEREIRA • *Quantitative Imaging and Nanobiophysics Group, MRC Laboratory for Molecular Cell Biology, Department of Cell and Developmental Biology, University College London, London, UK*
SHIRA PEREZ • *The Azrieli Faculty of Medicine, Bar-Ilan University, Safed, Israel*
ALEXANDER PLOSS • *Lewis Thomas Laboratory, Department of Molecular Biology, Princeton University, Princeton, NJ, USA*
JANNICK PRENTOE • *Copenhagen Hepatitis C Program (CO-HEP), Department of Infectious Diseases, Hvidovre Hospital, Hvidovre, Denmark; Department of Immunology and Microbiology, Faculty of Health and Medical Sciences, University of Copenhagen, Copenhagen, Denmark*
HANGFEI QI • *Department of Molecular and Medical Pharmacology, University of California, Los Angeles, CA, USA*

LUHUA QIAO • *Department of Microbiology, University of Alabama at Birmingham School of Medicine, Birmingham, AL, USA*

GLENN RANDALL • *Department of Microbiology, The University of Chicago, Chicago, IL, USA*

RANJIT RAY • *Department of Internal Medicine, Saint Louis University, St. Louis, MO, USA; Department of Molecular Microbiology and Immunology, Saint Louis University, St. Louis, MO, USA*

LAURA RIVA • *CIIL–Centre d'Infection et d'Immunité de Lille, Institut Pasteur de Lille, U1019-UMR 8204, Univ. Lille, CNRS, Inserm, CHU Lille, Lille, France*

ALLISON E. RODER • *Department of Molecular Genetics and Microbiology, Duke University Medical Center, Durham, NC, USA*

RICHARD H. SCHEUERMANN • *J. Craig Venter Institute, La Jolla, CA, USA; Department of Pathology, University of California, San Diego, CA, USA; Division of Vaccine Discovery, La Jolla Institute for Allergy and Immunology, La Jolla, CA, USA*

GEORGE M. SHAW • *Perelman School of Medicine, University of Pennsylvania, Philadelphia, PA, USA*

EUI-CHEOL SHIN • *Laboratory of Immunology and Infectious Diseases, Graduate School of Medical Science and Engineering, Korea Advanced Institute of Science and Technology (KAIST), Daejeon, Republic of Korea*

GUANGYU SUN • *Vecna Technologies, Greenbelt, MD, USA*

ALEXANDER W. TARR • *School of Life Sciences, The University of Nottingham, Nottingham, UK; NIHR Biomedical Research Unit in Gastrointestinal and Liver Diseases, Nottingham University Hospitals NHS Trust, The University of Nottingham, Nottingham, UK*

VIET LOAN DAO THI • *Center for Integrative Infectious Disease Research, University Hospital Heidelberg, Heidelberg, Germany; Laboratory of Virology and Infectious Diseases, Center for the Study of Hepatitis C, The Rockefeller University, New York, NY, USA*

RICHARD A. URBANOWICZ • *School of Life Sciences, The University of Nottingham, Nottingham, UK; NIHR Biomedical Research Unit in Gastrointestinal and Liver Diseases, Nottingham University Hospitals NHS Trust, The University of Nottingham, Nottingham, UK*

THOMAS VON HAHN • *Department of Gastroenterology, Hepatology und Endocrinology, Hannover Medical School, Hannover, Germany; Institute for Molecular Biology, Hannover Medical School, Hannover, Germany; German Center for Infection Research (DZIF), Hannover-Braunschweig Site, Hannover, Germany*

TAKAJI WAKITA • *Department of Virology II, National Institute of Infectious Diseases, Tokyo, Japan*

SHUYI WANG • *Perelman School of Medicine, University of Pennsylvania, Philadelphia, PA, USA*

YONG WANG • *Department of Pathology, Stanford University School of Medicine, Stanford, CA, USA*

YUANYUAN WANG • *Department of Chemistry and Chemical Biology, Center for Advanced Biotechnology and Medicine, Rutgers University, Piscataway, NJ, USA; Laboratory of Infectious Diseases, National Institute of Allergy and Infectious Diseases, National Institute of Health, Bethesda, MD, USA*

JILLIAN WHIDBY • *Department of Chemistry and Chemical Biology, Center for Advanced Biotechnology and Medicine, Rutgers University, Piscataway, NJ, USA*

DAVID WOLSKI • *Gastrointestinal Unit and Liver Center, Massachusetts General Hospital and Harvard Medical School, Boston, MA, USA*

NICHOLAS C. WU • *Department of Integrative Structural and Computational Biology, The Scripps Research Institute, La Jolla, CA, USA*
XIANFANG WU • *Laboratory of Virology and Infectious Diseases, Center for the Study of Hepatitis C, The Rockefeller University, New York, NY, USA*
SAMANTHA A. YOST • *Department of Chemistry and Chemical Biology, Center for Advanced Biotechnology and Medicine, Rutgers University, Piscataway, NJ, USA*
MUHAMMAD N. ZAHID • *Perelman School of Medicine, University of Pennsylvania, Philadelphia, PA, USA*
YUN ZHANG • *J. Craig Venter Institute, La Jolla, CA, USA*
CHRISTIAN ZMASEK • *J. Craig Venter Institute, La Jolla, CA, USA*

Part I

Review of Hepatitis C Virus and Bioinformatic Tools

Chapter 1

Overview of Direct-Acting Antiviral Drugs and Drug Resistance of Hepatitis C Virus

Darrick K. Li and Raymond T. Chung

Abstract

The advent of direct-acting antivirals (DAAs) has brought about a sudden renaissance in the treatment of chronic hepatitis C virus (HCV) infection with SVR rates now routinely >90%. However, due to the error-prone nature of the HCV RNA polymerase, resistance-associated substitutions (RASs) to DAAs may be present at baseline and can result in a significant effect on treatment outcomes and hamper the achievement of sustained virologic response. By further understanding the patterns and nature of these RASs, it is anticipated that the incidence of treatment failure will continue to decrease in frequency with the development of drug regimens with increasing potency, barrier to resistance, and genotypic efficacy. This review summarizes our current knowledge of RASs associated with HCV infection as well as the clinical effect of RASs on treatment with currently available DAA regimens.

Key words Direct-acting antiviral, Hepatitis C virus, Sustained virologic response, Resistance-associated substitution

1 HCV Virology

HCV is a member of the *Flavivirus* family (which also includes the yellow fever and dengue viruses) and is an enveloped (+)-strand RNA virus. The genome is approximately 9.6 kb in length and encodes a single large polyprotein which is ultimately cleaved to form ten proteins by cellular and viral proteases. These include three structural proteins, the nucleocapsid protein (C), and two envelope proteins (E1 and E2), as well as seven nonstructural proteins which include two proteins required for virion production (p7 and NS2) as well as five proteins that form the cytoplasmic viral replication complex (NS3, NS4A, NS4B, NS5A, and NS5B) (Fig. 1).

The following model synthesizes much of what is known to date [1]. HCV virions enter the hepatocyte via interaction with a number of co-receptors including CD81, claudin-1, occludin, and SR-B1 and are endocytosed into the cell. Following entry, the

Mansun Law (ed.), *Hepatitis C Virus Protocols*, Methods in Molecular Biology, vol. 1911,
https://doi.org/10.1007/978-1-4939-8976-8_1,

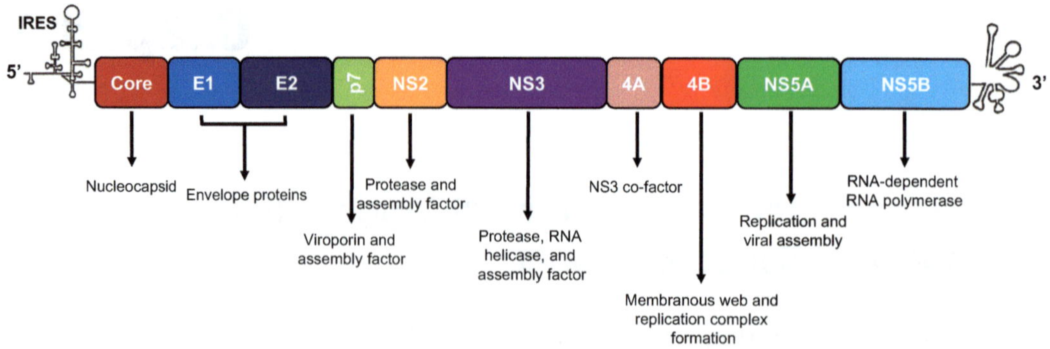

Fig. 1 Genomic organization of the hepatitis C virus. *IRES*, internal ribosome entry site, *NS* nonstructural protein

endosome becomes acidified which changes the conformation of the envelope proteins, releasing the viral (+)-strand RNA genome into the cytoplasm, which become associated with the ER. The RNA then becomes the template for the production of viral proteins. The envelope proteins are secreted into the lumen of the ER, while the core protein remains cytoplasmic. The replication complex of NS3, NS4A, NS4B, NS5A, and NS5B then forms "membranous webs" derived from the ER membrane and directs transcription of a (−)-strand genome which then becomes the template for further production of (+)-strand genomes, which are then packaged with the structural proteins to form mature virions, which are then released.

Given its central role in the viral life cycle, a number of the protein components of the viral replication complex have been a target for many of the effective antivirals that have recently been developed, in particular NS3, NS5A, and NS5B, all of which will be further described later. In brief, the NS3 protein functions as the key viral protease and is responsible for a number of the polypeptide processing events including the cleavage of the NS3/NS4A, NS4A/NS4B, NS4B/NS5A, and NS5A/NS5B junctions. This activity requires NS4A as a cofactor. NS5A is a membrane-bound RNA-binding protein whose precise role remains unclear but appears to play multiple essential roles in the regulation of viral replication, assembly, and exit. NS5B is the viral RNA-dependent RNA polymerase and the catalytic core of the viral replication machinery.

The NS5B polymerase lacks proofreading capability [2], and as such, this has led to the rapid accumulation of genetic diversity and the rise of at least six separate HCV genotypes (GTs). These GTs have important consequences for treatment as there are emerging differences in the response rate of various GTs to various antiviral regimens. The GTs also have geographic variability—genotype 1 (GT1) is the most widespread and is predominantly seen in

North America and is further split into two subtypes, GT1a and GT1b. GT2 is also widespread and is found principally in Central and West Africa. GT3 is found primarily in Asia; GT4 is found in the Middle East and Northern Africa; GT5 and GT6 are rare and can be found in regions of Africa and Asia [3].

2 A Brief History of HCV Treatment

The treatment of HCV has undergone a revolution. Indeed, until recent times, interferon (IFN)-based therapy had been the backbone of HCV therapy (Fig. 2). In fact, the first evidence of therapeutic efficacy for IFN-based therapy for HCV was performed even prior to its identification in a pilot study of ten patients for what was termed "non-A, non-B hepatitis," an entity originally described in 1976 [4–6]. Concurrent with the identification of HCV in 1989 [7], the first two randomized controlled trials for the use of IFN in HCV treatment were performed [8, 9]. In these trials, recombinant IFNα was given three times a week for 24 weeks, and treatment response was measured by a sustained normalization in alanine aminotransferase (ALT) levels in the serum. Only 10–25% of patients achieved a treatment response in these trials. IFN monotherapy was the standard of care for the next decade until the late 1990s, during which combination therapy of IFN-α and ribavirin led to the next step in the treatment of HCV. In a landmark study, 912 patients were randomized to subcutaneously injected IFNα2b with daily oral administration of ribavirin achieved on SVR (as measured by undetectable HCV RNA viral loads) in 38% of treated patients undergoing 48 weeks of therapy compared with 13% with IFNα2b monotherapy [10]. The introduction of

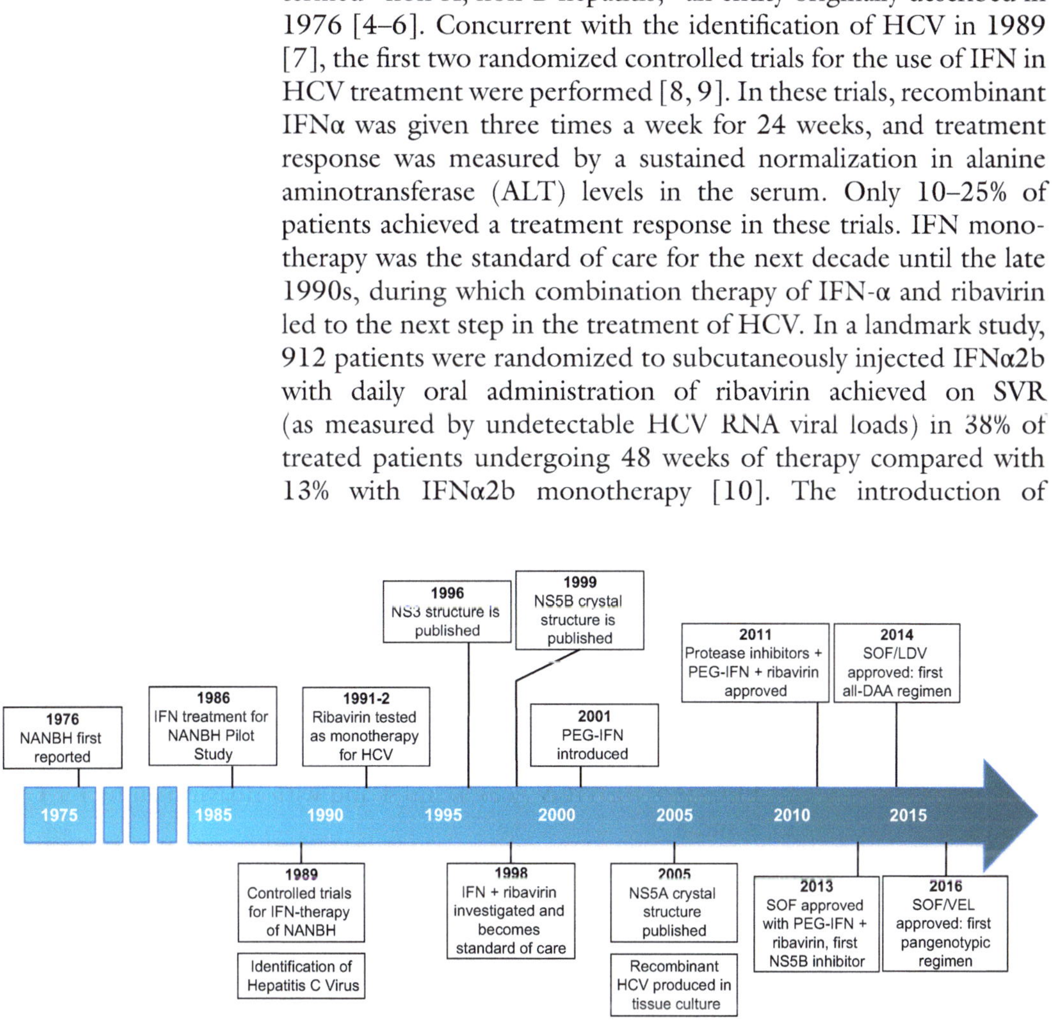

Fig. 2 Timeline of milestones in hepatitis C virus treatment. *NANBH* non-A, non-B hepatitis, *NS* nonstructural protein, *PEG-IFN* pegylated interferon; *SOF* sofosbuvir, *LDV* ledipasvir, *VEL* velpatasvir

pegylated IFN (PEG-IFN) in 2001, which had a longer half-life and more favorable pharmacokinetics in combination with ribavirin, led to additional incremental improvements in SVR rates to approximately 55% [11, 12]. For the next decade, treatment of HCV with PEG-IFN and ribavirin was the standard of care, though the treatment regimen continued to be plagued by variation in SVR rates with GT and viral load and significant side effects including asthenia, neutropenia, flu-like illness, cytopenias, and depression [13].

Significant advances in our understanding of the molecular and structural virology, life cycle, and pathogenesis of HCV as well as the ability to produce recombinant infectious HCV by tissue culture led directly to the development of the first DAAs [14]. These agents were the first-generation NS3/4A protease inhibitors, telaprevir (TVR) and boceprevir (BOC), which achieved SVR rates of 65–75% when used together with PEG-IFN and ribavirin [15, 16]. As such, they were approved by the FDA for use in "triple therapy" for HCV GT1 in 2011. Their approval was followed by a flurry of activity, notable for the development of several compounds targeting other stages of the HCV life cycle. Simeprevir (SMV), a once-daily NS3/4A protease inhibitor, was approved in 2013 to be used in combination with PEG-IFN and ribavirin for treatment of GT1, achieving comparable SVR rates as its predecessors, with better tolerability [17]. A major advance was the development of an NS5B polymerase inhibitor, sofosbuvir (SOF). SOF is a member of a family of nucleotide analogues that work by causing early chain termination after being incorporated into newly synthesized viral RNA [18]. Given its mechanism of action and the conservation of the NS5B RNA polymerase active site, it is active against all HCV GTs and has a high barrier to resistance, selecting only for viral mutants with exceedingly low replication fitness. As such, in a landmark trial enrolling individuals with predominantly HCV GT1 or GT4, SOF-anchored triple therapy was found to achieve SVR rates of 90% after 12 weeks of therapy (SVR12) [19]. Moreover, SVR12 rates of 95% and 82% were attained with SOF and ribavirin alone in treatment-naïve and treatment-experienced persons, respectively, with HCV GT2 or GT3 [19, 20]. Accordingly, in 2013, the FDA approved SOF for use as part of triple therapy with PEG-IFN for HCV GTs 1 and 4 and with ribavirin alone for GTs 2 and 3.

Of particular interest has been the development of all-oral IFN-free regimens utilizing two or more classes of DAAs to achieve the dual goal of rapid viral suppression and prevention of selection of resistant variants. This concept has been realized with the approval of SOF and ledipasvir (LDV, an NS5A inhibitor) by the FDA in October 2014 as a once-daily co-formulation for the treatment of HCV GT1. This was done on the basis of three pivotal trials that studied this combination with and without ribavirin in both treatment-naïve and treatment-experienced patients. These clinical studies found that irrespective of ribavirin use, individuals treated with this combination achieved SVR12 rates of 94–99% [21–23]. In addition, SOF/SMV for the treatment of HCV GT1 was approved by the FDA in November 2014, based on results from the COSMOS trial, which demonstrated >90% SVR12 rates and good safety and tolerability profiles [24]. Finally, the combination regimen of ombitasvir, ritonavir-boosted paritaprevir, and dasabuvir ± ribavirin was approved in December 2014 on the basis of several trials showing SVR12 rates >90% [25, 26]. Most recently, there has been a surge of approvals for a new generation of DAA regimens with increased antiviral potency and pan-genotypic efficacy including the approval of SOF and velpatasvir (VEL, a NS5A inhibitor). A list of currently approved IFN-sparing DAA regimens and additional DAAs that are currently in development can be found in Fig. 3

Published data regarding real-world experience with the new DAA regimens are rapidly accumulating, and preliminary findings have been encouraging. For instance, individuals with HCV GT1 treated with SOF/SMV ± ribavirin for 12 to 16 weeks also

Class	Name	Manufacturer	Status
NS3-4A inhibitors ("**-previr**")	Telaprevir	Jannsen, Mitsubishi	Approved (2011, now discontinued)
	Boceprevir	Merck	Approved (2011, now discontinued)
	Simeprevir	Janssen	Approved (2013)
	Vaniprevir	Merck	Approved (2014, only in Japan)
	Paritaprevir	AbbVie	Approved (2015)
	Asunaprevir	Bristol-Myers Squibb	Approved (2015, only in Asia, Middle East)
	Grazoprevir	Merck	Approved (2016)
	Glecaprevir	AbbVie	Approved (2017)
	Voxilaprevir	Gilead Sciences	Approved (2017)
NS5A inhibitors ("**-asvir**")	Ledipasvir	Gilead Sciences	Approved (2014)
	Ombitasvir	AbbVie	Approved (2014)
	Daclatasvir	Bristol-Myers Squibb	Approved (2015)
	Elbasvir	Merck	Approved (2016)
	Velpatasvir	Gilead Sciences	Approved (2016)
	Pibrentasvir	AbbVie	Approved (2017)
	Odalasvir (ACH-3102)	Janssen	Phase II
	Ravidasvir (PPI-668)	Presidio	Phase II
	MK-8408	Merck	Phase II
NS5B inhibitors ("**-buvir**")			
Nucleos(t)ide inhibitors	Sofosbuvir	Gilead Sciences	Approved (2013)
	MK-3682	Merck	Phase II
	VX-135	Vertex	Phase II
	ACH-3422	Achillion/Janssen	Phase I
	ALS-335	Janssen	Phase I
Non-nucleos(t)ide inhibitors	Dasabuvir	Abbvie	Approved (2014)
	Beclabuvir (BMS-791325)	Bristol-Myers Squibb	Phase III
	ABT-072	Abbvie	Phase II
	GS-9669	Gilead Sciences	Phase II
	TMC647055	Tibotec	Phase II
	MBX-700	Microbiotix/Merck	Phase I

Fig. 3 Approved direct-acting antivirals and current pipeline agents undergoing evaluation for chronic HCV infection. Unless otherwise indicated, approved drugs have been approved in the United States and the European Union

experienced high rates of SVR that were slightly decreased (84%, 675/802 patients) from what was seen in clinical trials, but with low rates of treatment discontinuation and serious adverse effects [27]. Recently, 2099 real-world patients with GT1 infection who received treatment with SOF/LDV ± ribavirin for 8 to 24 weeks were prospectively followed and were found to have extremely high SVR rates (≥95%) regardless of treatment duration [28]. Robust SVR rates have also been seen in real-world patients treated with various SOF-based regimens with GT2 [29] or GT3 infection [30] as well as those with advanced liver disease [31].

3 Principles of HCV Resistance to DAAs

As detailed above, the evolution of IFN-sparing, DAA-based treatment regimens for HCV has progressed at an incredible pace with encouraging real-world results. However, though most treated patients are able to achieve SVR, the phenomenon of DAA resistance has become increasingly appreciated. This has led to considerable interest in identifying common resistance-associated mutations, in understanding the biochemical mechanisms underlying viral resistance, and in developing new treatment strategies to treat individuals who fail initial therapy. The next sections of this chapter will highlight and discuss this rapidly evolving field in the study of HCV.

A key concept that underlies our emerging understanding of DAA resistance is the concept of a "quasispecies." The low fidelity of the HCV RNA polymerase combined with the high replication rate results in an extremely high number of different but closely related circulating HCV variants that can be observed in the plasma or liver at any time [32, 33]. The in vivo mixture of different but closely related variants is termed a "quasispecies." Each viral population that emerges from the process of random mutagenesis is then subject to selection based on the effect of the mutation(s) on overall viral fitness. Quasispecies theory stipulates that at any given timepoint, the exact distribution of viral populations within the quasispecies reflects an equilibrium between the replicative fitness of each variant, the continued generation of new variants, and the positive selective pressure applied by the environment [34]. Moreover, the quasispecies structure allows for a considerable evolutionary advantage by allowing the virus to rapidly adapt to a constantly changing milieu of stressors.

A virus's ability to evolve is thought to be determined by at least five interconnected parameters [35]. First, the development and emergence of RASs depend on the average mutation rate during viral genome replication. The HCV NS5B RNA polymerase is remarkably error-prone and does not have proofreading capacity, leading to an estimated 10^{-4} substitutions per site and round of replication [36]. In contrast, high replicative fidelity (e.g., with DNA polymerases with an estimated 10^{-8} to 10^{-11} substitutions per site) would result in a far more homogenous and genetically static viral population and would not allow the virus to quickly explore the sequence space. The development of resistant variants is also determined by the replication rate of the virus which, in the case of HCV, is estimated at 10^{12} virions per day [37]. The extremely high rate of replication combined with the high mutation rate allows the HCV to explore the sequence space available to it at a faster rate. A third factor that contributes to the development of resistant variants is the genetic barrier to drug resistance, which involves the number and type of mutations that are needed for the emergence of a RAS. Fourth, the fitness of the resistant variant populations is critical as it determines the likelihood that any resistant variant persist within the larger viral population. Finally, the emergence of viral resistance is determined by the level of drug exposure. Indeed, exposure to suboptimal concentrations of antiviral agents will result in the selection of RASs by allowing for the maintenance of a viral load in the presence of a mild selective pressure.

As has been discussed recently, the language to describe these mutations and viral variants should be standardized [38]. Most recently, it has been proposed that the amino acid substitutions that confer resistance will be called *resistance-associated substitutions* (*RASs*) and the viral populations that carry these RASs will be called *resistant variants* (*RAVs*). These resistant variants can also acquire additional mutations, termed compensatory or fitness-associated substitutions, which may increase their fitness. This can lead to their rapid outgrowth during the course of treatment, which is termed a *breakthrough*, or after treatment, which is termed a *relapse*.

3.1 Identification of Resistance-Associated Variants

With our developing understanding of DAA resistance, there has been increasing interest in identifying pre-existing RASs that exist within HCV quasispecies. To do so, several methods have been used to perform sequencing of varying depth to identify populations of viral variants within the larger quasispecies cloud [39, 40]. In most studies, the identification of pre-existing RASs is performed using population sequencing via the traditional Sanger method. While an excellent strategy to identify major sequences present within the quasispecies, its primary weakness is its lack of sensitivity, as it is generally unable to detect viral populations that are present at proportions lower than 10–25% of the total population [41]. However, in recent years, there has been incredible advancement in the development of high-throughput, next-generation sequencing technologies (e.g., Illumina, 454, Ion Torrent, PacBio, etc.), which has rapidly improved our ability to detect viral subpopulations that are present in ever smaller proportions within the quasispecies, even those comprising just ~0.1–1% [42, 43]. The highest sensitivities for the detection of minor viral population in

NGS studies typically occur when the analysis focuses on a specific gene or short region of a gene, though the benefits of high sensitivity must be weighed against the possibility of detecting false positives as a result of the amplification and sequencing steps. Therefore, minor variants that are present as less than 0.5% of the viral population are typically excluded in these studies [42]. In addition to focused sequencing techniques, increasingly sensitive methods for whole-genome sequencing have also been devised recently that allow for the detection of minor population of RASs and mixed GT/subtype infections that may be relevant for treatment response [44].

In general, it has been found that RASs that are present in low proportions (<15%) do not significantly affect treatment outcomes, whereas RASs existing as a greater than 15% proportion of the overall population are more associated with treatment failure. As such, there has been agreement that a 15% cutoff should be used in all clinical trials and studies of real-world patients in the reporting of RASs by population and next-generation sequencing [38].

4 Drug Resistance to DAAs

4.1 NS3/4A Protease Inhibitors

The NS3/4A viral protease is a heterodimer complex in which the NS3 protein contains the proteolytic cleavage site and NS4A functions as a cofactor. NS3/4A inhibitors block the NS3 catalytic site or inhibit NS3/4A interaction, thereby preventing the cleavage of the HCV polyprotein. The catalytic site of the NS3 protease is located within a shallow groove which has made the design of small inhibitor molecules challenging. Moreover, given the paucity of binding sites available for small molecules to the catalytic site, there is a relatively low barrier to genetic resistance to NS3/4A inhibitors [45].

The first-generation protease inhibitors (PIs) telaprevir (TVR) and boceprevir (BOC) were the first two DAAs approved for the treatment of HCV GT1 in conjunction with interferon and ribavirin [46, 47] and were part of a structural class known as linear ketoamides [45, 48]. The second wave of first-generation PIs are characterized by their increased potency and include the macrocyclic compounds: simeprevir (SMV), asunaprevir (ASV), and vaniprevir (VAN). SMV and ASV are currently approved for HCV treatment in the United States, while VAN is only approved in Japan. Given the development of these increasingly potent protease inhibitors, the production of TVR and BOC was terminated in 2014. Most recently, the second-generation protease inhibitors have been developed with a particular focus on not only continuing to increase potency and reduce susceptibility to resistance mutations but also optimizing broad genotypic activity [49]. Approved members of this class include paritaprevir (PTV), grazoprevir (GZR), glecaprevir (GLE), and voxilaprevir (VOX).

NS3 RASs can be detected at low levels in infected individuals even prior to DAA treatment. Early studies that relied on population sequencing of individuals who did not achieve SVR after treatment with a TVR-based regimen in phase II and III trials found that a number of RASs were preferentially enriched, including V36A/M, T54A/S, R155K/T, A156S/T, and D168N [50]. A subsequent study evaluating the presence of baseline RASs in individuals who did not achieve SVR after treatment with a SMV-based regimen in phase IIb and III trials found some similar mutations to that seen with the first-generation PI-based regimens, including those at positions 155, 156, and 168, along with a number of unique RASs including Q80K and S122R [49, 51]. RASs at positions 155, 156, and 168 have also been identified in individuals who were treated with PTV- or GZR-containing regimens [52–54]. However, in comparison to second-wave, first-generation PIs, in vitro replicon studies demonstrate that these mutations are associated with significantly less resistance toward second-generation PIs [52, 53]. These resistance-associated positions are all located around the catalytic site of the NS3 protease domain, and mutations at these sites lead to reduction in the ability of the inhibitor molecule to bind effectively to the active site. A complete list of known NS3 RASs to date that result in >twofold increase in resistance can be found in Fig. 4.

NS3 RASs are generally found at low levels (0.1–3.1%) at baseline because many of them incur a significant replicative cost [64]. The one exception to this is Q80K, which does not significantly impair replicative fitness. In one study of patients with GT1 infection, the Q80K RAS was identified in 13.6% of cases, with nearly all the cases being present in patients with GT1a infection [49].

After withdrawal of treatment, NS3 RASs gradually disappear with time as the environmental pressure that selected for these subpopulations is removed. This has been observed in patients treated with all generations of NS3/4A inhibitors. In one study that investigated the evolutionary dynamics of treatment-emergent RASs in 1797 patients treated with TVR, PEG-IFN, and ribavirin, 77% of those who did not achieve SVR harbored RASs at time of treatment failure, but these were lost over time after treatment cessation [65]. The median time to reversal to wild-type viruses predominating was 10.6 months for GT1a and 0.9 months for GT1b, but all patients had lost detectable RASs by 17 months and 13 months, respectively [65]. Another study of 197 patients treated with SMV, PEG-IFN, and ribavirin had similar results. 91% of those who did not achieve SVR had the presence of RASs at the time of treatment failure, and the median time until the return to dominance of the wild-type virus was 9 months for individuals with GT1a infection and 6 months for those with GT1b infection [49]. More recently, a study of patients treated with GZR/ELB

NS3 RASs (>2-fold resistance)		Genotype						
Position	Drug Resistance	1a	1b	2	3	4	5	6
36	ALL	V36A	V36A					V36A
	BOC/PTV/TVR	V36C	V36C					
	BOC/PTV/TVR	V36G	V36G					
	ASV/VOX	V36L						
	ASV/BOC/PTV/SMV/TVR	V36M	V36M					
43	SMV		F43I					
	SMV		F43V					
	ASV/GZR/SMV		F43S					
	PTV	F43L						
54	BOC/TVR	T54A	T54A	T54A	T54A			T54A
	BOC/TVR	T54S	T54S	T54S			T54S	
55	BOC	V55A	V55A		V55A			
56	PTV	Y56H						
80	ASV/PTV/SMV	Q80K	Q80K					
	ASV/GZR/SMV/GLE	Q80R	Q80R					
122	ASV/SMV	S122R	S122R					
132	VAN	**I132V**						
138	SMV		S138T					
155	ASV/BOC/GZR/SMV/TVR/VAN			R155A				
	BOC/GZR/PTV/SMV/TVR/VAN	R155G	R155G					
	PTV/TVR	R155I	R155I					
	ALL	**R155K**	R155K	R155K	R155K	R155K	R155K	R155K
	PTV/TVR	R155M	R155M					
	VAN	R155N						
	ALL	R155Q						
	PTV/TVR/VAN	R155S	R155S					
	ASV/BOC/PTV/SMV/TVR/VAN	R155T	R155T	R155T				R155T
	PTV/TVR	R155W	R155W					
156	ASV/BOC/TVR		A156F					
	ASV/BOC/GZR/PTV/SMV/VAN/GLE	A156G	A156G	A156G	A156G	A156G	A156G	A156G
	ASV/BOC/GZR/PTV/TVR/VAN	A156S	A156S	A156S	A156S	A156S	A156S	A156S
	ASV/BOC/GZR/SMV/TVR	A156T	A156T					
	ASV/BOC/GZR/SMV/TVR/VAN	A156V	A156V	A156V				
157	VAN			A157V				
158	BOC		V158I					
168	ASV/GZR/PTV/SMV/VAN	**D168A**	D168A	D168A	D168A		**D168A**	D168A
	ASV/GZR/PTV/SMV	D168C	D168C					
	ASV/GZR/PTV/SMV/VAN	D168E	D168E	D168E			D168E	D168E
	ASV/GZR/PTV/SMV/VAN	D168G	D168G	D168G				D168G
	ASV/GZR/PTV/SMV/VAN	D168H	D168H	D168H			**D168H**	
	SMV		**D168I**					
	ASV/GZR/PTV/SMV	D168K	D168K					
	ASV/GZR/PTV/SMV/VAN	D168N	D168N	D168N			D168N	D168N
	ASV/GZR/PTV/SMV	D168T	D168T					
	ASV/GZR/PTV/SMV/VAN	D168V	**D168V**	D168V			D168V	D168V
	ASV/GZR/PTV/SMV	**D168Y**	**D168Y**					
170	GZR	I170T						
	BOC/TVR		V170A					
175	BOC		M175L					

Fig. 4 Known NS3 resistance-associated substitutions (RASs). Listed RASs have been found to have >twofold increase in the EC_{50} when tested in in vitro replicon systems [48–63]. Bolded variants have been found to exhibit a very high level of resistance (>100-fold). Empty fields indicate that no data is available for the corresponding GT at that given amino acid position. *ASV* asunaprevir, *BOC* boceprevir, *GZR* grazoprevir, *PTV* paritaprevir, *SMV* simeprevir, *TVR* telaprevir, *VAN* vaniprevir, *GLE* glecaprevir, *VOX* voxilaprevir

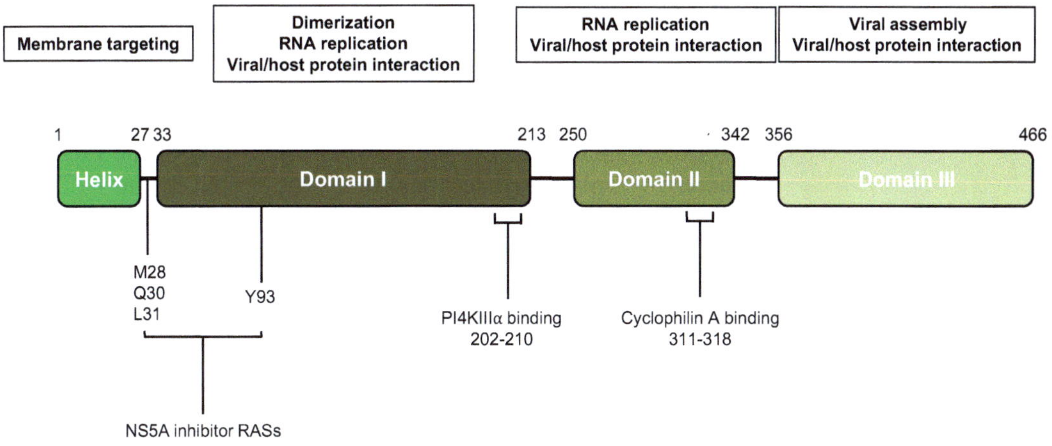

Fig. 5 Schematic of NS5A domain organization and known functions

found that the A156T RAS disappeared within several weeks posttreatment [66].

These findings show that NS3 RASs quickly lose their dominance in the viral quasispecies after treatment cessation. Whether physicians should wait until the NS3 RASs have disappeared prior to retreatment with a NS3/4A inhibitor remains an open question and active area of investigation.

4.2 NS5A Inhibitors

The NS5A protein is a multifunctional zinc-binding phosphoprotein [55] of still enigmatic function but is thought to be involved in multiple aspects of the HCV lifecycle, including replication, assembly, and egress of HCV virions [55, 67]. The protein itself is a ~450 amino acid phosphoprotein comprised of an N-terminal α-helix, which is critical for membrane targeting and helps to tether the protein to intracellular membrane bilayers [68], and three domains (D1–D3), depicted in Fig. 5. D1 is an RNA-binding region required for viral replication and is thought to function in several alternative dimerized states [69–71]. D2 and D3 are predicted to be largely unstructured domains involved in RNA replication and interaction with viral and/or host factors including cyclophilin A and phosphatidylinositol 4-kinase IIIα [72].

Current NS5A inhibitors that have been approved for HCV treatment include daclatasvir (DCV), ledipasvir (LDV), ombitasvir (OMV), elbasvir (ELB), velpatasvir (VEL), and pibrentasvir (PIB). Despite the rapid development of this class of DAAs, the exact mechanism of inhibition by DAAs against the NS5A protein remains incompletely understood. A number of biochemical studies have shown that NS5A inhibitors bind the dimerized D1 domain and stabilize it, thereby potentially locking it in position and preventing further structural rearrangements necessary for the protein to exert its myriad functions [73, 74], though the latter hypothesis remains to be substantiated. Treatment with NS5A inhibitors has been found to have global effects on NS5A, including alteration of phosphorylation patterns [75] and aberrant subcellular localization [76]. Recent data has further suggested that NS5A inhibitors may function by blocking the assembly of the membranous HCV replication factories that are essential in the HCV life cycle [73]. Though the exact mechanism by which NS5A inhibitors function to abrogate NS5A remains an active area of investigation, the multiple defects observed in the HCV replicative cycle after NS5A inhibition reflect the pleomorphic functions that the NS5A protein participates in the HCV life cycle.

As with NS3 RASs, NS5A RASs are frequently detected as natural variants in DAA-naïve infected individuals. Indeed, though NS5A inhibitors exhibit a broad pan-genotypic spectrum of action, likely the result of a more conserved interaction site within the NS5A protein [73, 77], they are also characterized by a low genetic barrier to resistance. The most clinically significant RASs that have been identified to date in GT1 infection include M28A/G/T, Q30D/E/G/H/K/L/R, L31F/M/V, and Y93C/H/N/S, all of which result in high-level resistance (>100-fold increase from wild type) to NS5A inhibitors. M28, Q30, and L31 map to the linker region between the N-terminal α-helix, while Y93 maps to deep within D1 on the putative dimer interface predicted by X-ray crystallography [55, 77]. Early studies of NS5A RAS prevalence in DAA-naïve populations performed via population sequencing found that between 0.3% and 2.8% of HCV-infected individuals harbored detectable RASs [78–83]. Individuals with GT1b infection showed higher prevalence of two mutations, L31M, which portends intermediate to high resistance depending on the NS5A inhibitor used and was found in upward of 6.3% of GT1b patients, and Y93H, which confers a very high level of resistance and was observed in upward of 14% of individuals with GT1b infection. Interestingly, L31M was observed to be extremely prevalent in GT2 (74–85%) and GT4 (92.5–100%) populations. More recent studies using more sensitive next-generation sequencing suggest that these previously reported numbers underestimate the true prevalence of NS5A RASs. A large analysis of 2144 participants in phase II or III with HCV GT1a or 1b infection who were treated with sofosbuvir/LDV discovered that by using a deep sequencing assay cutoff value of 1% (i.e., >1% of sequences need to harbor a RAS for inclusion), they found that 16.7% of patients harbored NS5A RASs [84]. Even when the threshold was increased to a cutoff value of 20%, 8.5% of patients were identified as having NS5A RASs. In GT1a patients, the most commonly identified RASs were K24R, L31M, Q30H, M28T, Y93H, and Q30R in descending order. In GT1b patients, the most commonly identified RASs were Y93H and L31M [84]. A complete list of known NS5A RASs to date that result in >twofold increase in resistance can be found in Fig. 6.

NS5A RASs (>2-fold resistance)		Genotype						
Position	Drug Resistance	1a	1b	2	3	4	5	6
24	LDV	K24G						
	LDV	K24N						
	LDV	K24R						
28	DCV/OMV			F28S				
	OMV			L28F				
	OMV						L28I	
	ELB		L28M					
	DCV/OMV		L28T					
	OMV					L28V		
	DCV/ELB/LDV	M28A						
	DCV/LDV	**M28G**						
	ALL	M28T			M28T			
	OMV	M28V						
30	DCV/ELB				A30K			
	DCV					**L30H**		
	VEL					L30R		
	DCV		R30H					
	ELB	**Q30D**						
	ALL	**Q30E**						
	LDV	**Q30G**						
	ALL	Q30H						
	LDV, VEL	**Q30K**						
	LDV	Q30L						
	ALL	**Q30R**						
	LDV	Q30T						
31	DCV/ELB/LDV/OMV	L31F	L31F		**L31F**		**L31F**	
	LDV/VEL	L31I						
	DCV/ELB/LDV/VEL	**L31M**	L31M	**L31M**	**L31M**			**L31M**
	ALL	**L31V**	L31V	**L31V**	**L31V**		**L31V**	L31V
32	DCV/LDV/VEL	**P32L**	P32L					**P32L**
	DCV/LDV/VEL	**P32S**	P32S					**P32S**
38	LDV	**S38F**						
58	ALL	**H58D**						
	LDV		**P58D**					
	DCV/OMV							T58A
	DCV/OMV							**T58N**
	DCV/OMV							T58S
92	LDV		**A92K**					
	LDV	A92T						
	DCV			C92R				
93	ALL	**Y93C**						
	LDV/VEL	Y93F						
	ALL	**Y93H**	Y93H	**Y93H**	**Y93H**	**Y93H**		
	ALL	**Y93N**	Y93N			Y93N		
	DCV/VEL	Y93R				**Y93R**		
	LDV/VEL	Y93S						
	VEL	**Y93W**						

Fig. 6 Known NS5A resistance-associated substitutions (RASs). Listed RASs have been found to have >twofold increase in the EC_{50} when tested in in vitro replicon systems [68, 73, 75, 76, 79–92]. Bolded variants have been found to exhibit a very high level of resistance (>100-fold). Empty fields indicate that no data is available for the corresponding GT at that given amino acid position. *DCV* daclatasvir, *ELB* elbasvir, *LDV* ledipasvir, *OMV* ombitasvir, *VEL* velpatasvir, *PIB* pibrentasvir

Unlike NS3 RASs, NS5A RASs that emerge during treatment are extremely long-lasting and can remain as the dominant population for years after treatment failure. In one study of 31 patients with GT1b patients treated with DCV/ASV, treatment failure was associated with the emergence of NS5A-L31/Y93 and NS3-D168 RASs [93]. While the NS3 RAS was replaced by wild-type virus in all patients by posttreatment weeks 12–60, the NS5A RASs persisted even at high levels even at posttreatment week 170 [93]. This was also observed in a study of 436 patients with GT1b infection treated with OMV/PTV/ritonavir + DSV in which those who experienced treatment failure also had emergence of NS3-D168 and NS5-Y93 RASs [94]. Similarly, while the NS3 RASs disappeared over time, NS5A RASs persisted at high levels through posttreatment week 48 [94].

These results demonstrate that NS5A RASs persist for years after treatment failure, likely due to their inherently higher replicative fitness. As such, their presence will strongly affect the outcomes of retreatment.

4.3 NS5B Inhibitors

NS5B inhibitors currently approved or in development are divided into two categories: nucleos(t)ide inhibitors (NIs) and non-nucleos(t)ide inhibitors (NNIs) [95, 96]. Sofosbuvir (SOF) is the only NI that is approved, and dasabuvir (DSV) is the only NNI that has been approved for treatment. NIs function by incorporating themselves during active transcription by NS5B and preventing the further incorporation of nucleotides. Chain termination is thought to occur by steric hindrance exerted through the 2′-C-methyl or 2′-fluoro groups found on the drug molecule [97]. Given that the mechanism of action is exerted on an extremely conserved aspect of the HCV genome, NIs tend to have pan-genotypic activity. SOF is currently the most advanced example of a NI that is currently on the market, though several more are currently in phase I and phase II trials.

In contrast, NNIs function by binding to the NS5B molecule in a noncompetitive fashion at sites away from the active site. The crystal structure of NS5B revealed a canonical "right-hand" structure with "finger" and "thumb" subdomains that encircle a "palm" domain that houses the active site (Fig. 7) [98–100]. The polymerase shifts between both "open" and "closed" conformations that are critical for different steps in de novo HCV genome synthesis, the details of which are beyond the scope of this chapter but have been reviewed elsewhere [96]. NNIs bind to one of five allosteric sites located on either the thumb or palm domains of NS5B and interfere with the conformational changes that are important for polymerase activity. The only NNI that has been approved for use is DSV, though several more are currently being developed that target the other domains of the NS5B polymerase.

NS5B RASs (>2-fold resistance)		Genotype						
Position	Drug Resistance	1a	1b	2	3	4	5	6
282	SOF	S282T	S282T	S282T	S282T			
289	SOF			M289L				
314	DSV	L314H						
316	DSV	**C316H**	**C316H**					
	DSV	**C316Y**	**C316Y**					
	DSV		C316N					
320	SOF	L320F						
368	DSV		**S368T**					
411	DSV		N411S					
446	DSV	E446K						
	DSV	E446Q						
448	DSV	**Y448C**	**Y448C**					
	DSV	**Y448H**	Y448H					
553	DSV	**A553T**						
	DSV		**A553V**					
554	DSV	G554S	G554S					
556	DSV	S556G	S556G					
	DSV	**S556R**						
558	DSV	G558R						
559	DSV	**D559G**	**D559G**					
561	DSV	Y561H						

Fig. 7 Structure and organization of NS5B HCV RNA polymerase. Ribbon structure diagram adapted from [98]. *Blue* sofosbuvir-related RASs. *Red* dasabuvir-related RASs

SOF exhibits a very high barrier to resistance, and as such, naturally occurring RASs have infrequently been detected in infected patients. A complete list of known NS5B RASs to date that result in >twofold increase in resistance can be found in Fig. 8. In initial in vitro replicon studies, S282T was the only RAS that was identified, but this mutation has rarely been identified in clinical cases, consistent with marked impairment of replicative fitness due to this mutation [101]. In one study of 982 patients enrolled in phase III clinical studies that evaluated the efficacy of SOF/RBV or SOF/IFN/RBV regimens, no patients with treatment failure were found to harbor the S282T RAS [102]. Moreover, among 1645 patients who received SOF-containing regimens in the phase II and III SOF trials, only one patient who failed treatment had developed the S282T mutation [103, 104]. Additional position 282 RASs are being discovered with the advent of increasingly deep sequencing techniques that are present in very minor proportions [112], but the clinical significance of these substitutions remains to be clarified.

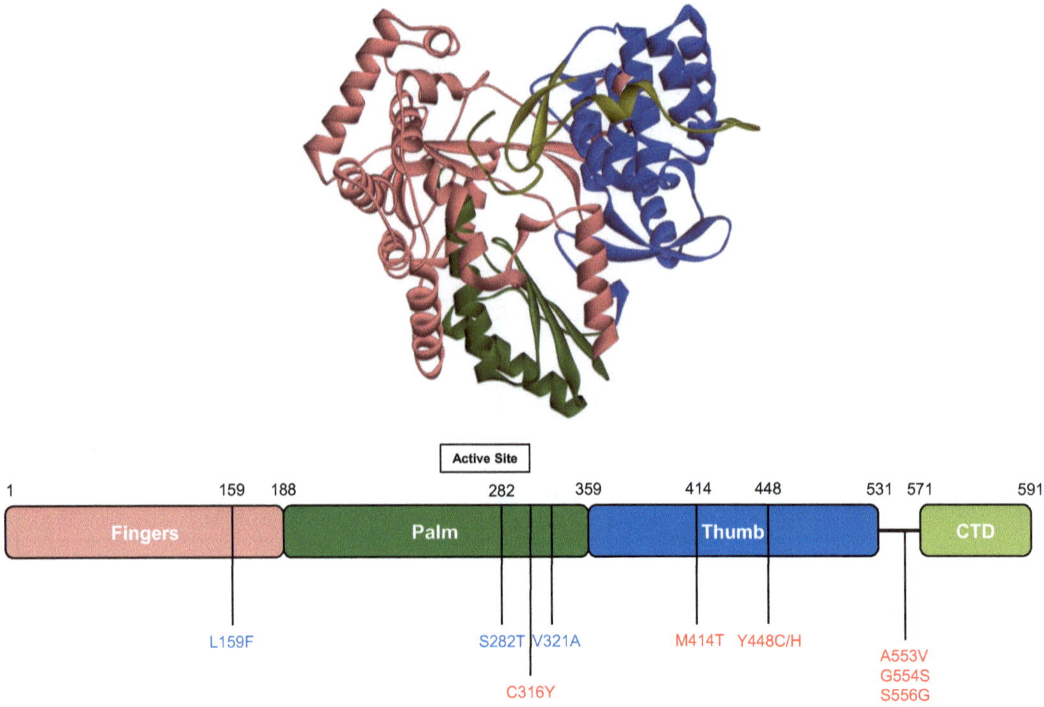

Fig. 8 Known NS5B resistance-associated substitutions (RASs). Listed RASs have been found to have >twofold increase in the EC_{50} when tested in in vitro replicon systems [25, 26, 91, 101–111]. Bolded variants have been found to exhibit a very high level of resistance (>100-fold). Empty fields indicate that no data is available for the corresponding GT at that given amino acid position. *SOF* sofosbuvir, *DSV* dasabuvir

In a large pan-genotypic analysis of phase II and III SOF clinical trials, L159F and V321A were identified as NS5B RASs that emerged after treatment [104]. However, these RASs only conferred minimally reduced susceptibility to SOF in vitro [104]. Moreover, in an analysis of individuals who developed L159F and V321A RASs in eight clinical trials involving SOF, these RASs were found not to have any impact on retreatment outcome with SOF, ribavirin, and PEG-IFN, and furthermore, the baseline presence of the L159F RAS in GT1 did not affect the treatment outcome with LDV/SOF [105]. Baseline C316N/H/F NS5B RASs were observed in seven patients with either treatment failure or relapse after treatment [102]. Other identified RASs include the S96T and L320I/F, but their impact on treatment outcomes remains unknown. It should be noted that most of these studies disproportionally focused on GT1 infections, and as such, the prevalence and significance of NS5B RASs in non-GT1 infection remain an area of particular interest.

In contrast to NIs, RASs related to NS5B NNIs are more common due to the relatively lower barrier of resistance associated with these agents. As expected, the identity of the substitutions that confer resistance to NNIs depends on the particular domain on which the NNI targets. In the case of DSV, a NNI that targets the NS5B palm domain, the most commonly observed RASs in GT1a patients who did not achieve SVR were M414T and S556G [25, 26, 106]. Interestingly, in a study of 1145 HCV isolates, the S556G RAS is naturally found in 100% of GT2 and GT3 strains as well as 97% of GT4 strains, though it is considerably less common in GT1a (0.6–3.1%) and GT1b (7–16%) viral strains [64, 94]. Other naturally occurring NNI-NS5B RASs that were identified in a screen of 1145 NS5B sequences isolated from NI/NNI treatment-naïve patients include M414L, M423I, C316N, A421V, C445F, I482L, V494A, and V499A [107]. Of these mutations, V499A deserves further mention as this RAS is found in over 96% of GT1a isolates, though in vitro, the mutation only results in a minor reduction in the potency of NNIs that target the thumb domain [113]. As additional NNIs are developed that target specific sites on NS5B and ultimately become implemented in clinical settings, new important RASs will be identified, and their impact on the outcome of treatment will be clarified.

5 Clinical Effect of RASs on Treatment Response

Given the relative abundance of RASs that exist in DAA-naïve individuals with HCV infection, a critical question garnering significant attention in the field is whether the pretreatment presence of RASs predicts treatment failure with IFN-free regimens. The following presents an overview of the current status of this rapidly evolving field of study. Only drug combinations that have been approved in the United States and Europe will be discussed.

5.1 Sofosbuvir/Ledipasvir

The combination of SOF/LDV with or without ribavirin has been studied in over 3000 patients in phase II/III studies [21–23, 114–119]. In the largest study to date specifically addressing the impact of RASs on treatment outcome with this combination, 2144 patients that were enrolled in these trials were evaluated, with the vast majority with GT1a or GT1b infection [84]. Among these, 2140 (99.8%) had NS5A gene sequencing performed, the majority with deep sequencing techniques. Using a cutoff value of 15%, 16.0% of patients were identified to have baseline NS5A RASs. There was a significantly decreased rate of SVR associated with the presence of NS5A RASs compared with those with no baseline NS5A RASs, though this was driven primarily by failures in GT1a treatment [84]. When further stratified by treatment duration and level of in vitro RAS resistance, a significant reduction in SVR rate was observed in treatment-naïve patients with NS5A RASs that conferred a high level of LDV resistance (>100-fold) in those who received 8 weeks of therapy compared to those who did not have baseline NS5A RASs (82.8% vs. 96.4%; $p = 0.011$) [84]. This was also seen in treatment-experienced individuals with baseline NS5A RASs that conferred a high level of resistance who received 8 weeks of therapy (64.7% vs. 97.4%; $p < 0.001$). This effect was not observed in treatment-naïve or treatment-experienced individuals who received 12 or 24 weeks of therapy. SVR12 rates ranged between 75% and 83% in those individuals with GT1a infection with high-resistance RASs (Q30H, Q30R, L31M, Y93H). In addition, the number of NS5A RASs at baseline directly associated with the risk of treatment failure [84].

NS5B sequencing was performed for 1692 patients (78.9%) in this study. Baseline NS5B RASs were exceedingly rare, with a total of 41 individuals with either L159F or N142T RASs, all of whom achieved SVR12 [84]. The NS5B RAS S282T was not detected in any of these patients. These clinical outcomes are in accordance with the very mild resistance that NS5B RASs have demonstrated in vitro, often at the expense of replicative fitness [105]. These results for NS5A and NS5B RASs were corroborated by similar findings in a Japanese cohort treated with SOF/LDV ± ribavirin [120].

Overall, these findings suggest that the presence of baseline NS5A RASs (at a cutoff of >15%) that confer high levels of in vitro resistance to LDV results in significant reductions in rates of SVR achieved with the combination of SOF/LDV in individuals with GT1a infection.

5.2 Sofosbuvir/Velpatasvir

The combination of SOF/VEL was approved for HCV treatment in 2016 and has pan-genotypic efficacy. In the three phase III ASTRAL registration trials which assessed the safety and efficacy of SOF/VEL combination therapy in treatment-naïve or treatment-experienced patients with or without compensated or decompensated cirrhosis, baseline resistance was assessed with deep sequencing with a cutoff of 1% [121–123]. In the ASTRAL-1 and 2 trials that evaluated treatment outcomes in all viral GTs except GT3, NS5A RASs were detected in 42% of patients (257/616), but only two experienced treatment failure [122]. The higher rate of detected RASs in this study was likely related to the lower cutoff for detection of viral RASs as well as the inclusion of treatment-experienced patients. The incidence of treatment failure was higher among individuals with baseline NS5A RASs with GT3 infection. Among 274 patients in the ASTRAL-3 trial with GT3 infection and deep sequencing data, 16% had baseline NS5A RASs (A30K, L31M, Y93H), and of these, 88% (38/43) achieved SVR compared to 97% (225/231) of those without pre-existing NS5A RASs [123]. As such, current recommendations are for treatment-naïve patients with cirrhosis to undergo testing for NS5A RASs, and if the Y93H RAS is present, then ribavirin should be added. A similar recommendation holds for

GT3-infected noncirrhotic patients with prior treatment failure, for whom addition of ribavirin enhances SVR rates in those harboring NS5A RASs [124].

The effect of baseline NS5A RASs on achievement of SVR with this drug combination was even more marked in individuals with decompensated cirrhosis (Child-Pugh class B). In the ASTRAL-4 trial, 83% of patients (75/90) treated with SOF/VEL achieved SVR12, but only 50% (7/14) of those infected with GT3 achieved SVR [121]. This was also observed even if the treatment duration was lengthened to 24 weeks. Interestingly, the addition of ribavirin attenuated this effect, resulting in 85% (11/13) SVR12 rates [121]. However, given the small numbers achieved in these initial trials, a definitive assessment of the effect of GT3 infection on treatment outcomes with SOF/VEL awaits larger studies. In all these trials, NS5B RASs (N142T, L159F, E237G, M289I) did not have any impact on treatment outcome as all patients with baseline NS5B RASs achieved SVR12.

Overall, these results suggest that baseline NS5A RASs do not result in significant effects on the treatment outcomes for a 12-week course of SOF/VEL in patients without cirrhosis or with compensated cirrhosis with the exception of those with GT3 infection. However, in the case of those with decompensated cirrhosis and GT3 infection, this effect may be mitigated with the addition of ribavirin.

5.3 Sofosbuvir/Daclatasvir

Overall, there has been considerably less data that has been published regarding the association between baseline prevalence of RASs and treatment outcomes with the combination of SOF/DCV compared to other drug combinations. In the early phase II study trial in which 167 noncirrhotic treatment-naïve or treatment-experienced patients with GT1, 2, or 3 HCV infection were treated with SOF/DCV ± ribavirin, 98% (169/171), 92% (24/26), and 89% (16/18) of GT1, 2, and 3 patients achieved SVR12 [125]. Ten patients were found to have baseline NS5A RASs, and nine achieved SVR. In a phase III trial of SOF/DCV ± ribavirin in HIV/HCV co-infected patients, SVR12 was achieved in 96.4% of patients with GT1 infection [126]. Of the 12 individuals who had relapse after therapy, only 6 were found to have pre-existing NS5A RASs though the treatment failures occurred in the setting of other complicating factors including cirrhosis and a shortened treatment regimen. In another study that investigated treatment outcomes in GT3 individuals, 16/152 patients did not achieve SVR12 [127]. Of these, three had baseline A30T/K mutations, and two had a baseline Y93H mutation. However, 11 of these patients who did not achieve SVR12 also had cirrhosis. Thus, given the small numbers involved in this trial and others, the true effect of NS5A RASs in treatment outcomes for SOF/DCV treatment remains unclear.

5.4 Sofosbuvir/Simeprevir

Studies that have evaluated the role of baseline RASs in treatment outcomes in individuals treated with SOF/SMV have focused on the Q80K RAS. As described before, the Q80K mutation is one of the most frequently observed baseline NS3 RASs observed in GT1a strains (15–20% prevalence), whereas it is rarely found in GT1b strains [49, 128]. In vitro studies have shown that Q80K decreases the susceptibility of the virus to SMV by nearly tenfold [49]. The combination of SOF/SMV ± ribavirin was first explored in the COSMOS trial, a small phase II study performed in patients with GT1 infection [24]. Though high SVR rates were achieved for either GT1a without Q80K (68/72, 94%) or with Q80K (51/58, 88%) in this trial, four out of six GT1a patients who experienced virologic relapse had baseline Q80K RASs, suggesting an important in vivo effect of this RAS. This initial observation was further explored in the phase III trials that led to the approval of the combination of SOF/SMV for the treatment of patients with HCV GT1 infection. In the OPTIMIST-1 trial, noncirrhotic patients with GT1 infection who were either treatment-naïve or had been previously treated with PEG-IFN-based regimens were treated with SOF/SMV. Of patients with GT1a infection, 97% (68/70) of those without Q80K and 96% (44/46) of those with Q80K achieved SVR12 [129]. However, in the OPTIMIST-2 trial, in which treatment-naïve or treatment-experienced patients with compensated cirrhosis (Child-Pugh class A), SVR12 was achieved in 92% (35/38) of those with GT1a infection without Q80K compared with 74% (25/34) of GT1a individuals with Q80K [130]. The majority of the treatment failures with Q80K occurred in treatment-experienced patients. These results demonstrate that the pretreatment presence of the Q80K RAS is associated with an increased risk of treatment failure at 12 weeks, particularly in those with underlying cirrhosis, GT1a infection, and prior treatment. In all these trials, no pre-existing NS5B RASs were identified. As a result of these findings, current recommendations are to screen for baseline presence of the Q80K RAS if the SOF/SMV regimen is to be used in a treatment-experienced individual with GT1a infection and cirrhosis. If Q80K is present, alternative regimens that do not contain SMV are recommended. If Q80K is not present, then a 24-week course is recommended [131].

SOF/SMV is also approved for the treatment of GT4 infection, but no data is available at this time in regard to the effect of baseline RASs on treatment outcome.

5.5 Ombitasvir/Paritaprevir/Ritonavir and Dasabuvir

Current published data regarding the impact of baseline RASs on treatment outcomes in individuals treated with the OMV/PTV/ritonavir and DSV combination comes principally from the AVIATOR study, a phase II study that enrolled 571 GT1-infected patients without cirrhosis who were either treatment-naïve or previously failed PEG-IFN-based therapies to various two- or three-drug combinations of the above medications [56]. RASs were identified via population sequencing. In GT1a-infected patients, Q80K was identified in 41% of available NS3 sequences, with the next common RASs including the NS5A RAS M28 V (6.0%) and the NS5B RAS S556G (3.1%). In GT1b-infected patients, the most frequent mutations were the NS5B RASs C316N (18.4%) and S556G (16.0%), followed by the NS5A RAS R30Q (8.5%) [108]. In this study, the majority of individuals who had an inability to achieve SVR24 were those with GT1a. Among these patients, the overall SVR rate was slightly decreased among those with baseline RASs in the NS3, NS5A, and/or the NS5B genes compared to those without (87% vs. 92%). Specifically in those with baseline Q80K variants, there was no significant difference in the rate of SVR24 in those with and without the variant (88% vs. 94%; $p = 0.14$).

A larger analysis was recently presented that evaluated over 2500 HCV GT1-infected patients who have been treated with this treatment regimen in two phase II and six phase III clinical trials [132]. NS3, NS5A, and NS5B RASs were identified by population sequencing and were available for over 700 individuals. Baseline NS3 RASs were rare in both GT1a and 1b; NS5A RASs were observed in 12.5% of GT1a and 7.5% of GT1b infections; and NS5B RASs in NS5B were observed in 5.2% of GT1a and 28.6% of GT1b infections. Despite the relatively high level of baseline RASs, none of these mutations had any impact on treatment outcome.

The effect of baseline polymorphisms in the treatment of GT4-infected patients with OMV/PTV/ritonavir and DSV was recently analyzed from the cohort of 135 patients enrolled in the PEARL-1 phase II clinical trial [133]. Of 132 patients for which sequencing analysis was available, 57.6% (76/132) had baseline RASs at the resistance-associated positions 28, 30, 31, 32, 58, and 93. The most common RAS was T58P. Despite the significant prevalence of baseline RASs, all patients with and without baseline variants who were treated with OMV/PTV/ritonavir and DSV achieved SVR12 [133].

These results demonstrate that baseline RASs do not have significant impact on treatment outcomes in individuals with GT1 or GT4 infection treated with this DAA regimen.

5.6 Grazoprevir/Elbasvir

The treatment combination of the NS3/4A protease inhibitor, GZR, and NS5A inhibitor, ELB, was approved in the United States in January 2016 and in Europe in July 2016. The approval stemmed in significant part from the C-EDGE phase III trial that enrolled 421 treatment-naïve cirrhotic and noncirrhotic individuals with GT1, 4, or 6 infection and randomized them to 12 weeks of treatment with GZR/ELB or to deferred therapy [134]. Though overall, the rates of SVR12 were high for the treatment group, the presence of baseline NS5A RASs had a significant effect on treatment outcomes in GT1a-infected patients. NS5A RASs were identified at baseline in 12% (19/154) of GT1a-infected patients. However, of those with baseline NS5A RASs, only 58% (11/19) achieved SVR12. Moreover, only 22% (2/9) of those with baseline NS5A RASs with >fivefold resistance to ELB and GT1a infection achieved SVR12. In comparison, the presence of NS5A RASs did not affect SVR12 in patients with GT1b infection. In addition, the presence of baseline NS3 RASs did not affect treatment outcome. In a recent study of treatment-experienced GT1-infected patients, only 68% (21/31) of GT1a-infected patients with baseline RASs at positions 28, 30, 31, or 93 identified by population sequencing (using a 25% sensitivity threshold cutoff) or next-generation sequencing (using a 15% cutoff) achieved SVR12 [135]. The most common baseline variants were Q30H, L31M, and Y93H/N. Baseline NS3 RASs had no impact on treatment outcome. Treatment of GT4- and GT6-infected patients appeared to be unaffected by the presence of baseline NS5 RASs, though the number of individuals with these GTs was very low. Similar findings were observed in a larger pooled analysis of population sequencing data from GT1a- and 1b-infected treatment-naïve or treatment-experienced cirrhotic or noncirrhotic patients who were treated with GZR/ELB in recent phase II or III trials [136]. Extension of treatment to 16 weeks and the addition of ribavirin result in significant improvement in SVR rates in these patients [137, 138].

These results suggest that baseline resistance testing for NS5A RASs may be important in all patients with GT1a infection that are considering treatment GZR/ELB. As such, current guidelines recommend that NS5A RASs be tested for prior to starting GZR/ELB and if present, to extend treatment to 16 weeks and add ribavirin [131].

5.7 Newest Approved Regimens: Glecaprevir/Pibrentasvir and Sofosbuvir/Velpatasvir/Voxilaprevir

The treatment combination of GLE/PIB was approved in the United States in August 2017 for the pan-genotypic treatment of HCV with or without cirrhosis, including patients with moderate to severe kidney disease and those who are on dialysis. It was also approved for adult patients with GT1 infection previously treated with a regimen either containing an NS5A inhibitor or an NS3/4A protease inhibitor, but not both, speaking to its higher threshold for the development of resistance compared to earlier treatment regimens. In a pooled analysis of eight phase II and III studies that evaluated the safety and efficacy of GLE/PIB, fewer than 1% (22/2256) experienced virologic failure [139]. Next-generation sequencing (using a cutoff of 15%) revealed that the presence of baseline polymorphisms in NS3 and/or NS5A did not have an impact on SVR12 rates for patients infected with GT1 or GT2 and no treatment failures were observed in those with GT4, 5, or 6 infection. The majority of treatment failures were in GT3

(18/22). Treatment-emergent substitutions were detected in both NS3 and NS5A in the majority of patients. A single 156G mutation conferred resistance to GLE, but multiple substitutions in NS5A were required to confer resistance to PIB (i.e., A30K + Y93H in GT3-infected patients) [139].

The SOF/VEL/VOX treatment regimen was approved in the United States in July 2017 for all HCV genotypes as a salvage regimen for those who had failed prior DAA treatment. A pooled analysis of four phase II trials revealed that the presence of baseline NS3, NS5A, or NS5B had no impact on treatment outcome for either DAA-naïve patients or DAA-experienced patients [140]. In the phase III POLARIS-1 trial, 97% of patients with baseline resistance to NS3 or NS5A inhibitors achieved SVR with a 12-week course of SOF/VEL/VOX. Of the six virologic failures, two individuals with GT1a infection had baseline NS3 Q80K RAS, and one patient with GT4 infection had treatment-emergent substitution of the NS5A Y93H RAS [141].

6 Conclusions

The advent of DAAs has brought about a sudden renaissance in HCV treatment with SVR rates now routinely eclipsing 90–95% in clinical trials. However, baseline RASs to DAAs can result in a significant effect on the treatment outcomes and prevent the achievement of SVR in a significant number of individuals. The incidence of treatment failure due to the existence of baseline RASs may continue to decrease in frequency with the continued development of drug regimens with increasing potency, barriers to resistance, and genotypic efficacy. However, effective treatment options in patients who initially fail an IFN-free DAA regimen have not been rigorously identified, and this represents an area of active investigation. Treatment alternatives that are currently being explored include lengthening duration, the addition of ribavirin, and the addition of PEG-IFN. This and other areas of research including the potential economic impact of baseline RAS testing prior to treatment promise to clarify and streamline our management of HCV to help prevent treatment failures. Moreover, further molecular understanding of the mechanisms of DAA resistance promises to uncover new and fundamental aspects of HCV virology as well as treatment strategies for flaviviruses on the whole.

References

1. Ray SC, Thomas DL (2015) Hepatitis C. In: Mandell, Douglas, and Bennett's principles and practice of infectious diseases, 8th edn. Elsevier Saunders, Philadelphia, pp 1904–1927
2. Powdrill MH, Tchesnokov EP, Kozak RA, Russell RS, Martin R, Svarovskaia ES et al (2011) Contribution of a mutational bias in hepatitis C virus replication to the genetic barrier in the development of drug resistance. Proc Natl Acad Sci U S A 108:20509–20513
3. Mohd Hanafiah K, Groeger J, Flaxman AD, Wiersma ST (2013) Global epidemiology of hepatitis C virus infection: new estimates of age-specific antibody to HCV seroprevalence. Hepatology 57:1333–1342
4. Alter HJ, Holland PV, Morrow AG, Purcell RH, Feinstone SM, Moritsugu Y (1975) Clinical and serological analysis of transfusion-associated hepatitis. Lancet 306:838–841
5. Feinstone SM, Kapikian AZ, Purcell RH, Alter HJ, Holland PV (1975) Transfusion-associated hepatitis not due to viral hepatitis type A or B. N Engl J Med 292:767–770
6. Hoofnagle JH, Mullen KD, Jones DB, Rustgi V, Di Bisceglie A, Peters M et al (1986) Treatment of chronic non-A, non-B hepatitis with recombinant human alpha interferon. N Engl J Med 315:1575–1578
7. Choo QL, Kuo G, Weiner AJ, Overby LR, Bradley DW, Houghton M (1989) Isolation of a cDNA clone derived from a blood-borne non-A, non-B viral hepatitis genome. Science 244:359–362
8. Davis GL, Balart LA, Schiff ER, Lindsay K, Bodenheimer HC, Perrillo RP et al (1989) Treatment of chronic hepatitis C with recombinant interferon alfa. N Engl J Med 321:1501–1506
9. Di Bisceglie AM, Martin P, Kassianides C, Lisker-Melman M, Murray L, Waggoner J et al (1989) Recombinant interferon alfa therapy for chronic hepatitis C. N Engl J Med 321:1506–1510
10. McHutchison JG, Gordon SC, Schiff ER, Shiffman ML, Lee WM, Rustgi VK et al (1998) Interferon alfa-2b alone or in combination with ribavirin as initial treatment for chronic hepatitis C. N Engl J Med 339:1485–1492
11. Fried MW, Shiffman ML, Reddy KR, Smith C, Marinos G, Gonçales FL et al (2002) Peginterferon alfa-2a plus ribavirin for chronic hepatitis C virus infection. N Engl J Med 347:975–982
12. Manns MP, McHutchison JG, Gordon SC, Rustgi VK, Shiffman M, Reindollar R et al (2001) Peginterferon alfa-2b plus ribavirin compared with interferon alfa-2b plus ribavirin for initial treatment of chronic hepatitis C: a randomised trial. Lancet 358:958–965
13. Poynard T, Bedossa P, Chevallier M, Mathurin P, Lemonnier C, Trepo C et al (1995) A comparison of three interferon alfa-2b regimens for the long-term treatment of chronic non-A, non-B hepatitis. N Engl J Med 332:1457–1463
14. Wakita T, Pietschmann T, Kato T, Date T, Miyamoto M, Zhao Z et al (2005) Production of infectious hepatitis C virus in tissue culture from a cloned viral genome. Nat Med 11:791–796
15. Jacobson IM, McHutchison JG, Dusheiko G, Di Bisceglie AM, Reddy KR, Bzowej NH et al (2011) Telaprevir for previously untreated chronic hepatitis C virus infection. N Engl J Med 364:2405–2416
16. Poordad F, McCone J, Bacon BR, Bruno S, Manns MP, Sulkowski MS et al (2011) Boceprevir for untreated chronic HCV genotype 1 infection. N Engl J Med 364:1195–1206
17. Manns M, Marcellin P, Poordad F, de Araujo ES, Buti M, Horsmans Y et al (2014) Simeprevir with pegylated interferon alfa 2a or 2b plus ribavirin in treatment-naive patients with chronic hepatitis C virus genotype 1 infection (QUEST-2): a randomised, double-blind, placebo-controlled phase 3 trial. Lancet 384:414–426
18. Membreno FE, Lawitz EJ (2011) The HCV NS5B nucleoside and non-nucleoside inhibitors. Clin Liver Dis 15:611–626
19. Lawitz E, Mangia A, Wyles D, Rodriguez-Torres M, Hassanein T, Gordon SC et al (2013) Sofosbuvir for previously untreated chronic hepatitis C infection. N Engl J Med 368:1878–1887
20. Jacobson IM, Gordon SC, Kowdley KV, Yoshida EM, Rodriguez-Torres M, Sulkowski MS et al (2013) Sofosbuvir for hepatitis C genotype 2 or 3 in patients without treatment options. N Engl J Med 368:1867–1877
21. Afdhal N, Reddy KR, Nelson DR, Lawitz E, Gordon SC, Schiff E et al (2014) Ledipasvir and sofosbuvir for previously treated HCV genotype 1 infection. N Engl J Med 370:1483–1493
22. Afdhal N, Zeuzem S, Kwo P, Chojkier M, Gitlin N, Puoti M et al (2014) Ledipasvir

and sofosbuvir for untreated HCV genotype 1 infection. N Engl J Med 370:1889–1898

23. Kowdley KV, Gordon SC, Reddy KR, Rossaro L, Bernstein DE, Lawitz E et al (2014) Ledipasvir and sofosbuvir for 8 or 12 weeks for chronic HCV without cirrhosis. N Engl J Med 370:1879–1888
24. Lawitz E, Sulkowski MS, Ghalib R, Rodriguez-Torres M, Younossi ZM, Corregidor A et al (2014) Simeprevir plus sofosbuvir, with or without ribavirin, to treat chronic infection with hepatitis C virus genotype 1 in non-responders to pegylated interferon and ribavirin and treatment-naive patients: the COSMOS randomised study. Lancet 384:1756–1765
25. Feld JJ, Kowdley KV, Coakley E, Sigal S, Nelson DR, Crawford D et al (2014) Treatment of HCV with ABT-450/r–ombitasvir and dasabuvir with ribavirin. N Engl J Med 370:1594–1603
26. Poordad F, Hezode C, Trinh R, Kowdley KV, Zeuzem S, Agarwal K et al (2014) ABT-450/r–ombitasvir and dasabuvir with ribavirin for hepatitis C with cirrhosis. N Engl J Med 370:1973–1982
27. Sulkowski MS, Vargas HE, Di Bisceglie AM, Kuo A, Reddy KR, Lim JK et al (2016) Effectiveness of simeprevir plus sofosbuvir, with or without ribavirin, in real-world patients with HCV genotype 1 infection. Gastroenterology 150:419–429
28. Terrault NA, Zeuzem S, Di Bisceglie AM, Lim JK, Pockros PJ, Frazier LM et al (2016) Effectiveness of ledipasvir-sofosbuvir combination in patients with hepatitis C virus infection and factors associated with sustained virologic response. Gastroenterology 151:1131–1140.e1135
29. Welzel TM, Nelson DR, Morelli G, Di Bisceglie A, Reddy RK, Kuo A et al (2016) Effectiveness and safety of sofosbuvir plus ribavirin for the treatment of HCV genotype 2 infection: results of the real-world, clinical practice HCV-TARGET study. Gut 66:1844–1852
30. Feld JJ, Maan R, Zeuzem S, Kuo A, Nelson DR, Di Bisceglie AM et al (2016) Effectiveness and safety of Sofosbuvir-based regimens for chronic HCV genotype 3 infection: results of the HCV-TARGET study. Clin Infect Dis 63:776–783
31. Reddy KR, Lim JK, Kuo A, Di Bisceglie AM, Galati JS, Morelli G et al (2016) All-oral direct-acting antiviral therapy in HCV-advanced liver disease is effective in real-world practice: observations through HCV-TARGET database. Aliment Pharmacol Ther 45:115–126
32. Laskus T, Wilkinson J, Gallegos-Orozco JF, Radkowski M, Adair DM, Nowicki M et al (2004) Analysis of hepatitis C virus quasispecies transmission and evolution in patients infected through blood transfusion. Gastroenterology 127:764–776
33. Martell M, Esteban JI, Quer J, Genesca J, Weiner A, Esteban R et al (1992) Hepatitis C virus (HCV) circulates as a population of different but closely related genomes: quasispecies nature of HCV genome distribution. J Virol 66:3225–3229
34. Andino R, Domingo E (2015) Viral quasispecies. Virology 479-480:46–51
35. Perales C, Quer J, Gregori J, Esteban J, Domingo E (2015) Resistance of hepatitis C virus to inhibitors: complexity and clinical implications. Viruses 7:5746–5766
36. Bartenschlager R, Lohmann V (2000) Replication of hepatitis C virus. J Gen Virol 81:1631–1648
37. Neumann AU (1998) Hepatitis C viral dynamics in vivo and the antiviral efficacy of interferon-therapy. Science 282:103–107
38. Pawlotsky J-M (2016) Hepatitis C virus resistance to direct-acting antiviral drugs in interferon-free regimens. Gastroenterology 151:70–86
39. Chevaliez S, Rodriguez C, Pawlotsky JM (2012) New virologic tools for management of chronic hepatitis B and C. Gastroenterology 142:1303–1313.e1301
40. Echeverria N, Moratorio G, Cristina J, Moreno P (2015) Hepatitis C virus genetic variability and evolution. World J Hepatol 7:831–845
41. Leitner T, Halapi E, Scarlatti G, Rossi P, Albert J, Fenyo EM et al (1993) Analysis of heterogeneous viral populations by direct DNA sequencing. BioTechniques 15:120–127
42. Fourati S, Pawlotsky J-M (2015) Virologic tools for HCV drug resistance testing. Viruses 7:6346–6359
43. Ninomiya M, Ueno Y, Funayama R, Nagashima T, Nishida Y, Kondo Y et al (2011) Use of Illumina deep sequencing technology to differentiate hepatitis C virus variants. J Clin Microbiol 50:857–866
44. Thomson E, Ip CLC, Badhan A, Christiansen MT, Adamson W, Ansari MA et al (2016) Comparison of next-generation sequencing technologies for comprehensive assessment of full-length hepatitis C viral genomes. J Clin Microbiol 54:2470–2484

45. Halfon P, Locarnini S (2011) Hepatitis C virus resistance to protease inhibitors. J Hepatol 55:192–206
46. Kwo PY, Lawitz EJ, McCone J, Schiff ER, Vierling JM, Pound D et al (2010) Efficacy of boceprevir, an NS3 protease inhibitor, in combination with peginterferon alfa-2b and ribavirin in treatment-naive patients with genotype 1 hepatitis C infection (SPRINT-1): an open-label, randomised, multicentre phase 2 trial. Lancet 376:705–716
47. McHutchison JG, Everson GT, Gordon SC, Jacobson IM, Sulkowski M, Kauffman R et al (2009) Telaprevir with peginterferon and ribavirin for chronic HCV genotype 1 infection. N Engl J Med 360:1827–1838
48. McCauley JA, Rudd MT (2016) Hepatitis C virus NS3/4a protease inhibitors. Curr Opin Pharmacol 30:84–92
49. Lenz O, Verbinnen T, Fevery B, Tambuyzer L, Vijgen L, Peeters M et al (2015) Virology analyses of HCV isolates from genotype 1-infected patients treated with simeprevir plus peginterferon/ribavirin in phase IIb/III studies. J Hepatol 62:1008–1014
50. Kieffer TL, De Meyer S, Bartels DJ, Sullivan JC, Zhang EZ, Tigges A et al (2012) Hepatitis C viral evolution in genotype 1 treatment-naïve and treatment-experienced patients receiving telaprevir-based therapy in clinical trials. PLoS One 7:e34372
51. Lenz O, Verbinnen T, Lin TI, Vijgen L, Cummings MD, Lindberg J et al (2010) In vitro resistance profile of the hepatitis C virus NS3/4A protease inhibitor TMC435. Antimicrob Agents Chemother 54:1878–1887
52. Howe AYM, Black S, Curry S, Ludmerer SW, Liu R, Barnard RJO et al (2014) Virologic resistance analysis from a phase 2 study of MK-5172 combined with pegylated interferon/ribavirin in treatment-naive patients with hepatitis C virus genotype 1 infection. Clin Infect Dis 59:1657–1665
53. Pilot-Matias T, Tripathi R, Cohen D, Gaultier I, Dekhtyar T, Lu L et al (2014) In vitro and in vivo antiviral activity and resistance profile of the hepatitis C virus NS3/4A protease inhibitor ABT-450. Antimicrob Agents Chemother 59:988–997
54. Summa V, Ludmerer SW, McCauley JA, Fandozzi C, Burlein C, Claudio G et al (2012) MK-5172, a selective inhibitor of hepatitis C virus NS3/4a protease with broad activity across genotypes and resistant variants. Antimicrob Agents Chemother 56:4161–4167
55. Moradpour D, Penin F (2013) Hepatitis C virus proteins: from structure to function. Curr Top Microbiol Immunol 369:113–142
56. Kowdley KV, Lawitz E, Poordad F, Cohen DE, Nelson DR, Zeuzem S et al (2014) Phase 2b trial of interferon-free therapy for hepatitis C virus genotype 1. N Engl J Med 370:222–232
57. Howe JA, Long J, Black S, Chase R, McMonagle P, Curry S et al (2014) Clinical implications of detectable baseline hepatitis C virus-genotype 1 NS3/4A-protease variants on the efficacy of boceprevir combined with peginterferon/ribavirin. Open Forum Infect Dis 1. https://doi.org/10.1093/ofid/ofu078
58. Jensen SB, Serre SBN, Humes DG, Ramirez S, Li Y-P, Bukh J et al (2015) Substitutions at NS3 residue 155, 156, or 168 of hepatitis C virus genotypes 2 to 6 induce complex patterns of protease inhibitor resistance. Antimicrob Agents Chemother 59:7426–7436
59. Jiang M, Mani N, Lin C, Ardzinski A, Nelson M, Reagan D et al (2013) In vitro phenotypic characterization of hepatitis C virus NS3 protease variants observed in clinical studies of telaprevir. Antimicrob Agents Chemother 57:6236–6245
60. Lawitz E, Sulkowski M, Jacobson I, Kraft WK, Maliakkal B, Al-Ibrahim M et al (2013) Characterization of vaniprevir, a hepatitis C virus NS3/4A protease inhibitor, in patients with HCV genotype 1 infection: safety, antiviral activity, resistance, and pharmacokinetics. Antivir Res 99:214–220
61. McPhee F, Friborg J, Levine S, Chen C, Falk P, Yu F et al (2012) Resistance analysis of the hepatitis C virus NS3 protease inhibitor asunaprevir. Antimicrob Agents Chemother 56:3670–3681
62. Serre SBN, Jensen SB, Ghanem L, Humes DG, Ramirez S, Li Y-P et al (2016) Hepatitis C virus genotype 1 to 6 protease inhibitor escape variants: in vitro selection, fitness, and resistance patterns in the context of the infectious viral life cycle. Antimicrob Agents Chemother 60:3563–3578
63. Verbinnen T, Fevery B, Vijgen L, Jacobs T, De Meyer S, Lenz O (2015) In vitro activity of simeprevir against hepatitis C virus genotype 1 clinical isolates and its correlation with NS3 sequence and site-directed mutants. Antimicrob Agents Chemother 59:7548–7557
64. Sarrazin C (2016) The importance of resistance to direct antiviral drugs in HCV

infection in clinical practice. J Hepatol 64:486–504
65. Sullivan JC, De Meyer S, Bartels DJ, Dierynck I, Zhang EZ, Spanks J et al (2013) Evolution of treatment-emergent resistant variants in telaprevir phase 3 clinical trials. Clin Infect Dis 57:221–229
66. Buti M, Gordon SC, Zuckerman E, Lawitz E, Calleja JL, Hofer H et al (2015) Grazoprevir, Elbasvir, and Ribavirin for chronic hepatitis C virus genotype 1 infection after failure of pegylated interferon and ribavirin with an earlier-generation protease inhibitor: final 24-week results from C-SALVAGE: table 1. Clin Infect Dis 62:32–36
67. Pawlotsky J-M (2013) NS5A inhibitors in the treatment of hepatitis C. J Hepatol 59:375–382
68. Penin F, Brass V, Appel N, Ramboarina S, Montserret R, Ficheux D et al (2004) Structure and function of the membrane anchor domain of hepatitis C virus nonstructural protein 5A. J Biol Chem 279:40835–40843
69. Lambert SM, Langley DR, Garnett JA, Angell R, Hedgethorne K, Meanwell NA et al (2014) The crystal structure of NS5A domain 1 from genotype 1a reveals new clues to the mechanism of action for dimeric HCV inhibitors. Protein Sci 23:723–734
70. Love RA, Brodsky O, Hickey MJ, Wells PA, Cronin CN (2009) Crystal structure of a novel dimeric form of NS5A domain I protein from hepatitis C virus. J Virol 83:4395–4403
71. Tellinghuisen TL, Marcotrigiano J, Rice CM (2005) Structure of the zinc-binding domain of an essential component of the hepatitis C virus replicase. Nature 435:374–379
72. de Chassey B, Navratil V, Tafforeau L, Andre P, Rabourdin-Combe C, Lotteau V (2008) Hepatitis C virus infection protein network. Int J Infect Dis 12:e175
73. Berger C, Romero-Brey I, Radujkovic D, Terreux R, Zayas M, Paul D et al (2014) Daclatasvir-like inhibitors of NS5A block early biogenesis of hepatitis C virus–induced membranous replication factories, independent of RNA replication. Gastroenterology 147:1094–1105.e1025
74. Gao M, Nettles RE, Belema M, Snyder LB, Nguyen VN, Fridell RA et al (2010) Chemical genetics strategy identifies an HCV NS5A inhibitor with a potent clinical effect. Nature 465:96–100
75. Fridell RA, Qiu D, Wang C, Valera L, Gao M (2010) Resistance analysis of the hepatitis C virus NS5A inhibitor BMS-790052 in an in vitro replicon system. Antimicrob Agents Chemother 54:3641–3650
76. Lee C, Ma H, Hang JQ, Leveque V, Sklan EH, Elazar M et al (2011) The hepatitis C virus NS5A inhibitor (BMS-790052) alters the subcellular localization of the NS5A non-structural viral protein. Virology 414:10–18
77. Bartenschlager R, Lohmann V, Penin F (2013) The molecular and structural basis of advanced antiviral therapy for hepatitis C virus infection. Nat Rev Microbiol 11:482–496
78. Bartels DJ, Sullivan JC, Zhang EZ, Tigges AM, Dorrian JL, De Meyer S et al (2012) Hepatitis C virus variants with decreased sensitivity to direct-acting antivirals (DAAs) were rarely observed in DAA-naive patients prior to treatment. J Virol 87:1544–1553
79. Plaza Z, Soriano V, Vispo E, del Mar Gonzalez M, Barreiro P, Seclén E et al (2012) Prevalence of natural polymorphisms at the HCV NS5A gene associated with resistance to daclatasvir, an NS5A inhibitor. Antivir Ther 17:921–926
80. Wang C, Huang H, Valera L, Sun JH, O'Boyle DR, Nower PT et al (2012) Hepatitis C virus RNA elimination and development of resistance in replicon cells treated with BMS-790052. Antimicrob Agents Chemother 56:1350–1358
81. Wang C, Jia L, Huang H, Qiu D, Valera L, Huang X et al (2011) In vitro activity of BMS-790052 on hepatitis C virus genotype 4 NS5A. Antimicrob Agents Chemother 56:1588–1590
82. Wang C, Jia L, O'Boyle DR, Sun JH, Rigat K, Valera L et al (2014) Comparison of daclatasvir resistance barriers on NS5A from hepatitis C virus genotypes 1 to 6: implications for cross-genotype activity. Antimicrob Agents Chemother 58:5155–5163
83. Wang C, Valera L, Jia L, Kirk MJ, Gao M, Fridell RA (2012) In vitro activity of daclatasvir on hepatitis C virus genotype 3 NS5A. Antimicrob Agents Chemother 57:611–613
84. Sarrazin C, Dvory-Sobol H, Svarovskaia ES, Doehle BP, Pang PS, Chuang S-M et al (2016) Prevalence of resistance-associated substitutions in HCV NS5A, NS5B, or NS3 and outcomes of treatment with ledipasvir and sofosbuvir. Gastroenterology 151:501–512.e501
85. Cheng G, Tian Y, Doehle B, Peng B, Corsa A, Lee Y-J et al (2016) In vitro antiviral activity and resistance profile characterization of the hepatitis C virus NS5A inhibitor ledipasvir.

Antimicrob Agents Chemother 60:1847–1853
86. Krishnan P, Beyer J, Mistry N, Koev G, Reisch T, DeGoey D et al (2014) In vitro and in vivo antiviral activity and resistance profile of ombitasvir, an inhibitor of hepatitis C virus NS5A. Antimicrob Agents Chemother 59:979–987
87. Kuntzen T, Timm J, Berical A, Lennon N, Berlin AM, Young SK et al (2008) Naturally occurring dominant resistance mutations to hepatitis C virus protease and polymerase inhibitors in treatment-naïve patients. Hepatology 48:1769–1778
88. Lahser FC, Bystol K, Curry S, McMonagle P, Xia E, Ingravallo P et al (2016) The combination of grazoprevir, a hepatitis C virus (HCV) NS3/4A protease inhibitor, and elbasvir, an HCV NS5A inhibitor, demonstrates a high genetic barrier to resistance in HCV genotype 1a replicons. Antimicrob Agents Chemother 60:2954–2964
89. Lawitz EJ, Dvory-Sobol H, Doehle BP, Worth AS, McNally J, Brainard DM et al (2016) Clinical resistance to velpatasvir (GS-5816), a novel pan-genotypic inhibitor of the hepatitis C virus NS5A protein. Antimicrob Agents Chemother 60:5368–5378
90. Liu R, Curry S, McMonagle P, Yeh WW, Ludmerer SW, Jumes PA et al (2015) Susceptibilities of genotype 1a, 1b, and 3 hepatitis C virus variants to the NS5A inhibitor elbasvir. Antimicrob Agents Chemother 59:6922–6929
91. Paolucci S, Fiorina L, Mariani B, Gulminetti R, Novati S, Barbarini G et al (2013) Naturally occurring resistance mutations to inhibitors of HCV NS5A region and NS5B polymerase in DAA treatment-naïve patients. Virol J 10:355
92. Wong KA, Worth A, Martin R, Svarovskaia E, Brainard DM, Lawitz E et al (2013) Characterization of hepatitis C virus resistance from a multiple-dose clinical trial of the novel NS5A inhibitor GS-5885. Antimicrob Agents Chemother 57:6333–6340
93. Yoshimi S, Imamura M, Murakami E, Hiraga N, Tsuge M, Kawakami Y et al (2015) Long term persistence of NS5A inhibitor-resistant hepatitis C virus in patients who failed daclatasvir and asunaprevir therapy. J Med Virol 87:1913–1920
94. Krishnan P, Schnell G, Tripathi R, Beyer J, Reisch T, Zhang X et al (2015) Analysis of hepatitis C virus genotype 1b resistance variants in Japanese patients treated with paritaprevir-ritonavir and ombitasvir. Antimicrob Agents Chemother 60:1106–1113
95. Eltahla A, Luciani F, White P, Lloyd A, Bull R (2015) Inhibitors of the hepatitis C virus polymerase; mode of action and resistance. Viruses 7:5206–5224
96. Soriano V, Vispo E, de Mendoza C, Labarga P, Fernandez-Montero JV, Poveda E et al (2013) Hepatitis C therapy with HCV NS5B polymerase inhibitors. Expert Opin Pharmacother 14:1161–1170
97. De Clercq E (2015) Curious (old and new) antiviral nucleoside analogues with intriguing therapeutic potential. Curr Med Chem 22:3866–3880
98. Ago H, Adachi T, Yoshida A, Yamamoto M, Habuka N, Yatsunami K et al (1999) Crystal structure of the RNA-dependent RNA polymerase of hepatitis C virus. Structure 7:1417–1426
99. Bressanelli S, Tomei L, Roussel A, Incitti I, Vitale RL, Mathieu M et al (1999) Crystal structure of the RNA-dependent RNA polymerase of hepatitis C virus. Proc Natl Acad Sci U S A 96:13034–13039
100. Lesburg CA, Cable MB, Ferrari E, Hong Z, Mannarino AF, Weber PC (1999) Crystal structure of the RNA-dependent RNA polymerase from hepatitis C virus reveals a fully encircled active site. Nat Struct Biol 6:937–943
101. Lam AM, Espiritu C, Bansal S, Micolochick Steuer HM, Niu C, Zennou V et al (2012) Genotype and subtype profiling of PSI-7977 as a nucleotide inhibitor of hepatitis C virus. Antimicrob Agents Chemother 56:3359–3368
102. Donaldson EF, Harrington PR, O'Rear JJ, Naeger LK (2014) Clinical evidence and bioinformatics characterization of potential hepatitis C virus resistance pathways for sofosbuvir. Hepatology 61:56–65
103. Lawitz E, Lalezari JP, Hassanein T, Kowdley KV, Poordad FF, Sheikh AM et al (2013) Sofosbuvir in combination with peginterferon alfa-2a and ribavirin for non-cirrhotic, treatment-naive patients with genotypes 1, 2, and 3 hepatitis C infection: a randomised, double-blind, phase 2 trial. Lancet Infect Dis 13:401–408
104. Svarovskaia ES, Dvory-Sobol H, Parkin N, Hebner C, Gontcharova V, Martin R et al (2014) Infrequent development of resistance in genotype 1–6 hepatitis C virus–infected subjects treated with sofosbuvir in phase 2 and 3 clinical trials. Clin Infect Dis 59:1666–1674
105. Svarovskaia ES, Gane E, Dvory-Sobol H, Martin R, Doehle B, Hedskog C et al

(2015) L159F and V321A sofosbuvir-associated hepatitis C virus NS5B substitutions. J Infect Dis 213:1240–1247

106. Zeuzem S, Jacobson IM, Baykal T, Marinho RT, Poordad F, Bourlière M et al (2014) Retreatment of HCV with ABT-450/r–ombitasvir and dasabuvir with ribavirin. N Engl J Med 370:1604–1614

107. Di Maio VC, Cento V, Mirabelli C, Artese A, Costa G, Alcaro S et al (2014) Hepatitis C virus genetic variability and the presence of NS5B resistance-associated mutations as natural polymorphisms in selected genotypes could affect the response to NS5B inhibitors. Antimicrob Agents Chemother 58:2781–2797

108. Krishnan P, Tripathi R, Schnell G, Reisch T, Beyer J, Irvin M et al (2015) Resistance analysis of baseline and treatment-emergent variants in hepatitis C virus genotype 1 in the AVIATOR study with paritaprevir-ritonavir, ombitasvir, and dasabuvir. Antimicrob Agents Chemother 59:5445–5454

109. Berger KL, Sarrazin C, Nelson DR, Scherer J, Sha N, Marquis M et al (2016) Resistance analyses of HCV NS3/4A protease and NS5B polymerase from clinical studies of deleobuvir and faldaprevir. PLoS One 11: e0160668

110. Kati W, Koev G, Irvin M, Beyer J, Liu Y, Krishnan P et al (2014) In vitro activity and resistance profile of dasabuvir, a nonnucleoside hepatitis C virus polymerase inhibitor. Antimicrob Agents Chemother 59:1505–1511

111. Lu L, Dekhtyar T, Masse S, Pithawalla R, Krishnan P, He W et al (2007) Identification and characterization of mutations conferring resistance to an HCV RNA dependent RNA polymerase inhibitor in vitro. Antivir Res 76:93–97

112. Ji H, Kozak RA, Biondi MJ, Pilon R, Vallee D, Liang BB et al (2015) Next generation sequencing of the hepatitis C virus NS5B gene reveals potential novel S282 drug resistance mutations. Virology 477:1–9

113. Kukolj G, McGibbon GA, McKercher G, Marquis M, Lefèbvre S, Thauvette L et al (2005) Binding site characterization and resistance to a class of non-nucleoside inhibitors of the hepatitis C virus NS5B polymerase. J Biol Chem 280:39260–39267

114. Abergel A, Asselah T, Metivier S, Kersey K, Jiang D, Mo H et al (2016) Ledipasvir-sofosbuvir in patients with hepatitis C virus genotype 5 infection: an open-label, multicentre, single-arm, phase 2 study. Lancet Infect Dis 16:459–464

115. Abergel A, Metivier S, Samuel D, Jiang D, Kersey K, Pang PS et al (2016) Ledipasvir plus sofosbuvir for 12 weeks in patients with hepatitis C genotype 4 infection. Hepatology 64:1049–1056

116. Bourlière M, Bronowicki J-P, de Ledinghen V, Hézode C, Zoulim F, Mathurin P et al (2015) Ledipasvir-sofosbuvir with or without ribavirin to treat patients with HCV genotype 1 infection and cirrhosis non-responsive to previous protease-inhibitor therapy: a randomised, double-blind, phase 2 trial (SIRIUS). Lancet Infect Dis 15:397–404

117. Charlton M, Everson GT, Flamm SL, Kumar P, Landis C, Brown RS Jr et al (2015) Ledipasvir and sofosbuvir plus ribavirin for treatment of HCV infection in patients with advanced liver disease. Gastroenterology 149:649–659

118. Gane EJ, Hyland RH, An D, Svarovskaia E, Pang PS, Brainard D et al (2015) Efficacy of ledipasvir and sofosbuvir, with or without ribavirin, for 12 weeks in patients with HCV genotype 3 or 6 infection. Gastroenterology 149:1454–1461.e1451

119. Lawitz E, Poordad FF, Pang PS, Hyland RH, Ding X, Mo H et al (2014) Sofosbuvir and ledipasvir fixed-dose combination with and without ribavirin in treatment-naive and previously treated patients with genotype 1 hepatitis C virus infection (LONESTAR): an open-label, randomised, phase 2 trial. Lancet 383:515–523

120. Mizokami M, Dvory-Sobol H, Izumi N, Nishiguchi S, Doehle B, Svarovskaia ES et al (2016) Resistance analyses of Japanese hepatitis C-infected patients receiving sofosbuvir or ledipasvir/sofosbuvir containing regimens in phase 3 studies. J Viral Hepat 23:780–788

121. Curry MP, O'Leary JG, Bzowej N, Muir AJ, Korenblat KM, Fenkel JM et al (2015) Sofosbuvir and velpatasvir for HCV in patients with decompensated cirrhosis. N Engl J Med 373:2618–2628

122. Feld JJ, Jacobson IM, Hézode C, Asselah T, Ruane PJ, Gruener N et al (2015) Sofosbuvir and velpatasvir for HCV genotype 1, 2, 4, 5, and 6 infection. N Engl J Med 373:2599–2607

123. Foster GR, Afdhal N, Roberts SK, Bräu N, Gane EJ, Pianko S et al (2015) Sofosbuvir and velpatasvir for HCV genotype 2 and 3 infection. N Engl J Med 373:2608–2617

124. Pianko S, Flamm SL, Shiffman ML, Kumar S, Strasser SI, Dore GJ et al (2015) Sofosbuvir plus velpatasvir combination therapy for treatment-experienced patients with

genotype 1 or 3 hepatitis C virus infection. Ann Intern Med 163:809

125. Sulkowski MS, Gardiner DF, Rodriguez-Torres M, Reddy KR, Hassanein T, Jacobson I et al (2014) Daclatasvir plus sofosbuvir for previously treated or untreated chronic HCV infection. N Engl J Med 370:211–221
126. Wyles DL, Ruane PJ, Sulkowski MS, Dieterich D, Luetkemeyer A, Morgan TR et al (2015) Daclatasvir plus sofosbuvir for HCV in patients coinfected with HIV-1. N Engl J Med 373:714–725
127. Leroy V, Angus P, Bronowicki J-P, Dore GJ, Hezode C, Pianko S et al (2016) Daclatasvir, sofosbuvir, and ribavirin for hepatitis C virus genotype 3 and advanced liver disease: a randomized phase III study (ALLY-3+). Hepatology 63:1430–1441
128. Sarrazin C, Lathouwers E, Peeters M, Daems B, Buelens A, Witek J et al (2015) Prevalence of the hepatitis C virus NS3 polymorphism Q80K in genotype 1 patients in the European region. Antivir Res 116:10–16
129. Kwo P, Gitlin N, Nahass R, Bernstein D, Etzkorn K, Rojter S et al (2016) Simeprevir plus sofosbuvir (12 and 8 weeks) in hepatitis C virus genotype 1-infected patients without cirrhosis: OPTIMIST-1, a phase 3, randomized study. Hepatology 64:370–380
130. Lawitz E, Matusow G, DeJesus E, Yoshida EM, Felizarta F, Ghalib R et al (2016) Simeprevir plus sofosbuvir in patients with chronic hepatitis C virus genotype 1 infection and cirrhosis: a phase 3 study (OPTIMIST-2). Hepatology 64:360–369
131. AASLD/IDSA HCV Guidance Panel (2015) Hepatitis C guidance: AASLD-IDSA recommendations for testing, managing, and treating adults infected with hepatitis C virus. Hepatology 62:932–954
132. Krishnan P, Tripathi R, Schnell G, Reisch T, Beyer J, Irvin M et al (2014) Pooled analysis of resistance in patients treated with ombitasvir/ABT-450/r and dasabuvir with or without ribavirin in Phase 2 and Phase 3 clinical trials. Hepatology 60:1134a–1135a
133. Schnell G, Tripathi R, Beyer J, Reisch T, Krishnan P, Lu L et al (2015) Hepatitis C virus genotype 4 resistance and subtype demographic characterization of patients treated with ombitasvir plus paritaprevir/ritonavir. Antimicrob Agents Chemother 59:6807–6815
134. Zeuzem S, Ghalib R, Reddy KR, Pockros PJ, Ari ZB, Zhao Y et al (2015) Grazoprevir–Elbasvir combination therapy for treatment-naive cirrhotic and noncirrhotic patients with chronic hepatitis C virus genotype 1, 4, or 6 infection. Ann Intern Med 163:1–13
135. Kwo P, Gane EJ, Peng C-Y, Pearlman B, Vierling JM, Serfaty L et al (2017) Effectiveness of elbasvir and grazoprevir combination, with or without ribavirin, for treatment-experienced patients with chronic hepatitis C infection. Gastroenterology 152:164–175.e164
136. Jacobson IM, Asante-Appiah E, Wong P, Black TA, Howe AYM, Wahl J et al (2015) Prevalence and impact of baseline NSA resistance associated variants (RAVs) on the efficacy of elbasvir/grazoprevir (EBR/GZR) against GT1a infection. Hepatology 62:1393A
137. Komatsu TE, Boyd S, Sherwat A, Tracy L, Naeger LK, O'Rear JJ et al (2017) Regulatory analysis of effects of hepatitis C virus NS5A polymorphisms on efficacy of elbasvir and grazoprevir. Gastroenterology 152:586–597
138. Zeuzem S, Rockstroh JK, Kwo P, Roth D, Lawitz E, Sulkowski M et al (2015) Predictors of response to grazoprevir/elbasvir among HCV genotype 1 (GT1)–infected patients: integrated analysis of phase 2-3 trials. Hepatology 62:554A
139. Krishnan P, Schnell G, Tripathi R, Ng T, Reisch T, Beyer J et al (2017) Pooled resistance analysis in HCV genotype 1–6-infected patients treated with glecaprevir/pibrentasvir in phase 2 and 3 clinical trials. J Hepatol 66: S500
140. Reau N, Nguyen MH, Kowdley KV, Gane E, Dvory-Sobol H, Svarovskaia E et al (2016) Integrated resistance analyses of HCV-infected patients treated with sofosbuvir, velpatasvir, and voxilaprevir for 8 and 12 weeks from phase 2 studies. Hepatology 64:415A–416A
141. Bourliere M, Gordon SC, Flamm SL, Cooper CL, Ramji A, Tong M et al (2017) Sofosbuvir, velpatasvir, and voxilaprevir for previously treated HCV infection. N Engl J Med 376:2134–2146

Chapter 2

Similarities and Differences Between HCV Pseudoparticle (HCVpp) and Cell Culture HCV (HCVcc) in the Study of HCV

Laura Riva and Jean Dubuisson

Abstract

For a long time, the study of the HCV infectious cycle has been a major challenge for researchers because of the difficulties in generating an efficient cell culture system leading to a productive viral infection. The development of HCVpp and later on HCVcc model allowing for functional studies of HCV in cell culture completely revolutionized HCV research. The aim of this review is to provide the reader with a brief overview of the development of these two models. We describe the advantages of each model as well as their limitations in the study of the HCV life cycle, with a particular emphasis on virus entry. A comparison between these two models is presented in terms of virion composition and their use as tools for the characterization of entry factors, envelope glycoprotein functions, and antibody neutralization. We also compare the production and biosafety level of these two types of viral particles. Globally, this review provides a general description of the most adequate applications for HCVpp and HCVcc in HCV research.

Key words Hepatitis C virus, HCVpp, HCVcc

1 Introduction

The identification of HCV genome, which was reported for the first time in 1989 [1], led to a new era in molecular and biochemical research on HCV. However, for a long time, it remained impossible to propagate HCV in cell culture, and this limited the understanding of virus-host interactions at the cellular and molecular levels. To circumvent these limitations, several groups have tried to develop surrogate models to study HCV entry or replication. Before the development of a cell culture system for HCV (HCVcc) in 2005 [2–4], selectable replicon systems [5] and retrovirus-based pseudotyped particles (HCVpp) [6–8] were major tools used to investigate HCV genomic replication and virus entry, respectively. In this review, we briefly present HCVpp and HCVcc systems, and we discuss the similarities and differences between these two systems.

Mansun Law (ed.), *Hepatitis C Virus Protocols*, Methods in Molecular Biology, vol. 1911,
https://doi.org/10.1007/978-1-4939-8976-8_2,

1.1 The HCVpp System

The development of the HCVpp system enabled for the first time to functionally investigate HCV entry. This model is based on the production of retroviral capsid particles, which incorporate in their lipid envelope unmodified HCV glycoproteins E1 and E2. This system was first described in 2003 by two different groups [6, 8], and since then, it has been extensively used to investigate the mechanisms of HCV entry and characterize antibody neutralization.

HCVpp are generated through the co-transfection of 293T cells with three plasmids, respectively, coding for (1) a retroviral genome expressing a reporter gene, (2) the gag-pol proteins from human immunodeficiency virus (HIV) or murine leukemia virus (MLV), and (3) HCV E1 and E2 glycoproteins. The presence of a reporter gene, such as GFP or luciferase, allows for the quantification of pseudoparticle entry into target cells (Fig. 1).

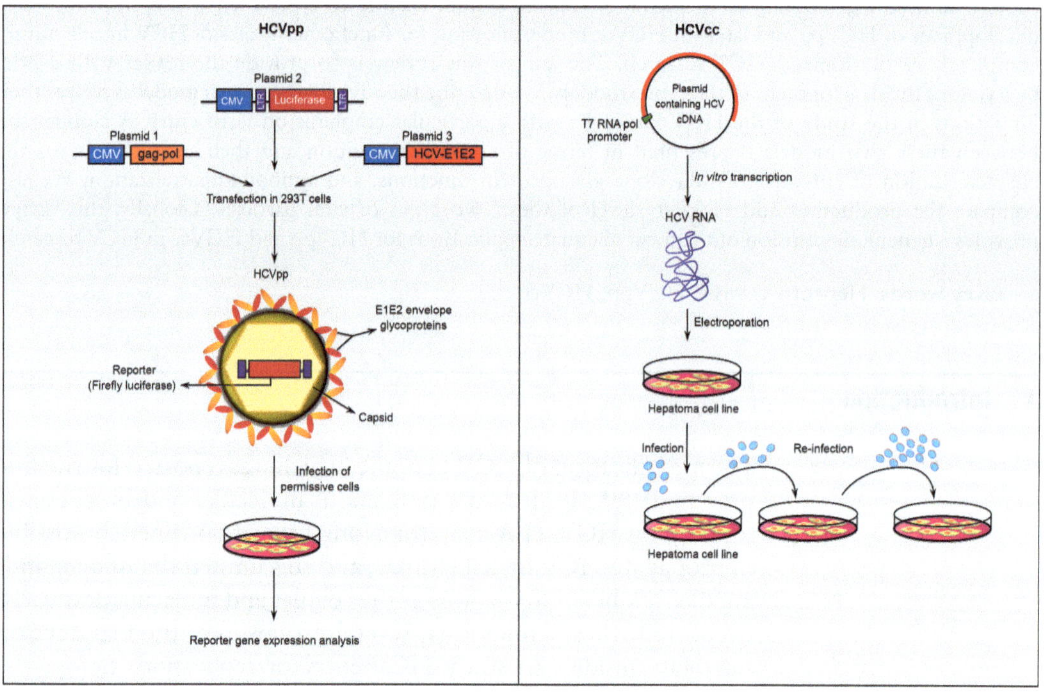

Fig. 1 *HCVpp.* HCVpp are produced through the co-transfection of 293T cells with three plasmids, respectively, coding for the HIV or MLV gag-pol, the reporter gene Firefly Luciferase and HCV glycoproteins E1 and E2 from a specific genotype. The co-transfection drives the assembly of pseudoparticles constituted by a viral capsid containing the mini-genome coding for the reporter gene and an envelope in which are inserted the HCV glycoproteins. These pseudoparticles are released by the cells and can be collected and used to infect permissive cell lines, allowing the quantification of the entry process through measurement of the reporter gene activity. *HCVcc.* A plasmid coding for the HCV cDNA is transcribed in vitro, thanks to the presence of a T7 RNA polymerase promoter. The synthesized HCV RNA is then electroporated in hepatoma cells (Huh-7 or Huh-7.5 cells), giving rise to the translation of all the viral proteins and to the generation of a complete infection cycle. The virions secreted can be collected and used to infect new hepatoma cells. Sequential passages of the supernatant on new permissive cells enable the amplification of the virus and the increase of the viral titer

Even if HCV morphogenesis strongly differs from retroviral/lentiviral assembly and the cell surface is supposed to be the main location where retrovirus/lentivirus budding occurs, retroviral assembly has also been observed in the late endosomes/multivesicular bodies [9]. Moreover, overexpression of the HCV envelope glycoproteins in 293T cells leads to the export of a fraction of these proteins at the plasma membrane, which otherwise are retained in the endoplasmic reticulum [6–8]. Both these observations can explain the production of retroviral/lentiviral pseudoparticles containing HCV glycoproteins. The presence of HCV envelope proteins at the surface of HCVpp mediates interactions with HCV-specific cellular entry factors as well as fusion with host membranes, whereas the following steps are mediated by the retroviral/lentiviral components of this chimeric particle. Therefore, HCVpp mimic HCV entry independently of the other steps of the HCV life cycle.

1.2 The HCVcc System

After the identification of HCV genome in 1989 [1] and its characterization [10], the development of a cell culture system for this virus has been a slow process involving several crucial steps, which started with the identification of the 3′ noncoding region of the genome [11]. This led to the construction of full-length cDNA clones of genotype 1a that were transcribed in vitro and tested for viral production. However, faced with a lack of replication in cell culture, productive viral replication had to be validated by intrahepatic inoculation of chimpanzees with genomic RNA generated in vitro, after the construction of a consensus clone [12, 13]. These in vivo experiments encouraged the scientific community to continue the search for in vitro replication models.

The first development of a functional in vitro system for HCV replication appeared in 1999, with the construction of the first subgenomic replicon of genotype 1b [5]. This subgenomic replicon is characterized by a bi-cistronic construct containing an internal ribosome entry site (IRES) from encephalomyocarditis virus (EMCV). In this system, the HCV IRES controls the expression of neomycin phosphotransferase as a selection gene, while the EMCV IRES drives the expression of HCV nonstructural (NS) proteins from NS3 to NS5B. All genes are under the control of a T7 promoter, necessary for the in vitro RNA transcription with a T7 RNA polymerase, in order to generate the viral RNA that once electroporated in Huh-7 cells can give rise to self-replication [5]. The addition of G418 allows for the selection of Huh-7 cells electroporated with the viral RNA, in order to isolate cellular clones replicating HCV. The development of the subgenomic replicon system enabled the study of HCV replication, leading to the identification of viral proteins as well as cellular proteins participating in this step of the viral life cycle.

Replication represents an important step during the viral life cycle, and it is an interesting target for the identification of antiviral compounds. Therefore, the establishment of subgenomic replicons gave a great impulse to the development of antiviral drugs directed against nonstructural proteins. The currently used direct acting antivirals (DAAs), mainly targeting NS3, NS5A, and NS5B, were validated in vitro on subgenomic replicons (reviewed in [14]).

After the assembly of functional subgenomic replicons, full-length genomes were reconstituted based on functional replicons. Unfortunately, transfection of hepatoma cells with such full-length replicons still did not lead to virion production [15–18]. This is due to the fact that efficient amplification of these replicons required replication-enhancing mutations that impaired virus assembly [19]. Indeed, in addition to structural components, HCV non-structural proteins are also involved in virion assembly [20].

A major breakthrough in the development of a cell culture system for HCV came with the construction of a genotype 2a subgenomic replicon which replicated with very high efficiency without the requirement of adaptive mutations [21]. This subgenomic replicon was cloned from a Japanese patient suffering from a fulminant hepatitis, and it was designated "Japanese fulminant hepatitis 1" (JFH1) virus [22]. Later on, three groups reported that the complete wild-type JFH1 genome or chimeras consisting of the JFH1 replicase genes NS3-NS5B and Core to NS2 regions of alternative HCV genomes replicated efficiently in Huh-7 cells and produced infectious viral particles [2–4]. These particles were designated HCVcc for cell culture-derived HCV, and they are now routinely used to study each step of the viral life cycle (Fig. 1). During the following years, this system was further optimized thanks to the selection of adaptive mutations, which strongly increased the viral titers of the released virions [23–25].

2 Differences in Virion Composition Between HCVpp and HCVcc

HCVcc particles are composed of the single-stranded RNA genome interacting with the core protein to form the nucleocapsid that is surrounded by a cell-derived lipid membrane in which the envelope glycoproteins E1 and E2 are anchored. In addition, HCV particles also contain host-derived components. Indeed, a specific feature of HCV virion is its interaction with very-low-density lipoproteins (VLDL) or low-density lipoproteins (LDL), resulting in the formation of a lipo-viro-particle [26]. As a consequence, apolipoproteins such as apoE, apoB, apoA1, and apoC1 can also be found in association with HCVcc particles [27–29]. In addition, neutral lipids such as triglycerides and cholesteryl esters are also incorporated into HCVcc virions [28]. However, differences in densities between HCVcc and serum-derived virions suggest that the lipid composition of HCVcc does not totally reflect that of native viral particles [30].

The composition of HCVpp is very different from that of HCVcc. Indeed, the only HCV-specific components of HCVpp are the envelope glycoproteins, which are anchored in a host-derived membrane. The rest of the particle is composed of the retroviral/lentiviral core proteins as well as a packaging-competent retrovirus-/lentivirus-derived genome harboring a marker gene. Therefore, only entry functions attributed to HCV envelope glycoproteins can be investigated with the HCVpp system. Furthermore, because of their production in kidney cells and their mode of assembly, HCVpp are not associated with lipoproteins, typically synthesized in hepatic cells. Moreover, ApoE is not found to be associated with HCVpp [31]. This therefore represents a major limitation for investigating the role of lipoprotein or lipid receptors in HCV entry.

Differences in maturation are also observed between envelope glycoproteins associated with HCVpp and HCVcc. Indeed, differences in the glycosylation profiles were observed between HCVpp and HCVcc, suggesting that the accessibility of HCV glycans to Golgi glycosidases and/or glycosyltransferases is not the same for the two types of particles, due to differences in the assembly process. Furthermore, HCVcc-associated E1 and E2 envelope glycoproteins form large covalent complexes stabilized by disulfide bridges [32], whereas mainly non-covalent E1E2 heterodimers are found to be associated with HCVpp [33]. It has also to be pointed out that a higher degree of heterogeneity has also been observed in the envelope glycoproteins associated with HCVpp [34].

2.1 Characterization of HCV Entry Factors with HCVpp and HCVcc Systems

HCVpp have been extensively used for the characterization of HCV-specific host entry factors. Cellular entry factors like tetraspanin CD81 [35] and scavenger receptor class B type I (SR-BI) [36], previously identified by a biochemical approach using a truncated form of E2 as a bait, were confirmed to play a functional role in HCV entry with the help of HCVpp [37, 38]. Later on, these entry factors were also confirmed with the HCVcc system [2, 3, 39]. HCVpp were also used in different large-scale screening approaches to identify new entry factors. Thanks to these screenings, the tight-junction proteins claudin-1 [40] and occludin [41] were identified and characterized as essential players during the entry process. More recently, with the help of HCVpp, E-cadherin was shown to be an additional HCV-specific host entry factor [42]. Since its entry functions are uncoupled from other steps of the HCV life cycle, HCVpp have also been a major tool to confirm the role of novel HCV-specific host entry factors identified using the HCVcc system, as shown for the transferrin receptor 1 [43], the VLDL receptor [44], as well as the epidermal growth factor receptor (EGFR) and ephrin receptor A2 (EphA2) [45].

In contrast, some HCV-specific host entry factors could not be confirmed in the HCVpp system. This is the case for Niemann-Pick C1-like 1 (NPC1L1) [46] and serum response factor binding protein 1 (SRFBP1) [47]. The lack of HCVpp dependence on these HCV-specific host entry factors could be explained by differences in virion composition between HCVpp and HCVcc. Indeed, as discussed above, infectious HCV particles are associated with lipoproteins and contain neutral lipids and apolipoproteins that can potentially play a role during the entry process. Since they are not associated with these host components, HCVpp cannot be used to validate the functions of these host factors in HCV entry. For instance, due to the role of ApoE in virion attachment to heparan sulfate proteoglycans on host cells [48, 49], HCVpp cannot be used to study the early steps of virus entry. Similarly, the role of the low-density lipoprotein receptor (LDL-R) could not be investigated with the HCVpp model [50]. Investigators need therefore to be aware of this major limitation when studying HCV entry.

2.2 Characterization of HCV E1E2 Envelope Glycoprotein Functions with HCVpp and HCVcc Systems

In addition to their contribution to the characterization of HCV-specific host entry factors, HCVpp have also often been used to investigate HCV envelope glycoprotein functions. Since these proteins are produced from a plasmid expressing only E1 and E2 envelope glycoproteins, site-directed mutagenesis can be easily performed, and a large number of mutants can be simultaneously investigated. HCVpp have indeed been extensively used to characterize functional regions in HCV envelope glycoproteins (reviewed in [51]). Another interesting aspect in using HCVpp for the characterization of HCV envelope glycoproteins is that mutations introduced by site-directed mutagenesis are stable, whereas in the context of HCVcc reversion or second-site mutations can potentially appear. It is worth noting that the entry functions of glycoprotein mutants having a defect in virion assembly cannot be investigated in the HCVcc system; however some of them can be investigated with the help of HCVpp [52]. One needs however to be careful in the interpretation of phenotypes observed with HCVpp. For instance, glycosylation mutants that are infectious in the HCVcc system are not necessarily functional in the HCVpp model [53].

One limitation in investigating the functional role of the viral envelope glycoproteins with the HCVpp and HCVcc systems is that not all the envelope proteins from patient isolates are functional in these systems. Furthermore, in a recent study, no correlation in relative specific infectivity of E1E2-matched HCVcc and HCVpp was observed [31]. This might in part be due to different amounts of E2 on HCVcc virions compared to HCVpp, as well as the presence of ApoE on HCVcc and not on HCVpp [31]. It has been suggested that multiple technical factors can influence production of functional HCVpp, including plasmid transfection ratios and use of either MLV or HIV reporter constructs [54].

An interesting approach that is often used to study the function of viral envelope glycoproteins consists in propagating the virus in serial passages in the presence of an entry inhibitor. This method has been successfully used to identify functional regions in HCV envelope glycoproteins [55–57]. However, this type of approach can only be performed with HCVcc. Similarly, HCVcc adapted to alternative receptors can also be selected to identify cross-talks between HCV envelope glycoproteins and receptors [58, 59].

2.3 Characterization of Antibody Neutralization with HCVpp and HCVcc Systems

The presence of E1 and E2 at the surface of HCVpp is compatible with the characterization of neutralizing antibodies targeting the two glycoproteins [8, 60] as well as cross-neutralizing antibodies [6, 38, 61, 62]. Recently, large panels of HCVpp have been produced that encompass more of the diversity of circulating HCV variants, which is critical for accurate measurement of antibody neutralizing breadth [54, 63].

When the HCVcc model was developed, one of its major limitations to study antibody neutralization was the lack of HCV sequence diversity available. However, lately, chimeric full-length viruses, which express diverse E1E2 variants in the context of JFH1-based genome background, have been produced [64–66]. Such panels of HCVcc expressing E1E2 from multiple genotypes have been used to measure neutralizing breadth of broadly neutralizing monoclonal antibodies [67, 68]. More recently, full-length infectious clones from other genotypes have also been developed. They include genotype 1a [69, 70], genotype 2b [71], and genotype 3a [72]. These viruses could also be used for the characterization of the neutralizing antibody response. More importantly, they will be essential for the analysis of genotype differences in terms of sensitivity to antivirals.

Very recently, Wasilewski and coworkers generated panels of E1E2-matched HCVcc and HCVpp to compare sensitivity to neutralizing antibodies [31]. Interestingly, they observed a very strong positive correlation between relative neutralization resistance of these same E1E2-matched HCVcc and HCVpp variants, indicating that both systems can be used for the characterization of neutralizing antibodies. However, the HCVpp system is more flexible for this type of study. Comparisons between the two models also showed that HCVpp have a tendency to be more sensitive than HCVcc to antibody neutralization [31, 54]. This quantitative difference in HCVpp and HCVcc neutralization may be explained by the observation that ApoE increases HCVcc neutralization resistance [73].

When studying antibody neutralization, it is important to assess viral evolution in the presence of specific monoclonal antibodies to determine their potency to prevent the formation of escape variants. However, this can only be performed with the HCVcc system which allows the virus to be serially passaged in the presence of antibodies [74]. In the same line, the HCVcc system can also be used to identify mutations leading to neutralization escape [75]. An additional advantage of the HCVcc system is that the viral particles produced are infectious in uPA-SCID mice transplanted with human hepatocytes, chimpanzees [30], and genetically humanized mice [76, 77]. This makes them better tools to study antibody neutralization in preclinical models [62, 78].

3 Comparison Between HCVpp and HCVcc in Terms of Production and Biosafety

In addition to all the aspects described in the previous paragraphs, these two viral models present some differences concerning their production both in terms of quantity of virion produced and biosafety level for their manipulation.

As previously described, HCVpp are generated by a single co-transfection of three different plasmids. This represents a limitation in the quantity of pseudoparticles that can be produced. In fact, the concomitant presence of the three plasmids in the same cell and their respective expression are necessary for assembly and release of HCVpp. Conversely, HCVcc virions can be produced in stocks with higher titers than HCVpp, since they can be generated by propagating them in serial passages of infected cells rather than by a single transfection step (Fig. 1).

On the other hand, even if it can be a limitation in terms of quantity, the lack of generation of a complete viral cycle for HCVpp represents an advantage in terms of biosafety. Specifically, the lack of HCV genome in HCVpp makes the safety level for manipulation less strict for HCVpp than HCVcc. As a consequence, the European legislation authorizes to handle HCVpp in a BSL2 facility, whereas experiments with HCVcc need to be performed in a BSL3 laboratory. To circumvent this problem, one can use a system that supports particle production based on trans-packaging of subgenomic viral RNA which produces viruslike particles that are similar to HCVcc virions [79].

In summary, HCVpp can be produced and handled in a BSL2 facility with a limited quantity of particles produced, while HCVcc can be produced and amplified to reach higher titers, but in more stringent biosafety conditions.

4 Conclusions

The establishment of HCVpp and HCVcc systems radically changed the study of the HCV life cycle and represented major advances to investigate virus-host interactions at the cellular level.

Table 1
HCVpp and HCVcc properties

	HCVpp	*HCVcc*
VLDL/LDL association	No	Yes
Apolipoprotein association	No	Yes
El and E2 envelope glycoproteins	Non-covalent E1E2 heterodimers	Large covalent complexes stabilized by disulfide bridges
Degree of heterogeneity in E1E2	High	Low
Study of E1E2 glycoproteins	Entry	Entry, assembly, and secretion
Stability of mutations	Stable	Possibility of reversion or second-site mutation
Characterization of entry factors	E1E2-dependent interactions	E1E2-dependent interactions Lipid receptors and apolipoprotein-dependent interactions
Selection of E1E2 adaptation to alternative receptors	No	Yes
Selection of drug-resistant variants	No	Yes
Selection of neutralization escape mutants	No	Yes
Sensitivity to antibody neutralization	Highly sensitive	Sensitive
Study of antibody neutralization in animal models	No	Yes
Viral production	Limited to one co-transfection	Can be amplified by sequential passages in cell culture
Biosafety level[a]	BSL2	BSL3

[a]According to the European Union legislation

These two models represent complementary techniques (Table 1) for the study of HCV entry and were extensively used during the past decade for the identification of cellular entry receptors, the characterization of E1E2 glycoproteins, and the study of neutralizing antibodies (Table 1). It is worth noting that besides these investigations, HCVcc have also been extensively used to study HCV genome replication and assembly. When using HCVcc and HCVpp systems, one needs to keep in mind that the virion composition of these two models is different due to the association of HCVcc particles with lipoproteins. As a consequence, experiments performed with these two models can sometimes lead to divergent results. However, in the context of antibody-mediated virus neutralization, the two models seem to produce similar results [31].

In summary, albeit HCVcc represents a more complete tool to study the whole HCV life cycle, HCVpp still represent a useful system to investigate the entry steps directly depending on E1 and E2 glycoproteins in a simpler and easier way. The complementary use of these two models represents thus a powerful approach for the characterization of the complex HCV entry steps.

Acknowledgments

The authors were supported by the French National Agency for Research on AIDS and Viral Hepatitis (ANRS) and the ANR through ERA-NET Infect-ERA program (ANR-13-IFEC-0002-01).

References

1. Choo QL, Kuo G, Weiner AJ, Overby LR, Bradley DW, Houghton M (1989) Isolation of a cDNA clone derived from a blood-borne non-A, non-B viral hepatitis genome. Science 244:359–362
2. Wakita T, Pietschmann T, Kato T, Date T, Miyamoto M, Zhao Z et al (2005) Production of infectious hepatitis C virus in tissue culture from a cloned viral genome. Nat Med 11:791–796
3. Lindenbach BD, Evans MJ, Syder AJ, Wolk B, Tellinghuisen TL, Liu CC et al (2005) Complete replication of hepatitis C virus in cell culture. Science 309:623–626
4. Zhong J, Gastaminza P, Cheng G, Kapadia S, Kato T, Burton DR et al (2005) Robust hepatitis C virus infection in vitro. Proc Natl Acad Sci U S A 102:9294–9299
5. Lohmann V, Korner F, Koch J, Herian U, Theilmann L, Bartenschlager R (1999) Replication of subgenomic hepatitis C virus RNAs in a hepatoma cell line. Science 285:110–113
6. Bartosch B, Dubuisson J, Cosset F-L (2003) Infectious hepatitis C virus pseudo-particles containing functional E1-E2 envelope protein complexes. J Exp Med 197:633–642
7. Drummer HE, Maerz A, Poumbourios P (2003) Cell surface expression of functional hepatitis C virus E1 and E2 glycoproteins. FEBS Lett 546:385–390
8. Hsu M, Zhang J, Flint M, Logvinoff C, Cheng-Mayer C, Rice CM et al (2003) Hepatitis C virus glycoproteins mediate pH-dependent cell entry of pseudotyped retroviral particles. Proc Natl Acad Sci U S A 100:7271–7276
9. Sherer NM, Lehmann MJ, Jimenez-Soto LF, Ingmundson A, Horner SM, Cicchetti G et al (2003) Visualization of retroviral replication in living cells reveals budding into multivesicular bodies. Traffic 4:785–801
10. Choo QL, Richman KH, Han JH, Berger K, Lee C, Dong C et al (1991) Genetic organization and diversity of the hepatitis C virus. Proc Natl Acad Sci U S A 88:2451–2455
11. Kolykhalov AA, Feinstone SM, Rice CM (1996) Identification of a highly conserved sequence element at the 3′ terminus of hepatitis C virus genome RNA. J Virol 70:3363–3371
12. Kolykhalov AA, Agapov EV, Blight KJ, Mihalik K, Feinstone SM, Rice CM (1997) Transmission of hepatitis C by intrahepatic inoculation with transcribed RNA. Science 277:570–574
13. Yanagi M, Purcell RH, Emerson SU, Bukh J (1997) Transcripts from a single full-length cDNA clone of hepatitis C virus are infectious when directly transfected into the liver of a chimpanzee. Proc Natl Acad Sci U S A 94:8738–8743
14. Schinazi R, Halfon P, Marcellin P, Asselah T (2014) HCV direct-acting antiviral agents: the best interferon-free combinations. Liver Int 34 (Suppl 1):69–78
15. Blight KJ, McKeating JA, Rice CM (2002) Highly permissive cell lines for subgenomic and genomic hepatitis C virus RNA replication. J Virol 76:13001–13014
16. Blight KJ, McKeating JA, Marcotrigiano J, Rice CM (2003) Efficient replication of hepatitis C virus genotype 1a RNAs in cell culture. J Virol 77:3181–3190
17. Ikeda M, Yi M, Li K, Lemon SM (2002) Selectable subgenomic and genome-length dicistronic RNAs derived from an infectious molecular clone of the HCV-N strain of hepatitis C virus

replicate efficiently in cultured Huh7 cells. J Virol 76:2997–3006
18. Pietschmann T, Lohmann V, Kaul A, Krieger N, Rinck G, Rutter G et al (2002) Persistent and transient replication of full-length hepatitis C virus genomes in cell culture. J Virol 76:4008–4021
19. Pietschmann T, Zayas M, Meuleman P, Long G, Appel N, Koutsoudakis G et al (2009) Production of infectious genotype 1b virus particles in cell culture and impairment by replication enhancing mutations. PLoS Pathog 5:e1000475
20. Murray CL, Jones CT, Rice CM (2008) Architects of assembly: roles of Flaviviridae non-structural proteins in virion morphogenesis. Nat Rev Microbiol 6:699–708
21. Kato T, Date T, Miyamoto M, Furusaka A, Tokushige K, Mizokami M et al (2003) Efficient replication of the genotype 2a hepatitis C virus subgenomic replicon. Gastroenterology 125:1808–1817
22. Kato T, Furusaka A, Miyamoto M, Date T, Yasui K, Hiramoto J et al (2001) Sequence analysis of hepatitis C virus isolated from a fulminant hepatitis patient. J Med Virol 64:334–339
23. Zhong J, Gastaminza P, Chung J, Stamataki Z, Isogawa M, Cheng G et al (2006) Persistent hepatitis C virus infection in vitro: coevolution of virus and host. J Virol 80:11082–11093
24. Kaul A, Woerz I, Meuleman P, Leroux-Roels G, Bartenschlager R (2007) Cell culture adaptation of hepatitis C virus and in vivo viability of an adapted variant. J Virol 81:13168–13179
25. Russell RS, Meunier J-C, Takikawa S, Faulk K, Engle RE, Bukh J et al (2008) Advantages of a single-cycle production assay to study cell culture-adaptive mutations of hepatitis C virus. Proc Natl Acad Sci U S A 105:4370–4375
26. Andre P, Komurian-Pradel F, Deforges S, Perret M, Berland JL, Sodoyer M et al (2002) Characterization of low- and very-low-density hepatitis C virus RNA-containing particles. J Virol 76:6919–6928
27. Meunier J-C, Russell RS, Engle RE, Faulk KN, Purcell RH, Emerson SU (2008) Apolipoprotein c1 association with hepatitis C virus. J Virol 82:9647–9656
28. Merz A, Long G, Hiet M-S, Brugger B, Chlanda P, Andre P et al (2011) Biochemical and morphological properties of hepatitis C virus particles and determination of their lipidome. J Biol Chem 286:3018–3032
29. Catanese MT, Uryu K, Kopp M, Edwards TJ, Andrus L, Rice WJ et al (2013) Ultrastructural analysis of hepatitis C virus particles. Proc Natl Acad Sci U S A 110:9505–9510
30. Lindenbach BD, Meuleman P, Ploss A, Vanwolleghem T, Syder AJ, McKeating JA et al (2006) Cell culture-grown hepatitis C virus is infectious in vivo and can be recultured in vitro. Proc Natl Acad Sci U S A 103:3805–3809
31. Wasilewski L, Ray S, Bailey JR (2016) Hepatitis C virus resistance to broadly neutralizing antibodies measured using replication competent virus and pseudoparticles. J Gen Virol 97:2883–2893
32. Vieyres G, Thomas X, Descamps V, Duverlie G, Patel AH, Dubuisson J (2010) Characterization of the envelope glycoproteins associated with infectious hepatitis C virus. J Virol 84:10159–10168
33. Op De Beeck A, Voisset C, Bartosch B, Ciczora Y, Cocquerel L, Keck Z et al (2004) Characterization of functional hepatitis C virus envelope glycoproteins. J Virol 78:2994–3002
34. Flint M, Logvinoff C, Rice CM, McKeating JA (2004) Characterization of infectious retroviral pseudotype particles bearing hepatitis C virus glycoproteins. J Virol 78:6875–6882
35. Pileri P, Uematsu Y, Campagnoli S, Galli G, Falugi F, Petracca R et al (1998) Binding of hepatitis C virus to CD81. Science 282:938–941
36. Scarselli E, Ansuini H, Cerino R, Roccasecca RM, Acali S, Filocamo G et al (2002) The human scavenger receptor class B type I is a novel candidate receptor for the hepatitis C virus. EMBO J 21:5017–5025
37. Zhang J, Randall G, Higginbottom A, Monk P, Rice CM, McKeating JA (2004) CD81 is required for hepatitis C virus glycoprotein-mediated viral infection. J Virol 78:1448–1455
38. Bartosch B, Vitelli A, Granier C, Goujon C, Dubuisson J, Pascale S et al (2003) Cell entry of hepatitis C virus requires a set of co-receptors that include the CD81 tetraspanin and the SR-B1 scavenger receptor. J Biol Chem 278:41624–41630
39. Zeisel MB, Koutsoudakis G, Schnober EK, Haberstroh A, Blum HE, Cosset F-L et al (2007) Scavenger receptor class B type I is a key host factor for hepatitis C virus infection required for an entry step closely linked to CD81. Hepatology 46:1722–1731
40. Evans MJ, von Hahn T, Tscherne DM, Syder AJ, Panis M, Wolk B et al (2007) Claudin-1 is a hepatitis C virus co-receptor required for a late step in entry. Nature 446:801–805
41. Ploss A, Evans MJ, Gaysinskaya VA, Panis M, You H, de Jong YP et al (2009) Human

occludin is a hepatitis C virus entry factor required for infection of mouse cells. Nature 457:882–886

42. Li Q, Sodroski C, Lowey B, Schweitzer CJ, Cha H, Zhang F et al (2016) Hepatitis C virus depends on E-cadherin as an entry factor and regulates its expression in epithelial-to-mesenchymal transition. Proc Natl Acad Sci U S A 113:7620–7625
43. Martin DN, Uprichard SL (2013) Identification of transferrin receptor 1 as a hepatitis C virus entry factor. Proc Natl Acad Sci U S A 110:10777–10782
44. Ujino S, Nishitsuji H, Hishiki T, Sugiyama K, Takaku H, Shimotohno K (2016) Hepatitis C virus utilizes VLDLR as a novel entry pathway. Proc Natl Acad Sci U S A 113:188–193
45. Lupberger J, Zeisel MB, Xiao F, Thumann C, Fofana I, Zona L et al (2011) EGFR and EphA2 are host factors for hepatitis C virus entry and possible targets for antiviral therapy. Nat Med 17:589–595
46. Sainz BJ, Barretto N, Martin DN, Hiraga N, Imamura M, Hussain S et al (2012) Identification of the Niemann-Pick C1-like 1 cholesterol absorption receptor as a new hepatitis C virus entry factor. Nat Med 18:281–285
47. Gerold G, Meissner F, Bruening J, Welsch K, Perin PM, Baumert TF et al (2015) Quantitative proteomics identifies serum response factor binding protein 1 as a host factor for hepatitis C virus entry. Cell Rep 12:864–878
48. Jiang J, Cun W, Wu X, Shi Q, Tang H, Luo G (2012) Hepatitis C virus attachment mediated by apolipoprotein E binding to cell surface heparan sulfate. J Virol 86:7256–7267
49. Xu Y, Martinez P, Seron K, Luo G, Allain F, Dubuisson J et al (2015) Characterization of hepatitis C virus interaction with heparan sulfate proteoglycans. J Virol 89:3846–3858
50. Albecka A, Belouzard S, Op de Beeck A, Descamps V, Goueslain L, Bertrand-Michel J et al (2012) Role of low-density lipoprotein receptor in the hepatitis C virus life cycle. Hepatology 55:998–1007
51. Vieyres G, Dubuisson J, Pietschmann T (2014) Incorporation of hepatitis C virus E1 and E2 glycoproteins: the keystones on a peculiar virion. Viruses 6:1149–1187
52. Falson P, Bartosch B, Alsaleh K, Tews BA, Loquet A, Ciczora Y et al (2015) Hepatitis C virus envelope glycoprotein E1 forms trimers at the surface of the virion. J Virol 89:10333–10346
53. Helle F, Vieyres G, Elkrief L, Popescu C-I, Wychowski C, Descamps V et al (2010) Role of N-linked glycans in the functions of hepatitis C virus envelope proteins incorporated into infectious virions. J Virol 84:11905–11915
54. Urbanowicz RA, McClure CP, King B, Mason CP, Ball JK, Tarr AW (2016) Novel functional hepatitis C virus glycoprotein isolates identified using an optimized viral pseudotype entry assay. J Gen Virol 97:2265–2279
55. Fénéant L, Potel J, François C, Sané F, Douam F, Belouzard S et al (2015) New insights into the understanding of hepatitis C virus entry and cell-to-cell transmission by using the ionophore monensin a. J Virol 89:8346–8364
56. Vausselin T, Seron K, Lavie M, Mesalam AA, Lemasson M, Belouzard S et al (2016) Identification of a new benzimidazole derivative as an antiviral against hepatitis C virus. J Virol 90:8422–8434
57. Perin PM, Haid S, Brown RJP, Doerrbecker J, Schulze K, Zeilinger C et al (2016) Flunarizine prevents hepatitis C virus membrane fusion in a genotype-dependent manner by targeting the potential fusion peptide within E1. Hepatology 63:49–62
58. Bitzegeio J, Bankwitz D, Hueging K, Haid S, Brohm C, Zeisel MB et al (2010) Adaptation of hepatitis C virus to mouse CD81 permits infection of mouse cells in the absence of human entry factors. PLoS Pathog 6: e1000978
59. Hopcraft SE, Evans MJ (2015) Selection of a hepatitis C virus with altered entry factor requirements reveals a genetic interaction between the E1 glycoprotein and claudins. Hepatology 62:1059–1069
60. Cai Z, Zhang C, Chang K-S, Jiang J, Ahn B-C, Wakita T et al (2005) Robust production of infectious hepatitis C virus (HCV) from stably HCV cDNA-transfected human hepatoma cells. J Virol 79:13963–13973
61. Owsianka A, Tarr AW, Juttla VS, Lavillette D, Bartosch B, Cosset F-L et al (2005) Monoclonal antibody AP33 defines a broadly neutralizing epitope on the hepatitis C virus E2 envelope glycoprotein. J Virol 79:11095–11104
62. Law M, Maruyama T, Lewis J, Giang E, Tarr AW, Stamataki Z et al (2008) Broadly neutralizing antibodies protect against hepatitis C virus quasispecies challenge. Nat Med 14:25–27
63. Osburn WO, Snider AE, Wells BL, Latanich R, Bailey JR, Thomas DL et al (2014) Clearance of hepatitis C infection is associated with the early appearance of broad neutralizing antibody responses. Hepatology 59:2140–2151

64. Gottwein JM, Scheel TKH, Hoegh AM, Lademann JB, Eugen-Olsen J, Lisby G et al (2007) Robust hepatitis C genotype 3a cell culture releasing adapted intergenotypic 3a/2a (S52/JFH1) viruses. Gastroenterology 133:1614–1626
65. Gottwein JM, Scheel TKH, Jensen TB, Lademann JB, Prentoe JC, Knudsen ML et al (2009) Development and characterization of hepatitis C virus genotype 1-7 cell culture systems: role of CD81 and scavenger receptor class B type I and effect of antiviral drugs. Hepatology 49:364–377
66. Scheel TKH, Gottwein JM, Carlsen THR, Li Y-P, Jensen TB, Spengler U et al (2011) Efficient culture adaptation of hepatitis C virus recombinants with genotype-specific core-NS2 by using previously identified mutations. J Virol 85:2891–2906
67. Keck Z, Xia J, Wang Y, Wang W, Krey T, Prentoe J et al (2012) Human monoclonal antibodies to a novel cluster of conformational epitopes on HCV E2 with resistance to neutralization escape in a genotype 2a isolate. PLoS Pathog 8:e1002653
68. Carlsen THR, Pedersen J, Prentoe JC, Giang E, Keck Z-Y, Mikkelsen LS et al (2014) Breadth of neutralization and synergy of clinically relevant human monoclonal antibodies against HCV genotypes 1a, 1b, 2a, 2b, 2c, and 3a. Hepatology 60:1551–1562
69. Li Y-P, Ramirez S, Jensen SB, Purcell RH, Gottwein JM, Bukh J (2012) Highly efficient full-length hepatitis C virus genotype 1 (strain TN) infectious culture system. Proc Natl Acad Sci U S A 109:19757–19762
70. Li Y-P, Ramirez S, Mikkelsen L, Bukh J (2015) Efficient infectious cell culture systems of the hepatitis C virus (HCV) prototype strains HCV-1 and H77. J Virol 89:811–823
71. Ramirez S, Li Y-P, Jensen SB, Pedersen J, Gottwein JM, Bukh J (2014) Highly efficient infectious cell culture of three hepatitis C virus genotype 2b strains and sensitivity to lead protease, nonstructural protein 5A, and polymerase inhibitors. Hepatology 59:395–407
72. Ramirez S, Mikkelsen LS, Gottwein JM, Bukh J (2016) Robust HCV genotype 3a infectious cell culture system permits identification of escape variants with resistance to Sofosbuvir. Gastroenterology 151:973–985
73. Fauvelle C, Felmlee DJ, Crouchet E, Lee J, Heydmann L, Lefevre M et al (2016) Apolipoprotein E mediates evasion from hepatitis C virus neutralizing antibodies. Gastroenterology 150:206–217
74. Keck Z, Angus AGN, Wang W, Lau P, Wang Y, Gatherer D et al (2014) Non-random escape pathways from a broadly neutralizing human monoclonal antibody map to a highly conserved region on the hepatitis C virus E2 glycoprotein encompassing amino acids 412-423. PLoS Pathog 10:e1004297
75. Gal-Tanamy M, Keck Z-Y, Yi M, McKeating JA, Patel AH, Foung SKH et al (2008) In vitro selection of a neutralization-resistant hepatitis C virus escape mutant. Proc Natl Acad Sci U S A 105:19450–19455
76. Dorner M, Horwitz JA, Robbins JB, Barry WT, Feng Q, Mu K et al (2011) A genetically humanized mouse model for hepatitis C virus infection. Nature 474:208–211
77. Dorner M, Horwitz JA, Donovan BM, Labitt RN, Budell WC, Friling T et al (2013) Completion of the entire hepatitis C virus life cycle in genetically humanized mice. Nature 501:237–241
78. de Jong YP, Dorner M, Mommersteeg MC, Xiao JW, Balazs AB, Robbins JB et al (2014) Broadly neutralizing antibodies abrogate established hepatitis C virus infection. Sci Transl Med 6:254ra129
79. Steinmann E, Brohm C, Kallis S, Bartenschlager R, Pietschmann T (2008) Efficient trans-encapsidation of hepatitis C virus RNAs into infectious virus-like particles. J Virol 82:7034–7046

Chapter 3

Hepatitis C Virus Database and Bioinformatics Analysis Tools in the Virus Pathogen Resource (ViPR)

Yun Zhang, Christian Zmasek, Guangyu Sun, Christopher N. Larsen, and Richard H. Scheuermann

Abstract

The Virus Pathogen Resource (ViPR; www.viprbrc.org) is a US National Institute of Allergy and Infectious Diseases (NIAID)-sponsored Bioinformatics Resource Center providing bioinformatics support for major human viral pathogens. The hepatitis C virus (HCV) portal of ViPR facilitates basic research and development of diagnostics and therapeutics for HCV, by providing a comprehensive collection of HCV-related data integrated from various sources, a growing suite of analysis and visualization tools for data mining and hypothesis generation, and personal Workbench spaces for data storage and sharing. This chapter introduces the data and functionality provided by the ViPR HCV portal. It describes example workflows for (1) searching HCV genome and protein sequences, (2) conducting phylogenetic analysis, and (3) analyzing sequence variations using pattern search for amino acid substitutions in proteins, single nucleotide variation calculation, metadata-driven comparison, and sequence feature variant type analysis. All data and tools are freely available via the ViPR HCV portal at https://www.viprbrc.org/brc/home.spg?decorator=flavi_hcv.

Key words Virus Pathogen Resource, ViPR, Database, Annotation, Genotype, Mature peptides, Comparative genomics, Phylogenetics, Antiviral, Drug resistance

1 Introduction

With the advent of direct antiviral agents (DAA) for HCV, highly effective and safer therapeutic regimens are now available on the global market. However, these antivirals remain very expensive in many countries, resulting in limited access to these treatments [1]. Therefore, research and development of new therapeutics of HCV is ongoing. For patients who have access to diagnosis and treatment, treatment options are based on the virus genotype and patient cirrhosis status [1]. This necessitates genotype/subtype-related diagnostic analysis tools. In order to advance research on infectious disease pathogens, the National Institute of Allergy and Infectious Diseases (NIAID) at the US National Institutes of Health (NIH) is supporting the freely available, Web-based Virus

Mansun Law (ed.), *Hepatitis C Virus Protocols*, Methods in Molecular Biology, vol. 1911,
https://doi.org/10.1007/978-1-4939-8976-8_3,

Pathogen Resource (ViPR; www.viprbrc.org) [2] through the Bioinformatics Resource Centers Program. The objective of the ViPR resource is to provide a one-stop shop for human virus data and analysis tools to drive new discoveries about virus transmission, pathogenesis, and to develop novel strategies for diagnostics and therapeutics. Considering HCV as a major public health burden, the ViPR team has developed custom annotation and analysis tools for HCV and subsequently created a dedicated HCV portal within ViPR to support various bioinformatics needs. In this chapter, we demonstrate a selection of workflows for using the HCV portal to search and analyze HCV data.

2 Overview of ViPR HCV Portal

2.1 Data Provided in the HCV Portal of ViPR

A key feature of ViPR is to provide access to a wide array of data types on a single platform. The HCV portal of ViPR provides a comprehensive collection of HCV-related data integrated from public archives, data submitters, and ViPR in-house curation and annotation pipelines, with data types covering genome sequences, mature peptide annotations, immune epitopes, 3D protein structures, clinical metadata, curated sequence features, and antiviral drug data (Table 1). In particular, the ViPR curation pipeline provides standardized, controlled vocabularies for sequence metadata fields, such as host and country, in order to better support metadata-based search and analysis. The ViPR annotation pipelines also generate predicted annotations, including genotype/subtype classifications, mature peptide sequences, protein domains, predicted immune epitopes, and known protein structural and functional regions termed Sequence Features (SFs). These data are searchable via a series of custom search interfaces tailored for each data type.

2.2 Annotation and Analytical Tools Integrated into the HCV Portal

The ViPR HCV portal provides a growing suite of analysis and visualization tools (Table 2), for data contained within the ViPR resource and data provided by the user. In addition to providing commonly used bioinformatics analysis tools, such as BLAST for sequence similarity comparison, MUSCLE for multiple sequence alignment, and FastME/RaxML/PhyML for phylogenetic tree construction, the ViPR team has also focused on developing customized tools for HCV data analysis, including the HCV genotype/subtype annotation pipeline, the mature peptide prediction tool, and the Sequence Feature Variant Types (SFVT) component.

2.2.1 Genotype/Subtype Annotation Pipeline

HCV genotypes have great clinical significance and are found to be important factors impacting patients' response to various HCV treatments. To characterize the HCV genotypes/subtypes, the International Committee on Taxonomy of Viruses (ICTV)

Table 1
Summary of HCV Data in ViPR

Data type	Number of records[a]	Data source
Strains	149,949	GenBank
Strains with predicted genotypes/subtypes	83,546	ViPR: Genotyping Pipeline
Genomes	223,444	GenBank
Complete genomes	3130	GenBank
Proteins	542,717	UniProt
Mature peptides	348,489	ViPR: Mature Peptide Prediction Pipeline
Samples with clinical metadata	367	NIAID Genome Sequencing Center
Pfam domains	456,017	ViPR: InterProScan
Other domains/motifs	356,825	ViPR: InterProScan
Experimentally determined epitopes	7626	Immune Epitope Database
Proteins with predicted epitopes	364,981	ViPR: NetCTL algorithm
Sequence features	367	ViPR: curation
3D structures	384	Protein Data Bank
Antiviral drugs	12	Drugbank Drug and Target Database

[a]As of July 2018

Flaviviridae Study Group maintains a list of HCV genotype/subtype assignments, as well as a reference alignment of representative sequences [3]. To support these assignments, a genotype/subtype annotation pipeline has been developed in ViPR. This computational pipeline uses a phylogenetic tree approach to classify HCV genomes, and consists of the following major parts:

1. An expanded reference alignment, which is computed from the ICTV reference alignment [3] with additional reference sequences added to the alignment using MAFFT.
2. An HCV genotype/subtype reference tree, which is built from the expanded reference alignment.
3. A phylogenetic pipeline, wherein each HCV genome sequence in ViPR is aligned with the expanded reference alignment, and subsequently placed into the reference tree using pplacer [4], with the reference tree serving as a scaffold onto which the query sequence is placed.
4. A genotype inference engine named cladinator, which infers the genotype/subtype for the query sequence based on its local environment within the phylogenetic tree.

Table 2
Annotation and Analytical Tools Integrated into the HCV Portal of ViPR

Tool name	Function	Source
Antiviral Resistance Risk Assessment	Determine if input sequences contain substitutions associated with altered response to antiviral drugs	ViPR
Archaeopteryx.js, Archaeopteryx.java	Visualize phylogenetic trees	ViPR
BLAST	Identify similar sequences in ViPR-compiled databases or custom datasets	External
FastME, PhyML, RAxML	Construct phylogenetic trees	External
Genome Annotation Transfer Utility	Predict mature peptides for user-provided sequences	External
Genotyping/subtyping tool	Annotate genome sequences with genotype/subtype assignments	ViPR
JalView (being replaced by MSAViewer)	Visualize sequence alignments	External
Jmol, JSmol	Visualize 3D structures	External
Mature Peptide Prediction	Annotate genome sequences with mature peptides	ViPR
Metadata-driven comparative analysis tool for sequences (meta-CATS)	Statistically compare groups of sequences to identify positions that significantly differ between them	ViPR
ModelCompare	Determine the optimal evolutionary model for phylogenetic trees	ViPR
MUSCLE	Calculate multiple sequence alignments	ViPR
Primer3	Design PCR primers	External
ReadSeq	Convert between sequence formats	External
Sequence Feature Variant Types	Compute sequence variations within curated Sequence Feature regions	ViPR
Sequence Variation Calculator	Compute entropy scores at each position of input sequences	External
Short Peptide Identification	Find short amino acid strings in target proteins	External

5. Last, a database of genotypes/subtypes is available, where the output genotype/subtype annotations are stored and made available for query and download.

A complete description of the mature peptide prediction method is described in a standard operating procedure (SOP) document [5] on the ViPR website.

2.2.2 Mature Peptide Prediction Tool

Mature peptide annotations for HCV genomes in GenBank are often incomplete. To support gene and protein level sequence search and analysis, a mature peptide prediction tool has been developed and used to annotate all HCV sequences in ViPR. This tool uses Refseq NC_004102 (GT1) as a reference sequence which has reference quality mature peptide annotations under ten gene symbols: C, E1, E2, p7, NS2, NS3, NS4A, NS4B, NS5A, and NS5B (*see* **Note 1**). The annotation process starts by aligning a query polyprotein sequence with the reference polyprotein sequence. Protease cleavage sites in the reference sequence are then mapped to the query sequence by pairwise alignment. For all public HCV sequences in the ViPR database, the resulting coordinates of the cleavage sites and mature peptide sequences are stored in the ViPR database, and made available for query and download. A complete description of the mature peptide prediction method is described in a standard operating procedure (SOP) document [6] on the ViPR website.

2.2.3 Metadata-Driven Comparative Analysis Tool for Sequences (Meta-CATS)

Previously, when researchers wanted to identify specific genetic determinants that could be responsible for viral characteristic changes, they often had to compare sequences manually. The meta-CATS method provides an automated and statistically rigorous approach for pinpointing the specific genetic variations associated with differences in virus characteristics [7]. This method consists of three parts:

1. A multiple sequence alignment using MUSCLE.
2. A chi-square goodness of fit test to identify positions of the multiple sequence alignment that significantly differ from the expected (random) distribution of residues between all metadata groups.
3. A Pearson's chi-square test to identify the specific pairs of metadata groups that contribute to the observed statistical difference.

A complete description of the meta-CATS method is described in an SOP document [8] on the ViPR website. This tool is discussed more in Subheading 3.4.3 below.

2.2.4 Sequence Feature Variant Type Component

The genetic determinants of important viral phenotypes, such as enzyme active sites, protein–protein interaction sites, and drug resistance sites, are determined by specific functional regions in viral proteins. To support search and analysis of these important regions, a Sequence Feature Variant Type (SFVT) component has been implemented in ViPR, based on similar work done for the influenza virus [9]. Each Sequence Feature (SF) defines the precise location of well-characterized regions in viral proteins. Information used to define SFs was obtained and curated from UniProt, the Immune Epitope Database (IEDB), Drugbank, and scientific publications. Once an SF has been defined, all sequence records of the designated taxon in ViPR are annotated with the SF location. Subsequently, strains carrying the same amino acid sequence combinations are grouped into a defined Variant Type (VT). The resulting SFVT component allows for rapid identification of important regions in viral proteins and genotypic polymorphisms within the SF region. This component is discussed more in Subheading 3.4.4 below.

3 Methods

For text format convention, please *see* **Note 2**.

3.1 Using the Hepatitis C Virus Portal in ViPR

The ViPR home page (https://www.viprbrc.org) is the main resource entry point for most users. The Hepatitis C Virus portal (Fig. 1) can be accessed by clicking **Hepatitis C Virus** in the 'Featured Virus' section. Alternatively, the Hepatitis C Virus portal is accessible from: https://www.viprbrc.org/brc/home.spg?

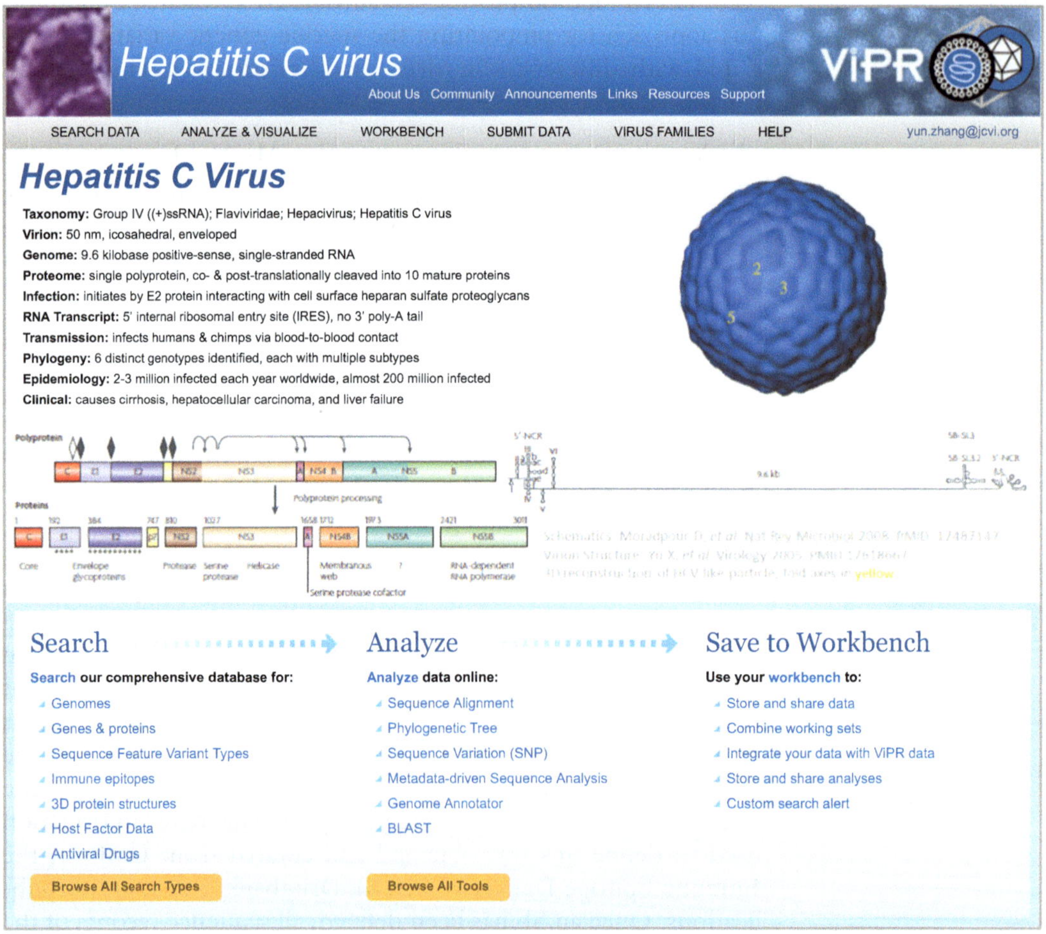

Fig. 1 The home page of the Hepatitis C Virus portal in ViPR—The portal serves as a gateway to search for a wide variety of data types, perform bioinformatics analyses, and save data and results in personal Workbenches

decorator=flavi_hcv (*see* **Note 3**). A gray navigation bar provides access to the following major functions:

1. Search for different types of data through tailored search interfaces accessible from the **Search Data** drop-down menu.
2. Analyze and visualize data contained within the portal, or provided by the user, using the tools in the **Analyze & Visualize** drop-down menu.
3. Save and share data and analysis results in the personal Workbench through the **Workbench** drop-down menu.
4. Access tutorials, computational protocols, and the ViPR help-desk from the **Help** drop-down menu.

3.2 Searching for Sequence Data

To make sequence data useful for basic and clinical research, it is necessary to ensure that these data are annotated in a correct, consistent, and enriched way. The comprehensive, consistent metadata and mature peptide annotations provided by ViPR serve as the foundation for a sequence search interface with structured taxonomy, standardized gene symbols, and metadata fields populated with controlled vocabularies (Fig. 2a). In this section, we briefly describe the basic search options and their usage.

1. To access the sequence search page, go to **Search Data** > **Search Sequences** > **Genome Sequences** or **Genes & Proteins** or **Strain Data**.
2. The sequence search page provides a variety of powerful search options:
   (a) The 'Select Virus(es) to Include in Search' section presents virus taxa in a hierarchy. Click the plus/minus sign to uncollapse/collapse the taxon (species, genotype, or subtype) and then click **Select All** for the desired taxon. As an example, drill down to **Hepatitis C virus** > **Genotype 1** > **Subtype 1b** and then click **Select All** for Subtype 1b. The count of the database record matching the selected criteria will update as new search criteria are added.
   (b) The 'Data To Return' section lists three data types available: Genome, Protein, and Strain. Choosing one data type will automatically update the search interface with additional search options specific to that data type. For example, choosing **Protein** will display the **Gene Symbol** selection list populated with **All, Polyprotein, C, F, NS5a**, etc. Many genome sequences in GenBank lack mature peptide annotations, therefore searching by gene symbol in GenBank often misses relevant strains. The gene symbol list on the HCV 'Gene/Protein Search' page is populated with standardized gene symbols generated from the ViPR mature peptide predictions, thereby

a

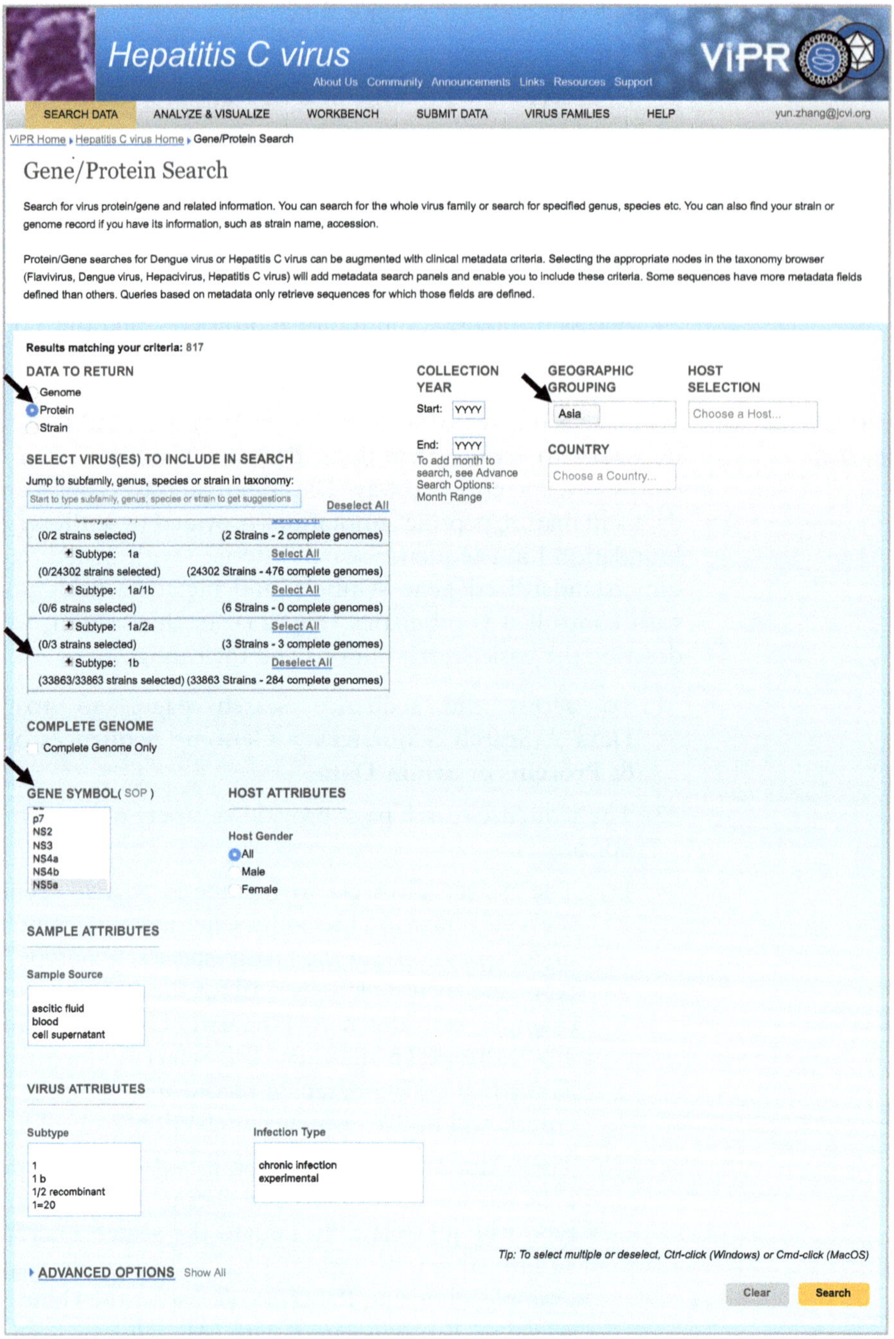

Fig. 2 Clinical metadata, genotype, and mature peptide annotations in ViPR—(**a**) The comprehensive, consistent genotype and mature peptide annotations in ViPR allow users to search for mature peptides based on genotypes, standardized gene symbols along with other standard metadata fields. (**b**) The 'Strain Details' page in ViPR presents virus characteristics, clinical metadata, sequence data, and sequence annotations in a single page. Clinical metadata is provided by the NIAID Genome Sequencing Center and displayed in the 'Host/Patient Sample Information' section when available. ViPR-generated genotype annotations are displayed in the 'Subtype/Genotype (ViPR)' field. Gene symbols and protein product names for ViPR-predicted mature peptides are displayed in a genome map and in a table

b

Strain Information

Strain Name:	HC-J8
Organism:	Hepatitis C virus
Taxonomy:	Flaviviridae -> Hepacivirus -> Hepatitis C virus -> Type 2b
Subtype/Genotype (ViPR) (SOP):	2b
Subtype/Genotype (Genbank):	Unknown
Host:	Unknown

Host/Patient Sample Information

Isolation Country:	Japan
Isolation Region:	Asia
HBV Coinfection:	N
HIV Coinfection:	N

Genome: D10988

GenBank Definition:	Hepatitis C virus genome.
Authors:	Mink,M.A., Benichou,S., Madaule,P., Tiollais,P., Prince,A.M. and Inchauspe,G. Okamoto,H., Kurai,K., Okada,S., Yamamoto,K., Lizuka,H., Tanaka,T., Fukuda,S., Tsuda,F. and Mishiro,S. Hotta,H., Doi,H., Hayashi,T., Purwanta,M., Soemarto,W., Mizokami,M., Ohba,K. and Homma,M.
GenBank Sequence Accession:	D10988
Sequence Length:	9511
Sequence Status:	Complete
Sequence:	View Nucleotide Sequence and design PCR primers
Number of Proteins:	11
Organism Name:	Hepatitis C virus
Mol Type:	genomic RNA
Host:	Unknown

Genome Image Map

Hide Show

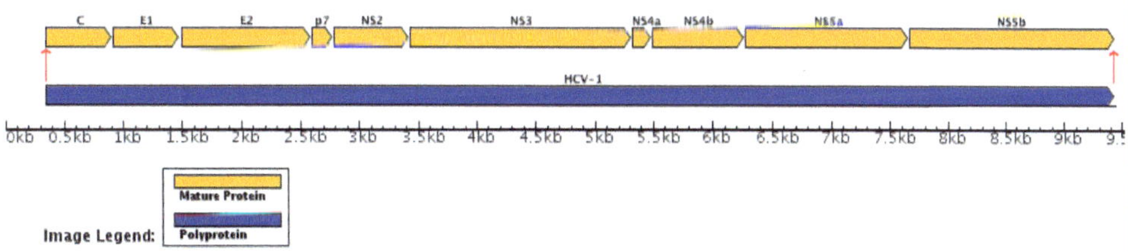

Protein Information (SOP)

	Gene Symbol	Protein Product Name	ViPR Locus ID	CDS Start	CDS End	NCBI Gene ID	Locus Name
GenBank							
View	-N/A-	polyprotein precursor	HCV-1	342	9443	-N/A-	-N/A-
ViPR-generated							
View	C	core protein	C	342	914	-N/A-	-N/A-
View	E1	E1 protein	E1	915	1490	-N/A-	-N/A-
View	E2	E2 protein	E2	1491	2591	-N/A-	-N/A-
View	p7	p7 protein	p7	2592	2780	-N/A-	-N/A-
View	NS2	NS2 protein	NS2	2781	3431	-N/A-	-N/A-
View	NS3	NS3 protease/helicase	NS3	3432	5324	-N/A-	-N/A-
View	NS4a	NS4A protein	NS4a	5325	5486	-N/A-	-N/A-
View	NS4b	NS4B protein	NS4b	5487	6269	-N/A-	-N/A-
View	NS5a	NS5A protein	NS5a	6270	7667	-N/A-	-N/A-
View	NS5b	NS5B RNA-dependent RNA polymerase	NS5b	7668	9440	-N/A-	-N/A-

Fig. 2 (continued)

providing an easy way of doing a comprehensive search for target proteins. For example, select **NS5a** to search for all strains with sequence coverage in the NS5a region (Fig. 2a).

(c) Other search fields include sequence length, collection year, geographic grouping, country, host, strain name, accession, and clinical metadata. These fields allow to add additional criteria if needed.

(d) Click **Search** at the bottom right of the page.

3. After processing, the 'Search Result' page will be loaded. The matching records are displayed in a tabular format (*see* **Note 4**). Above the search results table are a series of amber buttons that enable the following functions (*see* **Note 5**):

   (a) Customize the display format including the number of items per page, summary table columns, and sorting fields by using the **Display Settings** menu.

   (b) Store selected sequences as a working set in the user's Workbench so that the dataset can be retrieved later: select the desired records > **Add to Working Set**.

   (c) Save the search criteria to the user's personal Workbench by clicking **Save Search**. A 'Save Searches' lightbox will pop up. Users can choose to receive email alerts on newly released data that match the saved search criteria by checking the **Subscribe** option.

   (d) Select records and run an analysis on the selected records: **Run Analysis** > choose a desired option such as **Identify Similar Sequences (BLAST), Align Sequences (MSA), Identify Short Peptides in Proteins, Analyze Sequence Variation (SNP), Generate Phylogenetic Tree**, and others.

   (e) Download sequences in various formats by selecting sequences and then clicking **Download**.

4. To view the details of a sequence record, click **View** for the record to bring up the corresponding 'Details' page.

5. The 'Gene/Protein Details' page contains computationally improved annotations and curated information including sequence data, isoelectric point, molecular weight, protein domains and motifs, and predicted immune epitopes. These data are combined with the strain information on the page.

6. Clicking **View Strain Details** in the 'Strain Information' section loads the 'Strain Details' page, which contains clinical metadata if available, ViPR mature peptide annotations presented in a 'Genome Map' image, and a 'Protein Information' table (Fig. 2b). Selecting a protein in the image or table loads the 'Gene/Protein Details' page as described above.

3.3 Conducting Phylogenetic Analysis

Phylogenetic analysis is widely used to study virus evolution. A typical phylogenetic analysis workflow in the ViPR HCV portal involves the following steps:

1. Compile sequences. Typically, full-length genome sequences are used for virus phylogenetic analysis. The phylogenetic tree tool in ViPR accepts input sequences from a query result, an existing working set in the personal Workbench, or private sequences and metadata uploaded to the personal Workbench.
2. Infer a multiple sequence alignment. The user can then check the quality of the sequences and the alignment, remove any questionable sequences, and recompute the alignment if needed.
3. Build a phylogenetic tree using sequences that have passed the quality control step.
4. Visualize the tree to analyze the evolutionary relationships of the viruses in question.

In the following example, we briefly describe the workflow for performing phylogenetic inference between HCV strains from the years 2011 to 2012.

1. Go to **Search Data** > **Search Sequences** > **Genome Sequences**.
2. On the 'Genome Search' page, choose the following criteria and then click **Search**.

   Select Virus(es) to Include in Search: Hepatitis C virus—Select All

   Complete Genome: Complete Genome Only

   Collection Year: Start: "2011," End: "2012"

   Host Selection: Human
3. The 'Genome Search Result' page will be loaded. Review the sequence metadata listed in the table and check the strain details by clicking **View** for the corresponding record. Note that a few strains appear to represent viruses resulting from the recombination of two different subtypes.
4. Calculate a multiple sequence alignment on the retrieved genomes. To do so, select desired records by checking the corresponding checkboxes (or select all by checking the **Select all X genomes** checkbox above the table), and then go to **Run Analysis** > **Align Sequences (MSA)**.
5. On the 'Align Sequences (MSA)' page, keep all default selections and click **Run**.
6. On the 'Processing' page, users can save the analysis upon completion by providing an analysis name, (e.g., "HCV full

length genomes 2011–2012") and then clicking **Save to Workbench**, or request an email notification.

7. Once the analysis is finished, a 'Visualize Aligned Sequences' page will be loaded (*see* **Note 6**). Examine the alignment. Questionable sequences (e.g., sequences containing large insertions or deletions or regions of poor alignment) should be excluded from downstream analysis.
8. If satisfied with the alignment, construct a phylogenetic tree based on the alignment by clicking **Run Analysis** > **Generate Phylogenetic Tree** on the 'Aligned Sequences' page.
9. The 'Generate Phylogenetic Tree' page will be displayed. ViPR provides three algorithms for phylogenetic tree construction: FastME [10], RaXML [11], and PhyML [12] (*see* **Note 7**). As an example, choose **Quick Tree** which uses FastME. In the 'Label Tree Tips (Ends) With' section, choose desired fields for sequence labels (e.g., **Strain Name, Date**, and **Country**). Then click **Build Tree.**
10. Once the analysis is finished, the 'View Phylogenetic Tree' page will be loaded.
11. The tree can be visualized in two interactive tree viewers: Archaeopteryx.js and Archaeopteryx (Java Applet) (*see* **Note 8**). Choose **Archaeopteryx-js** and then click **View Tree.**
12. Sequence-associated metadata describing such features as genotype/subtype, host species, isolation date, or geographic location can be extremely helpful in understanding evolutionary dynamics. Toward this end, Archaeopteryx.js has been developed to support metadata-based visualization functionalities, including visualization of metadata as node shapes, sizes, fill colors, border colors, and automated sub-tree collapsing based on shared node metadata. Furthermore, the tree viewer allows for the display of values, rerooting of the tree, swapping of descendants, collapsing/expanding/displaying/removing of subtrees, downloading of node data and aligned sequences associated with a subtree. In addition, the program has powerful search functions, including logical *or* and *and*, as well as regular expression functionality. For this example, color sequence labels by subtype. To do so, in the Visualizations panel, choose 'Label Color': **Subtype/Genotype**. Then color tree nodes by geographic region by choosing 'Node Fill Color': **Region**.
13. All trees produced by the algorithms used in ViPR are rooted randomly. Reroot the tree by clicking the branching node for a genotype (e.g., genotype 1) and then clicking **Reroot** in the pop-up box.

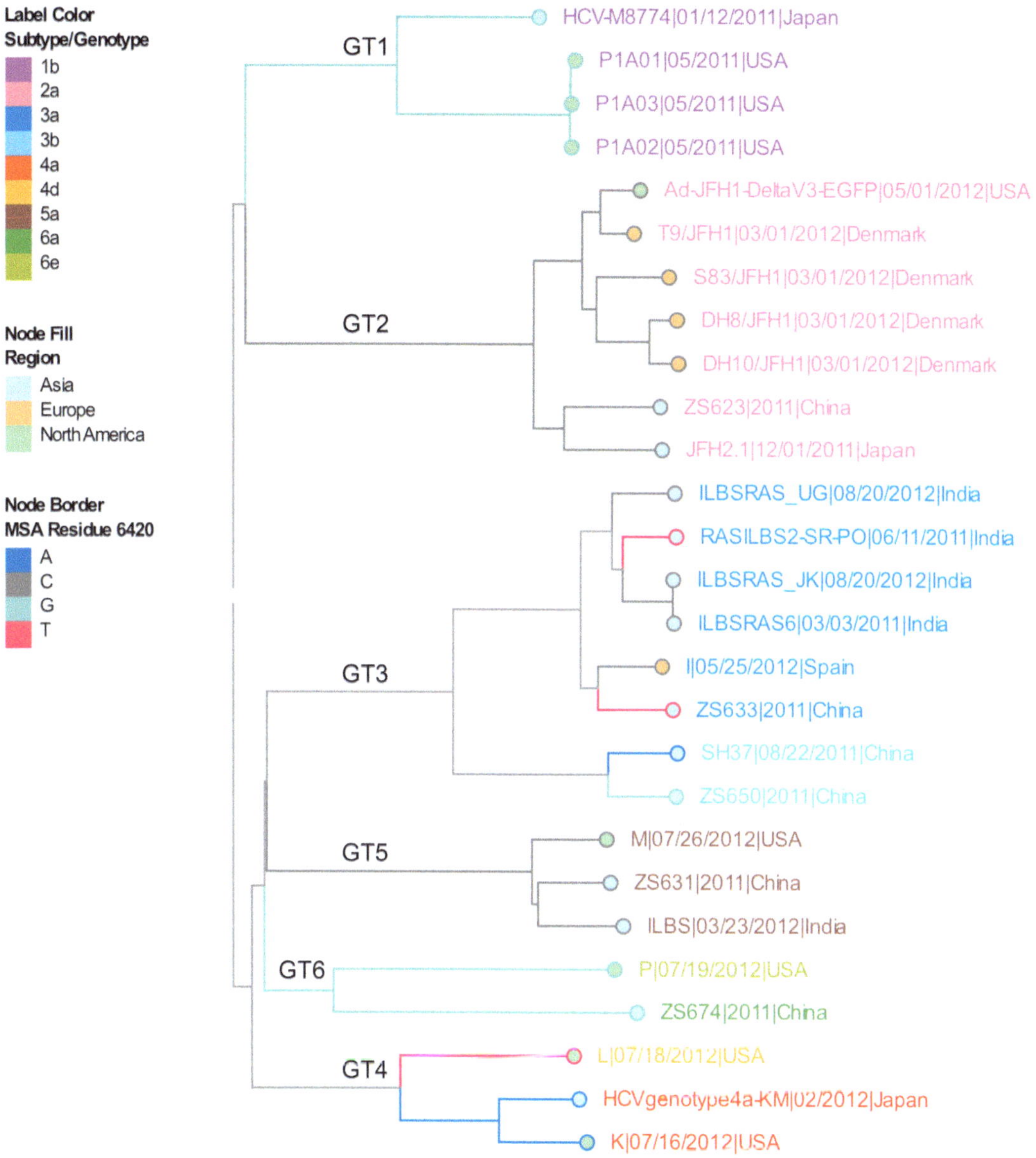

Fig. 3 Phylogenetic inference between HCV strains—The phylogenetic tree was constructed using selected full length HCV genomes from 2011 to 2012, and then visualized in Archaeopteryx.js in ViPR. Sequence labels are colored by genotype/subtype, tree nodes are filled by geographic region, and tree branches and node borders are colored by residue at position 6420

14. The decorated tree shows clear correlation between the tree topology and virus subtype (Fig. 3).
15. Download the tree as a computer-readable text file (currently Archaeopteryx.js provides Newick and phyloXML formats) or as a decorated tree image (PNG, SVG formats) by choosing the appropriate format and then clicking the **Download** button at the bottom of the control panel.

3.4 Analyzing Sequence Variations

3.4.1 Identifying Resistant-Associated Substitutions (RASs) or Other Substitutions

Currently available direct-acting antiviral drugs (DAAs) for HCV treatment target the viral proteins NS3, NS4a, NS5a or NS5b. Laboratory and clinical studies have characterized resistance-associated substitutions (RASs) in these viral proteins. In particular, the NS5a RASs are located at positions 28, 30, 31, and 93 in the NS5a protein [13]. The clinical impact of these RASs is significant enough to warrant RAS testing prior to treatment with an NS5a inhibitor. To identify strains carrying any of the RASs or other substitutions, users can take advantage of the short peptide search tool in ViPR. The following example demonstrates how to identify strains of HCV GT1b subtype carrying the L31 M/V substitution in NS5a.

1. Go to **Search Data** > **Search Sequences** > **Genes & Proteins**.
2. On the 'Gene/Protein Search' page, choose the following criteria and then click **Search**.

   Select Virus(es) to Include in Search: Hepatitis C virus > Genotype: 1 > Subtype: 1b – Select All.

   Gene Symbol: NS5a.

   Host: Human
3. The 'Gene/Protein Search Result' page will be loaded. Select all protein records by clicking the **Select all X proteins** checkbox above the table and then go to **Run Analysis** > **Identify Short Peptides in Proteins**.
4. The search page provides three search types:
   (a) **Exact Match** finds only exact matches of the search string.
   (b) **Fuzzy Match** finds amino acid strings with over 50% identity to the search string.
   (c) **Pattern Match** finds amino acid strings with 100% pairwise identity to the input pattern.
5. The majority of GT1b strains carry 28LP[HKQR]LP32 [14], where [HKQR] means position 30 contains H, K, Q, or R. To use this tool to find the L31M/V substitution, choose **Pattern Match**, type "LP[HKQR][MV]P" in the string to find box, and click **Run**.
6. The 'Identify Short Peptides Result' page displays a list of strains matching the provided pattern, including the matched sequence, the coordinate range in the NS5a protein, and a hyperlink to the protein sequence (Fig. 4a).
7. Similarly, users can use this tool to find short peptides in ViPR sequences saved in a working set in their personal Workbench, private sequences uploaded directly in the tool page, private sequences and metadata uploaded to their personal Workbench, or ViPR precompiled viral protein databases.

a

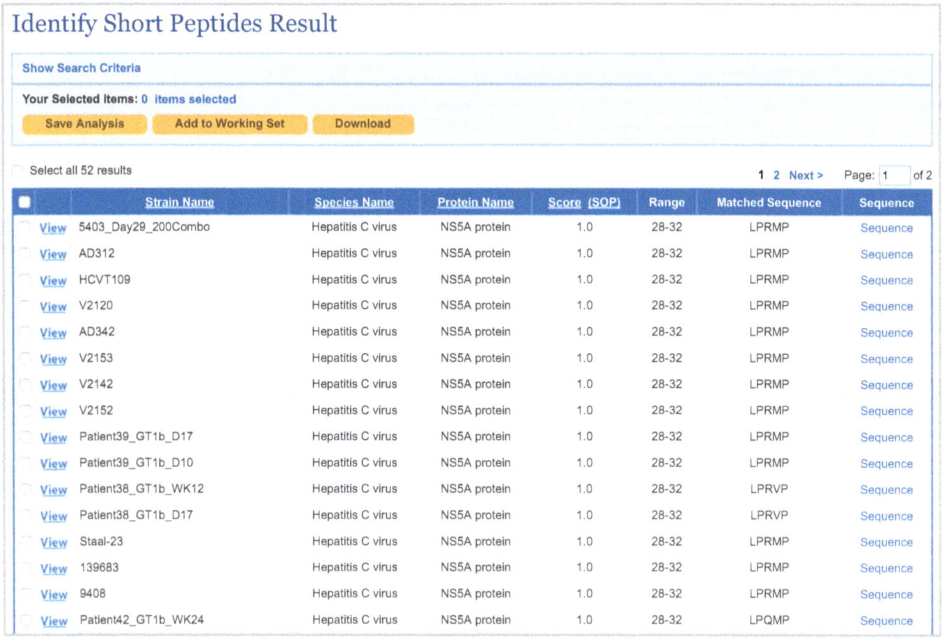

Identify Short Peptides Result

Show Search Criteria

Your Selected Items: 0 items selected

Save Analysis | Add to Working Set | Download

Select all 52 results

1 2 Next > Page: 1 of 2

		Strain Name	Species Name	Protein Name	Score (SOP)	Range	Matched Sequence	Sequence
	View	5403_Day29_200Combo	Hepatitis C virus	NS5A protein	1.0	28-32	LPRMP	Sequence
	View	AD312	Hepatitis C virus	NS5A protein	1.0	28-32	LPRMP	Sequence
	View	HCVT109	Hepatitis C virus	NS5A protein	1.0	28-32	LPRMP	Sequence
	View	V2120	Hepatitis C virus	NS5A protein	1.0	28-32	LPRMP	Sequence
	View	AD342	Hepatitis C virus	NS5A protein	1.0	28-32	LPRMP	Sequence
	View	V2153	Hepatitis C virus	NS5A protein	1.0	28-32	LPRMP	Sequence
	View	V2142	Hepatitis C virus	NS5A protein	1.0	28-32	LPRMP	Sequence
	View	V2152	Hepatitis C virus	NS5A protein	1.0	28-32	LPRMP	Sequence
	View	Patient39_GT1b_D17	Hepatitis C virus	NS5A protein	1.0	28-32	LPRMP	Sequence
	View	Patient39_GT1b_D10	Hepatitis C virus	NS5A protein	1.0	28-32	LPRMP	Sequence
	View	Patient38_GT1b_WK12	Hepatitis C virus	NS5A protein	1.0	28-32	LPRVP	Sequence
	View	Patient38_GT1b_D17	Hepatitis C virus	NS5A protein	1.0	28-32	LPRVP	Sequence
	View	Staal-23	Hepatitis C virus	NS5A protein	1.0	28-32	LPRMP	Sequence
	View	139683	Hepatitis C virus	NS5A protein	1.0	28-32	LPRMP	Sequence
	View	9408	Hepatitis C virus	NS5A protein	1.0	28-32	LPRMP	Sequence
	View	Patient42_GT1b_WK24	Hepatitis C virus	NS5A protein	1.0	28-32	LPQMP	Sequence

b

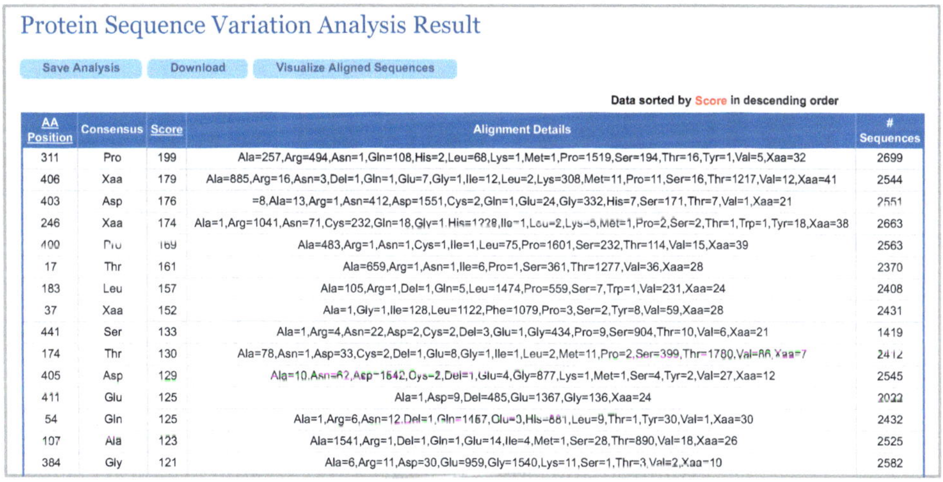

Protein Sequence Variation Analysis Result

Save Analysis | Download | Visualize Aligned Sequences

Data sorted by Score in descending order

AA Position	Consensus	Score	Alignment Details	# Sequences
311	Pro	199	Ala=257,Arg=494,Asn=1,Gln=108,His=2,Leu=68,Lys=1,Met=1,Pro=1519,Ser=194,Thr=16,Tyr=1,Val=5,Xaa=32	2699
406	Xaa	179	Ala=885,Arg=16,Asn=3,Del=1,Gln=1,Glu=7,Gly=1,Ile=12,Leu=2,Lys=308,Met=11,Pro=11,Ser=16,Thr=1217,Val=12,Xaa=41	2544
403	Asp	176	=8,Ala=13,Arg=1,Asn=412,Asp=1551,Cys=2,Gln=1,Glu=24,Gly=332,His=7,Ser=171,Thr=7,Val=1,Xaa=21	2551
246	Xaa	174	Ala=1,Arg=1041,Asn=71,Cys=232,Gln=18,Gly=1,His=1228,Ile=1,Leu=2,Lys=5,Met=1,Pro=2,Ser=2,Thr=1,Trp=1,Tyr=18,Xaa=38	2663
400	Pro	169	Ala=483,Arg=1,Asn=1,Cys=1,Ile=1,Leu=75,Pro=1601,Ser=232,Thr=114,Val=15,Xaa=39	2563
17	Thr	161	Ala=659,Arg=1,Asn=1,Ile=6,Pro=1,Ser=361,Thr=1277,Val=36,Xaa=28	2370
183	Leu	157	Ala=105,Arg=1,Del=1,Gln=5,Leu=1474,Pro=559,Ser=7,Trp=1,Val=231,Xaa=24	2408
37	Xaa	152	Ala=1,Gly=1,Ile=128,Leu=1122,Phe=1079,Pro=3,Ser=2,Tyr=8,Val=59,Xaa=28	2431
441	Ser	133	Ala=1,Arg=4,Asn=22,Asp=2,Cys=2,Del=3,Glu=1,Gly=434,Pro=9,Ser=904,Thr=10,Val=6,Xaa=21	1419
174	Thr	130	Ala=78,Asn=1,Asp=33,Cys=2,Del=1,Glu=8,Gly=1,Ile=1,Leu=2,Met=11,Pro=2,Ser=399,Thr=1780,Val=86,Xaa=7	2412
405	Asp	129	Ala=10,Asn=62,Asp=1542,Cys=2,Del=1,Glu=4,Gly=877,Lys=1,Met=1,Ser=4,Tyr=2,Val=27,Xaa=12	2545
411	Glu	125	Ala=1,Asp=9,Del=485,Glu=1367,Gly=136,Xaa=24	2022
54	Gln	125	Ala=1,Arg=6,Asn=12,Del=1,Gln=1467,Glu=3,His=881,Leu=9,Thr=1,Tyr=30,Val=1,Xaa=30	2432
107	Ala	123	Ala=1541,Arg=1,Del=1,Gln=1,Glu=14,Ile=4,Met=1,Ser=28,Thr=890,Val=18,Xaa=26	2525
384	Gly	121	Ala=6,Arg=11,Asp=30,Glu=959,Gly=1540,Lys=11,Ser=1,Thr=3,Val=2,Xaa=10	2582

Fig. 4 Sequence variation analysis tools in ViPR—(**a**) A section of an 'Identify Short Peptides Result' page showing strains matching pattern LP[HKQR][MV]P, with the intention of finding strains carrying the L31M/V substitution in NS5a. (**b**) A portion of a sequence variation SNP report showing the highly variable positions in the NS5a protein of subtype GT1b strains. (**c**) A report of the meta-CATS analysis comparing the GT1b strains between pretreatment and posttreatment groups from a GS-5816 clinical trial study [15]. The table contains two significant positions identified in the NS5a protein between the two groups: L31M/V/X and Y93H/X. (**d**) A section of a multiple sequence alignment generated from the meta-CATS report, including L31M/V/X and Y93H/X found in the meta-CATS analysis. (**e**) The 'Sequence Feature Details' page for a resistance-associated substitution (RAS) in the NS5a protein, named Hepatitis C virus_NS5a_RAS_31(1), covering position 31 of the NS5a protein. The 'Variant Types' (VTs) table shows a list of 'Phenotypic Variant Types' (31M: VT-2, 31V: VT-3, 31F: VT-9) with the strain count number hyperlinked to all strains carrying the corresponding VT. These Phenotypic Variant Types confer resistance to NS5a inhibitor treatment for certain genotype infections

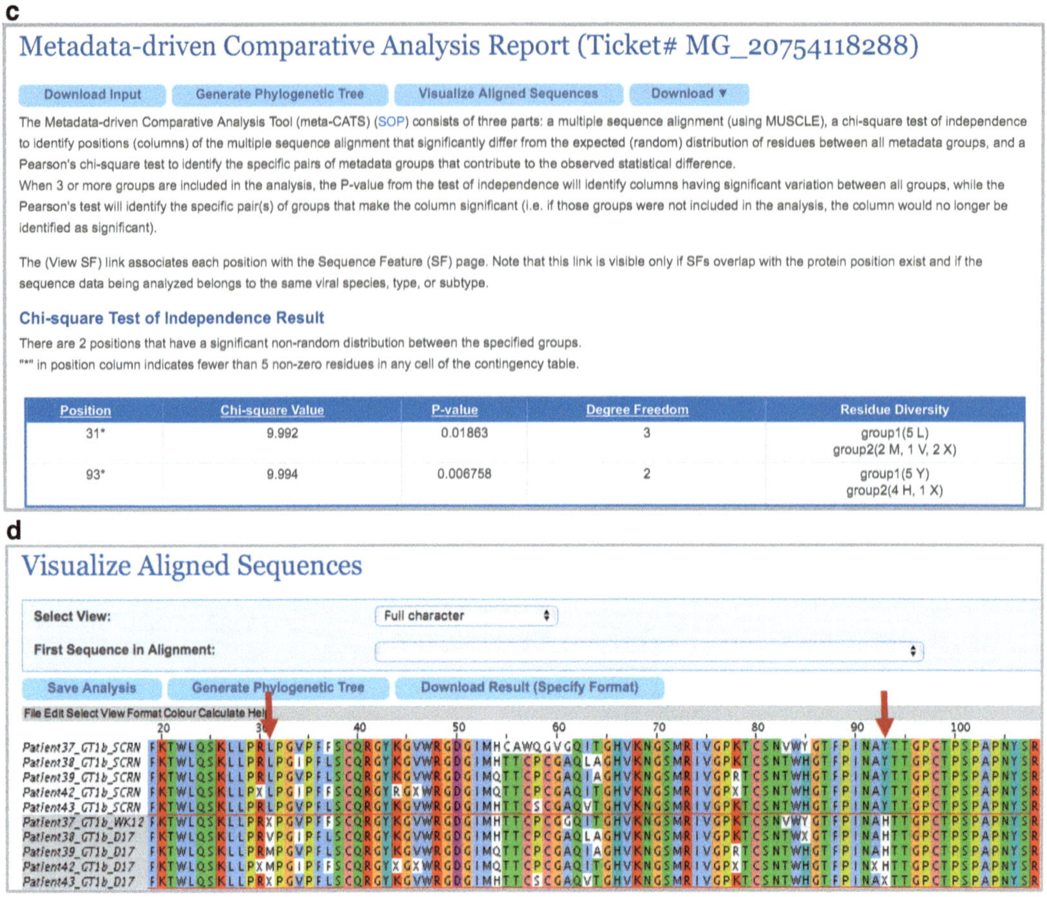

Fig. 4 (continued)

3.4.2 Identifying SNPs

Similar to other RNA viruses, the HCV viral RNA-dependent RNA polymerase (RdRp) lacks proofreading activity and therefore generates distinct viral variants at a rapid pace. Identifying sequence variations in viral genomes is important for understanding virus evolution and how variations affect treatment efficacy. One of the analytical tools that can detect sequence variations is the Analyze Sequence Variation tool. The ViPR implementation of the Analyze Sequence Variation tool allows users to analyze both nucleotide and amino acid sequences from a query result, an existing working set in their personal Workbench, or private sequences and metadata uploaded to their personal Workbench. Instructions below demonstrate a sequence variation analysis of NS5a protein sequences from GT1b subtype viruses.

1. Go to **Search Data** > **Search Sequences** > **Genes & Proteins**.
2. On the 'Gene/Protein Search' page, choose the following criteria and then click **Search**:

e

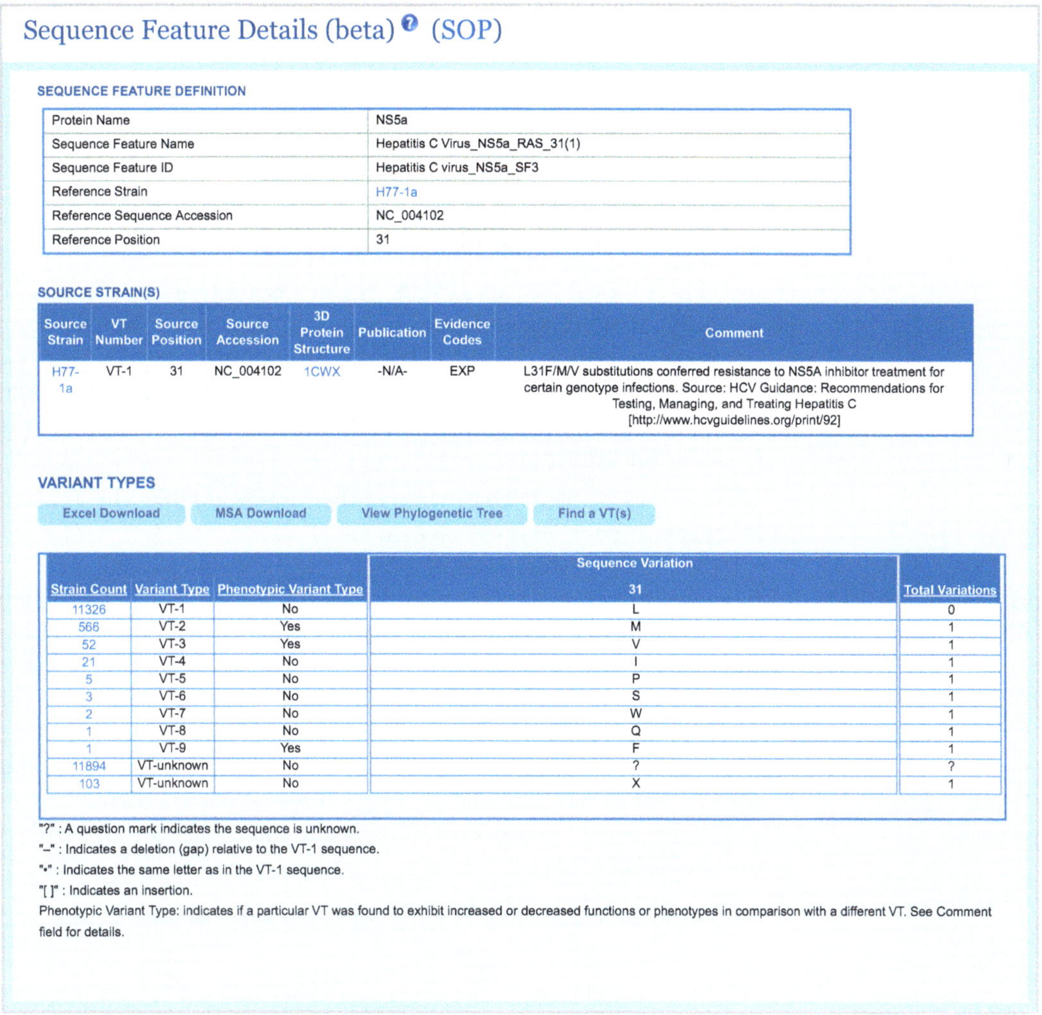

Fig. 4 (continued)

Select Virus(es) to Include in Search: Hepatitis C virus > Genotype: 1 > Subtype: 1b – Select All.
Gene Symbol: NS5a.
Host: Human

3. The 'Gene/Protein Search Result' page will be loaded. Select all protein records by clicking the **Select all X proteins** checkbox above the table and then go to **Run Analysis** > **Analyze Sequence Variation (SNP).**
4. In the 'Select Sequence Type' lightbox, choose **Amino Acid (AA)**. Then click **Continue**.
5. The 'Analyze Sequence Variation (SNP)' page will be loaded. Details of the SNP tool are provided. It begins with a multiple

sequence alignment. At each position (column) of the alignment, a consensus residue is computed based on a majority rule, while a variation score is calculated using a normalized entropy score. Click **Run** to continue.

6. Once the analysis is finished, the 'Protein Sequence Variation Analysis Result' page will be displayed (Fig. 4b). The table shows the amino acid position, consensus residue, variation score, and counts for each amino acid at each position. The consensus residue is the one with a frequency greater than 50%. If no residue exceeds 50%, N (for nucleotide) or Xaa (for amino acid) is used to indicate ambiguity. The score ranges from 0 (no variation) to 200 (highest variation). The table can be sorted based on the various column headings (e.g., the variation score to rank positions by variability) by clicking the column header.
7. Users can choose to download the displayed table, the consensus sequence, and the alignment.

3.4.3 Comparing Groups of Viral Genomes

In analyzing sequence variations in viral genomes, researchers often need to examine correlations between groups of virus sequences and other virus characteristics, such as finding specific genetic variations that distinguish different subtypes, or detecting the genetic determinants responsible for antiviral drug resistance. Meta-CATS [7] is an automated, statistical analysis tool for finding genetic variations associated with different virus characteristics using groups of sequences. Utilizing the meta-CATS tool involves the following steps:

1. Compile sequences. For function-related analyses, amino acid sequences are often used.
2. Input sequences to the meta-CATS tool in ViPR from a query result, an existing working set in a personal Workbench, or private sequences and metadata uploaded to a personal Workbench.
3. Segregate input sequences into groups manually, or using the auto-grouping option that segregates sequences based on metadata values and other characteristics stored in the ViPR database (*see* **Note 9**).
4. Run the meta-CATS tool.
5. The meta-CATS report then provides a list of positions that significantly differ between the sequence groups.

The example below demonstrates an analysis comparing the GT1b subtype strains between the pretreatment and posttreatment groups from a GS-5816 clinical trial study (PMID: 27353271) [15].

1. Go to **Search Data** > **Search Sequences** > **Genes & Proteins**.
2. On the 'Gene/Protein Search' page, choose the following criteria and then click **Search**.

   Select Virus(es) to Include in Search: Hepatitis C virus > Genotype 1 > Subtype 1b – Select All.
   Gene Symbol: NS5a.
   Advanced Options—Public Identifier Search—**PubMed ID**: "27353271".
3. The 'Gene/Protein Search Result' page will be loaded. Select all protein records by clicking the **Select all X proteins** checkbox above the table and then click **Add to Working Set**.
4. In the 'Add to Working Set' lightbox, choose **Create a new working set with the selected items**, provide a name for the working set (e.g., "PMID_27353271_1b") and then click **Add to Working Set**.
5. Go to **Workbench** > **View Your Workbench** > **View** for the saved working set 'PMID_27353271_1b'.
6. This dataset contains the NS5a sequences from eight patients. Based on the study, five patients (Patients 37, 38, 39, 42, and 43) were treated with a 3-day monotherapy of GS-5816.
7. Next, use meta-CATS to compare the NS5a protein sequences from the above five patients between pretreatment and posttreatment. For each of the five patients, select the sequence for the pretreatment time point (strain name with a '_SCRN' suffix) and the sequence for a posttreatment time point (strain name with a '_D17' suffix, with the exception of '_WK12' for Patient 37). A total of ten sequences will be selected. Then go to **Run Analysis** > **Metadata-driven Comparative Analysis Tool.**
8. In the Select Sequence Type lightbox, choose **Amino Acid (AA)** and then click **Continue**.
9. On the 'Metadata-driven Comparative Analysis Tool (meta-CATS)' page, choose **Manual Grouping: 2** and then click **Continue**.
10. This will bring up the 'Setup Subset' page. Select all sequences with a '_SCRN' suffix using Ctrl+ mouse button and add them to group 1. Then select the remaining sequences ('_D17' or '_WK12') in the main list of sequences and add them to group 2. The resulting group 1 will have NS5a sequences at the screening time point from five patients, while group 2 will have sequences from the D17 or WK12 time point from the same cohort. Click **Run**.
11. Once the analysis is finished, a 'Metadata-driven Comparative Analysis Report' page will be loaded (Fig. 4c). The report has

two parts: a chi-square test of independence result table listing the positions that have a significant nonrandom distribution between your specified sequence groups, and a Pearson's chi-square test result table listing the specific pairs of groups that contribute to the observed statistical difference. Review the chi-square test results to find the positions that differ significantly between the pretreatment and posttreatment groups. The residue diversity column lists the counts for each residue within a group. All five patients have RASs at positions 31 and 93 in posttreatment sequences.

12. Finally, view the protein sequence alignment to confirm the meta-CATS results. Click **Visualize Aligned Sequences** at the top of the page.
13. In the Label Sequence By lightbox, choose the **Strain Name** option, and then click **Run**.
14. The 'Visualize Aligned Sequences' page will be loaded (Fig. 4d). Users can manually adjust the alignment order by clicking a sequence name and using the up and down arrow to move the sequence within the alignment. Scroll right to the 30–100 region and examine positions 31 and 93. L31M/V/X and Y93H/X are observed in the posttreatment sequences.

3.4.4 Analyzing Sequence Variations in Curated Sequence Features

The ViPR team has curated interesting protein regions with known structural or functional properties as Sequence Features (SFs). To infer the functional implications of specific substitutions, users can search for SFs that cover the positions of interest and then examine the precomputed sequence polymorphisms in the SF region. The following example shows how to find and analyze SFs covering position 31 identified by the previous meta-CATS analysis.

1. Go to **Search Data > Sequence Feature Variant Types**.
2. On the 'Sequence Feature Variant Types' page, users can either browse the list of all SFs by clicking **Go to Sequence Feature List**, or search for SFs by protein, Sequence Feature type, amino acid coordinates, or keyword. For example, select the following criteria and then click **Search**.

   Select (Mature) Proteins: NS5a.

   Sequence Feature Type: Functional
3. The 'Sequence Feature Variant Types' page lists all functional SFs in the NS5a protein. The SF naming convention is: taxon_protein_SF-description_starting-position(SF-length). For example, 'Hepatitis C Virus_NS5a_RAS_31(1)' defines a resistance associated substitution (RAS) at position 31 in the NS5a protein. This position was identified during the previous meta-CATS analysis. The comments field provides a brief description

of this SF: L31F/M/V substitutions conferred resistance to NS5A inhibitor treatment for certain genotype infections.

4. Click **View** for this SF.
5. This will bring up the 'Sequence Feature Details' page (Fig. 4e). 'The Sequence Feature Definition' section lists the reference sequence used to define the coordinate system for HCV SFs. The 'Source Strain' section displays the strain where the Sequence Feature was first reported. The 'Variant Types' table shows all unique substitutions in the SF region, termed Variant Types (VTs). The 'Phenotypic Variant Type' column indicates whether a particular VT has been found to give rise to the phenotypic effect described in the 'Comment' field, in this case, resistance to the NS5a inhibitor.
6. Click the strain count for VT-3 (31V) .
7. This will bring up all strains carrying this substitution.

4 Notes

1. Recently, RefSeq sequences for other subtypes have become available, including NC_009823 (GT2), NC_009824 (GT3), NC_009825 (GT4), NC_009826 (GT5), and NC_009827 (GT6). The protease cleavage sites in these RefSeq sequences match those in NC_004102 (GT1). The mature peptide prediction tool is being augmented with the addition of protein F and these additional RefSeq sequences.
2. The text is formatted using the following convention: **Bold text**—action buttons or options on the website that need to be clicked or selected; 'text in single quotes'—page or section names on the website; "text in double quotes"—text that needs to be entered into an input box on the website.
3. Information presented in this chapter is based on the July 2018 release of the ViPR resource. Bimonthly updates and enhancements may be made to the user interfaces, database contents, and analysis tools; therefore, functions and interfaces may have changed since publication.
4. Results are displayed in a data-dependent manner. For example, an immune epitope search results table contains IEDB ID, Epitope Sequence, Proteins Matching Sequence, Protein Names, Host, Assay Type Category, Assay Result, etc.
5. The analysis options change depending on the type of data being displayed.
6. Currently, the JalView program is used for MSA visualization and editing. Since this program has limited browser compatibility, we are in the process of replacing it with MSAViewer which runs on all major browsers.

7. If the user is logged into the personal Workbench and a very large dataset is submitted to the tree construction tool, a recommendation for running the job in the CIPRES high performance computing environment will appear. The user can then choose the CIRPES option to compute the tree. Once the analysis is completed, the result will be saved to the personal Workbench.
8. Java Applets are becoming increasingly obsolete, so the ViPR team has pioneered Archaeopteryx-js, a JavaScript based tree viewer, which is faster under most circumstances and more robust.
9. Users can upload a sequence FASTA file and a sequence group assignment file (csv, xls, or xlsx) to automatically assign sequences to groups.

5 Summary

The HCV portal of ViPR provides a unique resource, combining (1) high quality annotations, (2) integrated data from both external sources and internal computational pipelines, and (3) a suite of analysis and visualization tools. With data updated and new features released every 2 months, the portal continues to expand to support HCV data mining and hypothesis generation. The development of the ViPR resource is driven by community needs. The ViPR team welcomes user feedback and provides a 'Contact Us' form to assist users with specific questions.

Acknowledgment

This work was supported by the National Institutes of Health/ National Institute for Allergy and Infectious Diseases [HHSN272201400028C].

References

1. WHO (2017) Hepatitis C. http://www.who.int/mediacentre/factsheets/fs164/en/. Accessed 11 June 2017
2. Pickett BE, Sadat EL, Zhang Y et al (2012) ViPR: an open bioinformatics database and analysis resource for virology research. Nucleic Acids Res 40:D593–D598
3. ICTV Flaviviridae Study Group (2017) Updated alignment (FASTA) of HCV genotypes and subtypes 1.6.17.FST. https://talk.ictvonline.org/ictv_wikis/flaviviridae/w/sg_flavi/57/hcv-reference-sequence-alignments. Accessed 11 June 2017
4. Matsen FA, Kodner RB, Armbrust EV (2010) pplacer: linear time maximum-likelihood and Bayesian phylogenetic placement of sequences onto a fixed reference tree. BMC Bioinformatics 11:538
5. ViPR (2017) Protocol for HCV typing. https://www.viprbrc.org/brcDocs/documents/SOP_HCV_typing.pdf
6. ViPR (2017) Protocol for virus mature peptide prediction. https://www.viprbrc.org/

brcDocs/documents/VIPR_MAT_PEPTIDE_SOP.pdf. Accessed 11 June 2017

7. Pickett BE, Liu M, Sadat EL et al (2013) Metadata-driven comparative analysis tool for sequences (meta-CATS): an automated process for identifying significant sequence variations that correlate with virus attributes. Virology 447:45–51
8. ViPR (2017) Protocol for meta-CATS analysis tool. https://www.viprbrc.org/brcDocs/documents/MGC_Protocol.pdf. Accessed 11 June 2017
9. Noronha JM, Liu M, Squires RB et al (2012) Influenza virus sequence feature variant type analysis: evidence of a role for NS1 in influenza virus host range restriction. J Virol 86:5857–5866
10. Desper R, Gascuel O (2002) Fast and accurate phylogeny reconstruction algorithms based on the minimum-evolution principle. J Comput Biol 9:687–705
11. Stamatakis A (2014) RAxML version 8: a tool for phylogenetic analysis and post-analysis of large phylogenies. Bioinformatics 30:1312–1313
12. Guindon S, Gascuel O (2003) A simple, fast, and accurate algorithm to estimate large phylogenies by maximum likelihood. Syst Biol 52:696–704
13. AASLD-IDSA (2017) Monitoring patients who are starting Hepatitis C treatment, are on treatment, or have completed therapy. Recommendations for testing, managing, and treating hepatitis C. http://www.hcvguidelines.org/print/92. Accessed 11 June 2017
14. Nakamoto S, Kanda T, Wu S et al (2014) Hepatitis C virus NS5A inhibitors and drug resistance mutations. World J Gastroenterol 20:2902–2912
15. Lawitz EJ, Dvory-Sobol H, Doehle BP et al (2016) Clinical resistance to velpatasvir (GS-5816), a novel pan-genotypic inhibitor of the hepatitis C virus NS5A protein. Antimicrob Agents Chemother 60:5368–5378

Part II

Methods for Virus Cloning, Culture, and Purification

Chapter 4

Establishment of Replication-Competent HCV Strain with Minimum Modifications

Takanobu Kato, Asako Murayama, and Takaji Wakita

Abstract

The HCV cell culture system, consisting of the JFH-1 strain and HuH-7 cells, has been broadly used to assess the complete HCV life cycle in cultured cells. However, being able to use multiple HCV strains in such a system is vital for future studies of this virus. We recently established a novel HCV cell culture system using another HCV genotype 2a strain, J6CF, which replicates in chimpanzees but not in cultured cells. We identified effective cell culture-adaptive mutations and established a replication-competent J6CF strain with minimum modifications in cultured cells. The strategy of how we established the replication-competent HCV strain and how we identified the effective cell culture-adaptive mutations is described here and could prove useful for establishing other replication-competent HCV strains.

Key words Replication, Infectious virus production, Adaptive mutation, Cell culture

1 Introduction

After the discovery of hepatitis C virus (HCV) in 1989 [1], HCV research was hindered by the lack of a cell culture system. Several HCV clones have been reported to be infectious in chimpanzees, but most of these clones could not replicate in cell culture [2–5]. In 1999, however, the HCV subgenomic replicon system was invented [6]. By replacing the HCV structural region with a neomycin resistance gene, this system allowed researchers to observe HCV replication in cultured cells. However, even in such a system, most HCV strains required adaptive mutations in order to replicate efficiently. Subsequently, a genome-length replicon with a structural region was reported. This system was expected to allow visualization of the infection step of this virus in cell culture, but it was still difficult to observe active infection because of poor replication efficiency and infectious particle production [7–9].

In 2005, we established a HCV cell culture system using the JFH-1 strain [10]. This strain was isolated from a fulminant hepatitis patient and belongs to the genotype 2a [11]. The subgenomic

Mansun Law (ed.), *Hepatitis C Virus Protocols*, Methods in Molecular Biology, vol. 1911,
https://doi.org/10.1007/978-1-4939-8976-8_4,

replicon of this strain is capable of replication without adaptive mutation, and the efficiency of colony formation is higher than that of the prototype subgenomic replicon of the Con1 strain [11]. This JFH-1 subgenomic replicon replicates not only in HuH-7 and its derived cell lines but also in other hepatocyte-derived cell lines, including HepG2 and IMY-N9 cells and the non-hepatocyte-derived cell lines, HeLa and 293 cells [12, 13]. Because subgenomic replicon form of this strain could replicate efficiently in cultured cells, we reasoned that genome-length form of this strain might be able to replicate efficiently in cultured cells. To test this idea, we constructed a full-length cDNA construct of JFH-1 and assessed replication in cell culture after transfection of in vitro synthesized genome-length RNA. As expected, the JFH-1 strain could replicate in HuH-7 cells and produced viral particles that were able to infect cultured cells and chimpanzees [10, 14]. This established system using the JFH-1 strain is now broadly used as a platform to study the entire HCV lifecycle, including infection, replication, assembly of infectious viruses, and virus secretion. To date, several other HCV strains have also been reported as competent for infection and replication in cell culture. However, these strains, which are different from JFH-1, require many adaptive mutations [15–21]. These HCV cell culture systems are useful for the high-throughput screening of antiviral reagents and have been used to identify novel direct-acting antivirals (DAAs) that efficiently inhibit HCV replication. However, for accurate and reliable characterization of drug resistance, a HCV cell culture system using multiple HCV strains that can replicate with few or no adaptive mutations is essential, because even just one amino acid mutation can reduce the effect of DAAs.

Recently, we established a novel HCV cell culture system using the HCV J6CF strain. This strain was constructed from the consensus sequence of HCV isolated from a chimpanzee inoculated with an HCV patient's plasma [22]. This strain also belongs to genotype 2a and is similar to the JFH-1 strain at the nucleotide (89% identity) and amino acid (91% identity) levels. Propagation of J6CF in vivo can be observed in chimpanzees after intrahepatic inoculation with in vitro synthesized genome-length HCV RNA [4]. However, J6CF does not replicate in cultured cells after transfection of genome-length HCV RNA [23]. We identified the minimum number of mutations essential for replication of this strain in cell culture after long-term culturing of a chimeric J6CF strain with low replication efficiency. Based on these experiments, we provide an overview here of how to establish this replication-competent HCV strain.

As a first step, we identified the important region of the JFH-1 genome by constructing chimeric subgenomic replicons between strains JFH-1 and J6CF [23]. The nonstructural regions of NS3 (protease and helicase domains), NS4AB, NS5A, and NS5B-to-3'X of the JFH-1 subgenomic replicon were replaced with those of J6CF individually or in combination, and replication efficiencies were tested in cultured cells. We found that the replacement of the J6CF NS3 helicase (N3H) and NS5B-to-3'X (N5BX) regions with those of JFH-1 confers replication competence to J6CF. Efficient replication and infectious virus production by replacement of these regions were also confirmed by full-length chimeric constructs between JFH-1 and J6CF. As a next step, we identified the responsible nucleotide or amino acid mutations in the N5BX region of J6CF [24]. We assessed several mutations introduced into the J6CF chimeric virus with JFH-1 N3H and found that the JFH-1-type amino acids in the NS5B region (A450S, R517K, and Y561F) and a nucleotide mutation at the 3′-untranslated region (UTR), C9458G, were responsible for the replication competency of this clone. Finally, to establish replication-competent J6CF strains with minimum modifications, we identified the responsible mutations instead of replacing the N3H region. We attempted to find the responsible amino acids in the N3H region by replacing a partial region of N3H but could not identify the responsible region or specific amino acids. Therefore, we exploited the strategy of long-term culture of chimeric constructs with low replication efficiencies [25]. For this purpose, we used the J6CF chimeric virus J6/5BSLX-JFH1, which carries JFH-1 predicted stem-loop structures in NS5B (5BSL)-to-3'X. This J6/5BSLX-JFH1 chimeric virus could replicate efficiently in cell culture when combined with a N3H replacement but exhibited very low replication efficiency without it. Fortunately, using this strategy, we could identify the essential amino acid mutations that conferred replication competency to J6CF in combination with the previously identified three amino acid mutations and single-nucleotide mutation. Surprisingly, the identified mutation was not in NS3 but in NS4A.

2 Materials

2.1 Cell Culture

1. Huh-7.5.1 cell line [26], which is derived from HuH-7 hepatoma cells [27].
2. Dulbecco's modified eagle's medium (DMEM) (high glucose; Sigma-Aldrich) supplemented with 10% fetal bovine serum, 0.1 mM MEM nonessential amino acids solution (Thermo Fisher Scientific), 2 mM L-glutamine, 10 mM HEPES pH 7.4, 1 mM sodium pyruvate, 100 U/mL penicillin, and 100 μg/mL streptomycin.

2.2 Plasmid DNA Preparation

1. pJ6/5BSLX-JFH1, the plasmid for a chimeric virus of J6/5BSLX-JFH1, was generated using pJFH1 [10] and pJ6CF [4]. The fragment containing the stem-loop structure in the

NS5B region and the entire 3'UTR of JFH-1 was amplified and fused with the NS5B region of J6CF by PCR. The generated fragment was ligated into the genome-length J6CF plasmid by digestion with SrfI (nt. 8843) and XbaI (nt. 9711) [25].

2.3 RNA Synthesis

1. XbaI (New England Biolabs).
2. Mung Bean Nuclease (New England Biolabs).
3. QIAquick PCR Purification Kit (QIAGEN).
4. MEGAscript T7 Transcription Kit (Thermo Fisher Scientific).
5. RNeasy Mini Kit (QIAGEN).

2.4 RNA Transfection

1. Opti-MEM I reduced-serum Medium (Thermo Fisher Scientific).
2. Cytomix buffer [28].
3. Gene Pulser Electroporation Cuvette (Bio-Rad).
4. Gene Pulser Xcell apparatus (Bio-Rad).

2.5 Quantification of HCV Core Antigen

1. Lumipulse Ortho HCV Antigen Kit (Ortho Clinical Diagnostics) [29].
2. LUMIPULSE G1200 apparatus (Fujirebio).

2.6 HCV Sequencing

1. QIAamp Viral RNA Mini Kit (QIAGEN).
2. SuperScript III reverse transcriptase (Thermo Fisher Scientific).
3. Random Primer (hexadeoxyribonucleotide mixture; TaKaRa Bio).
4. TaKaRa LA Taq polymerase (TaKaRa Bio).
5. Primers (Table 1).

2.7 Introduction of Amino Acid Mutations

1. Pyrobest DNA Polymerase (TaKaRa Bio).
2. Primers (Table 2).
3. QIAquick PCR Purification Kit (QIAGEN).
4. BbvCI (New England Biolabs).
5. EcoRI (New England Biolabs).
6. QIAquick Gel Extraction Kit (QIAGEN).
7. DH5α competent cells.
8. PureLink HiPure Plasmid DNA Purification Kit (Thermo Fisher Scientific).

2.8 Titration of Virus Infectivity

1. Millex-HV Syringe-Driven filter unit (PVDF, Φ13 mM, 0.45 μm pore size) (Merck Millipore).
2. Collagen-coated 96-well plates (Corning).

Table 1
Primers used for amplification of genome-length chimeric virus

Fragment		Primer	Sequence (5′–3′)
A	First PCR	44S-IH	CTGTGAGGAACTACTGTCTT
		3189R-IH	CCAGTCCACCTGCCAAGG
	Second PCR	C/S17	CGGGAGAGCCATAGTGG
		2445R-2a	TCCACGATGTTCTGGTGAAG
B	First PCR	2099S-2a	ACGGACTGTTTTAGGAAGCA
		4859R-2a	TGAGGCTCGCTCACCAGTGG
	Second PCR	2285S-2a	AACTTCACTCGTGGGGATCG
		4717R-IH	TTGCAGTCGATCACGGAGTCAAAGTC
C	First PCR	4199S-J6	CAACATTAGGACTGGAGTCA
		7271R-2a	GGGAGAGCACAGCCCGCAAC
	Second PCR	4607S-IH	AGAGGGTTGGACGTCTCCATAATACCA
		7220R-2a	TGTAATCAGGCCGTGCCCA
D	First PCR	6774S-IH	TCCGGGATGAGGTCTCGTTC
		3X-75R-2a	TACGGCACTCTCTGCAGTCA
	Second PCR	6881S-IH	ATTGATGTCCATGCTAACAG
		3X-54R-2a	GCGGCTCACGGACCTTTCAC

Table 2
Primers used for introduction of amino acid mutations

Fragment		Primer	Sequence (5′–3′)
Mut-1	First PCR	3471S-IH	TGGGCACCATAGTGGTGAG
		A1680Eas	CACCCGGTCtCCAGGCAATACGCGGCGACG
Mut-2	First PCR	A1680Es	ATTGCCTGGaGACCGGGTGTGTTTGCATCA
		6542R-IH	CGCACTGGCCCTCCGTGTA
Mut-3	Second PCR	3471S-IH	TGGGCACCATAGTGGTGAG
		6542R-IH	CGCACTGGCCCTCCGTGTA

3. Anti-HCV core antibody (clone C7–50; Abcam).
4. Alexa Fluor 488 Goat Anti-mouse IgG (Thermo Fisher Scientific).

3 Methods

3.1 RNA Synthesis

1. Digest the pJ6/5BSLX-JFH1 plasmid with XbaI.
2. Treat the XbaI-digested plasmid with Mung Bean Nuclease to remove a four-nucleotide overhang, ensuring the correct 3′ end of the HCV cDNA.
3. Purify the digested plasmid DNA using the QIAquick PCR Purification Kit.
4. By using the purified plasmid DNA as a template, synthesize genome-length HCV RNA with the MEGAscript T7 Transcription Kit following the manufacturer's instructions.
5. Purify the synthesized RNA using the RNeasy Mini Kit.
6. To determine the RNA concentration, measure the optical density at OD260 and OD280. Determine the purity and size of the synthesized HCV RNA by agarose gel electrophoresis.

3.2 RNA Transfection

1. To prepare Huh-7.5.1 cells for transfection, trypsinize the cells, wash them with Opti-MEM I reduced-serum medium, and resuspend them at a concentration of 7.5×10^6 cells/mL with Cytomix buffer.
2. Add 10 μg of J6/5BSLX-JFH1 RNA to the cell suspension (3×10^6 cells/400 μL of Cytomix buffer).
3. Transfer the cell mixture into the Gene Pulser electroporation Cuvette.
4. Pulse the cells at 260 V and 950 μF using the Gene Pulser Xcell apparatus.
5. Transfer the electroporated cells immediately to a 10 cm culture dish containing 10 mL of culture medium.

3.3 Long-Term Culture of HCV RNA-Transfected Cells

1. Maintain the HCV RNA-transfected cells by passaging them every 2–3 days prior to the cells becoming confluent.
2. Monitor the HCV replication by measuring the concentration of HCV core antigen (Ag) in the culture medium (*see* **Note 1**). In our experiments, HCV core Ag in the culture medium started to increase at day 28 and peaked at day 49 after transfection (Fig. 1).

3.4 Identification of Adaptive Mutations

1. Extract the total RNA from the culture medium that was collected at the time of peak HCV core Ag level using the QIAamp Viral RNA Mini Kit.
2. Synthesize cDNA using the extracted RNA with SuperScript III reverse transcriptase and a random primer.

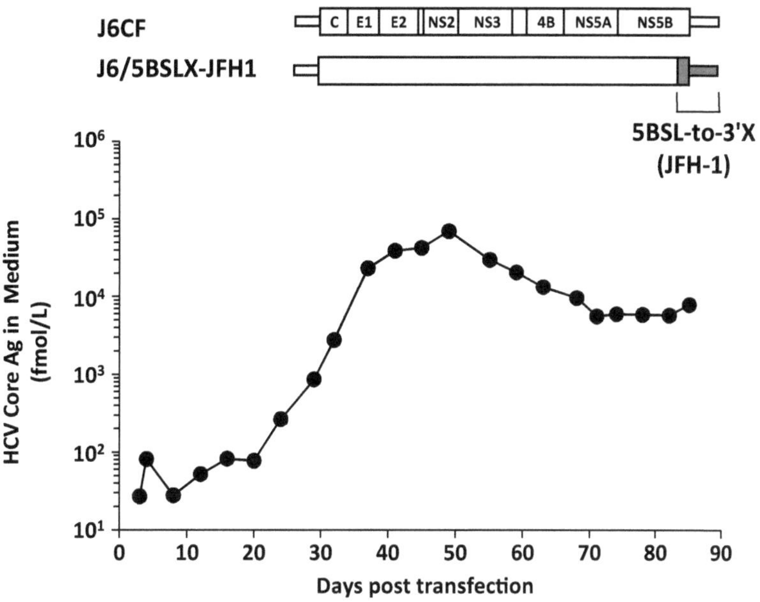

Fig. 1 Long-term culturing of genome-length J6/5BSLX-JFH1 RNA-transfected cells. HCV core Ag levels were measured in the culture media of genome-length J6/5BSLX-JFH1 RNA-transfected cells. The schematic structure of J6/5BSLX-JFH1 is indicated in the upper panel

3. Amplify the fragments of the HCV genome covering the entire open reading frame by PCR using TaKaRa LA Taq polymerase and the primers indicated in Table 1.
4. To determine if the introduced adaptive mutations are present, sequence amplified fragments directly using the appropriate primers. In our case, one synonymous mutation was detected in NS4A region (nt. C5379A; aa. A1680E).

3.5 Introduction of Adaptive Mutation into the J6CF Genome

1. To generate the mutation-introduced fragment, perform mutagenesis by amplifying the target region in the J6CF genome using Pyrobest DNA Polymerase and mutation-introducing primers, as indicated in Table 2.
2. Purify the amplified fragment using the QIAquick PCR Purification Kit.
3. Digest the pJ6/5BSLX-JFH1 plasmid and purified fragment with the appropriate restriction enzymes. In our case, restriction enzymes BbvCI (nt. 3663) and EcoRI (nt. 6006) were used.
4. Purify the digested fragment by gel electrophoresis using the QIAquick Gel Extraction Kit.
5. Ligate the digested fragment with the digested plasmid, and transform the resulting product into DH5α competent cells.

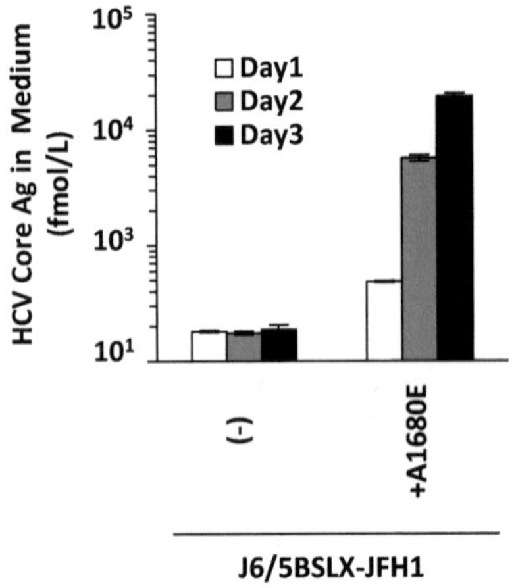

Fig. 2 Enhanced replication of adaptive mutation-introduced J6/5BSLX-JFH1. HCV core Ag levels were measured in the culture media of genome-length J6/5BSLX-JFH1 RNA- and A1680E-introduced J6/5BSLX-JFH1 RNA-transfected cells

6. Isolate the resulting plasmid using the HiPure Plasmid DNA Purification Kit.
7. Confirm the introduced mutation by sequencing the isolated plasmid.

3.6 Confirmation of the Advantages of the Adaptive Mutation

1. Synthesize genome-length HCV RNA using mutation-introduced J6/5BSLX-JFH1 similarly to synthesizing J6/5BSLX-JFH1.
2. Transfect the genome-length HCV RNA of mutation-introduced J6/5BSLX-JFH1 into Huh-7.5.1 cells by electroporation as described in Subheading 3.2.
3. Confirm the enhanced replication of mutation-introduced J6/5BSLX-JFH1 by measuring the HCV core Ag in the culture medium of the transfected cells (Fig. 2).

3.7 Construction of J6CF with Minimum Modifications

1. Introduce the adaptive mutation, A1680E, into the pJ6CF plasmid, which contains three previously identified amino acid mutations, A450S, R517K, and Y561F, in NS5B and a nucleotide mutation, C9458G, within the 3’-UTR (resulting plasmid called pJ6-mem).
2. Prepare the plasmid of pJ6-mem using the PureLink HiPure Plasmid DNA Purification Kit.
3. Synthesize genome-length HCV RNA from J6-mem as described in Subheading 3.1.
4. Transfect the genome-length HCV RNA of J6-mem into Huh-7.5.1 cells by electroporation as described in Subheading 3.2.

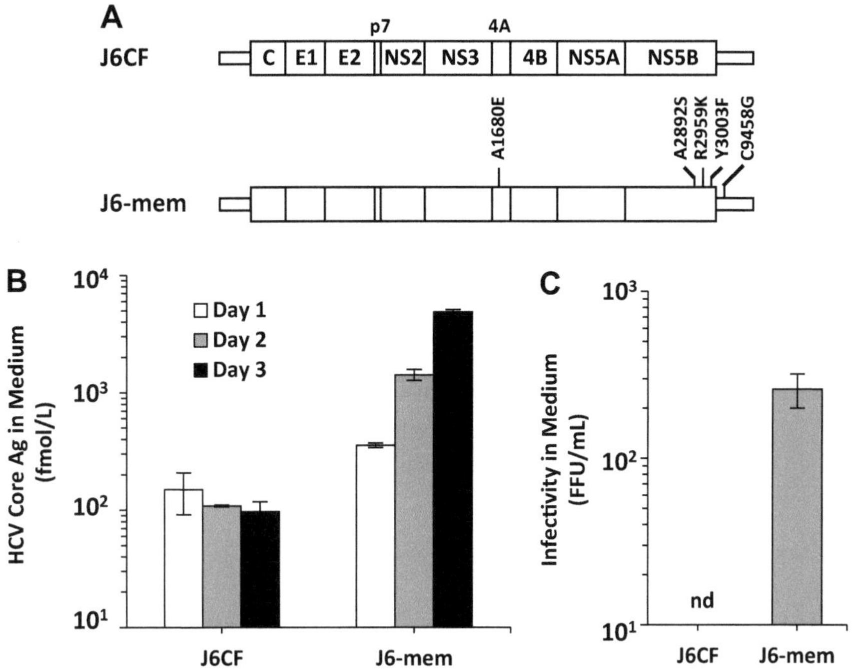

Fig. 3 Replication and infectious virus production of J6-mem. (**a**) Position of introduced mutations in the J6-mem genome. (**b**) HCV core Ag levels in the culture media of genome-length J6CF RNA- and J6-mem RNA-transfected cells. (**c**) Infectivity titer of culture media of genome-length J6CF RNA- and J6-mem RNA-transfected cells

5. Confirm HCV replication by measuring the HCV core Ag levels in the culture medium of the HCV RNA-transfected cells (Fig. 3).

3.8 Determination of Infectivity

1. Seed Huh-7.5.1 cells onto collagen-coated 96-well plates at a density of 1×10^4 cell/well 1 day prior to infection.
2. Pass the culture medium containing the virus through a Millex-HV syringe-driven filter unit.
3. Prepare inocula with tenfold serial dilutions of culture medium containing virus.
4. Inoculate the prepared inocula after aspiration of the culture medium.
5. Replace the medium with fresh medium after a 5 h incubation of the infected cells.
6. After 3 days of culture, fix the infected cells with 100% methanol for 20 min at −20 °C.
7. Stain the cells with anti-HCV core antibody (C7–50, 1 μg/mL) and dye-conjugated anti-mouse antibody (Alexa Fluor 488 Goat Anti-mouse IgG; 1 μg/mL) (*see* **Note 2**).

8. Count the stained foci using a fluorescence microscope, and multiply the number of infected foci by the dilution factor. The infectious titer is represented as focus-forming units (FFU)/mL (Fig. 3) (*see* **Note 3**).

4 Notes

1. Other HCV core Ag kits, such as ARCHITECT HCV antigen assay kit (Abbott Laboratories) or HCV RNA measuring kits, are applicable [29].
2. Final concentration is provided.
3. Determination of TCID50 following the Reed-Muench method is also available [30].

References

1. Choo QL, Kuo G, Weiner AJ, Overby LR, Bradley DW, Houghton M (1989) Isolation of a cDNA clone derived from a blood-borne non-A, non-B viral hepatitis genome. Science 244:359–362
2. Yanagi M, Purcell RH, Emerson SU, Bukh J (1997) Transcripts from a single full-length cDNA clone of hepatitis C virus are infectious when directly transfected into the liver of a chimpanzee. Proc Natl Acad Sci U S A 94:8738–8743
3. Lanford RE, Lee H, Chavez D, Guerra B, Brasky KM (2001) Infectious cDNA clone of the hepatitis C virus genotype 1 prototype sequence. J Gen Virol 82:1291–1297
4. Yanagi M, Purcell RH, Emerson SU, Bukh J (1999) Hepatitis C virus: an infectious molecular clone of a second major genotype (2a) and lack of viability of intertypic 1a and 2a chimeras. Virology 262:250–263
5. Bukh J, Pietschmann T, Lohmann V, Krieger N, Faulk K, Engle RE et al (2002) Mutations that permit efficient replication of hepatitis C virus RNA in Huh-7 cells prevent productive replication in chimpanzees. Proc Natl Acad Sci U S A 99:14416–14421
6. Lohmann V, Korner F, Koch J, Herian U, Theilmann L, Bartenschlager R (1999) Replication of subgenomic hepatitis C virus RNAs in a hepatoma cell line. Science 285:110–113
7. Ikeda M, Yi M, Li K, Lemon SM (2002) Selectable subgenomic and genome-length dicistronic RNAs derived from an infectious molecular clone of the HCV-N strain of hepatitis C virus replicate efficiently in cultured Huh7 cells. J Virol 76:2997–3006
8. Pietschmann T, Lohmann V, Kaul A, Krieger N, Rinck G, Rutter G et al (2002) Persistent and transient replication of full-length hepatitis C virus genomes in cell culture. J Virol 76:4008–4021
9. Blight KJ, McKeating JA, Marcotrigiano J, Rice CM (2003) Efficient replication of hepatitis C virus genotype 1a RNAs in cell culture. J Virol 77:3181–3190
10. Wakita T, Pietschmann T, Kato T, Date T, Miyamoto M, Zhao Z et al (2005) Production of infectious hepatitis C virus in tissue culture from a cloned viral genome. Nat Med 11:791–796
11. Kato T, Date T, Miyamoto M, Furusaka A, Tokushige K, Mizokami M et al (2003) Efficient replication of the genotype 2a hepatitis C virus subgenomic replicon. Gastroenterology 125:1808–1817
12. Date T, Kato T, Miyamoto M, Zhao Z, Yasui K, Mizokami M et al (2004) Genotype 2a hepatitis C virus subgenomic replicon can replicate in HepG2 and IMY-N9 cells. J Biol Chem 279:22371–22376
13. Kato T, Date T, Miyamoto M, Zhao Z, Mizokami M, Wakita T (2005) Nonhepatic cell lines HeLa and 293 support efficient replication of the hepatitis C virus genotype 2a subgenomic replicon. J Virol 79:592–596
14. Kato T, Choi Y, Elmowalid G, Sapp RK, Barth H, Furusaka A et al (2008) Hepatitis C virus JFH-1 strain infection in chimpanzees is associated with low pathogenicity and emergence of an adaptive mutation. Hepatology 48:732–740

15. Li YP, Ramirez S, Jensen SB, Purcell RH, Gottwein JM, Bukh J (2012) Highly efficient full-length hepatitis C virus genotype 1 (strain TN) infectious culture system. Proc Natl Acad Sci U S A 109:19757–19762
16. Li YP, Ramirez S, Mikkelsen L, Bukh J (2015) Efficient infectious cell culture systems of the hepatitis C virus (HCV) prototype strains HCV-1 and H77. J Virol 89:811–823
17. Date T, Morikawa K, Tanaka Y, Tanaka-Kaneko K, Sata T, Mizokami M et al (2012) Replication and infectivity of a novel genotype 1b hepatitis C virus clone. Microbiol Immunol 56:308–317
18. Li YP, Ramirez S, Gottwein JM, Scheel TK, Mikkelsen L, Purcell RH et al (2012) Robust full-length hepatitis C virus genotype 2a and 2b infectious cultures using mutations identified by a systematic approach applicable to patient strains. Proc Natl Acad Sci U S A 109: E1101–E1110
19. Date T, Kato T, Kato J, Takahashi H, Morikawa K, Akazawa D et al (2012) Novel cell culture-adapted genotype 2a hepatitis C virus infectious clone. J Virol 86:10805–10820
20. Ramirez S, Li YP, Jensen SB, Pedersen J, Gottwein JM, Bukh J (2014) Highly efficient infectious cell culture of three hepatitis C virus genotype 2b strains and sensitivity to lead protease, nonstructural protein 5A, and polymerase inhibitors. Hepatology 59:395–407
21. Kim S, Date T, Yokokawa H, Kono T, Aizaki H, Maurel P et al (2014) Development of hepatitis C virus genotype 3a cell culture system. Hepatology 60:1838–1850
22. Okamoto H, Okada S, Sugiyama Y, Kurai K, Iizuka H, Machida A et al (1991) Nucleotide sequence of the genomic RNA of hepatitis C virus isolated from a human carrier: comparison with reported isolates for conserved and divergent regions. J Gen Virol 72:2697–2704
23. Murayama A, Date T, Morikawa K, Akazawa D, Miyamoto M, Kaga M et al (2007) The NS3 helicase and NS5B-to-3'X regions are important for efficient hepatitis C virus strain JFH-1 replication in Huh7 cells. J Virol 81:8030–8040
24. Murayama A, Weng L, Date T, Akazawa D, Tian X, Suzuki T et al (2010) RNA polymerase activity and specific RNA structure are required for efficient HCV replication in cultured cells. PLoS Pathog 6:e1000885
25. Murayama A, Sugiyama N, Suzuki R, Moriyama M, Nakamura N, Mochizuki H et al (2017) Amino acid mutations in the NS4A region of hepatitis C virus contribute to viral replication and infectious virus production. J Virol 91(4):e02124-16
26. Zhong J, Gastaminza P, Cheng G, Kapadia S, Kato T, Burton DR et al (2005) Robust hepatitis C virus infection in vitro. Proc Natl Acad Sci U S A 102:9294–9299
27. Nakabayashi H, Taketa K, Miyano K, Yamane T, Sato J (1982) Growth of human hepatoma cells lines with differentiated functions in chemically defined medium. Cancer Res 42:3858–3863
28. van den Hoff MJ, Moorman AF, Lamers WH (1992) Electroporation in 'intracellular' buffer increases cell survival. Nucleic Acids Res 20:2902
29. Murayama A, Sugiyama N, Watashi K, Masaki T, Suzuki R, Aizaki H et al (2012) Japanese reference panel of blood specimens for evaluation of hepatitis C virus RNA and core antigen quantitative assays. J Clin Microbiol 50:1943–1949
30. Lindenbach BD (2009) Measuring HCV infectivity produced in cell culture and in vivo. Methods Mol Biol 510:329–336

Chapter 5

Full-Length Open Reading Frame Amplification of Hepatitis C Virus

Ulrik Fahnøe and Jens Bukh

Abstract

The purpose of this method is to amplify the full coding sequence of hepatitis C virus (HCV) by a single round reverse transcriptase-polymerase chain reaction (RT-PCR) approach. Our method relies on a highly robust and sensitive RNA extraction procedure and cutting-edge RT-PCR enzymes, all of which have been rigorously tested and optimized. This will not only allow for robust amplification of the entire open reading frame (ORF) of HCV for sequencing by Sanger or next-generation sequencing (NGS), but can also be used for cloning of the ORF of uncharacterized samples and for linkage analysis of mutations on individual genomes spanning the entire ORF. The method has been validated on a variety of samples, including sera from HCV patients and cell-culture supernatants.

Key words Hepatitis C virus, HCV, RNA extraction, Long PCR, Full-length ORF RT-PCR, Next-generation sequencing, Sanger sequencing, Cloning

1 Introduction

HCV is present in the sera of patients and in supernatants from cell-culture infections as an approximately 9.6 kb ssRNA virus [1]. This makes full-length amplification within reach of modern long PCR techniques [2]. Since the introduction of the long PCR technique, the method has been applied to liver-tropic viruses such as HAV, HBV, and HCV [3, 4]. Several other viruses have been amplified like the *Pestivirus* genus [5–7] and norovirus [8]. The advantages of having a long amplification strategy for HCV are that the 5′ and 3′ end untranslated regions (UTRs) are highly conserved within genotype variants permitting the use of one set of primers for each subtype or genotype. This eliminates the need for primer walking or prior knowledge of the coding sequence that sometimes hampers the multiple amplicon approach because of mismatches in the primer regions. For HCV the poly-U/UC tract in the 3′ UTR severely inhibits efficient RT-PCR amplification. Therefore the reverse primer is usually situated upstream of the poly-U/UC and

Mansun Law (ed.), *Hepatitis C Virus Protocols*, Methods in Molecular Biology, vol. 1911,
https://doi.org/10.1007/978-1-4939-8976-8_5, © Springer Science+Business Media, LLC, part of Springer Nature 2019

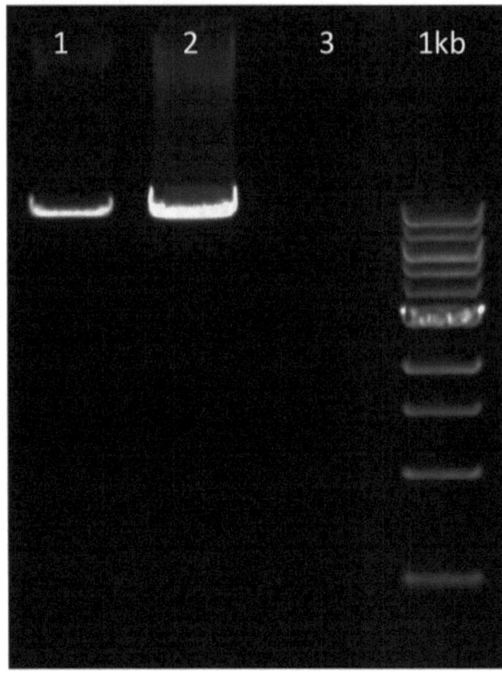

Fig. 1 Gel electrophoresis of full-length ORF HCV TNcc 1a RT-PCR. Lanes 1 and 2 are products amplified from two cell-culture supernatants and lane 3 is the negative control. Sample 1 had an RNA titer of 4.5 $\log_{10}$ IU per ml and sample 2 had a titer of 6.3 $\log_{10}$ IU per ml. The 1 kb ladder (NEB) was used as the size marker

just downstream of the stop codon in the 3′ UTR. However, the 5′ UTR has a stretch of extremely conserved sequence among all genotypes making a true pan inter-genotypic forward primer possible. The long PCR product obtained from uncharacterized patient or culture samples can be used for sequencing by NGS and Sanger or by cloning permitting mutation linkage analysis of single genomes covering the entire ORF to look for epistasis of important adaptive mutations, for example, in response to direct-acting antiviral treatment. Here we present a modified version of HCV full ORF RT-PCR amplification that is highly sensitive and robust for genotype 1a [9]. In addition, we describe a very robust RNA extraction procedure in combination with the RT-PCR technique for high performance (*see* Fig. 1).

2 Materials

2.1 RNA Extraction

1. Virus sample (serum, plasma, cell-culture supernatant).
2. Phase Lock Gel Heavy 2 ml tubes (5′, 2302830).
3. RNase-free 1.5 ml tubes.

4. TRIzol LS (Thermo Scientific, 10296-010).
5. RNA Clean & Concentrator-5 including buffers and Zymo-Spin IC Columns (Zymo research, R1015).
6. Chloroform.
7. 100% Ethanol (EtOH).
8. Nuclease-free water.
9. Refrigerated centrifuge.
10. Chemical hood.

2.2 Reverse Transcription

1. Maxima H Minus Reverse Transcriptase including RT enzyme, 5× RT buffer (Thermo Scientific, EP0751).
2. RNasin Plus RNase inhibitor (Promega, N2611).
3. dNTPs (10 mM).
4. Thermo cycler with heated lid.
5. RNase H (Thermo Scientific, EN0201).
6. 1a RT primer 2 μM (5′ TAA GAG GCC GGA GTG TTT AC 3′).

2.3 PCR Amplification

1. Q5 Hot start High-Fidelity DNA Polymerase including 5× Buffer and GC enhancer (New England Biolabs, M0493S).
2. dNTPs (10 mM).
3. Thermo cycler with heated lid.
4. Nuclease-free water.
5. Agarose.
6. Electrophoresis gel chamber.
7. Power supply.
8. Gel Loading buffer.
9. 1a PCR forward primer 10 μM (5′ TGC CTG ATA GGG TGC TTG CG 3′).
10. 1a PCR reverse primer 10 μM (5′ AGG CCG GAG TGT TTA CCC CA 3′).

3 Methods

3.1 RNA Extraction

1. Prepare Phase Lock Gel Heavy tubes by centrifuging 15 s (>10,000 × *g*) and subsequently adding 750 μl TRIzol LS.
2. Add 250 μl of sample to each tube and invert several times to mix (*see* **Note 1**).
3. Incubate 5 min at room temperature.

4. Add 200 μl of chloroform and shake the tube horizontally for 15 s.
5. Incubate 3 min at room temperature.
6. Centrifuge at 12,000 × *g* for 15 min at 2–8 °C.
7. Add 1 volume (approx. 450 μl) 100% EtOH to a new tube.
8. Transfer water phase to EtOH and mix briefly by pipetting (*see* **Note 1**).
9. Transfer sequentially 700 μl to the Zymo-Spin IC Column in a Collection Tube and centrifuge for 30 s (>10,000 × *g*). Discard the flow-through in the chemical hood.
10. Add 400 μl RNA Prep Buffer and centrifuge 30 s (>10,000 × *g*). Discard the flow-through.
11. Add 700 μl RNA wash buffer and centrifuge 30 s (>10,000 × *g*). Discard the flow-through.
12. Add 400 μl RNA wash buffer and centrifuge 30 s (>10,000 × *g*).
13. Place immediately the RNeasy column in a new 2 ml collection tube and centrifuge 5 min with open lid (>10,000 × *g*).
14. Place immediately the RNeasy column in a new RNase-free tube.
15. Add 15 μl RNase-free water directly to the spin column membrane.
16. Spin 1 min at >10,000 × *g*. The tube with eluted RNA is placed on ice and stored at −80 °C as soon as possible.
17. Repeat **step 16** to get two sequential elution of the column in the same tube if needed. Do not pass the first eluate on the column a second time (*see* **Note 2**).

3.2 Reverse Transcription of HCV RNA for Full-Length ORF PCR Amplification

1. Thaw 5× RT buffer, 10 mM dNTPs, and RT primer and place on ice.
2. Combine the following in a 0.2 ml thin-walled PCR tube:

RNA from eluate	12 μl
2 μM RT primer	1 μl
RNasin Plus RNase inhibitor	1 μl
10 mM dNTPs	1 μl
	15 μl

3. Incubate at 70 °C for 5 min and then immediately place on ice for at least 1 min.

4. Combine the cDNA synthesis mix:

5× RT buffer	4 μl
Maxima H minus RT enzyme	1 μl*
	5 μl

*Use 0.1 or 0.5 μl per reaction for standard cell-culture supernatant; add H_2O for total volume of 5 μl (*see* **Note 3**).

5. Add 5 μl cDNA synthesis mix to each RNA/primer mix.
6. Mix gently by pipetting 10 times.
7. Incubate in a PCR cycler for 120 min at 50 °C (place tube only after the cycler has reached 50 °C), followed by 5 min at 85 °C. Place on ice.
8. Briefly centrifuge and add 1 μl of RNase H to each tube and mix.
9. Incubate 20 min at 37 °C. Place on ice.
10. Store at −20 °C or proceed directly to full-length PCR (*see* **Note 4**).

3.3 PCR Amplification of Full-Length ORF

1. Thaw 5× Q5 Reaction buffer, primers, dNTPs, and GC enhancer and place on ice.
2. Combine the following in a 0.2 ml thin-walled PCR tube:

5× Q5 reaction buffer	10 μl
5× Q5 High GC Enhancer	10 μl
Forward 5' UTR primer (10 μM)	2.5 μl
Reverse 3' UTR primer (10 μM)	2.5 μl
dNTPs (10 mM)	1 μl
cDNA template	2 μl
Q5 Hot start High-Fidelity DNA Polymerase	0.5 μl
dH_2O	21.5 μl
	50 μl

3. Mix gently with pipette. If necessary briefly centrifuge.
4. Run the samples in a thermocycler using the following program (use heated lid):

98 °C	30 s
35 cycles of:	
98 °C	10 s
65 °C	10 s
72 °C	8 min

(continued)

72 °C	8 min
4 °C	∞

5. Analyze 5 μl on a 1% TAE gel and store the rest at −20 °C.
6. Full-length ORF HCV PCR product should have a size of approximately 9.1 kb (*see* Fig. 1). The protocol can be modified for other HCV genotypes or viruses (*see* **Note 5**) and the final PCR product can be used for several downstream applications (*see* **Notes 6–8**).

4 Notes

1. Remember to mix after adding sample to TRIzol by repeated inversions of the tubes. In addition, mix EtOH and water phase after phase separation by pipetting 5–10 times to ensure proper binding of the RNA to the column matrix.
2. Do not put the first eluate on the matrix again because it will severely decrease RT-PCR efficiency.
3. When amplifying routine cell-culture supernatants, 0.1 μl RT enzyme is sufficient and will save operational funds. However, when RNA titer is below 5 $\log_{10}$ IU per ml, we recommend using 0.5–1 μl RT enzyme.
4. Proceed to PCR after RT step directly to enhance sensitivity as the cDNA will slowly decay over time and give less full-length ORF products.
5. The protocol can be expanded to amplify other genotypes and subtypes without major modifications. The 5′ UTR forward primer can be made pan genotypic, whereas the 3′ UTR reverse primer has to be genotype- and sometimes subtype- specific because of the less conserved sequence at the 3′ UTR variable region. In addition, we have shown that the protocol can be applied to a related hepacivirus [10].
6. The ORF product can be used for direct sanger sequencing after purification by a DNA purification column, giving high-quality long reads.
7. The product can also be applied to NGS sequencing with reads mapping almost 100% to the HCV genome, compared to RNA-seq where normally only a small fraction of the reads are viral.
8. Using the long RT-PCR amplicon, the ORF can be cloned in one piece into a vector and thereby studied for linkage of mutations over the entire ORF. Thus, this approach can be used to generate ORF clones of HCV patient samples and from HCV cultures to study evolution in greater detail.

References

1. Bukh J (2016) The history of hepatitis C virus (HCV): basic research reveals unique features in phylogeny, evolution and the viral life cycle with new perspectives for epidemic control. J Hepatol 65:S2–S21
2. Barnes WM (1994) PCR amplification of up to 35-kb DNA with high fidelity and high yield from lambda bacteriophage templates. Proc Natl Acad Sci 91:2216–2220
3. Tellier R, Bukh J, Emerson SU et al (1996) Long PCR and its application to hepatitis viruses: amplification of hepatitis A, hepatitis B, and hepatitis C virus genomes. J Clin Microbiol 34:3085–3091
4. Yanagi M, Purcell RH, Emerson SU, Bukh J (1997) Transcripts from a single full-length cDNA clone of hepatitis C virus are infectious when directly transfected into the liver of a chimpanzee. Proc Natl Acad Sci U S A 94:8738–8743
5. Rasmussen TB, Reimann I, Hoffmann B et al (2008) Direct recovery of infectious pestivirus from a full-length RT-PCR amplicon. J Virol Methods 149:330–333
6. Fahnøe U, Pedersen AG, Dräger C et al (2015) Creation of functional viruses from non-functional cDNA clones obtained from an RNA virus population by the use of ancestral reconstruction. PLoS One 10:1–17
7. Fahnøe U, Pedersen AG, Risager PC et al (2014) Rescue of the highly virulent classical swine fever virus strain "Koslov" from cloned cDNA and first insights into genome variations relevant for virulence. Virology 468:379–387
8. Kostela J, Ayers M, Nishikawa J et al (2008) Amplification by long RT-PCR of near full-length norovirus genomes. J Virol Methods 149:226–230
9. Li Y-P, Ramirez S, Jensen SB et al (2012) Highly efficient full-length hepatitis C virus genotype 1 (strain TN) infectious culture system. Proc Natl Acad Sci U S A 109:19757–19762
10. Billerbeck E, Wolfisberg R, Fahnøe U et al (2017) Mouse models of acute and chronic hepacivirus infection. Science 357 (80):204–208

Chapter 6

InFusion Cloning for the Generation of Biologically Relevant HCV Chimeric Molecular Clones

Barnabas King, Richard Urbanowicz, Alexander W. Tarr, Jonathan K. Ball, and C. Patrick McClure

Abstract

This chapter describes how to generate chimeric molecular cassettes that are ready to receive PCR-amplified E1/E2 genes using new DNA cloning technology. The method is divided into three sections: (1) generation of a ΔCore-NS2 cassette based upon the full-length JFH-1 molecular clone; (2) insertion of a "structural gene" fragment encoding the Core, p7, and NS2 genes of a given genotype reference sequence, to generate a ΔE1/E2 cassette; and (3) insertion of patient-isolated E1/E2 genes that are genotype-matched to the structural genes. The final assembled chimeric genomes can then be analyzed in the HCV cell culture system. These cassettes allow characterization of the extensive in vivo viral diversity without the need to isolate and clone whole virus genomes. This method can be readily applied to the study of other HCV genes and other viruses.

Key words Hepatitis C virus, Chimeric virus, Full-length molecular clones, Patient-isolated E1/E2

1 Introduction

Investigations into the properties of authentic hepatitis C virus (HCV) strains circulating in vivo are hampered by the limited cell culture models to study in vitro replication. The NS3-5B genes of HCV are capable of self-replicating the HCV genome when in the form of a sub-genomic replicon [1]. However, attempts at isolating full-length infectious clones, capable of replicating in vitro, have had limited success, apart from the genotype 2a isolate, JFH-1 [2–4]. The generation of viable chimeric viruses, containing the Core NS2 genes from one isolate with the NS3-NS5B genes of another isolate, has only been possible when the NS3-5B genes are taken from the JFH-1 isolate [5, 6]. In addition, preliminary evidence indicates that genotype matching of the E1/E2 genes with the Core-p7-NS2 genes is essential for the generation of infectious molecular clones, in vitro [6, 7]. For studies of the properties of HCV virions, these chimeras provide a novel and valuable approach

Mansun Law (ed.), *Hepatitis C Virus Protocols*, Methods in Molecular Biology, vol. 1911,
https://doi.org/10.1007/978-1-4939-8976-8_6, © Springer Science+Business Media, LLC, part of Springer Nature 2019

to investigating entry and inhibition of genetically diverse isolates. However, existing methods to create viral chimeras are cumbersome and time consuming.

Here we describe methods established in our laboratory that facilitate rapid production of viral chimeras encoding the E1/E2 genes of a range of genetically diverse patient-derived isolates [7]. Utilizing the JFH-1 NS3-5B backbone and UTRs, we describe methods with which to create suitable cloning cassettes and simple methods for inserting recombinant structural gene constructs using InFusion® cloning. These approaches could be applied to any molecular viral clone and significantly improve on existing methods for creating chimeras.

2 Materials

1. Full-length cloned JFH-1 HCV parental cDNA (pJFH-1—a gift from Takaji Wakita) [3].
2. cDNA encoding E1E2 genes (*see* Subheading 3.3, **step 1**).
3. Nuclease-free (n.f.) H_2O.
4. 100 μM dNTP (1:1:1:1 mixture from 100 μM stocks of each constituent dNTP) (Sigma).
5. Primers (*see* Table 1) were obtained by commercial synthesis and reconstituted at 100 μM followed by 1:10 dilutions for use as 10 μM working dilutions. Further dilutions were made of

Table 1
PCR primers used for the generation and confirmation of full-length, inter-genotype HCVcc chimeras encoding patient-derived genotype 3 E1/E2 genes

Primer number	Primer name	Primer sequence (5′–3′)
1	J6_ΔCore-NS2_*AfeI*_SDMf	GCTCCCATCACTGCTTATGCCCAGCAAACAC
2	J6_ΔCore-NS2_*AfeI*_SDMr	GCTCATGGTGCACGGTCT
3	JFH_C-NS2_SCRf	ACTCTATGCCCGGCCATTT
4	JFH_C-NS2_SCRr	GGCCTGTTCTGTCCTGTCA
5	G3_E1/E2InF_f	AGGGAACTTGCCCGGTTGCTCCTTTTCTATCTTCC
6	G3_E1/E2InF_r	AGCATCAGCCAAAGGGCAACGCACACGC
7	G3_E1/E2Scr_f	CGTAGGAGGCGTCGCAAG
8	G3_E1/E2Scr_r	AGCCAATACCATGTGTCCCA

primers used for Sanger sequencing to 3.3 μM. All primer dilutions were made in nuclease-free (n.f.) H_2O.

6. Plasmid DNA miniprep kit.
7. Chemically competent *E. coli* cells (NEB).
8. Luria-Bertani (LB) media and LB-agar plates, containing 100 μg/mL ampicillin.
9. Q5 site-directed mutagenesis (SDM) kit (NEB).
10. InFusion HD cloning kit (Clontech).
11. Cloning Enhancer (Clontech).
12. *AfeI* and *FseI* restriction endonucleases (REs) and corresponding buffers.

3 Methods

This method is specifically designed for use with InFusion cloning technology (Clontech). However the main principles are also applicable to Gibson Assembly cloning technology (NEB).

Carry out all procedures at room temperature (18–22 °C) unless otherwise stated.

3.1 Generation of a JFH-1 Δ Core-NS2 Cassette with a Novel AfeI Site by Site-Directed Mutagenesis (SDM)

Deletion of the Core-NS2 genes from the full-length JFH-1 parental clone is achieved using a single site-directed mutagenesis (SDM) reaction. Specifically, this reaction removes the JFH-1 sequence (accession: AB047639) from nucleotide (nt) 347-3430 (codon 3 of Core to the C-terminus of NS2) followed by re-circularization of the plasmid, resulting in the formation of a novel *AfeI* site (AGC˅GCT) at the point of ligation (Fig. 1a, b). This *AfeI* site is then unique within the plasmid and at the precise junction point for insertion of heterologous Core-NS2 genes.

1. Set up SDM PCR mixture in a 0.2 mL thin-walled PCR tube adding 0.62 μL of 10 μM Primers #1 and #2, 0.25 μL of 100 μM dNTP mixture, 5–50 ng of full-length JFH-1 plasmid cDNA, 2.5 μL NEB Q5 5× Buffer, and 0.125 μL NEB Q5 DNA polymerase. Make up the reaction volume to 12.5 μL with nuclease-free (n.f.) H_2O and mix by pipetting.
2. Place the tube in a thermal cycler using the following program:
   (a) 98 °C for 30 s.
   (b) 25 cycles of:
   - 98 °C for 10 s
   - 61 °C for 20 s (*see* **Note 1**).
   - 72 °C for 6 min.

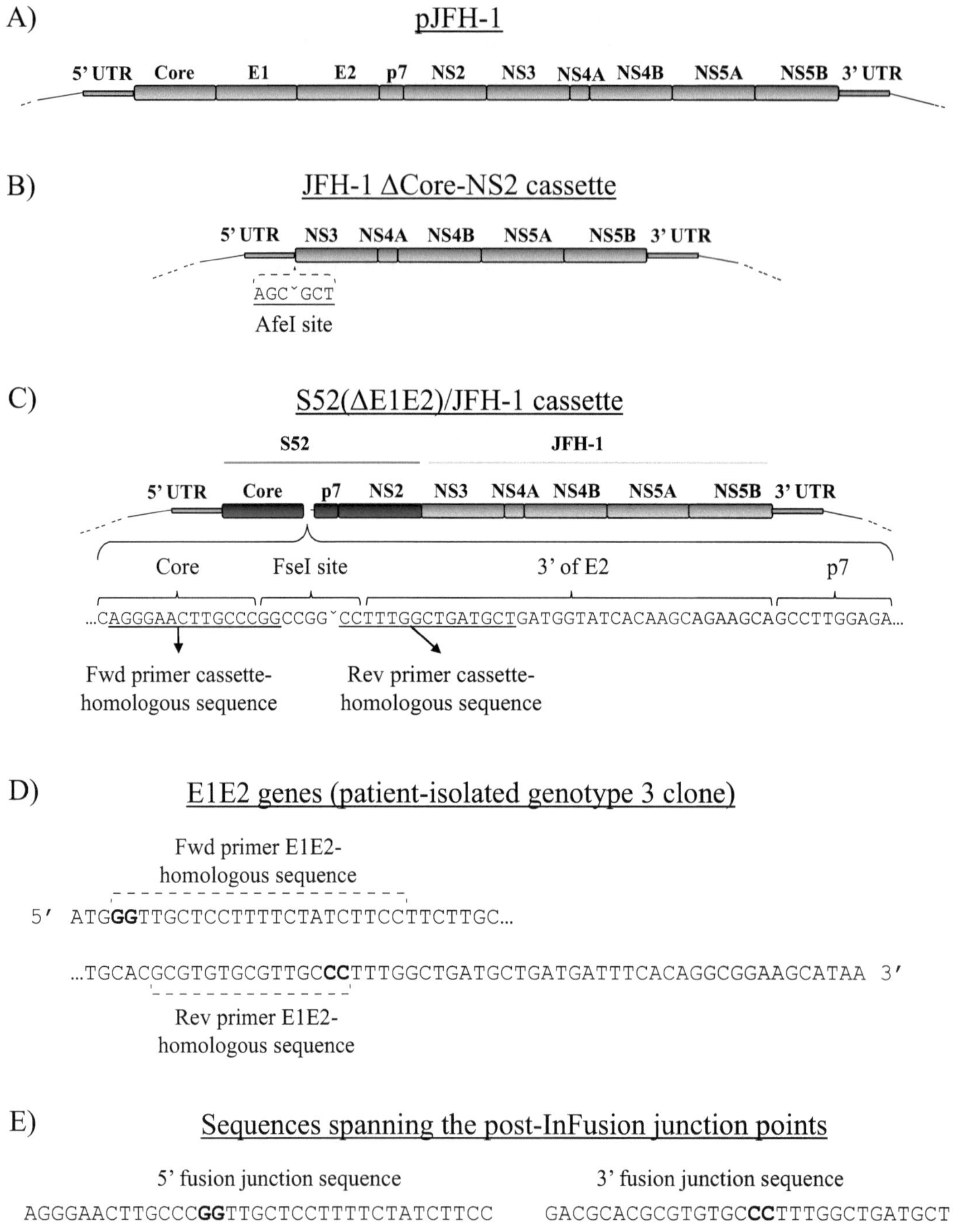

Fig. 1 Schematic diagram of the HCV DNA cassettes and E1/E2 InFusion primers design. (**a**) The genome arrangement of the HCV genes in the pJFH-1 plasmid used as the backbone for the generation of HCV chimeric viruses. (**b**) The JFH-1 ΔCore-NS2 cassette generated following the SDM reaction using Primers #1 and #2. The reaction results in the formation of a unique *AfeI* site at the point of Core-NS2 deletion. (**c**) The S52(ΔE1/E2)/JFH-1 cassette is generated following insertion of the S52_Core-(ΔE1/E2[*FseI*])-NS2 sequence into the *AfeI*-linearized JFH-1 ΔCore-NS2 cassette. The sequence proximal to the *FseI* site is shown. Underlined

(c) 72 °C for 10 min.

(d) Hold at 8 °C.

3. Setup KLD reaction in 0.2 mL thin-walled PCR tube on ice (*see* **Notes 2** and **3**). Add 2.5 μL of NEB 2× KLD reaction buffer, 0.5 μL NEB KLD Enzyme Mix, 0.5 μL SDM PCR from **step 2**, and 1.5 μL n.f. H_2O. Mix by pipetting and incubate at 22 °C for 5–15 min in a thermal cycler (*see* **Note 2**).
4. Transform 0.5–2 μL of the KLD reaction into an aliquot of high-efficiency (>1 × 10^8 cfu/μg) chemically competent *E. coli* cells, following the manufacturer's instructions, and plate out transformed cells onto LB-agar plates containing 100 μg/mL ampicillin. Incubate plate overnight at 37 °C.
5. Perform colony screening PCR to confirm the removal of the Core-NS2 coding region using Primers #3 and #4.
6. Analyze PCR products by agarose gel DNA electrophoresis. Successful SDM reactions will give a small product of 210 base pairs (bp). Unsuccessful SDM reactions will give an amplicon of ~3.3 kb as it includes the entire Core-NS2 coding region.
7. Confirm the sequence of positive clones by Sanger sequencing. Initially, a positive PCR product can be diluted 1:10 in n.f. H_2O and sequenced using Primer #3 to determine correct sequence around the ligation site and the establishment of a complete *AfeI* site, as this is the most likely area of PCR-introduced error.
8. Following initial identification of a ΔCore-NS2 clone with correct ligation site, generate a DNA miniprep, and Sanger sequence the entirety of the remaining HCV genome (including UTRs) to ensure no off-target PCR errors have been introduced.
9. The fully sequenced plasmid clone with intact *AfeI* site is referred to as the "JFH-1 ΔCore-NS2 cassette."

Fig. 1 (continued) nucleotides represent the minimum sequence of homology that must be encoded by the primers used to PCR-amplify the E1/E2 genes to enable insertion using InFusion cloning. (**d**) An example of an E1/E2 gene sequence suitable for analysis in the HCV pseudoparticle system (*see* Subheading 3.3, **step 7**) from the start (ATG) and stop codons (TAA). Underlined nucleotides represent the homologous sequence encoded by the primers used to PCR-amplify the E1/E2 genes. The nucleotides of *FseI* site retained following InFusion are highlighted in bold (**GG** and **CC**). (**e**) The desired sequences spanning the post-InFusion junction points were compiled from the cassette and E1/E2 gene sequences. These junction sequences were used to define the E1/E2 InFusion primer sequences, ensuring a minimum 15 nt of homology to the cassette sequence for InFusion cloning and a minimum 17 nt of homology to the E1/E2 genes for PCR amplification. The forward primer sequence being identical to the 5′ InFusion junction sequence, the reverse primer sequence is the reverse complement of the 3′ InFusion junction sequence. ˇ denotes point of digest in RE site on positive strand by corresponding RE

10. In order to insert heterologous Core-NS2 genes by InFusion cloning, the cassette must be linearized at the point of insertion (the *AfeI* site). Digest 5–10 μg of the JFH-1 ΔCore-NS2 cassette in a reaction containing 3 μL *AfeI* restriction enzyme along with the corresponding buffer, and make to a final volume of 30 μL with n.f. H_2O. Mix the reaction by pipetting and incubate overnight at 37 °C in a thermal cycler (*see* **Note 4**).
11. Column-purify the digest using a commercial PCR-purification kit and elute in a final volume of 30 μL n.f. H_2O.
12. Determine DNA concentration using a microvolume spectrophotometer, and analyze by agarose gel DNA electrophoresis alongside the undigested plasmid to confirm complete linearization.
13. This cassette is ready to accept Core-NS2 gene fragments using InFusion cloning and is stable for >1 year at −20 °C.

3.2 Generation of Inter-Genotype HCV (Δ E1/E2) Chimeras in the JFH-1 Δ Core-NS2 Cassette

The current method describes the generation of a second-generation cassette containing the Core, p7, and NS2 genes (but lacking the E1/E2 genes) of a heterologous genotype to enable the analysis of a diverse population of genotype-matched E1/E2 genes (Fig. 1c). However, the linearized JFH-1 ΔCore-NS2 cassette is also suitable for the insertion and study of whole Core-NS2 gene sequences or to generate a similar cassette lacking a different gene, e.g., p7.

In order to generate a ΔE1/E2 cassette for the analysis of genotype 3 E1/E2 genes, the Core-(ΔE1/E2)-p7-NS2 genes from the S52 isolate (genotype 3, accession: GU814263) are InFusion-cloned into the JFH-1 ΔCore-NS2 cassette. Although InFusion and Gibson Assembly cloning approaches differ mechanistically, they both allow the seamless insertion of a DNA molecule (i.e., PCR product) into a linearized vector. The major technical requirement to enable fusion is that the 5′ and 3′ termini of the insert possess 15 bps of homology to the sequence immediately proximal to the point of vector linearization.

1. Generate in silico sequences for the "JFH-1 ΔCore-NS2 cassette" in the plasmid backbone and "JFH-1 ΔCore-NS2 cassette" containing the complete Core-NS2 sequence of S52, referred to as "pS52/JFH-1" (*see* **Note 5**).
2. Analyze the pS52/JFH-1 sequence to identify any RE sites that are not present within the whole sequence. This list is then narrowed to exclude REs which leave a 5′ overhang following digestion, as this reduces the post-InFusion footprint left by the RE site (*see* **Note 6**). Restriction endonucleases (REs) that cut after the third nucleotide of the site were also excluded as these add an increased restriction on the placement of the

InFusion junction, as all nucleotides 3′ to the cleavage site have to be a naturally occurring motif in the cassette sequence to prevent incorporation of Non-native coding sequence into the final molecule. If possible, select a commonly used RE with a standard 6-base pair recognition site that is digested between the fifth and sixth bp of the site (e.g., SacI) (*see* **Note 7**). In this method, *FseI* (RE site: 5′—GGCCGG˅CC cutting between the sixth and seventh bases) was chosen. *FseI* requires that the 5′ and 3′ InFusion junctions of the parental cloned genome end with "GG" and start with "CC," respectively.

3. The pS52/JFH-1 sequence was analyzed for a "GG" motif proximal to the Core/E1 junction and a "CC" motif proximal to the E2/p7 junction (*see* **Note 8**). The 5′ of the InFusion junction was chosen to be "GG" nt 511–512 of Core gene and "CC" nt 1071-2 of E2.
4. Generate an in silico DNA sequence comprising nt 1-512 of S52 Core gene followed by the nucleotides "CCGG" (additional bases to form a complete *FseI* site), nt 1071–1107 of S52 E2 gene, and the complete S52 p7 and NS2 genes. Add the last 9 nt of the JFH-1 IRES (CCGTGCACC) to the 5′ of this in silico sequence, and add the first 15 nt of JFH-1 NS3 (GCTCCCATCACTGCT) to the 3′ of this sequence. These two additional DNA sequences establish 15 bp motifs that are homologous to the JFH-1 ΔCore-NS2 cassette sequence immediately proximal to the *AfeI* site. These regions of homology allow InFusion cloning into the JFH-1 ΔCore-NS2 cassette. The final in silico DNA sequence, referred to as "S52_Core-(ΔE1/E2[*FseI*])-NS2," should be generated using commercially available DNA synthesis platforms. The sequence abounding the *FseI* site is shown in Fig. 1d.
5. Resuspend the synthesized S52_Core-(ΔE1/E2[*FseI*])-NS2 molecule to a concentration of 20 ng/μL in n.f. H_2O.
6. Set up an InFusion reaction on ice in 0.2 mL thin-walled PCR tube (*see* **Note 9**). Mix together 0.5 μL 5x InFusion HD reaction mix, 0.5 μL *AfeI*-linearized, purified JFH-1 ΔCore-NS2 cassette (recommended to use 12.5–50 ng for a 2.5 μL reaction), 1.0 μL S52_Core-(ΔE1/E2[*FseI*])-NS2 (20 ng μL^{-1}), and 0.5 μL n.f. H_2O. Incubate at 50 °C for 15 min in a thermal cycler then transfer tube to ice (*see* **Note 2**).
7. Transform 2 μL of the InFusion reaction into an aliquot of high-efficiency ($>1 \times 10^8$ cfu/μg) chemically competent *E. coli* cells, following the manufacturer's instructions, and plate out transformed cells onto LB-agar plates containing 100 μg/mL ampicillin. Incubate plate overnight at 37 °C.

8. Perform colony screening PCR to confirm the insertion of the S52_Core-(ΔE1/E2[*FseI*])-NS2 region using Primers #3 and #4.
9. Analyze PCR products by agarose gel DNA electrophoresis. Successful SDM reactions will give an amplicon of ~1.6 kb. Unsuccessful SDM reactions will give an amplicon of 210 bp corresponding to religated or undigested parental cassette without any insert.
10. Produce plasmid minipreps of positive clones. Positive clones should be confirmed by Sanger sequencing of the entirety of the inserted Core-(ΔE1/E2[*FseI*])-NS2 fragment and the InFusion junctions using Primers #3 and #4 (*see* **Note 10**).
11. A correct clone with fully sequenced insert region, InFusion junctions, and an intact *FseI* site is referred to as the "S52 (ΔE1/E2)/JFH1 cassette."
12. In order to insert heterologous E1/E2 genes by InFusion cloning, the cassette must be linearized at the new point of insertion (the *FseI* site). Digest 5–10 μg S52(ΔE1/E2)/JFH1 cassette in a reaction containing 3 μL *FseI* RE along with the corresponding buffer, and make to a final volume of 30 μL with n.f. H_2O. Mix the reaction by pipetting and incubate overnight at 37 °C in a thermal cycler (*see* **Note 4**).
13. Column-purify the digest using a commercial PCR-purification kit and elute in a final volume of 30 μL n.f. H_2O.
14. Determine DNA concentration using a microvolume spectrophotometer, and analyze by agarose gel DNA electrophoresis alongside the undigested plasmid to confirm complete linearization.
15. This cassette is ready to accept genotype 3 E1/E2 genes using InFusion cloning and is stable for >1 year at −20 °C.

3.3 Insertion of Patient-Derived Genotype 3 E1/E2 Genes into the S52 (Δ E1/E2)/JFH-1 Cassette

The following section is specifically for the construction of full-length HCV molecular clones containing JFH-1 (genotype 2a) NS3-5B genes and UTRs; the S52 (genotype 3a) Core, p7, and NS2 genes; and the E1/E2 genes (UKNP3.1.2; accession: KU285215) isolated from the serum of a patient with chronic HCV (genotype 3) infection.

1. Extraction and cloning of E1/E2 cDNA from patient samples is described elsewhere in this book (*see* Subheading 3.3, **step** 7). Alternatively, E1/E2 genes can be obtained by commercial synthesis in a form immediately compatible for InFusion cloning (*see* **Notes 5**, **11** and **12**). If E1/E2 genes are being synthesized, resuspend DNA in n.f. H_2O to give a final concentration of 100–200 ng/μL and proceed to Subheading 3.3.

2. Design primers to amplify the E1/E2 genes for InFusion® cloning into the cassette (*see* Fig. 1c–e).
   (a) The forward primer should be designed such that the 5′ end is homologous to the 15 nt upstream of the *FseI* site in the cassette and the 3′ is homologous to the 5′ of the E1/E2 genes to be inserted.
   (b) The reverse primer should be similarly designed such that the 5′ 15 nt of the primer are homologous to the 3′ of the E1/E2 genes to be inserted and the 3′ 15 nt of the primer are homologous to the sequence downstream of the *FseI* site in the cassette.
3. Generation of E1/E2 genes for InFusion cloning by PCR amplification from cDNA should be done as follows: In a 0.2 mL thin-walled PCR tube, add 0.62 μL of 10 μM primers #5 and #6 (*see* **Note 12**), 0.25 μL of 100 μM dNTP mixture, 5–50 ng of E1/E2 cDNA, 2.5 μL NEB Q5 5× Buffer, and 0.125 μL NEB Q5 DNA polymerase. Make up the reaction volume to 12.5 μL with nuclease-free (n.f.) H_2O and mix by pipetting.
4. Place the tube in thermal cycler using the following program:
   (a) 98 °C for 30 s.
   (b) 25 cycles of:
      - 98 °C for 20 s.
      - 61 °C for 20 s (*see* **Note 1**).
      - 72 °C for 2 min.
   (c) 72 °C for 5 min.
   (d) Hold at 8 °C.
5. Analyze 5 μL of PCR products by agarose gel DNA electrophoresis. These primers ought to produce a single amplicon of ~1.8 kb in length.
6. Prepare successfully amplified E1/E2 genes by treatment with Cloning Enhancer (C.E.). Add 1 μL of C.E. to 2.5 μL PCR product in a 0.2 mL thin-walled PCR tube and incubate at 37 °C for 15 min, followed by inactivation at 80 °C at 15 min.
7. Set up an InFusion reaction on ice in a 0.2 mL thin-walled PCR tube (*see* **Note 9**). Mix together 0.5 μL 5× InFusion HD reaction mix, 0.5 μL of C.E.-treated E1/E2 (or synthesised E1/E2 genes resuspended to 100–200 ng/μL), 1.0 μL *FseI*-linearised S52(ΔE1/E2)/JFH-1 cassette (20 ng/μL), and 0.5 μL n.f. H_2O. Incubate at 50 °C for 15 min in a thermal cycler then transfer tube to ice (*see* **Note 2**).

8. Transform 2 μL of the InFusion reaction into an aliquot of high-efficiency (>1 × 10^8 cfu/μg) chemically competent *E. coli* cells, following the manufacturer's instructions, and plate out transformed cells onto LB-agar plates containing 100 μg/mL ampicillin. Incubate plate overnight at 37 °C.
9. Perform colony screening PCR to confirm the insertion of the E1/E2 genes using Primers #7 and #8.
10. Analyze PCR products by agarose gel DNA electrophoresis. Successful InFusion reactions will give an amplicon of ~2.0 kb. Unsuccessful InFusion reactions will give an amplicon of 185 bp corresponding to religation of the parental cassette without any insert.
11. Produce plasmid minipreps of positive clones. Positive clones should be confirmed by Sanger sequencing to confirm the integrity of the inserted E1/E2 genes and the InFusion junctions using Primers #7 and #8 (*see* **Note 10**).
12. The resulting molecular clone can now be linearized using *XbaI* (or *SbfI*; *see* **Note 5**) and used to generate an RNA transcript and electroporated into Huh7.5 cells [2].

4 Notes

1. When performing a PCR with new primers, a gradient PCR may be required to determine the most effective annealing temperature.
2. This sample can be frozen at −20 °C prior to undertaking the next step.
3. Analysis of the PCR product by gel electrophoresis is not recommended as this program and cycle number is intentionally designed to give low levels of PCR product in order not to saturate the template-DNA degradation step in the subsequent KLD reaction.
4. Many REs are available in a "fast digest" form which is suitable for 30-minute digestion reactions for analysis of digest patterns. The authors recommend overnight digests in order to minimize carryover of undigested material where even small amounts may impair subsequent steps.
5. It is important to note that the ultimate aim of this chapter is to generate RNA transcripts of the final full-length chimeric genomes for electroporation into mammalian cells. This process involves linearization of the cDNA template molecule using the *XbaI* RE, prior to RNA transcription, to prevent the production of concatenated transcripts. Therefore it is important to analyze any sequence being introduced into the JFH-1

backbone for the presence of *XbaI* sites. The S52 Core-p7-NS2 genes do not encode an *XbaI* site; however the S52 E1/E2 genes do encode an *XbaI* ER site. For these chimeras it is possible to use *SbfI* in place of *XbaI* to linearize cDNA templates prior to RNA transcription as *SbfI* has only a single RE site in the final full-length molecule which is proximal to the *XbaI* site.

6. Enzymes which generate a 3′ overhang generate the smallest footprint in InFusion cloning. Therefore, digestion with *FseI* (GGCCGG^CC) and subsequent InFusion at this site retains only the two overhanging bases (5′ GG, 3′ CC) of the RE site post-cloning at the chimeric junction.
7. If no appropriate and common enzyme is identified, then the search can be widened to include enzymes that ONLY cut within the E1/E2 genes as they will not be present in the final S52(ΔE1/E2)/JFH1 cassette. Alternatively, Gibson assembly can be used as this technology is most compatible with enzymes which leave a 5′ overhang (converse to InFusion technology).
8. This particular cassette is designed to allow for the insertion of E1/E2 genes isolated from patient samples to assess in the context of full-length clones. Therefore picking an InFusion junction that lies within the E1/E2 genes should only be done if absolutely necessary or if the terminal sequence is highly conserved. Otherwise primers used to amplify the E1/E2 genes for InFusion cloning will encode amino acids of E1/E2 which differ from those of the original patient-derived sequence. The 3′ region of E2 (encoding the trans-membrane domain) is conserved among the genotype 3 isolates E1/E2 clones in the Nottingham Panel; therefore it was possible to select a "CC" motif in this region (S52 E2 gene nt 1071-2) as the junction site. If this were not the case, then the "CC" motif in the S52 p7 gene (nt 2–3) would have been a more suitable junction point.
9. A negative control reaction, lacking the Core-(ΔE1/E2[*FseI*])-NS2 fragment, can be set up to assess frequency of cassette "re-ligation."
10. Some DNA synthesis techniques do not produce a uniform population of DNA molecules (e.g., DNA Strings—Thermo-Fisher); therefore it is essential to sequence the entire synthesized insert molecule for potential errors introduced during production. It is NOT necessary to sequence the entirety of the JFH-1 ΔCore-NS2 cassette genome as this has not been subjected to any error-prone reactions (such as PCR). The only points of mutation observed during InFusion cloning (by the authors) have been at the InFusion junction points.

11. If the E1/E2 genes are being generated by de novo synthesis, then they should be designed with the 15 bp sequences of homology corresponding to the cassette sequence, either side of the *FseI* junction. Thus, the synthesized constructs will be the same as if they had been produced by PCR using the method described in Subheading 3.3, **steps 2–4**.
12. This protocol uses the specific example of cloning the UKNP3.1.2 E1/E2 genes into the S52(ΔE1/E2)/JFH-1 cassette. The user will need to design primers with homologous sequences specific to both the cassette and E1/E2 genes being used. It should be noted that if the cassette junction point is at a highly conserved region across the genotype, then it may be possible to successfully amplify and clone many distinct E1/E2 genes using a single primer pair.

Acknowledgments

The authors would like to thank Takaji Wakita for plasmid pJFH-1 and Chris Lounds for molecular biology discussions. This work was supported by the Medical Research Council UK (G0801169) and by the EU FP7 Grant “HepaMAb” (305600).

References

1. Lohmann V, Korner F, Koch J, Herian U, Theilmann L, Bartenschlager R (1999) Replication of subgenomic hepatitis C virus RNAs in a hepatoma cell line. Science 285:110–113
2. Lindenbach BD, Evans MJ, Syder AJ, Wolk B, Tellinghuisen TL, Liu CC et al (2005) Complete replication of hepatitis C virus in cell culture. Science 309:623–626
3. Wakita T, Pietschmann T, Kato T, Date T, Miyamoto M, Zhao Z et al (2005) Production of infectious hepatitis C virus in tissue culture from a cloned viral genome. Nat Med 11:791–796
4. Zhong J, Gastaminza P, Cheng G, Kapadia S, Kato T, Burton DR et al (2005) Robust hepatitis C virus infection in vitro. Proc Natl Acad Sci U S A 102:9294–9299
5. Yi M, Villanueva RA, Thomas DL, Wakita T, Lemon SM (2006) Production of infectious genotype 1a hepatitis C virus (Hutchinson strain) in cultured human hepatoma cells. Proc Natl Acad Sci U S A 103:2310–2315
6. Pietschmann T, Kaul A, Koutsoudakis G, Shavinskaya A, Kallis S, Steinmann E et al (2006) Construction and characterization of infectious intragenotypic and intergenotypic hepatitis C virus chimeras. Proc Natl Acad Sci U S A 103:7408–7413
7. McClure CP, Urbanowicz RA, King BJ, Cano-Crespo S, Tarr AW, Ball JK (2016) Flexible and rapid construction of viral chimeras applied to hepatitis C virus. J Gen Virol 97:2187–2193

Chapter 7

Production and Purification of Cell Culture Hepatitis C Virus

Cynthia de la Fuente and Maria Teresa Catanese

Abstract

Hepatitis C virus (HCV) is a peculiar member of the *Flaviviridae* family, with features in between an enveloped virus and a human lipoprotein and, consequently, unusual biophysical properties that made its production and purification rather challenging.

Here we describe methods to generate HCV stocks in cell culture by electroporating in vitro transcribed viral RNA into permissive cell lines as well as downstream concentration and purification strategies.

Key words Hepatitis C virus, Virus production, In vitro transcription of HCV genomic RNA, Electroporation of viral genomes in mammalian cells, Virus purification

1 Introduction

The development of a robust cell culture system for HCV required a series of advances to allow for the complete virus replication cycle. The reader is referred to a number of excellent in depth reviews detailing the history of HCV cell culture systems [1–6]. Early attempts at propagating patient-derived HCV, either in cell lines or primary cultures, proved to be difficult with low levels of viral replication (reviewed in [7]). Similarly, the first cDNA molecular clones based on genotype 1, while infectious in chimpanzees, failed to replicate to detectable levels in tissue culture [8, 9]. The establishment of selectable subgenomic replicons (encoding for the HCV nonstructural (NS) proteins NS3-NS5B; [10]) and retroviral pseudoparticles exposing the HCV E1–E2 glycoproteins on the surface [11, 12] allowed for assessment of viral RNA replication and entry, respectively. Cell clones that were highly permissive for HCV replication and viral adaptive mutations that synergistically enhanced viral fitness could be selected for within these replicon systems. Human hepatoma Huh-7 cell line and derivatives (i.e., Huh7-Lunet, Huh-7.5, and Huh-7.5.1) with increased permissiveness were generated by "curing" selected replicon cell clones with interferon or inhibitors [13–15]. Unfortunately, infectious virus

Mansun Law (ed.), *Hepatitis C Virus Protocols*, Methods in Molecular Biology, vol. 1911,
https://doi.org/10.1007/978-1-4939-8976-8_7,

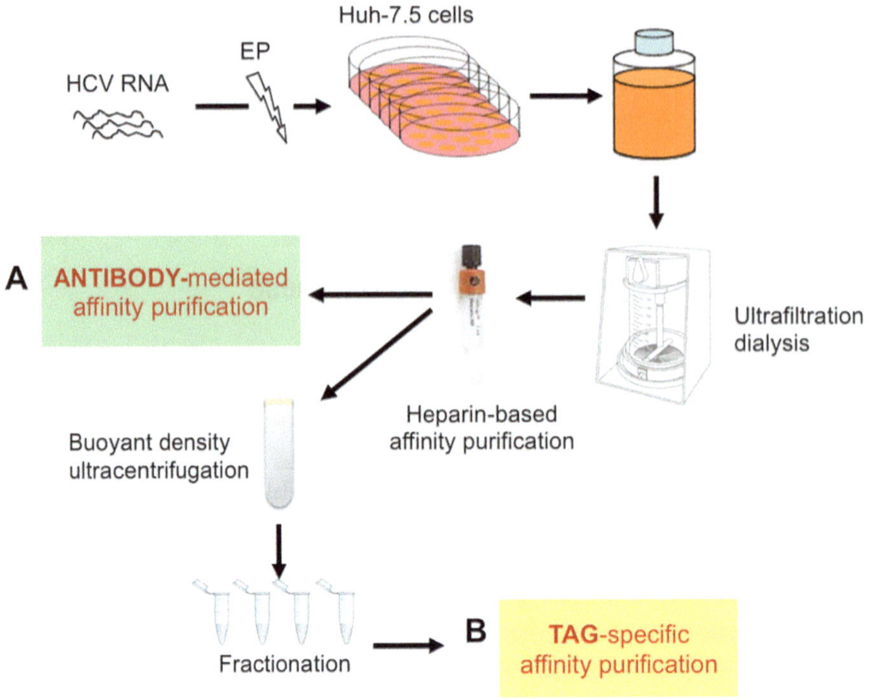

Fig. 1 Flowchart of HCV production and purification (adapted from Lussignol et al. [22])

production and spread was severely decreased in full-length genomic clones containing these viral adaptive mutations (Fig. 1).

In 2003, Kato and colleagues constructed a subgenomic replicon that could replicate to high levels without requiring selection or the development of viral enhancing mutations [16]. This was based on a viral isolate from a Japanese patient with *f*ulminant *h*epatitis termed JFH1 (genotype 2a). Full-length molecular clones either coding for the entire JFH1 genome or chimeras, in which the core through NS2 region belongs to a different HCV strain, were found to produce infectious particles to varying degrees [15, 17, 18]. Of these initial molecular clones, the intra-genotypic chimera J6/JFH1 had the highest viral titers, 100- to 1000-fold higher compared to the original JFH1 genome alone. Over the years, additional improvements, including alternative crossover points, selection for adaptive mutations that enhanced virus production and spread, as well as insertion of fluorescent proteins or luciferase reporters, have broadened the variety of HCV chimeric clones. Some of these ultimately enabled detection of HCV infection in primary hepatocytes and small animal models [1–4].

Several inherent features of HCV have posed significant barriers to its production in cell culture, including poor yields, purity, and stability (half-life ~3.5 h). Furthermore, the heterogeneity both in size (diameter, 45–100 nm) and density (1.03 to 1.25 g/mL) of the infectious particles and poor exposure of viral glycoproteins made the default viral sedimentation and purification approaches quite challenging [19–21].

As more information on the biology of HCV assembly became available, many of the unusual properties of this virus could be explained by its close interaction with very low-density lipoproteins (VLDL) and low-density lipoproteins (LDL), leading to the formation of chimeric lipo-viral particles (LVPs), which are pleomorphic and lack obvious symmetry [19].

Thus, a number of different strategies were required for overcoming these challenges. The use of highly fit genomes with adaptive mutations that enhanced virus production and spread allowed for higher virus yields in large-scale preparations. Frequent harvests and purification at 4 °C minimized stability issues. During HCV production, cells were switched to low fetal bovine serum (FBS) or serum-free media (SFM), and cell culture supernatants were concentrated in a nitrogen-pressurized ultrafiltration cell with low protein-binding membranes. This approach helped to retain infectivity while removing a large percentage of highly abundant serum proteins, such as albumin. Additional issues such as low specific infectivity, poor exposure of the glycoproteins on the surface of HCV particles, and aggregation required additional purification steps. Two sequential affinity purifications were performed post-concentration [22]. The first was a heparin-based affinity purification that increased specific infectivity from 1:5000 to 1:5 and helped to remove 95% of contaminants from the concentrated cell culture supernatant. The second purification step relied either on HCV E2 antibody-specific or tag-specific affinity purification. The use of a tandem tag inserted within the hypervariable region of E2 proved advantageous for structural studies and overcame limitations due to accessibility of E2 [19]. Finally, while not completely resolved, the constant stirring during concentration and increase in salt concentration has helped to mitigate aggregation observed during purification.

2 Materials

Below are indicated reagents, solutions, and instruments required for HVC production (in vitro transcription of viral genomic RNA + electroporation), concentration (to achieve high-titer stocks), affinity purification (via heparin, glycoprotein-specific antibodies, or genetically encoded tags), and fractionation (by buoyant density ultracentrifugation).

Diligently follow all waste disposal regulations for infectious material according to national and institutional guidelines as they may differ significantly.

2.1 In Vitro Transcription of HCV Genomes

2.1.1 Linearization

1. Purified HCV-encoding plasmid (*see* **Note 1**).
2. *XbaI* or other appropriate restriction enzyme with reaction buffer (*see* **Note 2**).
3. Qiagen MinElute PCR purification kit.

2.1.2 RNA Transcription

1. Purified linearized HCV template.
2. Nuclease-free microcentrifuge tubes.
3. Nuclease-free water.
4. T7 in vitro transcription kit (Promega T7 RiboMAX Express Large Scale RNA Production System).
5. RNase AWAY.
6. Qiagen RNeasy Mini kit.
7. Qiagen RNase-free DNAse.
8. Ambion RNA storage solution.

2.2 Electroporation

1. Huh-7.5 human hepatoma cell line or derivatives (i.e., Huh-7.5.1; *see* **Notes 3** and **4**).
2. In vitro transcribed RNA of HCV cell culture infectious clone (*see* **Note 5**).
3. Dulbecco's modified Eagle's medium (DMEM) with 0.1 mM nonessential amino acids (NEAA) and 10% fetal bovine serum (FBS; *see* **Note 6**).
4. Sterile Dulbecco's phosphate-buffered saline solution without Ca^{2+}/Mg^{2+} (DPBS).
5. Corning 175 mm^2 culture flasks with vented caps.
6. 0.25% Trypsin-EDTA.
7. Conical 250 mL centrifuge bottles.
8. Sterile 0.2 or 0.45 μm filter flask.
9. Cell strainer, 100 μm.
10. Hemocytometer.
11. Electroporator (parameters and settings found in this protocol have been specifically optimized for the following instruments: Bio-Rad Gene Pulser Xcell and BTX Electro Square Porator ECM 830; *see* **Note** 7 and **8**).
12. Electroporation single- or multi-well cuvettes (*see* Table 1).
13. Sterile, V-bottom 96-well plates.
14. Multichannel pipettor.
15. 96-Well Deep well block (2 mL volume/well).

Table 1
Cell concentrations, volumes, and EP settings for BTX Electro Square Porator ECM 830 (a–d) and Bio-Rad Gene Pulser Xcell (b)

Cuvette	Cells/mL[a]	Volume per well (μL)	Cells/well
(a) 2 mm single-well	1.5×10^7	400	6×10^6
(b) 4 mm single-well	1.5×10^7	800	1.2×10^7
(c) 4 mm 25-well[a]	$1.5/2.2 \times 10^7$	200/230	$3/5 \times 10^6$
(d) 4 mm 96-well[a]	$1.5/2.2 \times 10^7$	200/230	$3/5 \times 10^6$

[a]For multi-well cuvettes, lower cell concentration is sufficient for mutant testing; use higher cell concentration for large-scale virus preparation

2.3 Virus Large-Scale Production and Concentration

1. Millipore Stirred Ultrafiltration Cell, Model # 8400.
2. Ultracel Amicon ultrafiltration membranes, 100 kDa molecular weight cutoff (MWCO).
3. Magnetic stirring table.
4. Protecting shield.

2.4 HCV Affinity Purification and Fractionation

1. HiTrap heparin columns (GE Healthcare).
2. Peristaltic pump and tubing.
3. Heparin equilibration buffer (0.15 M NaCl, 0.02 M Tris–HCl, pH 7.4).
4. Heparin washing buffer (0.2 M NaCl, 0.02 M Tris–HCl, pH 7.4).
5. Heparin elution buffer (0.5 M NaCl, 0.02 M Tris–HCl, pH 7.4).
6. PBS 1×.
7. Amicon Ultra Centrifugal Filter Units.
8. Gradient maker (Biocomp).
9. SW41 rotor (Beckman Coulter).
10. Tubes for ultracentrifugation.
11. OptiPrep.
12. Gradient fractionator (Biocomp; optional—alternatively manual collection of fractions from the top of the gradient).
13. ProtA-coated magnetic beads (Bio-Adembeads; Ademtech), conjugated to HCV glycoproteins-specific (i.e., AR4A) or irrelevant (i.e., HIV glycoproteins) antibodies.
14. His-Dynabeads (Invitrogen).
15. Imidazole (in PBS 1×, pH 7.7).
16. Protease inhibitors.

3 Methods

3.1 In Vitro Transcription of HCV Genomes

HCV RNA is prepared by in vitro transcription of a linearized cDNA template coding for the viral genomic RNA, using the bacteriophage T7 DNA-dependent RNA polymerase.

3.1.1 Linearization

1. Mix together in a microcentrifuge tube, 5 μg of HCV-encoding plasmid, 1 μL of *XbaI* restriction enzyme (20 U; *see* **Note 2**), and 10 μL of 10× restriction enzyme buffer and nuclease-free water. Final volume of reaction should be 100 μL. Incubate at 37 °C for 2 h.
2. Clean up and purify linearized plasmid template using MinElute PCR Purification Kit or other reaction cleanup kit. Follow manufacturer's directions with following exceptions: (1) perform two washes with 500 μL of PE buffer, (2) leave first PE buffer wash on column for 3–5 min to improve removal of salts, and (3) elute in 15 μL of EB buffer. Quantify template by reading the absorbance at 260 nm using a spectrophotometer.

3.1.2 Transcription of HCV Genomic RNA

1. Assemble transcription reactions using the T7 RiboMAX Express Large Scale RNA Production System and 0.5–1 μg of linearized cDNA HCV template. Incubate reactions for 1 h at 37 °C.
2. Add 2 U RNase-free DNase to degrade the cDNA template, and incubate for 15 min at 37 °C.
3. Purify HCV RNA using RNeasy Mini kit following manufacturer's direction for RNA cleanup. Elute purified RNA in 50 μL of nuclease-free water or RNA storage solution.
4. Quantify RNA by measuring the absorbance at 260 nm with a spectrophotometer. Verify RNA integrity by gel electrophoresis.
5. Aliquot 5 μg of purified RNA into RNase-free microcentrifuge tubes, and store at −80 °C.

3.2 Electroporation

In this section, purified viral RNA will be transfected into permissive human hepatoma (Huh-7.5) cells by electroporation (EP) to allow translation and replication of the incoming viral genomes and secretion of infectious HCV particles in the cell culture supernatant.

3.2.1 Cell Preparation

1. Two days prior to EP, Huh-7.5 cells (85–90% confluence) are expanded by 1:3 to 1:5 split in 175 mm^2 flasks (*see* **Note 4**).
2. On the day of EP, aspirate media from sub-confluent cells, wash each flask twice with 10 mL D-PBS, and incubate with 5 mL of 0.25% Trypsin-EDTA for 5 min at 37 °C.

3. Add 5 mL of complete DMEM media to neutralize trypsin and resuspend cells by gently pipetting up and down. To break up cell clumps, filter cell mixture using 100 μm cell strainer and pool cells in 50 mL conical tubes.
4. Centrifuge cells for 5 min at 1500 rpm (514 × *g*) at 4 °C.
5. After pelleting cells, remove supernatant, and resuspend cells in 30 mL cold DPBS. It is critical to keep bottle of DPBS on ice or at 4 °C between uses and all centrifugation steps at 4 °C. Centrifuge cells using same conditions as **step 4**.
6. Remove supernatant and resuspend again the cell pellet in 30 mL cold DPBS. Mix well and remove 200–500 μL aliquot for counting. Spin 5 min at 1500 rpm (514 × *g*) at 4 °C.
7. During this wash step, count cells using a hemocytometer (or Beckman Coulter cell counter) to achieve an accurate cell count.
8. Remove supernatant and resuspend the cell pellet in ice-cold 1× DPBS according to Table 1. Pipette gently to avoid cell shearing. Keep cells in 50 mL conical tube in ice-filled bucket, or in sterile reservoir for plate handler electroporations. Ensure the cell suspension remains homogenous throughout the EP process by swirling or pipetting periodically.

3.2.2 Transfection of HCV Genomic RNA (Single-Well Cuvette Directions)

1. Add cell suspension to a microcentrifuge tube containing HCV genomic RNA (use between 2 and 10 μg per EP; *see* **Note 6**). Pipette up and down twice, immediately transfer the mix to an EP cuvette and pulse, in order to minimize the degradation of viral RNA by residual RNases in the cell suspension. Note the pulse length and output voltage to ensure consistency across EP. Allow cells to stand 5 to 10 min for recovery.
2. Use plastic Pasteur pipettes to transfer cells from cuvettes to 50 mL tubes containing 10–30 mL of complete DMEM media. Mix gently by pipetting.
3. Plate cells as quickly as possible (within 30 min). The following plating conditions are recommended (*see* Table 2). The indicated cell numbers refer to the initial cell count pre-EP and therefore may be slightly overestimated due to cell death. Cells should be confluent 48–72 h post-EP.

3.2.3 Transfection of HCV Genomic RNA (Multi-Well Cuvette Directions)

For the high-throughput (HT) multi-well plates, all wells must be filled within a column in order to transmit the electrical current. If unable to fill all wells with cell suspension, then pre-fill unused wells with 230 μL of DPBS.

1. Aliquot HCV genomic RNA into V-bottom 96-well plate before trypsinizing Huh-7.5 cells. Keep on ice until cells are ready.

Table 2
Plating conditions for electroporated cells[a]

Tissue culture dish	Final volume (mL)	Number of cells (pre-EP count)
T75	10–15	$8\text{–}10 \times 10^6$
6-Well	2–3	4×10^5
24-Well	0.5–1	1×10^5
96-Well	0.1	1.5×10^4

[a]Cell numbers are indicated for growth in serum-containing media. Increase plating numbers if cells will be shifted to serum-free media for harvests

2. Place cells from Subheading 3.2.1, **step 8** in a sterile reservoir. Using multichannel pipettor, remove appropriate amount of cell suspension, and mix gently with HCV genomic RNA. Transfer quickly the cell-RNA mixture to the multi-well EP plate.
3. Seal top with enclosed sterile plate tape. Ensure that pins are in the correct orientation depending on which plate format you are using (*see* manufacturer's directions). Place plate in high-throughput plate handler (25-well plate has adapter). Select column and electroporate by pressing START (*see* **Note 7**). When pulsing is done, switch to next column and press START again. Continue until all columns containing samples are pulsed.
4. Allow cells to stand for 5 min at room temperature for recovery. Using multichannel pipettor, transfer cells to 2 mL deep-well block containing 1.7 mL of complete DMEM media per well. Plate as directed in Table 2.

3.2.4 Harvest and Storage of Virus Stocks

1. Harvest viral supernatants when viral kinetics have plateaued (typically between 24 and 96 h post-EP for most HCV genomes). Pool supernatants and clarify by centrifugation at 3000 rpm for 10 min at 4 °C. Given the poor stability of extracellular HCV particles at 37 °C [23, 24], virus-containing media can be collected every 3–6 h in order to maximize viral titers.
2. Sterile filter supernatants using either 0.22 μm filter unit or syringe filters attached to 10 mL syringe. Aliquot supernatants into single-use cryotubes, and store at −80 °C until needed.

3.2.5 Measure EP Efficiency and Viral Kinetics

When setting up EP conditions with a new electroporator, it is important to vary a series of parameters to find the best balance between efficient delivery of the viral RNA and cell survival. These include cell number, pulsing mode (namely, exponential decay

Table 3
Parameters varied to optimize EP efficiency of HCV genomic RNA using Bio-Rad Gene Pulser Xcell[a]

EP	Cuvette gap	Cells per well[a]	Volume per well (μL)	EP settings
1	4 mm	6×10^6	400	270 V, 950 μF,100 Ω, exponential
2	4 mm	6×10^6	400	270 V, 25 ms,square wave: 1 pulse
3	4 mm	6×10^6	400	500 V, 0.1 ms, square wave: 5 pulses, interval 1.1 ms
4	4 mm	1.2×10^7	800	860 V, 5 ms, square wave: 2 pulses, interval 5 s
5	2 mm	6×10^6	400	140 V, 950 μF, exponential
6	4 mm	1.2×10^7	800	270 V, 950 μF,100 Ω, exponential

[a]For all conditions tested, a cell concentration of 1.5×10^7/mL was used

versus square wave pulse), voltage, and, for square wave pulses, the number of pulses and length of intervals.

EP efficiency and viral kinetics can then be assessed by quantifying viral RNA copies, infectious titers in the cell culture supernatants, or percentage (%) of HCV-infected cells by quantitative RT-PCR, limiting dilution/$TCID_{50}$ assay or flow cytometry, respectively. If EP efficiency is assessed by flow cytometry (FACS), intracellular staining with an HCV-specific antibody will enable to precisely determine how many cells were successfully transfected. A commonly used antibody (Ab) is the clone 9E10 that recognizes the HCV nonstructural protein NS5A [17]. Alternatively, reporter genomes encoding fluorescent proteins can prove very helpful [4]. In Table 3, we report an example of the variables tested to optimize EP conditions utilizing the Bio-Rad Gene Pulser Xcell electroporator. The results are shown in Fig. 2, using both Ab staining and detection of fluorescent reporter as FACS readouts in a side-by-side analysis.

3.3 Large-Scale Production of Virus and Concentration

High-titer HCV stocks can be produced by switching electroporated cells to low-serum (1.5% FBS) or serum-free media (SFM) not earlier than 12 h post-EP, followed by serial collections and concentration of the resulting infectious culture supernatants. This paragraph describes parameters for concentration of HCV by ultrafiltration in a pressurized stirring chamber and/or Amicon centricon tubes using 100 MWCO low protein-binding membranes made of regenerated cellulose. With this step, >85% of the viral infectivity in the starting supernatant is retained. Using the highly infectious J6/JFH clone 2, final concentrations of up to 1×10^8 $TCID_{50}$/mL can be achieved [22].

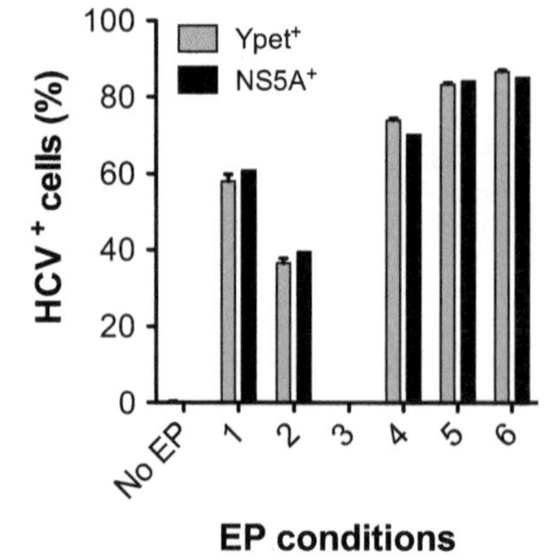

Fig. 2 Electroporation (EP) efficiency of HCV genomic RNA as measured by flow cytometry. An HCV reporter genome encoding a Ypet fluorescent protein between NS5A and NS5B (Clone 2-5AB-Ypet) was electroporated using the conditions indicated in Table 3 (each one in duplicate). At 72 h post-EP, cells were harvested for flow cytometric analysis. The percentage (%) of HCV-electroporated cells was determined either by intracellular staining with the NS5A antibody (9E10; black bars) or by detection of the Ypet fluorescent signal (gray bars). 9E10 was directly conjugated to Alexa Fluor 647 and used at a 1:4000 dilution in FACS buffer [22]. The highest cell survival, as noted by confluency at time of harvest, and infectious titers, as determined by $TCID_{50}$ limiting dilution assay, were achieved with condition 6

1. Depending on the replication kinetics of the isolate, cell culture supernatants can be harvested as often as every 3–6 h up to 5 days post-EP. Stop harvesting at early signs of cytopathic effect.
2. Centrifuge supernatants for 5 min at 1300 rpm, 4 °C using conical 250 mL bottles.
3. Put clarified supernatant in sterile bottles, and keep at 4 °C until harvesting is complete.
4. The concentration step can start immediately after harvesting. Harvests can be stored at 4 °C for a couple of days before concentration.
5. Depending on the starting volume and the desired concentration factor, either use Ultrafiltration Cell 8400 (max capacity 400 mL, concentrated vol. 25 mL) or tubes (Amicon 15, max capacity 15 mL, concentrated vol. 250 μL; Amicon 4, max capacity 4 mL, concentrated vol. 50 μL).
6. The amount of time required for this step will vary, depending on your starting volume and the protein content of the sample.

3.4 Virus Concentration by Ultrafiltration

1. To concentrate virus by ultrafiltration, assemble the ultrafiltration cell according to manufacturer's instruction (https://www.nist.gov/sites/default/files/documents/ncnr/StirredUltrafiltrationCell_Millipore.pdf).

2. Insert a membrane with a 100 kDa MWCO at the base of the cell by placing the shiny side up (*see* **Note 9**).
3. Pour the HCV-containing culture supernatant into the cell and place the cell on the magnetic stirring table. Place outlet into a disposable flask for flow-through collection.
4. Connect the inlet to a regulated nitrogen gas pressure source, and open the valve to inject the gas into the cell and pressurize.
5. Follow the instructions in the membrane package to determine optimal operating pressure. For 100 kDa MWCO membranes, use pressure between 30 and 50 psi. Do NOT exceed the cell pressure limit of 75 psi.
6. To avoid membrane damage, make sure the cell is pressurized before turning on the stirring table. Place a protecting shield in front of the ultrafiltration cell to protect yourself and your labmates during the concentration.
7. To refill the chamber and retrieve the concentrated HCV-containing supernatant, turn off the stirring table and then the nitrogen pressure source. Slowly vent the pressure inside the cell, by turning the pressure-relief valve knob to the horizontal position. Push the cap down, and then slide the cell out from the retaining stand.
8. With the ultrafiltration cell, HCV supernatants can be concentrated down to 50 mL.
9. Amicon Ultra tubes with 100 kDa MWCO can be used at 2800 rpm, 4 °C, to concentrate further the retentate from the ultrafiltration cell. Alternatively, they can be used for (1) concentration of smaller volumes of HCV supernatants and (2) to equilibrate the concentrated virus preparations to DPBS or other desired buffers.
10. Concentrated high-titer HCV stocks can be stored into cryovials and frozen at −80 °C for later use.

3.5 HCV Affinity Purification and Fractionation

3.5.1 Heparin-Based Affinity Purification

Viral supernatants (concentrated or not) can be affinity purified over heparin columns (HiTrap heparin; GE Healthcare), according to manufacturer's instructions. This purification step increases the specific infectivity [infectious units per genome RNAs] of the sample by eliminating noninfectious particles (~50% of input RNA copies) and possibly by removing HCV inhibitors contained in the cell culture supernatant. Moreover, the heparin column removes >95% of the protein contaminants in the concentrated culture supernatant.

1. Perform all steps at 4 °C; make sure column is not drying out.
2. Equilibrate heparin column with 50 mL equilibration buffer.
3. Apply virus sample by pumping it onto the column.

4. Wash heparin column twice with 50 mL wash buffer.
5. Elute virus sample off the column with 50 mL elution buffer.
6. Virus can then be further concentrated/dialyzed with Amicon Ultra Centrifugal Filter Units.

3.5.2 Buoyant Density Ultracentrifugation

HCV can be separated according to its density with the following ultracentrifugation and fractionation method. Using iodixanol buoyant density gradients, >90% of the input infectivity can be recovered in few fractions (*see* **Note 10**).

1. 500 μL of a high-titer HCV stock is carefully loaded onto an ultracentrifuge tube containing a 10–40% (wt/vol) iodixanol gradient solution (OptiPrep; Sigma).
2. Spin samples for 8 h at 39,000 rpm (187813 × *g*) at 4 °C, using a SW41 rotor.
3. Collect 16 fractions of 600 μL each from the top of the gradient (this can be done manually or with a fractionator).
4. Test each fraction to determine the number of HCV RNA copies and infectious particles in order to identify the peak of infectivity.
5. Fractions with the highest titers can be pooled and used for experiments requiring high multiplicity of infection (MOI), dialyzed to D-PBS using 100 kDa Amicon Ultra Centrifugal Filter Units, stored at −80 °C, or purified further as described below.

3.5.3 Tag- or Antibody-Based Affinity Purification

Depending on the starting viral titers, desired degree of purity, and final amount of infectious particles needed, HCV can be affinity purified (from cell culture supernatants, after heparin columns as well as post-fractionation) using antibodies specific to the viral glycoproteins (E1, E2) and/or to tags (if genetically modified clones with tagged E1/E2 are available). These antibodies can be adsorbed to electron microscopy (EM) grids or to beads for biochemical approaches (i.e., compositional analyses [22]; virus-host interaction studies).

1. For coating of antibodies to EM grids and ultrastructural studies of HCV virions, refer to the methods described in [19].
2. For coupling of antibodies to protein A- or protein G-coated magnetic beads (Bio-Adembeads; Ademtech), incubate 20 μL of resin overnight at 4 °C with 2 μg of α-HCV (i.e., AR4A) or α-tag (i.e., his; HA; Flag) antibodies in 500 μL of D-PBS + protease inhibitors.
3. Wash bead-antibody complexes three times with 1 mL D-PBS, and incubate with 500 μL of HCV-containing supernatant for 1 h at RT in the presence of protease inhibitors. In case of

His-tagged HCV, 20 mM imidazole can be added during this step to increase the stringency of binding and reduce background.

4. Wash bead-bound virions six times with 1 mL D-PBS or 50 mM imidazole in D-PBS for pull downs of His-tagged HCV.
5. Pellet HCV particles for 1 min on a magnet, and elute with 0.9 M imidazole in PBS1×, pH 7.7, or heat at 70 °C for 10 min in 1× SDS loading buffer.

4 Notes

1. HCV cDNA clones of the JFH1 isolates and its derivatives are generally maintained in a high-copy ampicillin-resistant backbone and propagated in *DH5-alpha* bacteria. Plasmid DNA is isolated by Qiagen Plasmid Maxi kit to ensure high yields and purity for downstream steps.
2. JFH1-based plasmids rely on *XbaI* cleavage to generate appropriate 3′ ends for runoff transcription reactions. For other genomes, a different restriction enzyme may be necessary.
3. A critical parameter affecting electroporation efficiency is the total time Huh-7 cells are outside of the 37 °C incubator (i.e., from cell trypsinization to plating). The entire process should be accomplished in less than 2 h, and multiple small batches (8–10 single-cuvette EPs) are therefore preferable. To facilitate large-scale preparations of cell culture HCV, HT plate handler can be connected to the BTX electroporator, and multi-well (25 or 96) cuvette plates may be purchased for synchronous multiple electroporations.
4. Since the number and health of cells electroporated is of critical importance, use cells between passage 20–40, and do not reduce cell concentration below 1.5×10^7 cells/mL. Avoid splitting cells 24 h prior to electroporation, and ensure that they are in the exponential growth phase (i.e., not overgrown). When first starting out, prepare ~2× the number of cells you will need. You can lower this as you become more comfortable with the protocol and minimize cell loss through the harvest and wash steps.
5. High-quality RNA is essential to obtaining high virus titers. Nuclease-free microcentrifuge tubes and water should be used. Although optional, barrier tips are recommended to minimize contamination with RNases that are highly stable. 0.5% SDS solution or commercial products such as RNase AWAY can be used on bench area. Pipettes should be wiped down before use. Verify RNA integrity by agarose gel electrophoresis, and store at −80 °C in single-use aliquots ready for electroporation.

6. Changing FBS lots may impact on cell health, growth, and survival rate after electroporation and therefore affect HCV titers. It is recommended to test different FBS batches to ensure optimal production of infectious HCV particles.
7. BTX HT plate handler: both the voltage generator and the plate holder need to be turned on when electroporating. Pin configuration is different depending on well format, and 25-well plates utilize an adaptor; consult manufacturer's directions to ensure correct setup. If arcing occurs during pulsing, check the electrode pins. If the pins are black, they must be changed out.
8. Even if users have access to the same electroporators mentioned in this protocol, optimal settings need to be verified for each instrument. It is recommended to vary the parameters described in our own optimization experiments with the Bio-Rad instrument (*see* Fig. 2 and Table 3).
9. Handle the ultrafiltration membrane by its edges, to avoid scratching or contaminating the surface.
10. Several gradient solutions have been tested to achieve best separation of HCV virions from protein contaminants while preserving their infectivity. While able to recover most of the input HCV RNA copies, both sucrose (10–30% for 1.5 h at 39,000 rpm) and potassium tartrate (5–25% for 2 h at 39,000 rpm) sedimentation velocity gradients drastically reduced the initial infectivity of the viral preparations (by ~1 log). This could be explained at least in part by the significant aggregation and damage to the viral particles as assessed by electron microscopy [22].

Acknowledgments

The authors would like to thank Dr. Charles M. Rice and the members of his laboratory at Rockefeller University for contributing to the development of the protocol described here, in particular Dr. Martina Kopp. This work builds on extensive knowledge that was created by the HCV community far beyond the selected references cited here for space limitation. Our appreciation goes to all the researchers that have contributed to advancing our understanding of the HCV biology. We are grateful to Dr. Kunihiro Uryu, Prof. Richard Kuhn, and Dr. Brian Chait for constant guidance and training on ultrastructural and proteomics studies of HCV particles. A sincere thanks to Dr. Marion Lussignol and Dr. Susan John for helping with the setup of the EP conditions for the Bio-Rad Gene Pulser Xcell at King's College London. This work was supported by King's Health Partners grant to MT Catanese.

References

1. Lohmann V, Bartenschlager R (2014) On the history of hepatitis C virus cell culture systems. J Med Chem 57:1627–1642
2. Steinmann E, Pietschmann T (2013) Cell culture systems for hepatitis C virus. Curr Top Microbiol Immunol 369:17–48
3. Bukh J (2016) The history of hepatitis C virus (HCV): basic research reveals unique features in phylogeny, evolution and the viral life cycle with new perspectives for epidemic control. J Hepatol 65:S2–S21
4. Catanese MT, Dorner M (2015) Advances in experimental systems to study hepatitis C virus in vitro and in vivo. Virology 479-480:221–233
5. Blight KJ, Norgard EA (2006) HCV replicon systems. In: Tan SL (ed) Hepatitis C viruses: genomes and molecular biology. Horizon Bioscience, Norfolk
6. Wakita T, Kato T (2006) Development of an infectious HCV cell culture system. In: Tan SL (ed) Hepatitis C viruses: genomes and molecular biology. Horizon Bioscience, Norfolk
7. Bartenschlager R, Lohmann V (2000) Replication of hepatitis C virus. J Gen Virol 81:1631–1648
8. Kolykhalov AA, Agapov EV, Blight KJ, Mihalik K, Feinstone SM, Rice CM (1997) Transmission of hepatitis C by intrahepatic inoculation with transcribed RNA. Science 277:570–574
9. Yanagi M, Purcell RH, Emerson SU, Bukh J (1997) Transcripts from a single full-length cDNA clone of hepatitis C virus are infectious when directly transfected into the liver of a chimpanzee. Proc Natl Acad Sci U S A 94:8738–8743
10. Lohmann V, Korner F, Koch J, Herian U, Theilmann L, Bartenschlager R (1999) Replication of subgenomic hepatitis C virus RNAs in a hepatoma cell line. Science 285:110–113
11. Bartosch B, Dubuisson J, Cosset FL (2003) Infectious hepatitis C virus pseudo-particles containing functional E1-E2 envelope protein complexes. J Exp Med 197:633–642
12. Hsu M, Zhang J, Flint M, Logvinoff C, Cheng-Mayer C, Rice CM et al (2003) Hepatitis C virus glycoproteins mediate pH-dependent cell entry of pseudotyped retroviral particles. Proc Natl Acad Sci U S A 100:7271–7276
13. Blight KJ, McKeating JA, Rice CM (2002) Highly permissive cell lines for subgenomic and genomic hepatitis C virus RNA replication. J Virol 76:13001–13014
14. Friebe P, Boudet J, Simorre JP, Bartenschlager R (2005) Kissing-loop interaction in the 3′ end of the hepatitis C virus genome essential for RNA replication. J Virol 79:380–392
15. Zhong J, Gastaminza P, Cheng G, Kapadia S, Kato T, Burton DR et al (2005) Robust hepatitis C virus infection in vitro. Proc Natl Acad Sci U S A 102:9294–9299
16. Kato T, Date T, Miyamoto M, Furusaka A, Tokushige K, Mizokami M et al (2003) Efficient replication of the genotype 2a hepatitis C virus subgenomic replicon. Gastroenterology 125:1808–1817
17. Lindenbach BD, Evans MJ, Syder AJ, Wolk B, Tellinghuisen TL, Liu CC et al (2005) Complete replication of hepatitis C virus in cell culture. Science 309:623–626
18. Wakita T, Pietschmann T, Kato T, Date T, Miyamoto M, Zhao Z et al (2005) Production of infectious hepatitis C virus in tissue culture from a cloned viral genome. Nat Med 11:791–796
19. Catanese MT, Uryu K, Kopp M, Edwards TJ, Andrus L, Rice WJ et al (2013) Ultrastructural analysis of hepatitis C virus particles. Proc Natl Acad Sci U S A 110(23):9505–9510
20. Gastaminza P, Dryden KA, Boyd B, Wood MR, Law M, Yeager M et al (2010) Ultrastructural and biophysical characterization of hepatitis C virus particles produced in cell culture. J Virol 84:10999–11009
21. Merz A, Long G, Hiet MS, Brugger B, Chlanda P, Andre P et al (2011) Biochemical and morphological properties of hepatitis C virus particles and determination of their lipidome. J Biol Chem 286:3018–3032
22. Lussignol M, Kopp M, Molloy K, Vizcay-Barrena G, Fleck RA, Dorner M et al (2016) Proteomics of HCV virions reveals an essential role for the nucleoporin Nup98 in virus morphogenesis. Proc Natl Acad Sci U S A 113:2484–2489
23. Ciesek S, Friesland M, Steinmann J, Becker B, Wedemeyer H, Manns MP et al (2010) How stable is the hepatitis C virus (HCV)? Environmental stability of HCV and its susceptibility to chemical biocides. J Infect Dis 201:1859–1866
24. Song H, Li J, Shi S, Yan L, Zhuang H, Li K (2010) Thermal stability and inactivation of hepatitis C virus grown in cell culture. Virol J 7:40

Chapter 8

Embryonic or Induced Pluripotent Stem Cell-Derived Hepatocellular Systems for HCV Culture

Xianfang Wu and Viet Loan Dao Thi

Abstract

Human-induced pluripotent stem cell-derived hepatocyte-like cells (iHeps) constitute a powerful tool for modeling hepatotropic pathogen infections in cell culture. Meanwhile, CRISPR-Cas9 technology enables precise editing of stem cell genomes to generate patient-specific disease models and thus development of personalized experimental systems. Here we present a detailed stepwise protocol for the differentiation of stem cells to hepatocyte-like cells for HCV studies in cell culture. We also outline the use of an inducible iCRISPR platform for the rapid and efficient modification of host factors of interest to better understand their function during HCV infection.

Key words Human embryonic/induced pluripotent stem cells, Hepatocyte-like cells, iCRISPR, Precision medicine

1 Introduction

Chronic infection with HCV affects 130–150 million people worldwide and is the primary cause of hepatocellular carcinoma in developed countries [1]. During the last decade, a better understanding of the HCV life cycle and viral genome has led to the development of direct-acting antivirals (DAA), and in turn, a high rate of viral clearance in the clinic. Although virologic cure improves liver fibrosis, some DAA-treated patients with advanced liver disease stage have been reported to experience unexpected early hepatocellular cancers and recurrence of infection [2, 3]. Therefore, in addition to viral factors, host genetic diversity appears to contribute to the spectrum of the disease and treatment outcome (Reviewed in [4]). For example, genome-wide association studies (GWAS) have identified host polymorphisms in the *IFNL4* locus that correlate with spontaneous HCV clearance and viral response to interferon (IFN)-based therapy [5]. These studies have also identified a single nucleotide polymorphism (SNP) in the *PNPLA3* gene that has been associated with steatosis and liver fibrosis in chronic HCV

Mansun Law (ed.), *Hepatitis C Virus Protocols*, Methods in Molecular Biology, vol. 1911,
https://doi.org/10.1007/978-1-4939-8976-8_8, © Springer Science+Business Media, LLC, part of Springer Nature 2019

patients [6]. To date, the mechanistic underpinnings of these correlations are still poorly understood.

Hepatoma cells have been invaluable across the history of HCV studies in cell culture. Their use however is limited for comprehensive studies of host genetic diversity on viral pathogenesis, as these cells are typically dedifferentiated and possess altered proliferative, metabolic, innate immune, and apoptotic responses. Human embryonic stem cells (hESC) or induced pluripotent stem cells (iPSC) offer an attractive alternative as they have the capacity to self-renew indefinitely while maintaining their potential to differentiate into specific cell types, such as iHeps. As mature iHeps are non-cancerous, they constitute a more physiologically relevant cell culture model than hepatoma cells. Our lab and others have demonstrated the utility of iHeps for studies of HCV and other hepatotropic viruses [7–15]. Strikingly, iHeps, unlike hepatoma cells, support infection of HCV clinical isolates [7]. In tandem with the capability to generate host-specific iPSC lines [16] and/or novel genome-editing technologies in stem cells, iHeps therefore enable the generation of patient-specific disease models (Fig. 1). These personalized models should advance the

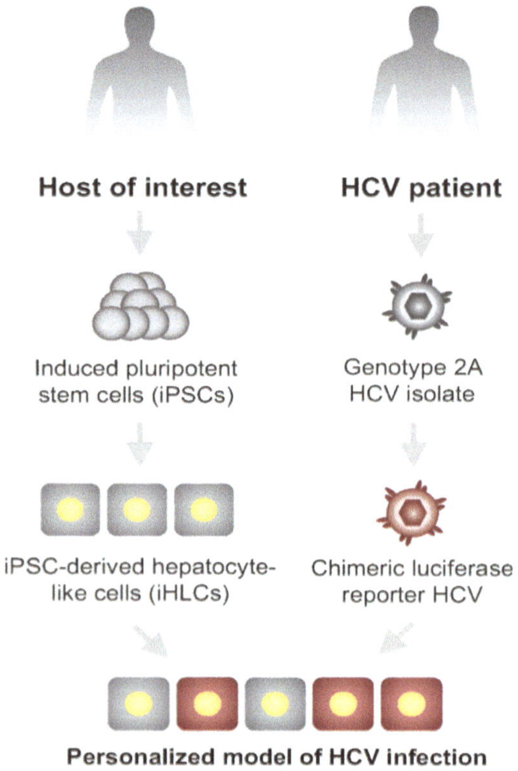

Fig. 1 Infection of iHeps derived from one donor with HCV isolate from the same or another donor for personalized disease modeling. Reprinted with permission from [8]

understanding of particularly those diseases that are associated with genetic variants, such as chronic HCV.

Here we describe an efficient and reproducible stem cell differentiation protocol detailing both the generation of iHeps and their use for HCV infections in cell culture. Moreover, we describe the application of a previously developed inducible CRISPR-Cas9 (iCRISPR) platform [17] for rapid and efficient genome editing of HCV host factors in iHeps to more generally explore the importance of specific host genes and pathways in HCV replication. These systems enable the generation of patient-specific models for the investigation of host-pathogen interactions and thus personalized and precise identification of drug targets and biomarkers for improving clinical practice, especially with regard to chronic patients who do not respond to available treatments.

2 Materials

2.1 Cell Lines

Human ESC (WA09) and iPSC (iPS.C3A) cells were obtained from the WiCell Research Institute and Stephen Duncan at Medical College of Wisconsin, respectively. iPSC line iPS.LVID2 was generated in the Rice lab. Irradiated mouse embryonic fibroblasts (iMEFs) were purchased from MTI-GlobalStem.

2.2 Reagents

Reagent	Supplier	Cat. number
Accutase	Innovative Cell Technologies	AT 104-500
Ascorbic acid	Sigma	A4544-25G
DMEM/F12 (1)	Invitrogen	A1370801
DMEM/F-12 (2)	Mediatech	15 090 CV
D-PBS	Invitrogen	14190250
ES-DMEM	MTI-GlobalStem	GSM-2001
FBS (ES-qualified)	Invitrogen	16141061
GlutaMAX-I	Invitrogen	35050-061
Holo-transferrin	Invitrogen	11107-018
Knockout serum replacement	Invitrogen	10828028
Lipofectamine RNAiMAX	Invitrogen	13778150
MEM-NEAA	Invitrogen	11140-050
mTeSR1	Stemcell Technologies	05850
Opti-MEM	Invitrogen	31985088

(continued)

Reagent	Supplier	Cat. number
Pen/strep	Sigma	P4333-100ML
Probumin	Millipore	81-068-3
ReLeSR	Stemcell Technologies	05872
Retinoic acid	Sigma	R2625-50MG
SB431542 hydrate	Sigma	S4317-5MG
T7EI endonuclease I	New England Biolabs	
Trace element A	Mediatech	25-021-CI
Trace element B	Mediatech	25-022-CI
Trace element C	Mediatech	25-023-CI
Trypan blue	Thermo Fisher Scientific	15250061
Y-27632	Stemcell Technologies	72308
β-Mercaptoethanol	Invitrogen	21985-023

2.3 Growth Factors

Growth factor	Supplier	Cat. number
Recombinant human EGF	Peprotech	AF-100-15
Recombinant human FGF-10	Peprotech	100-26
Recombinant human FGF-4	Peprotech	100-31
Recombinant human HGF	Peprotech	100-39
Recombinant human/mouse/rat activin A	R&D systems	338-AC-010
Recombinant mouse Wnt-3a	R&D systems	1324-WN-002
Recombinant human FGF-basic (AA 1-155)	Invitrogen	PHG0261

2.4 Media Compositions: Keep All Media Protected from Light

1. Differentiation supplement for basal media.

Component	Amount
Probumin	5 g
Ascorbic acid (stock: 100 mg/ml)	1.25 ml
Holo-transferrin (stock: 2 mg/ml)	12.5 ml
Trace elements (A/B/C)	2.5 ml + 2.5 ml + 2.5 ml
MEM-NEAA	25 ml
DMEM/F-12 (1)	Up to 50 ml

Mix the reagents thoroughly by shaking at 4 °C overnight, protected from light. Make 1 ml aliquots and store at −20 °C.

2. Differentiation basal media.

Component	Amount (ml)
Probumin (20%)	4.5
Supplement (2.5.1)	1
GlutaMAX-I	0.5
β-Mercaptoethanol	0.05
DMEM/F-12 (1)	Up to 50

3. Hepatocyte differentiation media.

Component	Medium-A	Medium-B	Medium-C	Medium-D	Medium-E	Notes
Activin-A	100 ng/ml	100 ng/ml				
FGF-basic	8 ng/ml	8 ng/ml				
Wnt-3a	25 ng/ml					
FGF-10			50 ng/ml	50 ng/ml		
SB431542				1 μM		
Retinoid acid				0.1 μM		Freshly made
EGF					50 ng/ml	
HGF					50 ng/ml	
FGF-4					30 ng/ml	

Prepare the media by supplementing the differentiation basal media with indicated growth factors at indicated concentrations and, if necessary, 1% Pen/Strep.

4. Cas9-induction medium: mTeSR1 with 2.5 μg/ml of doxycycline (Dox).
5. Transfection medium: mTeSR1 with 2.5 μg/ml of Dox and 10 μM Y-27632.
6. MEF-feeder medium.

Component	Amount (ml)
ES-DMEM	16.8
FBS (ES-qualified)	3
Pen/Strep	0.2

7. hPSC-feeder medium.

Component	Amount (ml)
DMEM/F12 (2)	38.5
GlutaMax (100×)	0.5
Pen/Strep	0.5
Non-essential AA	0.5
Knockout serum replacement (KSR)	10
β-Mercaptoethanol	0.05
FGF-β (10 μg/ml)	0.05

8. hPSC-freezing medium, prepare freshly: 50% hPSC-feeder medium, 40% knockout serum replacement, and 10% DMSO.

3 Methods

3.1 Stem Cell Culture

1. Preparation of Matrigel aliquots. Matrigel is frozen at −20 °C to −80 °C, liquid at 4 °C, and gels rapidly at room temperature. It is crucial to keep the Matrigel frozen until you are ready to aliquot it. We usually use Matrigel at 0.5 mg/6-well plate and prepare aliquots accordingly. Matrigel should be thawed overnight on ice in a cold room or refrigerator. Ensure that the neck of the bottle is not submerged in ice to reduce risks of contamination.
2. Thawing aliquots and coating plate(s) with Matrigel. Each Matrigel aliquot is intended for a single use and cannot be thawed and refrozen. In a sterile biosafety cabinet, place a sterile 15 ml conical tube on ice and add 6 ml of cold DMEM/F12 medium. Remove a 0.5 mg Matrigel aliquot from the freezer, and add 1 ml of the ice-cold DMEM/F12 medium to it. Gently pipette up and down to thaw and dissolve the Matrigel. Immediately transfer it to the 15 ml conical tube. Pipette to mix. Immediately plate 0.5 ml of the mix to one well of a 12-well plate. Adjust the volumes according to the surface area of culture plates. Gently shake the plate to ensure that the Matrigel covers the entire surface of the wells. Place the plates

into 37 °C incubator and wait for at least 1 h before use (*see* **Note 1**).

3. Maintenance of human pluripotent stem cells. It is critical to start with high a quality stem cell culture for efficient hepatic differentiation. Human pluripotent stem cells (hPSCs) are maintained in mTeSR1 medium, typically in Matrigel-coated 6-well plates. hPSCs grow as compact, multicellular colonies characterized by distinct borders and should be tightly packed, exhibit a high nuclear-to-cytoplasm ratio, and have prominent nucleoli. The cells are ready for passaging when approximately 70% confluency is reached.

3.2 Differentiation of Stem Cell-Derived Hepatocyte-Like Cells

This step describes the seeding of stem cells cultured in a 6-well plate to a Matrigel-coated 12-well plate. Adjust volumes if other plate types are used.

1. Coat one well of a 12-well plate with Matrigel as described in Subheading 3.1, **step 2**.
2. Warm (20–25 °C) sufficient volumes of mTeSR1, DMEM/F12, and Accutase for passaging. Prepare single-cell passaging medium by adding Y-27632 to mTeSR1 to reach a final concentration of 10 μM.
3. Transfer the old culture medium to a 15 ml conical tube.
4. Wash the cells with 1 ml of D-PBS.
5. Aspirate wash medium and add 0.5 ml Accutase.
6. Incubate at 37 °C for 3–5 min (*see* **Note 2**).
7. Transfer the old medium to the well and dislodge cells by pipetting up and down 3–5 times using a pipette with a p1000 tip. Ensure that all remaining cell aggregates are broken up into single cells.
8. Immediately transfer the cells back to the 15 ml conical tube. Wash the well once with 1 ml of DMEM/F12 to collect any remaining cells and transfer to the same tube.
9. Invert the tube up and down several times to ensure thorough mixing. Determine the number of live cells using Trypan blue.
10. Transfer 7.5–8 $\times$ 10^5 cells to a new 15 ml conical tube and centrifuge at 300 $\times$ g for 3–5 min (*see* **Note 3**).
11. Resuspend cells in 1 ml of single-cell passaging medium by pipetting up and down 5–10 times and plate cells onto a pre-coated well.
12. Incubate at 37 °C for 20–24 h.
13. On day 1, warm (37 °C) 3 ml of DMEM/F12 and 1 ml of differentiation basal medium. Prepare Medium-A (Subheading 2.4, **item 3**).

14. Aspirate culture medium and wash three times with 1 ml of DMEM/F12. Aspirate wash medium and replace with 1 ml of Medium-A.
15. Incubate at 37 °C for 24 h.
16. On day 2, warm (37 °C) 3 ml of DMEM/F12. Prepare sufficient Medium-B to be used on days 2–4, i.e., 3 ml in total per well. Warm up each day only the amount for immediate use.
17. Warm (37 °C) 1 ml of Medium-B. Aspirate old medium and wash three times with 1 ml of DMEM/F12. Aspirate wash medium and replace with 1 ml of Medium-B.
18. Incubate at 37 °C for 24 h.
19. On day 3, warm (37 °C) 1 ml of Medium-B. Aspirate old medium and replace with 1 ml of Medium-B.
20. Incubate at 37 °C for 24 h.
21. On day 4, warm (37 °C) 1 ml of Medium-B. Aspirate old medium and replace with 1 ml of Medium-B.
22. Incubate at 37 °C for 24 h.
23. On day 5 post-differentiation, definitive endoderm cells are ready for hepatic specification. Use ~20,000 cells to assess endoderm induction efficiency by flow cytometry and staining for endoderm markers CXCR4 and GATA4. Ideally, a properly maintained hPSC line should yield ≥90% $CXCR4^+GATA4^+$ cells (*see* **Note 4**).
24. Coat three wells of a 12-well plate with Matrigel as described in Subheading 3.1, **step 2**.
25. Warm (20–25 °C) sufficient volumes of DMEM/F12, differentiation basal medium, and Accutase for passaging.
26. Prepare 9 ml of Medium-C. Warm (37 °C) only the volume of Medium-C for day 5 use, i.e., 3 ml for three wells. Store remaining Medium-C at 4 °C and warm up the following days only the amount for immediate use. Add Y-27632 to pre-warmed Medium-C to reach a final concentration of 10 μM.
27. Aspirate culture medium and wash the cells with 1 ml of D-PBS. Aspirate wash medium and add 0.5 ml of Accutase. Incubate at 37 °C for 2–3 min.
28. Add 1 ml of differentiation basal medium to the well and dislodge cells by pipetting up and down 3–5 times using a pipette with a p1000 tip. Ensure that all remaining cell aggregates are broken up into single cells. Immediately transfer cells to a 15 ml conical tube. Wash the well once with 1 ml of DMEM/F12 to collect any remaining cells and transfer to the same tube. Centrifuge the tube at 300 × *g* for 3–5 min.

29. Resuspend cells in 1 ml of Medium-C supplemented with 10 μM Y-27632 by pipetting up and down 5–10 times. Add another 2 ml of Medium-C with Y-27632 to the same tube and invert the tube several times.
30. Divide the cells onto the three pre-coated wells, i.e., 1 ml per well.
31. Incubate at 37 °C for 24 h.
32. On day 6, warm (37 °C) 1 ml of Medium-C. Aspirate old medium and replace with 1 ml of Medium-C.
33. Incubate at 37 °C for 24 h.
34. On day 7, warm (37 °C) the remaining 1 ml Medium-C. Aspirate old medium and replace with 1 ml of Medium-C. Incubate at 37 °C for 24 h.
35. On day 8, prepare and warm 3 ml of complete Medium-D to 37 °C. Aspirate old medium and replace with 1 ml of Medium-D in each well.
36. Incubate at 37 °C for 24 h.
37. On day 9, prepare and warm 3 ml of the complete Medium-D to 37 °C. Aspirate old medium and replace with 1 ml of Medium-D in each well.
38. Incubate at 37 °C for 24 h.
39. On day 10, prepare and warm sufficient Medium-D as well as DMEM/F12 for passaging. Plan the experiments carefully at this moment, since splitting is not possible after this day.
40. Coat sufficient number of wells of a new 12-well plate with Matrigel as described in Subheading 3.1, **step 2**. Transfer the old culture medium to a 15 ml conical tube. Wash the cells with 1 ml of D-PBS. Aspirate wash medium and add 0.5 ml of Accutase. Incubate at 37 °C for 3–5 min.
41. Transfer the old medium back to the same well and dislodge cells by pipetting up and down 3–5 times using a pipette with a p1000 tip. Immediately transfer cells back to the 15 ml conical tube. Wash the well once with 1 ml of DMEM/F12 to collect any remaining cells and transfer to the same tube.
42. Centrifuge the tube at 300 × *g* for 3–5 min. Resuspend cells in 1 ml of complete Medium-D by pipetting up and down 5–10 times. Add pre-warmed Medium-D to the same tube. Divide the cells onto the pre-coated 12-well plate. For this passage step, the split ratio is usually around 1:5 but varies depending on the cell line. Adjust the ratio so that the cells are approximately 70–80% confluent the next day.

43. Incubate at 37 °C for 24 h.
44. During the next 10 days, the cells will be cultured in maturation Medium-E, which will be replaced every other day. Complete maturation Medium-E is stable for 2 weeks if properly stored at 4 °C. Warm (37 °C) only the volume of Medium-E required for each medium replacement.

3.3 Infection of iHeps with HCV

1. Cell culture-grown HCV (HCVcc) infection: HCVcc particles are produced in Huh-7.5 cells as previously described [18]. iHeps are incubated with virus inoculum for 4–6 h at 37 °C and then thoroughly washed three times with DMEM/F12. After the washes, infected iHeps are kept in maturation Medium-E. Infections are usually assessed 48–60 h post-media change depending on the assay.
2. iHep cell lysates can be harvested for quantification of HCV RNA genome copies and/or virus protein detection. HCV infections can be monitored in live iHeps using the RFP-NLS-IPS reporter system [19] (Fig. 2a). We usually transduce the iHeps at day 11 of the differentiation protocol with lentiviral vectors encoding the reporter and perform HCV infections at day 13 (*see* **Note 5**). When using Gaussia luciferase (GLuc) reporter virus for infections, harvest iHep supernatant at different time points and store at −20 °C for further analysis (*see* **Note 6**).
3. HCV isolate infection: One of the main advantages of using iHeps over hepatoma cells is their ability to support infection of HCV clinical isolates [7]. For infection, we usually make a serial dilution of the patient serum with maturation Medium-E, ranging from 1:5 to 1:100, and incubate with iHeps overnight. iHeps are then thoroughly washed and kept in Medium-E. Similar to HCVcc, infections are usually assessed 60–72 h post-medium change by immunoblotting and/or the RFP-NLS-IPS reporter system (Fig. 2b) (*see* **Note 7**).

3.4 Genetic Editing Using iCRISPR-Cas9

1. Generation of hPSCs with inducible Cas9 expression (iCas9): Based on a recent study [17], we generated iCas9-hPSCs by TALEN-mediated gene targeting into the "safe harbor" of *AVVS1* (also known as *PPP1R12C*) locus. In the iCas9-hPSC lines, Cas9 expression is induced by doxycycline (Dox) treatment. Single guide RNAs (sgRNA) are designed using the CRISPR design tool developed by Dr. Feng Zhang's group at MIT (http://crispr.mit.edu/). sgRNAs are cloned into a target vector, amplified, and in vitro transcribed following conventional protocols. Here we describe a stepwise protocol to generate knockout cell lines. If the knock-in of a specific mutation is desired, a single-stranded oligo DNA nucleotide with the

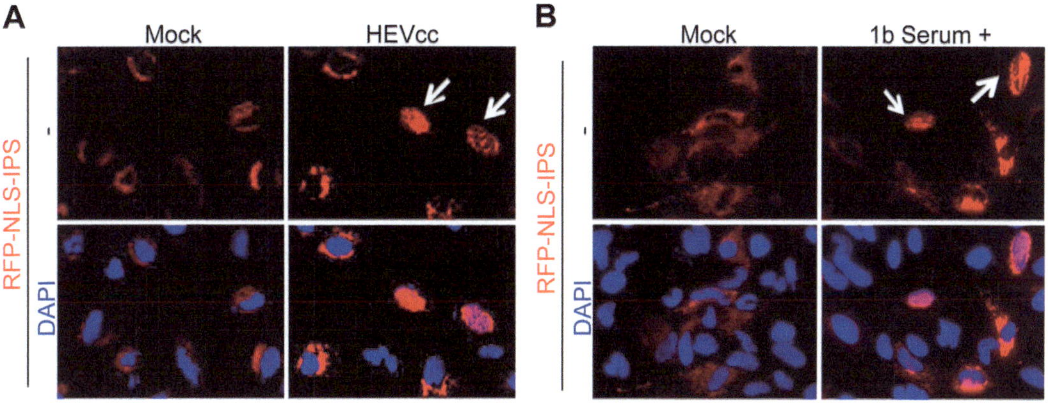

Fig. 2 Infection of iHeps with cell culture-grown HCV and HCV clinical isolates. (**a**) Fluorescence images of iHeps expressing RFP-NLS-IPS in the absence (Mock) or presence of JFH-1-based cell culture-grown HCV (HCVcc). (**b**) Fluorescence images of iHeps expressing RFP-NLS-IPS in the absence (Mock) or presence of genotype 1b HCV serum (1b Serum+). White arrows indicate infection events as visualized by translocation of the RFP-NLS-IPS reporter (red) into cell nuclei, stained with DAPI (blue). Reproduced from [7] with permission from PLoS Pathogens

sequence of interest should be co-transfected as DNA repair template.

2. Transfection of iCas9-hPSC with single or multiple sgRNAs: iCas9-hPSC cells are maintained in mTeSR1 medium, as detailed in Subheading 3.1. It is very critical to start with a high-quality cell culture for generating knockout lines. Cells are ready for passage when cultures are approximately 70% confluent.
3. On day 0, split cells and plate onto two Matrigel-coated wells of a 6-well plate, at a ratio of 1:8. Replace the medium with warm mTeSR1 daily.
4. On day 3, prepare 6 ml of Cas9-induction medium. Warm 3 ml of the induction medium to 37 °C. Keep the remaining induction medium at 4 °C in the dark. Aspirate old medium and replace with 1.5 ml of Cas9-induction medium.
5. On day 4, replace the old medium with 1.5 ml of new induction medium. Coat four wells with Matrigel of a 24-well plate as described in Subheading 3.1, **step 2**.
6. On day 5, 2 days post-Dox treatment, prepare 4–6 ml of the transfection medium by supplementing mTeSR1 with 2.5 μg/ml of Dox and 10 μM Y-27632. Warm 2 ml of transfection medium and keep the remaining medium at 4 °C. Warm sufficient volumes of DMEM/F12 and Accutase.
7. Prepare transfection mixes in two tubes (*see* **Note 8**): Mix 1: 100 μl Opti-MEM + 3 μl sgRNA (1000 ng); Mix 2: 100 μl

Opti-MEM + 6 μl Lipofectamine RNAiMAX. Incubate at room temperature for 5 min, then add mix 2 to 1, and incubate for 5–10 min. The final concentration of the sgRNA is 10 nM.

8. During the incubation, dissociate hPSC into single cells with Accutase. Count cells, transfer 1.5–1.8 × 10^5 cells to a new 15 ml conical tube, and centrifuge at 300 × *g* for 3–5 min.
9. Resuspend cells with 2 ml of transfection medium and plate cells into four-coated wells of the 24-well plate. Add 50 μl of the transfection mix dropwise into each of the four wells. Shake the plate gently back and forth, left and right.
10. Incubate the plate at 37 °C.
11. Replace the medium at 4 h post transfection with fresh transfection medium.
12. On day 6, perform a secondary transfection following the same protocol.
13. Preparation of mouse embryonic fibroblasts (MEFs) feeder cells: On day 7, plate irradiated MEFs in three wells of a 6-well plate with MEF-feeder medium. A wide range of MEF plating densities (ranging from 2–5 × 10^4 cells/cm^2) is used for human pluripotent stem cell culture. The density depends on the iPSC lines you are using. Therefore, we highly recommend you to optimize the density for your specific stem cell line. Transfer the plate with MEFs into a 37 °C incubator (*see* **Note 9**).
14. On day 8, prepare hPSC-feeder medium. The complete medium is stable for 2 weeks if being stored properly at 4 °C. Warm up to 37 °C only the volume of hPSC-feeder medium required for each medium replacement.
15. Warm (37 °C) sufficient volumes of DMEM/F12 and Accutase. Prepare 5 ml of first-day hPSC-feeder medium by adding Y-27632 to reach a final concentration of 10 μM and warm it to 37 °C. Dissociate cells from three wells of the 24-well plate into single cells, using Accutase supplemented with 10 μM Y-27632. Determine the number of live cells using Trypan blue.
16. Transfer 1.5–2.5 × 10^3 cells to a new 15 ml conical tube and centrifuge the tube at 300 × *g* for 3–5 min. The seeding density varies depending on the cell lines that are used. Therefore we recommend seeding cells at different densities to achieve an optimal separation between different colonies.
17. Resuspend cells with 4.5 ml of first-day hPSC-feeder medium and plate cells onto the three wells pre-seeded with irradiated MEFs. Shake the plate gently back and forth, left and right.
18. Place the plate into 37 °C incubator. Replace the medium daily.

19. Cell colony expansion: The cells are allowed to grow until colonies reach 1.5–2 mm in diameter. This usually takes about 10–14 days with daily medium changes. Single cell colonies are picked and plated on irradiated MEFs seeded in a 24-well plate in hPSC-feeder medium supplemented with 10 μM Y-27632. Once confluency is reached, which takes about 5–7 days, the cells will be detached with ReLeSR and replated in one well of a 12-well plate pre-coated with Matrigel.
20. hPSC clones are ready for further expansion into 6-well plates when cultures are approximately 60–70% confluent. Once this confluency is reached, detach the cells using ReLeSR and freeze down half of the cells with hPSC-freezing medium. The other half can be divided into two parts and stored at −80 °C for downstream analysis, i.e., colony screening by Western blot or target site sequencing (*see* **Note 10**).
21. Cell colony screening: We routinely screen for mutant clones through direct sequencing of PCR products. This procedure involves a simple protocol to extract genomic DNA, a PCR to amplify the regions of interest, subcloning into a TOPO vector, and sequencing. Alternatively, hPSC clones can be screened using Western blot analysis, when the corresponding protein of interest is expressed to a level that can be easily detected. When expression is low or a SNP is introduced, a surveyor nuclease assay to facilitate the screen should be employed.

4 Notes

1. If the Matrigel-coated plates will not be used immediately after preparation, add an additional 1 ml of DMEM/F12 medium to each well to prevent drying. Store plates in a 37 °C incubator or wrapped in parafilm at 2–8 °C. Both storage techniques require use of the plates within 7–10 days after preparation. If stored at 2–8 °C, plates should be warmed before use. If any portion of the well dries out, do not use the well.
2. The incubation time varies depending on the cell line; therefore dissociation should be monitored under the microscope until the optimal time is determined.
3. Adjust density for different cell lines, so that the cells are approximately 90–100% confluent the next day.
4. Very consistent and reproducible definitive endoderm can also be generated using the STEMdiff Definitive Endoderm Kit (Stemcell Technologies) following the manufacturer's instructions.

5. Due to autofluorescence properties of hepatocytes, we find that the RFP-NLS-IPS reporter gives much cleaner signals compared to the EGFP-NLS-IPS reporter.
6. Typically iHeps are susceptible to HCV infection from day 11 to 16 of the differentiation protocol [7]. To increase infection efficiency and to remove residual FBS in the HCVcc inoculum, viral particles can be purified and concentrated with PEG-8000 precipitation. Of note, it is usually not necessary to remove the residual FBS for infections at a mature iHep stage. Highly efficient infection can be achieved through ablation of innate immunity [20].
7. Supplementing the medium with epidermal growth factor (EGF, [21]) to a final concentration of 100 ng/ml during virus inoculum incubation enhances infection efficiency.
8. If multiplexed sgRNA transfection is desired, divide the total amount of 250 ng of transfected RNA per well by equal division between the sgRNAs.
9. MEF feeder cells should be plated 24 h prior to plating hPSCs. MEFs should not be used for longer than 10–14 days.
10. ReLeSR will detach both, the hPSC cells and MEFs. After replating into Matrigel-coated 12-well plate, some of the irradiated MEFs will survive and attach to the plate. However, during the expansion from 12-well to 6-well plates, these MEFs will be completely eliminated from the hPSC cells. We usually do not collect cells for colony screening unless all the MEFs have been removed from the hPSC cells, since the remaining MEFs may interfere with the analysis.

References

1. Goossens N, Hoshida Y (2015) Personalized management of hepatocellular carcinoma based on molecular information: future prospects. Clin Liver Dis 5:132–135
2. Conti F, Buonfiglioli F, Scuteri A, Crespi C, Bolondi L, Caraceni P et al (2016) Early occurrence and recurrence of hepatocellular carcinoma in HCV-related cirrhosis treated with direct-acting antivirals. J Hepatol 65:727–733
3. Reig M, Marino Z, Perello C, Inarrairaegui M, Ribeiro A, Lens S et al (2016) Unexpected high rate of early tumor recurrence in patients with HCV-related HCC undergoing interferon-free therapy. J Hepatol 65:719–726
4. Asselah T, Bieche I, Sabbagh A, Bedossa P, Moreau R, Valla D et al (2009) Gene expression and hepatitis C virus infection. Gut 58:846–858
5. Heim MH (2013) 25 years of interferon-based treatment of chronic hepatitis C: an epoch coming to an end. Nat Rev Immunol 13:535–542
6. Patin E, Kutalik Z, Guergnon J, Bibert S, Nalpas B, Jouanguy E et al (2012) Genome-wide association study identifies variants associated with progression of liver fibrosis from HCV infection. Gastroenterology 143:1244–1252
7. Wu X, Robotham JM, Lee E, Dalton S, Kneteman NM, Gilbert DM et al (2012) Productive hepatitis C virus infection of stem cell-derived hepatocytes reveals a critical transition to viral permissiveness during differentiation. PLoS Pathog 8:e1002617
8. Schwartz RE, Trehan K, Andrus L, Sheahan TP, Ploss A, Duncan SA et al (2012) Modeling hepatitis C virus infection using human

induced pluripotent stem cells. Proc Natl Acad Sci U S A 109:2544–2548

9. Shlomai A, Schwartz RE, Ramanan V, Bhatta A, de Jong YP, Bhatia SN et al (2014) Modeling host interactions with hepatitis B virus using primary and induced pluripotent stem cell-derived hepatocellular systems. Proc Natl Acad Sci U S A 111:12193–12198
10. Carpentier A, Tesfaye A, Chu V, Nimgaonkar I, Zhang F, Lee SB et al (2014) Engrafted human stem cell-derived hepatocytes establish an infectious HCV murine model. J Clin Invest 124:4953–4964
11. Paganelli M, Dallmeier K, Nyabi O, Scheers I, Kabamba B, Neyts J et al (2013) Differentiated umbilical cord matrix stem cells as a new in vitro model to study early events during hepatitis B virus infection. Hepatology 57:59–69
12. Roelandt P, Obeid S, Paeshuyse J, Vanhove J, Van Lommel A, Nahmias Y et al (2012) Human pluripotent stem cell-derived hepatocytes support complete replication of hepatitis C virus. J Hepatol 57:246–251
13. Yoshida T, Takayama K, Kondoh M, Sakurai F, Tani H, Sakamoto N et al (2011) Use of human hepatocyte-like cells derived from induced pluripotent stem cells as a model for hepatocytes in hepatitis C virus infection. Biochem Biophys Res Commun 416:119–124
14. Kaneko S, Kakinuma S, Asahina Y, Kamiya A, Miyoshi M, Tsunoda T et al (2016) Human induced pluripotent stem cell-derived hepatic cell lines as a new model for host interaction with hepatitis B virus. Sci Rep 6:29358
15. Pan-Genotype Hepatitis E Virus Replication in Stem Cell-Derived Hepatocellular Systems
16. Takahashi K, Tanabe K, Ohnuki M, Narita M, Ichisaka T, Tomoda K et al (2007) Induction of pluripotent stem cells from adult human fibroblasts by defined factors. Cell 131:861–872
17. Gonzalez F, Zhu Z, Shi ZD, Lelli K, Verma N, Li QV et al (2014) An iCRISPR platform for rapid, multiplexable, and inducible genome editing in human pluripotent stem cells. Cell Stem Cell 15:215–226
18. Lindenbach BD, Meuleman P, Ploss A, Vanwolleghem T, Syder AJ, McKeating JA et al (2006) Cell culture-grown hepatitis C virus is infectious in vivo and can be recultured in vitro. Proc Natl Acad Sci U S A 103:3805–3809
19. Jones CT, Catanese MT, Law LM, Khetani SR, Syder AJ, Ploss A et al (2010) Real-time imaging of hepatitis C virus infection using a fluorescent cell-based reporter system. Nat Biotechnol 28:167–171
20. Zhou X, Sun P, Lucendo-Villarin B, Angus AG, Szkolnicka D, Cameron K et al (2014) Modulating innate immunity improves hepatitis C virus infection and replication in stem cell-derived hepatocytes. Stem Cell Reports 3:204–214
21. Lupberger J, Zeisel MB, Xiao F, Thumann C, Fofana I, Zona L et al (2011) EGFR and EphA2 are host factors for hepatitis C virus entry and possible targets for antiviral therapy. Nat Med 17:589–595

Part III

Methods for the Study of HCV Life Cycle

Chapter 9

Molecular Identification of Transmitted/Founder Hepatitis C Viruses and Their Progeny by Single Genome Sequencing

Hui Li, Muhammad N. Zahid, Shuyi Wang, and George M. Shaw

Abstract

Chronic hepatitis C virus (HCV) infection exists as a complex mixture of genetically distinct viruses, commonly referred to as a "quasispecies." Quasispecies complexity can vary substantially during the course of natural infection as a consequence of viral population "bottlenecking." This occurs at the time of transmission from one individual to the next and during the course of chronic infection of an individual when adaptive immune responses eliminate certain viruses but allow others to escape and expand. Antiviral treatment with drugs that fail to eradicate virus can also lead to virus population bottlenecks and emergence of drug-resistant variants. Single genome sequencing (SGS) combined with mathematical modeling and phylogenetic inference is a recently described approach for characterizing the HCV quasispecies in unprecedented detail, allowing for the first time the retention of genetic linkage across genes and near full-length genomes and precise identification of transmitted/founder (T/F) genomes. Here, we describe the methodological approach to SGS and show how this strategy allows for the precise and unambiguous molecular identification of transmitted viruses as well as those that repopulate the body after drug or immune-mediated selective sweeps. This is an enabling experimental strategy that allows for a precise genetic, biologic, and antigenic characterization of HCV viruses that are responsible for transmission and persistence. Such an approach can be particularly valuable to future HCV vaccine design efforts, as it has been for human immunodeficiency virus type 1 (HIV-1).

Key words Single genome sequencing, HCV transmission, HCV persistence, HCV escape

1 Introduction

Hepatitis C virus (HCV) infects nearly 3% of the world's population and exhibits extraordinary genetic diversity reaching as much as 30% at the nucleotide level [1]. HCV replication depends on a virally encoded error-prone RNA-dependent RNA polymerase (RdRp) [2]. As a consequence of the extraordinary numbers of virions produced daily ($\leq 10^{12}$/day), the exceedingly short plasma viral half-life (<1 h), an RdRp error rate of 2.5×10^{-5} per nucleotide per genome replication [3], and significant virus-specific adaptive cellular and humoral immune responses, HCV generally exists as a complex mixture of genetically distinct variants in chronically

Mansun Law (ed.), *Hepatitis C Virus Protocols*, Methods in Molecular Biology, vol. 1911,
https://doi.org/10.1007/978-1-4939-8976-8_9, © Springer Science+Business Media, LLC, part of Springer Nature 2019

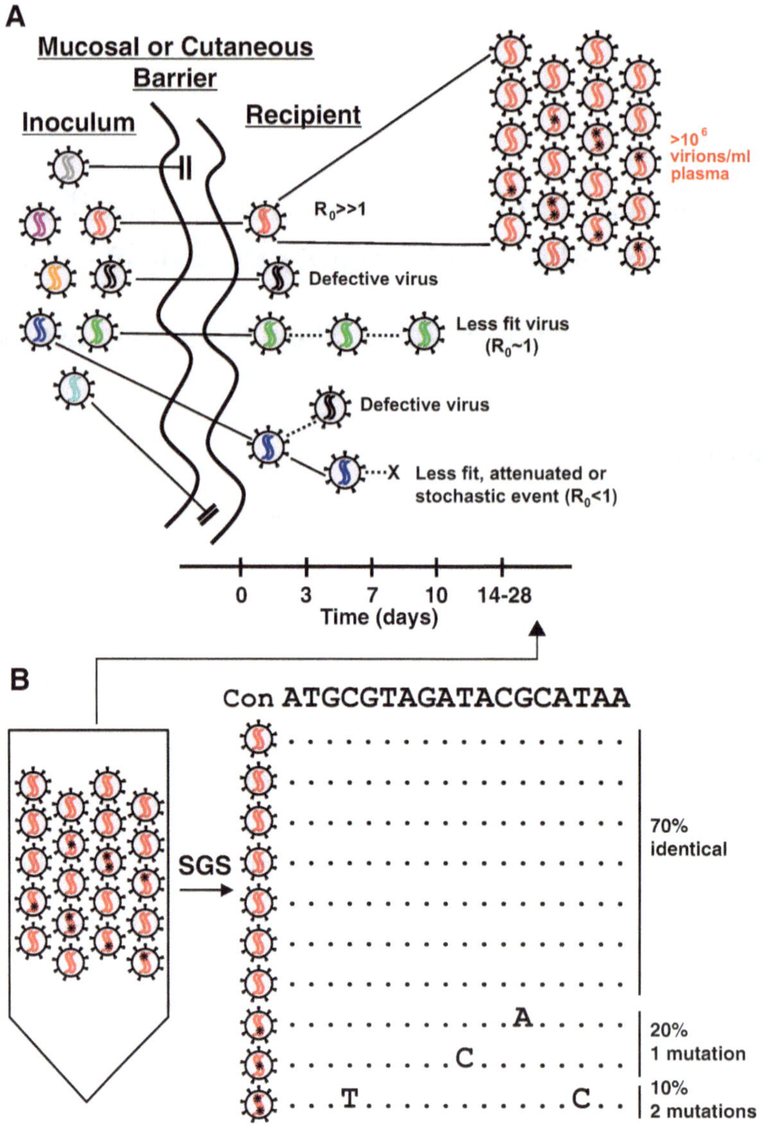

Fig. 1 HCV transmission model and single genome sequencing strategy. (**a**) Schematic illustration of HCV transmission bottleneck showing a complex genetically diverse quasispecies in the donor and much more restricted genetic diversity in an acutely infected recipient. Asterisks overlying viral RNA genomes denote single base-pair substitutions. (**b**) Sequences generated by SGS from plasma virion RNA proportionally represent virus present in plasma

infected individuals (Fig. 1). In contrast, HCV typically exhibits a stringent population bottleneck at or near the moment of transmission, since infection usually results from the acquisition of only a very limited number of viruses that expand exponentially and with essentially random early mutational events (Fig. 2) [4–6]. The most precise quantitative estimates of the transmission bottleneck to date

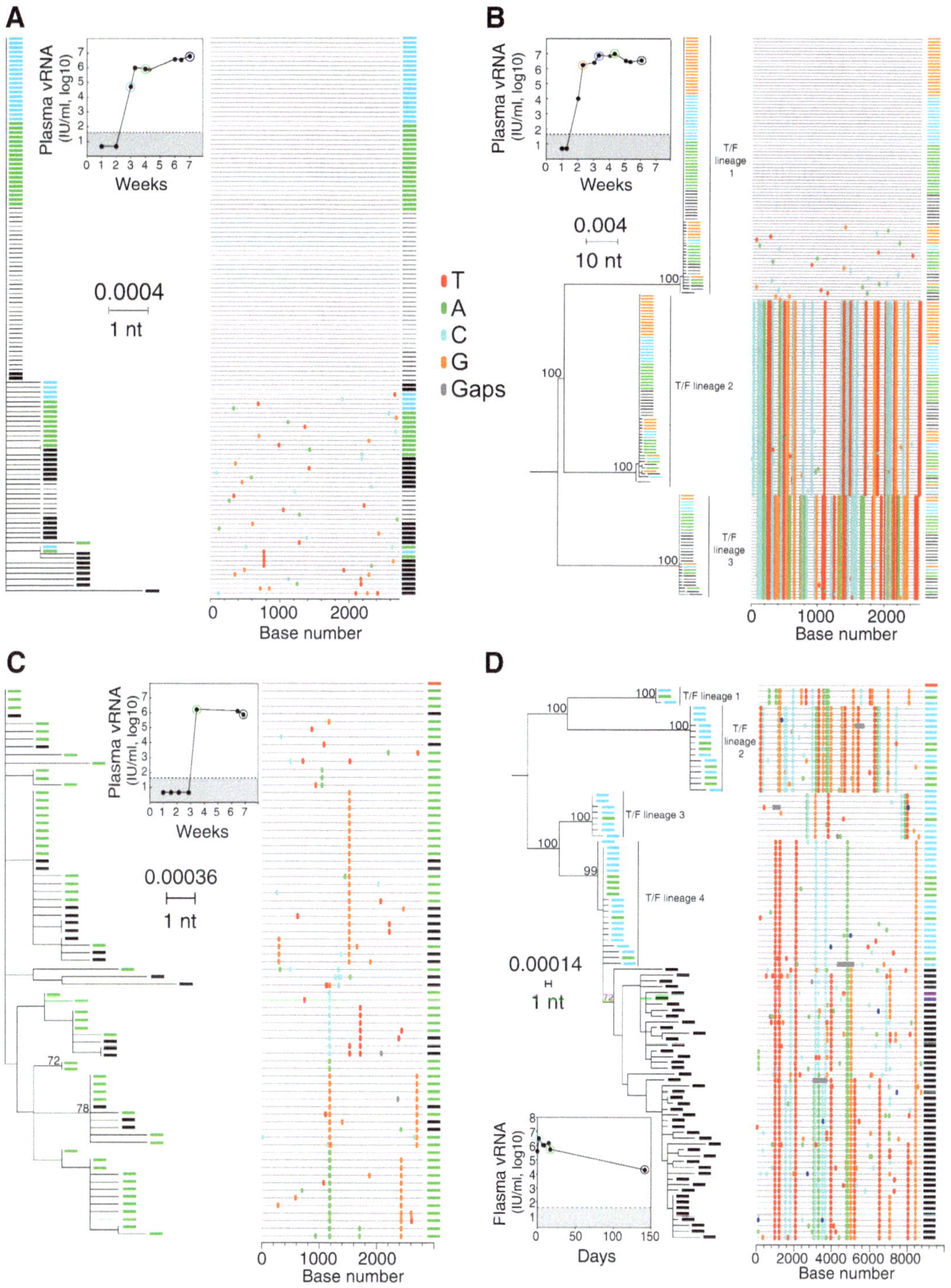

Fig. 2 Single genome HCV sequences from acutely infected human subjects. Shown are viral load kinetics, maximum likelihood phylogenetic trees, and *Highlighter* plots from four subjects previously reported [6, 25]. The time points sequenced are circled in different colors corresponding to color-coded sequences. (**a**) Single-variant transmission. (**b**) Transmission of three genetically diverse variants resulting in sequences comprising T/F lineages 1–3. (**c**) Acute-to-acute transmission of multiple closely related variants. (**d**) Transmission of four genetically diverse variants resulting in sequences comprising T/F lineages 1–4 followed by elimination of lineages 1–3 and evolution of lineage 4

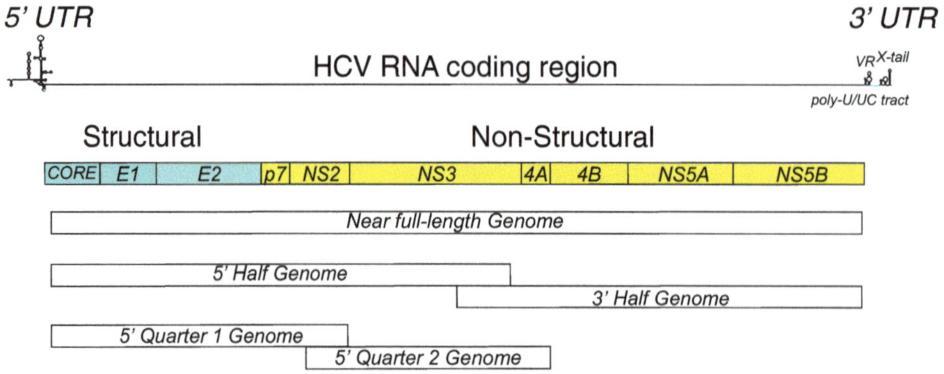

Fig. 3 Diagram of HCV RNA, gene organization of structural and nonstructural coding regions, and amplification strategy of 5′ half, 3′ half, 5′ quarter, and near full-length genomes

have been based on an approach first applied to HIV-1 that uses single genome sequencing (SGS), mathematical modeling, and phylogenetic inference to identify discrete transmitted viral sequence lineages [3, 6–11]. This strategy allows for unambiguous identification and enumeration of HIV-1 and HCV T/F viral genomes responsible for establishing productive clinical infection [8, 12, 13] and is based on two considerations: The first involves the SGS method itself. In SGS, cDNA templates are endpoint diluted prior to PCR amplification, which ensures that artifactural recombination (template switching) does not occur between heterotemplates. Next, uncloned DNA amplicons are directly sequenced without an intermediate cloning step, which ensures that *Taq* polymerase errors are not represented in finished sequences. Together, these methods ensure that SGS-derived sequences accurately and proportionately reflect vRNA sequences in the plasma (or other tissue or medium) while retaining genetic linkage across whole genes or genomes (Fig. 3) [8, 9]. The second consideration is the extremely short in vivo life span of plasma HIV-1 and HCV and of productively HIV-1-infected cells and HCV replication complexes in hepatocytes [8, 14–16]. This allows for plasma vRNA to act as a sensitive, real-time measure of viral genomes that are actively replicating in cells and tissues. Mathematical models underlying this strategy to infer T/F genomes assume a virus population evolving from each T/F genome that replicates exponentially infecting R_0 new cells at each generation with no selection pressure at variable sites and a constant mutation rate across lineages. Under these conditions, the frequency distribution of the Hamming distances (HD; the pairwise distance between two genetic virus strains) follows a Poisson distribution. If the expansion is sufficiently rapid, sequences of each lineage exhibit a star-like phylogeny, and the consensus sequence of each lineage corresponds to the actual sequence of T/F virus that was responsible for establishing productive clinical infection (Figs. 1 and 2) [8, 14–19].

To underpin this strategy for HIV-1, a mathematical model and Monte Carlo (MC) forward simulation of viral early evolution was developed aiming to quantitatively assess whether infections were established by single or multiple viruses. This model enables an assessment of the expected sequence diversity originating from a single T/F virus, estimation of the most recent common ancestor (MRCA) of the T/F viral lineage, and estimation of the time to MRCA. It also calculates the probability of the MRCA being the transmitted virus or an evolved variant [13]. The good agreement between model simulations and results from patient data [8] confirmed the model assumptions of random diversification of single T/F virus during acute infection [8, 12, 13].

Single genome sequencing strategies also enabled the identification of full-length T/F genomes and characterization of their subsequent evolutionary pathways, including population bottlenecks driven by host immune responses or antiviral drugs. It made possible for the first time a genome-wide analysis of cytotoxic T-cell responses against the complete infecting HIV-1 proteome [20] and provided a sensitive way to assess early low-titer neutralizing antibody responses against HIV-1 envelope [21]. It also allowed for dynamic assessment of the composite immune pressures exerted by both T and B cells on the T/F HIV-1 proteome and the evolving virus quasispecies [21, 22].

Because of the distinctive replication strategy of HCV, we developed two complementary mathematical models to analyze HCV sequence diversity [6]. The first model was the random evolution model employed to analyze HIV-1 diversification. Our studies demonstrated that during early infection, HCV genomes generally evolved according to this simple model of random evolution where the coalescent corresponded to the T/F sequence [3, 6, 7]. However, HCV replication differs from HIV-1 in multiple steps in the viral life cycle. Instead of integrating into chromosomal DNA and producing daughter progeny after a large burst of infected cell, HCV RNA forms as many as 40 cytosolic replication complexes (RC) that continue to produce viruses throughout the lifetime of the cell. The long-lived hepatocytes prior to the onset of HCV-specific cellular immune responses could result in accumulation of RCs from various generations. Unlike HIV-1 progeny that are produced only from the latest generation, HCV progeny are produced from replication complexes of all generations [3, 6, 23, 24]. These differences could lead to HCV diversification that may deviate from a star-like phylogeny and Poisson distribution of mutations in spite of similar error rates between HCV RdRp and HIV-1 RT [3, 6, 8]. To account for these differences, we developed an alternative model of HCV diversification that predicts occasional violations of previous model prediction with increasing probability as time goes on. The latter model calculation expects that HCV sequence diversity is basically random during virus linear growth phase but saturates when the viral load stabilizes and predicts about three times as many shared stochastic mutations in HCV compared to HIV-1 infection which could lead to more common deviations from a star-like phylogeny and Poisson distribution of mutations [6, 8]. This alternative model provided us with a mathematical and statistical basis for distinguishing closely related T/F HCV lineages from sequences that evolved from a single genome but shared early stochastic mutations. Based on this model, we also developed a phylogenetically based clustering method yielding a more conservative estimate of T/F HCV genome numbers [6]. Maximum likelihood (ML) phylogenetic analyses of sequences from sequential time points from acutely HCV-infected subjects revealed variable patterns of transmission and diversification [6]. Figure 2a depicts a typical pattern of HCV early diversification. Sequences sampled from sequential time points spanning early to late stage of acute infection formed a monophyletic discrete low-diversity lineage in the phylogenetic tree and exhibited random mutations in the *Highlighter* plot. Sequences sampled from the early exponential growth phase showed high identity or very close similarity and conformed well to the predictions of the simple model for Poisson distribution and star phylogeny of mutations. Sequences from later time points revealed that the percentage sequence identity decreased over time and shared mutations became evident at later time points but still conformed well to the alternative model prediction of single virus diversification. We thus concluded this pattern of evolution represents single virus transmission and the consensus sequence of the discrete lineage corresponds to the actual T/F genome [6]. Figure 2b reveals another pattern of sequence diversification, which exhibited three discrete low-diversity lineages with high inter-lineage diversity. Combined sequences from all lineages violated both model predictions, but when each sub-lineage as analyzed individually, it conformed to the model; we thus concluded the consensus sequence of each low-diversity lineage corresponds to a unique T/F HCV genome. Estimates of the numbers of T/F genomes are minimum estimates, limited by the sampling depth of SGS [6, 8]. Fig. 2c shows sequences from an acutely infected subject who exhibited many discrete low-diversity lineages with a small number of shared unique mutations, which violated model predictions of diversification from single T/F virus. We hypothesized that the multiple discrete lineages observed in this subject resulted from multiple virus acquisition from a donor individual who was recently infected by a single virus or by closely related viruses. Figure 2d shows an example of multivariant transmission (four T/F genomes) followed over a period of 5 months where the viral progeny of T/F genomes 1–3 were eliminated altogether leaving only the progeny of T/F virus 4 [6]. In the study of epidemiologically linked acute-to-acute HCV transmission pairs, we confirmed that inferred T/F sequences in recipients were identical to viral sequences in their respective donors [25].

Based on the SGS approach, we were able to generate full-length T/F HCV molecular clones containing variable-length poly(U/UC) tracts from genotypes 1a, 1b, 3a, and 4a. This was carried out by SGS of acute infection plasma vRNA to amplify large overlapping 5′ and 3′ half-gene segments that encode for entire structural and nonstructural polyprotein region (Fig. 3), followed by adapter-primed 5′- and 3′-terminal amplifications and a "bridging" SGS method that incorporated a molecular cloning step to span the poly(U/UC) tract. These full-length T/F clones contain all the essential viral elements that are necessary and sufficient for productive infection in human liver cells in vivo [7].

Most recently, we have used SGS to characterize HCV persistence and bottlenecking in subjects who developed broadly neutralizing antibodies and cleared viremia [26] and in other subjects treated with a potent HCV protease inhibitor [27]. In the former study, we found broadly neutralizing antibodies with few somatic mutations leading to hepatitis C virus clearance. We observed that bNAbs from two HCV-infected humans bound the same HCV Env epitope, were encoded by the same germline heavy chain variable-gene segment, and were encoded by antibody variable genes with sparse somatic mutations. For one of the most potent bNAbs, these somatic mutations were shown to be critical for antibody neutralizing breadth and for binding to autologous envelope variants circulating late in infection. However, somatic mutations were not necessary for binding of the bNAb unmutated ancestor to envelope proteins of early autologous transmitted/founder viruses. This study thus identified a public B-cell clonotype favoring early recognition of a conserved HCV epitope, proving that anti-HCV bNAbs can achieve significant neutralizing breadth with relatively few somatic mutations, and identified HCV envelope variants by SGS that favored selection and maturation of an anti-HCV bNAb in vivo. These data, based on SGS of full-length E1E2 and 5′ half-genome sequences, provided insight into the molecular mechanisms of immune-mediated clearance of HCV infection and a molecular road map to guide the development of vaccine candidates capable of stimulating anti-HCV bNAbs with a physiologic number of somatic mutations characteristic of vaccine responses. In the latter study, we used SGS to reveal HCV superinfection and cure of infected cells as novel mechanisms for hepatitis C virus expansion and persistence. We probed hepatitis C virus persistence by analyzing clinical samples taken from subjects who were treated with a novel protease inhibitor MK5172. Frequent longitudinal viral load determinations and SGS analyses revealed rapid antiviral resistance development and, surprisingly, dynamic turnover of dominant drug-resistant mutant populations long after treatment cessation. We fitted mathematical models to the data, and the results provided strong support for critical roles that superinfection and cure of infected cells play in facilitating the rapid turnover and persistence of viral populations. Based on these SGS and modeling analyses, we proposed a new theoretical framework integrating viral and molecular mechanisms to explain rapid HCV evolution, resistance, and persistence despite antiviral treatment and host immune responses. These studies highlight the kinds of novel insights that can be gleaned from SGS-based analysis of HCV plasma viral RNA.

2 Materials

2.1 RNA Extraction

1. This section provides two options for Viral RNA Extraction:
   (a) Using the Qiagen BioRobot EZ1 Workstation with EZ1 Virus Mini Kit v2.0 (Qiagen).
   (b) Using the QIAamp Viral Mini Kit (Qiagen).
2. Ethanol (70%).
3. Nuclease-free water (Ambion).

2.2 Reverse Transcription

1. Superscript III Reverse Transcriptase (Invitrogen).
   (Components: SSIII RT (200 U/μL), 5× First-Strand Buffer, 0.1 M DTT).
2. 10 mM dNTP mix (Invitrogen).
3. RNase OUT Recombinant Ribonuclease inhibitor (Invitrogen).
4. RNase H (Invitrogen).
5. HCV-specific reverse transcription primers (*see* Table 1).
6. Nuclease-free water (Ambion).

2.3 PCR Amplification of HCV-Specific Sequence

1. Platinum Taq DNA Polymerase High Fidelity (Invitrogen).
   (Components: HiFi Platinum Taq, 10× High Fidelity PCR buffer, 50 mM Magnesium Sulfate).
2. 10 mM dNTP mix (Invitrogen).
3. HCV-specific PCR primers (*see* Table 1).
4. MicroAmp Optical 96-well reaction plate (Applied Biosystems).
5. MicroAmp Clear Adhesive Film (Applied Biosystems).
6. MicroAmp PCR strip.
7. E-Gel Electrophoresis System (Invitrogen) (Components: Electrophoresis bases and E-Gel 96 1% Agarose).
8. E-Gel High Range DNA Marker (Invitrogen).

Table 1
Oligonucleotide primers for the amplification of HCV sequences

Geno-type	Genome[a]	Primer Name	Usage[b]	Position (nt)	Orienta-tion	Sequence (5′–3′)
1a	near full-length	1aCORE-F1	1	342–368	F	ATGAGCACGAATCCTAAACCTCAAAGA
		1aCORE-F2	2	362–391	F	TCAAAGAAAAACCAAACGTAACACCAACCG
		1aNS5B.R1	1 & RT	9354–9377	R	TCATCGGTTGGGGAGGAGGTAGAT
		1aNS5B.R2	2	9349–9374	R	TCGGTTGGGGAGGAGGTAGATGCCTA
	5′ half	1aCORE-F1	1	342–368	F	ATGAGCACGAATCCTAAACCTCAAAGA
		1aCORE-F2	2	362–391	F	TCAAAGAAAAACCAAACGTAACACCAACCG
		1aNS4A-R1	1 & RT	5451–5474	R	GCACTCTTCCATCTCATCGAACTC
		1aNS4A-R2	2	5297–5319	R	AGGTGCTCGTGACGACCTCCAGG
	3′ half	1aNS3-F1	1	5037–5065	F	ACTCATATAGACGCCCACTTTCTATCCCA
		1aNS3-F2	2	5146–5168	F	CATCGTGGGACCAGATGTGGAAG
		1aNS5B.R1	1 & RT	9354–9377	R	TCATCGGTTGGGGAGGAGGTAGAT
		1aNS5B.R2	2	9349–9374	R	TCGGTTGGGGAGGAGGTAGATGCCTA
	5′ quarter 1	1aCORE-F1	1	342–368	F	ATGAGCACGAATCCTAAACCTCAAAGA
		1aCORE-F2	2	362–391	F	TCAAAGAAAAACCAAACGTAACACCAACCG
		1aNS2.R1	1 & RT	3406–3433	R	GCCGTGATGGGCGCCAGCAACCTCCACC
		1aNS2.R2	2	3327–3354	R	GGGCAGAGACGGGCAAGCCGTTGATGAT
	5′ quarter 2	1aNS2.F2	1	2983–3008	F	ACCCGCTTTGGTATTTGACATCACC
		1aNS2.F3	2	3063–3084	F	AAAGTGCCCTACTTGTGCGCGT
		1aNS4B.R3	1 & RT	5634–5662	R	TATTGTATCCCACTGATGAAGTTCCACAT
		1aNS4B.R4	2	5516–5539	R	AGGGCCTTCTGCTTGAACTGCTC
1b	5′ half	1bCORE-F1	1	342–368	F	ATGAGCACGAATCCTAAACCTCAAAGA
		1bCORE-F2	2	362–391	F	TCAAAGAAAAACCAAACGTAACACCAACCG
		1bNS4A-R1	1 & RT	5451–5474	R	GCACTCTTCCATCTCATCGAACTC
		1bNS3-R2	2	5295–5318	R	GGTGCTCGTGACGACCTCCAGGTC
	3′ half	1bNS3.F1	1	4841–4861	F	CCTTCGGGCATGTTCGATTCCTC
		1bNS3.F2	2	4878–4899	F	TCCTCGGTCCTGTGTGAGTGCTATGA
		1bNS5B.R1	1 & RT	9354–9377	R	TCATCGATTGGGGAGCAGGTAGAT
		1bNS5B.R2	2	9325–9344	R	TACAGAAAGTAGGAGTAGGC

(continued)

Table 1
(continued)

Geno-type	Genome[a]	Primer Name	Usage[b]	Position (nt)	Orienta-tion	Sequence (5′–3′)
3a	5′ half	3a COR-F1	1	342–368	F	ATGAGCACACTTCCTAAACCTCAAAGA
		3a COR-F2	2	362–391	F	TCAAAGAAAAACCAAAAGAAACACCATCCG
		3aNS3-R2	1, 2 & RT	5287–5307	R	TTACTTCCAGATCAGCTGACA
	3′ half	3aNS3-F1	1	4642–4663	F	TTTGCGCTACTGACGCCCTCAT
		3aNS3-F2	2	4651–4671	F	CTGACGCCCTCATGACTGGAT
		3aNS5B-R	1, 2 & RT	9055–9074	R	GTACCCCAAGTTTCCTGAGG
4a	5′ half	4aCOR-F1	1	342–368	F	ATGAGCACACTTCCTAAACCTCAAAGA
		4aCOR-F2	2	362–391	F	TCAAAGAAAAACCAAACGTAACACCAACCG
		4aNS3-R2	1, 2 & RT	5299–5319	R	ACGTGCTCGTCACTACCTCGA
	3′ half	4aNS3-F1	1	4642–4663	F	TCTGCGCCACGGACGCCCTCAT
		4aNS3-F2	2	4681–4707	F	ACTTTGACTCAGTGATAGACTGCAA
		4aNS5B-R	1, 2 & RT	9312–9337	R	AGTAGGAGTAGGCACAGGAGTAAATA

[a] 5′ half—Partial Core, E1, E2, p7, NS2, and partial NS33′ half—partial NS3, NS4A, NS4B, NS5A, and partial NS5B5′ quarter 1—Partial Core, E1, E2, p7, and NS25′ quarter 2—Partial NS2, NS3, NS4A

[b] 1—first-round PCR2—second-round PCRRT—reverse transcription primer

3 Methods

3.1 RNA Extraction

1. Thaw plasma samples at room temperature (15–25 °C) and process immediately when they have equilibrated to room temperature. Microfuge briefly to ensure no infectious material is in the cap before proceeding to next step (*see* **Note 1**).
2. Store RNA elution on ice for immediate use or aliquot and freeze at −80 °C for future use.

3.1.1 Option A: Automatic Extraction Using Qiagen BioRobot EZ1 Workstation with EZ1 Virus Mini Kit

1. Clean up the EZ1 Workstation and the Biosafety hood with 70% ethanol.
2. Using pre-prepared carrier RNA stock solution (1.0 μg/μL), prepare carrier working solution by adding 3.6 μL stock solution to 56.4 μL AVE buffer or RNase-free water to total volume of 60 μL (*see* **Note 2**).
3. In 2 mL sample tubes (provided by EZ1 Virus Mini kit), normalize the 400 μL plasma sample to contain approximate 20,000 viral copies. For the high viral load samples, increase volume to 400 μL with PBS or AVE buffer. If the sample viral load is too low, concentrate by centrifugation for 1 h at 21,100 × *g*, then remove supernatant down to 400 μL and resuspend pellet.
4. Switch on the BioRobot EZ1. Set up the EZ1 worktable according to the manufacture instruction. Press "START" to start the protocol. Select 400 μL sample volume and 60 μL elution volume.

3.1.2 Option B: Manual Extraction Using QIAamp Viral RNA Mini Kit

1. Clean up the working area and microcentrifuge with 70% ethanol. Use QIAamp Viral RNA Mini Kit as written in the handbook with the following adaptations.
2. Normalize the 140 μL plasma sample to contain 20,000 viral copies. Increase volume to 140 μL with PBS. Or, if viral load is too low, concentrate by centrifugation for 1 h at 21,100 × *g*. Remove supernatant down to 140 μL and resuspend viral pellet.
3. Elute RNA from spin column with 60 μL elution buffer (stock elution buffer is diluted 1:5 in sterile RNase-free water).

3.2 Reverse Transcription

1. Preheat two heating blocks at 65 °C and 50 °C, respectively.
2. Thaw RNA samples (if frozen) and reagents for reverse transcription to room temperature. Mix each reagent by quick vortex twice and then centrifuge briefly at maximum speed to collect all the liquid to the bottom of the tube to avoid contamination.

3. Set up the reverse transcription reaction with the following volumes of components into 0.5 mL tube: 2 μL 20 μM primer, 3 μL 10 mM dNTP mix, 4 μL RNase-free H_2O, and 30 μL RNA template. The choice of primers varies by the amplification region and genotype (*see* Table 1).
4. Place reaction tube in 65 °C heat block for 5 min, remove to ice for at least 1 min. Centrifuge briefly to collect any condensation from the tube cap from heating.
5. After incubation, turn the heat block down to 70 °C for next step.
6. Add the following components to the reaction mix: 12 μL Buffer (5×), 3 μL 0.1 M DTT, 3 μL 40 U/μL RNase OUT, and 3 μL 200 U/μL Superscript III Reverse Transcriptase. Vortex twice and centrifuge briefly.
7. Incubate at 50 °C for 1 h and then increase to 55 °C for 1 h.
8. After incubation, turn the heat block down to 37 °C for next step.
9. Place reaction tube in 70 ° C heat block for 15 min to inactivate Superscript III Reverse Transcriptase.
10. Allow the reaction to cool to room temperature, then centrifuge briefly at maximum speed to collect the condensate, add 3 μL of RNaseH, and incubate at 37 °C for 20 min.
11. Store cDNA reaction on ice for immediate use or aliquot and freeze at −80 °C for future use.

3.3 PCR Amplification

1. Thaw cDNA samples (if frozen) and the reagents for PCR amplification.
2. The choice of primers varies by the amplification region and genotype (*see* Table 1). The PCR amplification mixture contents in 20 μL reaction are as follows: 2 μL 10× High Fidelity PCR buffer, 0.8 μL 50 mM Magnesium Sulfate, 0.4 μL 10 mM dNTP mix, 0.2 μL 20 μM forward primer, 0.2 μL 20 μM reverse primer, 15.28 μL RNase-free water, 0.12 μL Platinum Taq polymerase HiFi, and 1 μL of template (*see* **Note 3**).
3. Perform PCR amplification using the following conditions: First round, 94 °C for 2 min, 94 °C for 20 s, 55 °C for 30 s, 68 °C for 5.5 min, repeat **steps 2–4** for 37 more times, 68 °C for 7 min, end, and incubate at 4 °C; second round, 94 °C for 2 min, 94 °C for 20 s, 58 °C for 30 s, 68 °C for 5.5 min, repeat **steps 2–4** for 45 more times, 68 °C for 7 min, end, and incubate at 4 °C (*see* **Note 4**).
4. Analyze PCR results using high-throughput E-Gel electrophoresis system with 1% E-Gel 96 gels according to manufacture instruction. For HCV single genome amplification product,

load each sample well with mixture of 1 μL of PCR product and 9 μL of nuclease-free water and marker well with E-Gel High Range DNA Ladder.

3.3.1 cDNA Titration to Establish Appropriate Dilutions for SGS

This section describes the titration of cDNA to a point where, upon setting up multiple replicate PCR reactions with this dilution, the PCR product generated has a high probability of having been amplified from a single genome. In this case, according to Poisson distribution, if PCR positive wells constitute less than 30% of the total number of reactions, most wells contain amplicons derived from a single cDNA molecule.

1. Perform serial dilutions of cDNA to obtain <30% positivity. Dilutions of cDNA will vary from sample to sample and depend on viral load, sample integrity, amplicon size, number of free-ze–thaws, the efficiency of the RT and PCR primers, and more. At least six dilutions should be tested for each cDNA sample and do at least six replicates for each dilution. If 20,000 viral copies are extracted, the initial dilutions should be at 1:3, 1:9, 1:27, 1:81, 1:243, and 1:729 for HCV genome amplification.
2. Quickly vortex and briefly centrifuge the cDNA. Serial dilution of cDNA should be performed on ice. For example, to make serial dilutions at 1:3, 1:9, 1:27, 1:81, 1:243, and 1:729, prepare six nuclease-free tubes each with 8 μL nuclease-free water and add 4 μL of crude cDNA to the first tube to make 1:3 dilution. Mix well via pipetting and discard tip. Using clean tip transfer 4 μL of the 1:3 dilution to the next tube to make 1:9 dilution. Repeat until last dilution.
3. Set up the PCR master mix without cDNA template according to Subheading 3.3.2. In a MicroAmp Optical 96-well plate, aliquot master mix into 6 wells for each cDNA dilution and transfer 1 μL of the diluted cDNA into each well.
4. Seal the plate with MicroAmp Clear Adhesive Film tightly. Take to the PCR thermal cycler room. Spin the plates down and then run the first round PCR reaction according to Subheading 3.3. Centrifuge the first round PCR plate briefly and place in ductless recirculating fume hood after the cycles end at 4 °C.
5. In the clean room, prepare master mix without template for second round PCR on ice (*see* Subheading 3.3.2), vortex and centrifuge briefly the master mix, and pipet 19 μL of master mix into each well of MicroAmp Optical 96-well plate. Cover with MicroAmp Clear Adhesive Film lightly.
6. In PCR thermal cycler room in ductless recirculating fume hood, transfer 1 μL of first round PCR to the corresponding second round reaction using multichannel pipet (*see* **Note 5**). Seal second round PCR plate with a new film tightly and centrifuge briefly. Perform second round PCR amplification.

7. Analyze PCR results using 1% 96-well E-Gel. Determine the limiting cDNA dilution that results in <30% positivity. If too few/too many positive reactions, repeat PCR with lower/higher dilution.

3.3.2 Single Genome Amplification

1. Prepare PCR reaction master mix without cDNA template for 100 reactions (94 test + 2 negative controls + 4 extra). Add 19 μL of master mix plus 1 μL of nuclease-free water to two negative wells.
2. Based on the determined cDNA dilution (*see* Subheading 3.3.1), make 100 μL of cDNA dilution with nuclease-free water and add 98 μL to the master mix. Quickly vortex and briefly centrifuge the master mix. Distribute 20 μL to each well of the remaining 94 wells.
3. Follow Subheading 3.3.1, **steps 4–7** for the remaining steps for the first and second round PCR.
4. Select all positive wells. Sequence DNA directly and store remaining second round for future use. From all positive wells, save remaining first round in 0.5 mL tube at −80 °C for future envelope cloning.

3.4 Sequencing

This section provides two options for single genome amplicon sequencing. Inspection of individual amplicon sequencing results allows for the identification of amplicons derived from single versus multiple templates. The absence of mixed bases at each nucleotide position throughout the entire amplified gene is taken as evidence of single genome amplification from a single viral RNA/cDNA template. This quality control measure allows one to exclude from the analysis amplicons that result from PCR-generated in vitro recombination events or Taq polymerase errors and to obtain multiple individual sequences that proportionately represent those virions circulating in vivo.

3.4.1 Option A: Direct Sequencing Using BigDye Terminator Chemistry (Sanger)

Single genome amplicons are directly sequenced by cycle sequencing by using BigDye terminator chemistry (Applied Biosystems). Sequencing reaction products are analyzed with an ABI 3730xl genetic analyzer (Applied Biosystems). Both DNA strands should be sequenced by using partially overlapping fragments. Individual sequence fragments for each amplicon are assembled and edited by using the Sequencher program 5.4.1 (Gene Codes).

1. Dilute 1 μL of second round PCR product with 125 μL of nuclease-free water.
2. Per sequencing reaction add 5 μL of PCR product and 1 μL of 5 μM primer.

3. Inspect each individual chromatogram for double peaks at each nucleotide position. Exclude any sequence with ambiguities from the subsequent analyses.

3.4.2 Option B: Next-Generation Sequencing (NGS) Using MiSeq or MiniSeq (Illumina)

Single genome amplicons are sequenced by NGS platforms, MiSeq or MiniSeq (Illumina), which sequence DNA by synthesis technology. Raw reads of each amplicon are aligned de novo to generate a single contig using Geneious R10. The consensus sequence of each contig is used as the final sequence for each amplicon. Inspection of the base frequency at each nucleotide position allows for the identification of amplicons derived from single versus multiple templates and PCR errors. Any nucleotide position with more than one nucleotide base present and any at a frequency of >25% are treated as resulting from amplification of multiple vRNA/cDNA templates or PCR error and is excluded from further analysis.

1. Prepare MicroAmp strip with 9 μL of nuclease-free water in each tube.
2. Take 1 μL of second round PCR reaction from each positive well and mix well with 9 μL of nuclease-free water in the MicroAmp tube.
3. Nextera XT DNA sample preparation kit (Illumina) is used to construct libraries from PCR amplicons.
4. Sequence reads from the MiSeq/MiniSeq output will be first trimmed of the adapter sequences. Reads from each sample will then be separated by barcode.
5. Reads from each sample will be de novo assembled using SPAdes program.
6. Contigs are assembled by using Geneious R10 software. Inspect each individual contig for polymorphic sites at each nucleotide position. Exclude any sequence with single or multiple polymorphic sites from the subsequent analyses.

3.4.3 Sequence Analysis

1. Align SGS sequences using Geneious R10. Manually adjust the alignment to improve the codon alignment.
2. The alignment can be displayed with the sequence visualization tool, *Highlighter* (http://www.HIV.lanl.gov), that allows tracing of common ancestry between sequences based on individual nucleotide polymorphisms.
3. Select the best model for phylogenetic analysis by testing sequence alignment with jmodeltest-2.1.3. Generate maximum likelihood phylogenetic trees using PhyML 3.0 or other algorithms.

4 Notes

1. vRNA/cDNA should be handled in an isolated clean room to prevent contamination from double-strand PCR product or plasmid DNA. Extraction HCV RNA from plasma should always be performed under sterile biosafety cabinet in clean room with gloves, lab coat, and eye protection.
2. Dissolve the lyophilized carrier RNA (provided by EZ1 Virus Mini Kit v2.0) in 310 μL Buffer AVE, divide it into 12 μL aliquots, and store at −20 °C. Thawed aliquots of this stock solution can be stored at 2–8 °C for up to 6 weeks. Do not freeze–thaw the aliquots more than 3 times.
3. First round PCR reaction should be assembled in clean room.
4. PCR thermocyclers should be set up separately in different area than the clean room. Extension time should be modified depending on the amplicon size. A general rule for Taq polymerase: allow approximate 1 min per 1000 base pairs.
5. Transferring template from first round product to the second round reaction should be performed in PCR thermocycler room in order to avoid double-strand DNA contamination to clean room.

Acknowledgments

This work was supported by NIH grants U19 AI088791 and P30 AI045008.

References

1. Smith DB, Bukh J, Kuiken C, Muerhoff AS, Rice CM, Stapleton JT et al (2014) Expanded classification of hepatitis C virus into 7 genotypes and 67 subtypes: updated criteria and genotype assignment web resource. Hepatology 59:318–327
2. Moradpour D, Penin F, Rice CM (2007) Replication of hepatitis C virus. Nat Rev Microbiol 5:453–463
3. Ribeiro RM, Li H, Wang S, Stoddard MB, Learn GH, Korber BT et al (2012) Quantifying the diversification of hepatitis C virus (HCV) during primary infection: estimates of the in vivo mutation rate. PLoS Pathog 8: e1002881
4. Wang GP, Sherrill-Mix SA, Chang KM, Quince C, Bushman FD (2010) Hepatitis C virus transmission bottlenecks analyzed by deep sequencing. J Virol 84:6218–6228
5. Bull RA, Luciani F, McElroy K, Gaudieri S, Pham ST, Chopra A et al (2011) Sequential bottlenecks drive viral evolution in early acute hepatitis C virus infection. PLoS Pathog 7: e1002243
6. Li H, Stoddard MB, Wang S, Blair LM, Giorgi EE, Parrish EH et al (2012) Elucidation of hepatitis C virus transmission and early diversification by single genome sequencing. PLoS Pathog 8:e1002880
7. Stoddard MB, Li H, Wang S, Saeed M, Andrus L, Ding W et al (2015) Identification, molecular cloning, and analysis of full-length hepatitis C virus transmitted/founder genotypes 1, 3, and 4. MBio 6:e02518
8. Keele BF, Giorgi EE, Salazar-Gonzalez JF, Decker JM, Pham KT, Salazar MG et al (2008) Identification and characterization of transmitted and early founder virus envelopes

in primary HIV-1 infection. Proc Natl Acad Sci U S A 105:7552–7557

9. Salazar-Gonzalez JF, Salazar MG, Keele BF, Learn GH, Giorgi EE, Li H et al (2009) Genetic identity, biological phenotype, and evolutionary pathways of transmitted/founder viruses in acute and early HIV-1 infection. J Exp Med 206:1273–1289
10. Abrahams MR, Anderson JA, Giorgi EE, Seoighe C, Mlisana K, Ping LH et al (2009) Quantitating the multiplicity of infection with human immunodeficiency virus type 1 subtype C reveals a non-poisson distribution of transmitted variants. J Virol 83:3556–3567
11. Haaland RE, Hawkins PA, Salazar-Gonzalez J, Johnson A, Tichacek A, Karita E et al (2009) Inflammatory genital infections mitigate a severe genetic bottleneck in heterosexual transmission of subtype A and C HIV-1. PLoS Pathog 5:e1000274
12. Giorgi EE, Funkhouser B, Athreya G, Perelson AS, Korber BT, Bhattacharya T (2010) Estimating time since infection in early homogeneous HIV-1 samples using a poisson model. BMC Bioinformatics 11:532
13. Lee HY, Giorgi EE, Keele BF, Gaschen B, Athreya GS, Salazar-Gonzalez JF et al (2009) Modeling sequence evolution in acute HIV-1 infection. J Theor Biol 261:341–360
14. Ho DD, Neumann AU, Perelson AS, Chen W, Leonard JM, Markowitz M (1995) Rapid turnover of plasma virions and CD4 lymphocytes in HIV-1 infection. Nature 373:123–126
15. Markowitz M, Louie M, Hurley A, Sun E, Di Mascio M, Perelson AS et al (2003) A novel antiviral intervention results in more accurate assessment of human immunodeficiency virus type 1 replication dynamics and T-cell decay in vivo. J Virol 77:5037–5038
16. Wei X, Ghosh SK, Taylor ME, Johnson VA, Emini EA, Deutsch P et al (1995) Viral dynamics in human immunodeficiency virus type 1 infection. Nature 373:117–122
17. Stafford MA, Corey L, Cao Y, Daar ES, Ho DD, Perelson AS (2000) Modeling plasma virus concentration during primary HIV infection. J Theor Biol 203:285–301
18. Mansky LM, Temin HM (1995) Lower in vivo mutation rate of human immunodeficiency virus type 1 than that predicted from the fidelity of purified reverse transcriptase. J Virol 69:5087–5094
19. Slatkin M, Hudson RR (1991) Pairwise comparisons of mitochondrial DNA sequences in stable and exponentially growing populations. Genetics 129:555–562
20. Goonetilleke N, Liu MK, Salazar-Gonzalez JF, Ferrari G, Giorgi E, Ganusov VV et al (2009) The first T cell response to transmitted/founder virus contributes to the control of acute viremia in HIV-1 infection. J Exp Med 206:1253–1272
21. Bar KJ, Tsao CY, Iyer SS, Decker JM, Yang Y, Bonsignori M et al (2012) Early low-titer neutralizing antibodies impede HIV-1 replication and select for virus escape. PLoS Pathog 8: e1002721
22. Liu MK, Hawkins N, Ritchie AJ, Ganusov VV, Whale V, Brackenridge S et al (2013) Vertical T cell immunodominance and epitope entropy determine HIV-1 escape. J Clin Invest 123:380–393
23. Perelson AS, Neumann AU, Markowitz M, Leonard JM, Ho DD (1996) HIV-1 dynamics in vivo: virion clearance rate, infected cell lifespan, and viral generation time. Science 271:1582–1586
24. Quinkert D, Bartenschlager R, Lohmann V (2005) Quantitative analysis of the hepatitis C virus replication complex. J Virol 79:13594–13605
25. Li H, Stoddard MB, Wang S, Giorgi EE, Blair LM, Learn GH et al (2015) Single-genome sequencing of hepatitis C virus in donor-recipient pairs distinguishes modes and models of virus transmission and early diversification. J Virol 90:152–166
26. Bailey JR, Flyak AI, Cohen VJ, Li H, Wasilewski LN, Snider AE et al (2017) Broadly neutralizing antibodies with few somatic mutations and hepatitis C virus clearance. JCI Insight 2:e92872
27. Ke R, Li H, Wang S, Ding W, Ribeiro RM, Giorgi EE et al (2018) Superinfection and cure of infected cells as novel mechanisms contributing to rapid hepatitis C virus expansion and persistence. Proc Natl Acad Sci U S A 115 (30):E7139–E7148

Chapter 10

A Cell Culture Model for Persistent HCV Infection

Victoria Castro, Ginés Ávila-Pérez, Lidia Mingorance, and Pablo Gastaminza

Abstract

Chronic hepatitis C virus (HCV) infection affects millions of humans throughout the globe, causing liver disease and hepatocellular carcinoma. Persistence of the virus in the infected host can last for decades as a result of a faulty immune response that fails to clear the virus while constituting a major player in viral pathogenesis. In addition to evading immune responses, HCV has evolved intracellular survival strategies that enable persistent replication without directly killing the host cell.

After the generation of cell culture infection models for HCV, the knowledge about this virus and host-virus interactions acquired in the last decade has been greatly increased. Interestingly, persistent infection can also be established in cell culture. This model recapitulates persistent HCV RNA replication and viral protein expression as well as infectious progeny virus assembly and secretion and may be used to study not only these aspects of the virus replication cycle but also to study host-virus interactions in a model of prolonged HCV infection. In this chapter, we describe a methodology to generate persistently HCV-infected cultures and to monitor viral load and progeny virus production. Also, we provide generic protocols to study the impact of chemical compounds and host-targeting shRNAs to illustrate the applications of this model in the study of HCV infection in cell culture.

Key words Hepatitis C, Persistent infection, Cell culture, Antiviral, Host-virus interaction

1 Introduction

Hepatitis C virus chronically infects more than 150 million humans worldwide [1]. Chronic HCV infection, which is established in 85% of the infected individuals, is characterized by infection periods that can last for decades, in part due to a deficient immune response that not only is a major player in viral pathogenesis but also fails to eliminate infected cells [2, 3]. This, together with the fact that HCV infection is not cytolytic per se, results in the existence of fluctuating populations of persistently infected cells that replicate viral RNA, express viral antigens, and produce progeny virus during chronic infection [4].

Interestingly, HCV is capable of establishing persistent infections in cell culture [5]. Persistent HCV infection in cell culture is

Mansun Law (ed.), *Hepatitis C Virus Protocols*, Methods in Molecular Biology, vol. 1911,
https://doi.org/10.1007/978-1-4939-8976-8_10,

characterized by constant, albeit fluctuating genomic RNA replication, viral protein expression and progeny virus production that can be maintained for months in cell cultures [5]. Like other virus infection systems, selection of target cells and virus strain are critical to successfully establish these persistently infected cultures. Cells that are hyperpermissive for HCV infection, such as Huh-7.5 and derived clones [6], may undergo cell death due to excessive viral replication [5]. On the other hand, infection of permissive cells by cell culture-adapted HCV strains, such as Jc1 or similar chimeras, may also result in cell death [7, 8], thus hampering establishment of a persistent infection. In many studies, the experimental setup chosen by virologists is a combination of hyperpermissive cell lines and cell culture-adapted strains to obtain robust acute infection models, regardless of the outcome of the infection to cell viability. These constitute excellent models for the study of the viral replication cycle, antivirals, and factors that drive efficient HCV infection. Reduced viability of these cell cultures impedes the study of the long-term impact of the infection on cellular homeostasis and the impact that experimentally induced changes in the host cells (pharmacologically or by RNAi/overexpression) have on HCV infection in the long term.

The experimental system described herein is based on the use of susceptible human hepatoma cells (Huh-7) [9] established by culture of a human liver tumor and a recombinant infectious HCV molecular clone from genotype 2a generated from the consensus sequence obtained from an acutely infected patient [10]. Neither of these reagents is optimal in terms of viral replication and permissiveness, but they enable establishing a persistent infection model in cell culture.

This system by no means recapitulates chronic infection in infected patients, but it enables studying viral and cellular determinants that govern host-virus interactions during persistent HCV infection. Moreover, this system has been shown to be particularly useful for the study of late aspects of the HCV replication cycle, such as assembly and secretion of infectious virions [11, 12].

2 Materials

2.1 Biological Materials

1. Cells: Huh-7 cells were established from a primary culture of human hepatoma tumor tissue from a 57-year-old Asian male [9] (*see* **Note 1**). Huh-7-derived subclone Huh-7.5.1 clone 2, hereafter clone 2, was established by curing a Huh-7.5.1 cell [13] subclone bearing a subgenomic HCV replicon [14].
2. Viruses: Genotype 2a JFH-1 strain infectious molecular clone [10].

3. Antibodies: Human monoclonal anti-E2 (AR3A) (antibody provided by Mansun Law, The Scripps Research Institute; *see* **Note 2**) [15]. Fluorophore-conjugated secondary antibody.
4. Other reagents of biological origin: Fetal bovine serum. Bovine serum albumin (BSA).

2.2 Cell Culture Reagents

1. Complete Dulbecco's Modified Eagle's Medium supplemented with 10 mM HEPES, nonessential amino acids, 100 U/ml penicillin/streptomycin, and 10% fetal bovine serum (heat-inactivated at 56 °C for 30 min). Hereafter named complete DMEM.
2. Phosphate-buffered saline 10×: Dissolve 80 g NaCl, 2 g KCl, 26.8 g Na_2HPO_4-H_2O, and 2.4 g KH_2PO_4 in 800 ml H_2O. Adjust to pH 7.4 with HCl. Adjust volume to 1 l with H_2O.
3. Trypsin solution: 0.05% Trypsin, 0.48 mM EDTA, and 146 mM NaCl solution in PBS.
4. Fixation solution (3.7% formaldehyde solution): For 50 ml use 5 ml of 37% formaldehyde solution, 5 ml of 10× PBS, and 40 ml of double-distilled water.
5. Staining solution: 3% BSA and 0.3% Triton X100 in PBS.
6. Other chemicals: DAPI (4′,6-diamidino-2-phenylindole).

2.3 Equipment

1. Cell culture-treated sterile, disposable plasticware (filter tips, pipettes, vented cap flasks, and multiwell clusters).
2. CO_2 incubator.
3. BSL2 cell culture hood.
4. BSL2+ and above biological containment facility (*see* **Note 3**).
5. Fluorescence microscope (inverted fluorescence microscope with long-distance objectives is required for direct observation of cell culture plates).

3 Methods

3.1 Infectious Virus Rescue from Recombinant DNA

This procedure has been described by the original author in ref. 16.

3.2 JFH-1 Virus Stock Production

Although transfection or electroporation supernatants [16] can directly be used for inoculation of naive cells to produce persistently infected cell cultures, it is recommended to produce virus stocks to facilitate the process. The use of Huh-7.5.1 clone 2 cells provides the opportunity to obtain higher infectivity titers, which are particularly limiting when using JFH-1 strain.

1. Seed 4.5 × 10^6 clone 2 cells in an T162 flask 1 day before inoculation.
2. Incubate for 24 h in 5% CO_2 at 37 °C.
3. Dilute the starting JFH-1 stock (transfection supernatant [16]) in complete medium (20 ml/flask) to obtain a multiplicity of infection (MOI) of 0.01 FFU/cell (i.e., 4.5 × 10^4 focus-forming units, FFU, per flask).
4. Adding the virus dilution onto the cells.
5. Incubate cells in 5% CO_2 at 37 °C for 72 h.
6. Wash the cells once with warm PBS and discard.
7. Add 1 ml of trypsin/EDTA solution per flask.
8. Incubate for 3–5 min at 20–25 °C.
9. Add 9 ml of complete DMEM and carefully resuspend the cells.
10. Add 50 ml of complete DMEM and split cells into three T162 flasks (20 ml per flask).
11. Incubate the cells for additional 72 h in 5% CO_2 at 37 °C.
12. Collect cell supernatants; centrifuge for 1 min at 2000–2500 × *g* at 20–25 °C. collect supernatants and store at −80 °C in aliquots.
13. Replenish cells with complete DMEM (20 ml per T162 flask).
14. Incubate for 24 h in 5% CO_2 at 37 °C.
15. Repeat **steps 13** through **15** for 2–3 consecutive days. Discontinue if significant cell death is observed.
16. Determine infectivity titer (*see* Subheading 3.3) of the samples collected in **step 13**.

3.3 Virus Infectivity Titer Determination

Non-cytolytic virus infectivity may be titrated using immunofluorescence microscopy to identify and count infection foci, which are defined as clusters of viral antigen-expressing cells that are revealed after inoculation of naïve cells with an infectious supernatant. Viral infectivity titer is defined as the number of focus-forming units per milliliter in analogy to the plaque-forming units used for cytolytic viruses.

3.3.1 Intracellular Infectious Virion Sample Production

Infectivity titers are generally determined from extracellular (supernatants) but may also be determined for intracellular infectious HCV virions in whole-cell lysates [17]. These lysates are produced by resuspending infected cells in complete media and subjecting this suspension to four freeze-thawing cycles by consecutive immersions in ethanol-dry ice mixture and 37 °C water. Procedure for a 12-well plate with 10^5 cells per well:

1. Wash the infected cells once with warm PBS and discard.
2. Add 100 μl of trypsin/EDTA solution per well.

3. Incubate for 3–5 min at 20–25 °C.
4. Add 400 μl of complete DMEM and carefully resuspend the cells. Place the suspension (or a fraction of it) in a clean Eppendorf tube.
5. Prepare a container with a mixture of ethanol and dry ice.
6. Plunge the tubes into the freezing solution and wait until they are completely frozen.
7. Rapidly transfer the tubes into a 37 °C water bath, and incubate until the cell suspension melts.
8. Repeat **steps 6** and 7 four times.
9. Spin the lysates at 2000–2500 × *g* for 5 min at 4 °C.
10. Use supernatants as initial sample in Subheading 3.3.2, **step 5**.

3.3.2 Infectivity Titration Assay

1. Seed 10^4 clone 2 cells/well (*see* **Note 4**) in a final volume of 100 μl in 96-well flat bottom plates. Four to eight wells/sample are required, depending on the expected infectivity titer. Two replicates are strongly encouraged.
2. Incubate for 24 h in 5% CO_2 at 37 °C.
3. Thaw the infectious samples at 20–25 °C.
4. Serial sample dilutions are performed in clean 96-well plates (with no cells). Add 200 μl of complete medium to a 96-well plate, using 4–8 wells per sample. Set the wells sequentially in a lane to accommodate the different sample dilutions.
5. Add 50 μl of the sample onto the first well.
6. Discard the tip.
7. Mix meticulously the virus-medium dilution using the micropipette and a new filter tip, without creating bubbles.
8. Take 50 μl of the mixture, and carefully transfer into the next well.
9. Repeat **steps 5**–7 until reaching the desired dilution factor (*see* **Note 5**).
10. Remove the supernatant of the cells from **step 1**.
11. Add 100 μl of the different virus dilutions onto each well.
12. Incubate for 72 h in 5% CO_2 at 37 °C.
13. Discard cell supernatants.
14. Add 150 μl of the 3.7% formaldehyde fixation solution.
15. Incubate for 20 min at 20–25 °C.
16. Wash two times with PBS (200 μl).
17. Add 30 μl/well of staining solution.
18. Incubate for 1 h at 20–25 °C.

19. Prepare a 1 μg/ml dilution of the primary antibody in staining solution.
20. Discard staining solution from the 96-well plate.
21. Add 25 μl/well of the anti-E2 antibody dilution.
22. Incubate for 1 h at 20–25 °C.
23. Prepare a 1:500 dilution of the fluorophore-conjugated secondary antibody in staining solution. Add DAPI (1:1000; 1 μg/ml).
24. Wash the 96-well plate two times with PBS (200 μl).
25. Add 25 μl/well of the secondary antibody dilution.
26. Incubate for 1 h at 20–25 °C.
27. Wash the 96-well plate two times with PBS (200 μl).
28. Add 50 μl of PBS to the test wells. In case of using a glass coverslip, mount onto a microscope slide using Prolong™.
29. Observe under the fluorescence microscope. Count infection foci only in dilutions where they are discrete and well separated from one another, until reaching negative wells (endpoint dilution).
30. Calculate the infectivity titer in FFU/ml by multiplying the number of infection foci found in the different dilutions by the corresponding dilution factor. Multiply by 10 to obtain the foci number per ml (as 100 μl were used for inoculation). Average the infectivity titers estimated in the different dilutions (typically discrete foci are only observed in 2–3 wells; *see* **Note 6**).

3.4 Generation of Persistently Infected Cell Cultures

1. Seed 2×10^6 Huh-7 cells in a T75 flask.
2. Incubate for 24 h in 5% CO_2 at 37 °C.
3. Inoculate cells with JFH-1 at an MOI of 0.01 FFU/cell in a final volume of 10 ml per flask.
4. Incubate for a minimum of 12–14 days in 5% CO_2 at 37 °C.
5. Maintain subconfluent cultures at all times by splitting the cultures every 3–4 days. Wash cells with warm PBS and trypsinize cells as above. Add 5 ml of complete DMEM, and resuspend cells in a final volume of 6 ml. Plate 2 ml of cell suspension into a T75 flask, and add 8 ml of complete DMEM.
6. Detach the cells from the flask as described in **step 5** at day 12–14 postinoculation (*see* **Note 7**).
7. Collect 200 μl of cell suspension, and plate it onto a sterile glass coverslip in a M12 plate well. Add 800 μl of complete DMEM and incubate for 24 h. Proceed with this sample to **step 9**.
8. Seed the rest of the cells in a T162 flask. Proceed to **step 14**.

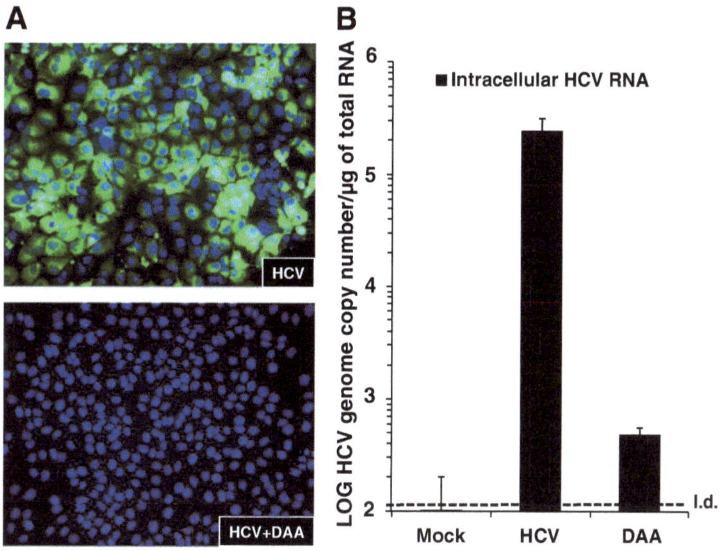

Fig. 1 Persistently infected cell populations express viral antigens and replicate HCV RNA. Treatment of the cells with a combination of direct-acting antivirals (DAA: sofosbuvir 1 μM and daclatasvir 100 nM [21]) for 3 weeks results in virtual loss of viral antigen expression and viral RNA accumulation. (Panel **a**) Immunofluorescence microscopy on cells prepared as described in Subheading 3.4 performed as described in Subheading 3.3.2, displaying E2 expression before and after treatment. (Panel **b**) HCV RNA abundance before and after treatment as determined by reverse-transcription and real-time PCR (l.d., limit of detection)

9. Remove supernatant from **step 8**, and add a 4% formaldehyde solution in PBS (500 μl) onto the glass coverslip.
10. Incubate for 20 min at 20–25 °C.
11. Wash the cells two times with PBS (1 ml).
12. Process the sample to perform immunofluorescence staining (*see* Subheading 3.3.2, **step 16**).
13. Determine the percentage of HCV antigen-positive cells by immunofluorescence microscopy. At day 12–14 postinoculation, nearly 100% of the cells should be HCV antigen-positive (Fig. 1a).
14. Once infection reaches approximately 100% of the cells, the persistently infected cell cultures are maintained subconfluent for a maximum of 6–8 weeks (*see* **Note 8**).

3.5 Study of the Impact of Antiviral Compounds on Persistent HCV Infection

1. Seed 10^5 persistently infected cells into M12 wells. Prepare at least triplicate wells for each experimental condition, including diluent control and a positive control of inhibition.
2. Incubate for 24 h in 5% CO_2 at 37 °C.
3. Prepare test compound dilutions as well as a negative control (solvent) and a positive control (e.g., 1 μM sofosbuvir or

10 μM 2′-C-methyladenosine) sufficient for triplicate samples (typically 3 ml).

4. Remove supernatant from **step 1**.
5. Add 1 ml of compound dilutions from **step 3** onto the cell cultures.
6. Incubate for 24 h in 5% CO_2 at 37 °C.
7. Harvest cell culture supernatants. Use this sample to determine extracellular infectivity titer (*see* **Note 9**).
8. Wash the cells once with warm PBS and discard.
9. Add 100 μl of trypsin/EDTA solution per well.
10. Incubate for 3–5 min at 20–25 °C.
11. Add 400 μl of complete DMEM and carefully resuspend the cells.
12. Divide the sample into 2 × 250 μl aliquots. Directly freeze (−80 °C) one of the samples for intracellular infectivity determination (*see* Subheading 3.3.1).
13. Spin the second 250 μl aliquot at 2000–2500 × *g* 5 min in a microfuge.
14. Remove supernatant, and freeze (−80 °C) the cell pellet for HCV RNA load determination [13] (*see* **Note 10**). The impact of antiviral on infectious progeny virus production or HCV RNA replication may be determined by infectivity titration (Subheading 3.3) or by reverse-transcription and real-time PCR (Fig. 2).

3.6 Study of the Impact of Host Gene Silencing by RNAi on Persistent HCV Infection

For RNAi experiments in persistently infected cells, we recommend the use of lentiviral vectors for shRNA delivery. Optimally, these vectors should enable expression of a selection marker to maximize silencing efficacy in the persistently infected cell population [18].

1. Seed 4 × 10^5 persistently infected cells into M6 wells. Prepare one well per each treatment, including a negative (no transduction) and a positive control (transduction with an irrelevant shRNA).
2. Incubate for 24 h in 5% CO_2 at 37 °C.
3. Inoculate with the adequate lentiviral vector dose to obtain 100% transduction efficiency.
4. Maintain subconfluent cultures at all times by splitting them every 3–4 days as above at a 1:3 ratio. The rest of the cells may be used to monitor target gene silencing efficiency by Western blot and to determine the optimal time for testing the impact of silencing on persistent HCV infection.
5. Once the optimal silencing time has been determined, seed 10^5 cells per well in triplicate M12 wells for each test condition.

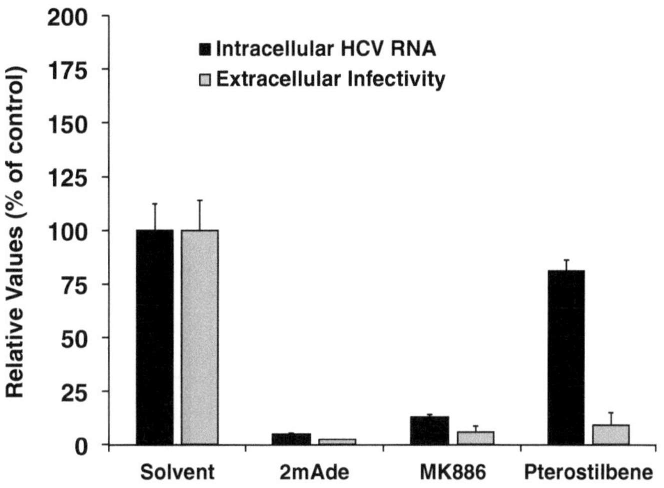

Fig. 2 Study of the impact of antiviral compounds on persistent HCV infection. Persistently infected cells (Subheading 3.4) were treated with solvent (DMSO) as negative control, the viral polymerase inhibitor (2′-C-m-adenosine; 10 μM) [22] as a positive control, MK886 (10 μM) or pterostilbene (10 μM), compounds that were identified as replication and assembly inhibitors after compound library screening [12]. Persistently infected cells were treated with 10 μM pterostilbene or DMSO as control and analyzed exactly as described in Subheading 3.5. The impact on infectious progeny virus production is shown by reduced levels of extracellular infectivity as determined by infectivity titration (Subheading 3.3). The impact of the compounds on persistent HCV RNA replication is shown by the relative intracellular HCV RNA levels as determined by reverse-transcription and real-time PCR

Seed an additional M6 plate to collect cell and/or supernatant samples to monitor silencing efficacy.

6. Incubate for 24 h in 5% CO_2 at 37 °C.
7. Harvest cell culture supernatants. Use this sample to determine extracellular infectivity titer.
8. Wash the cells once with warm PBS and discard.
9. Add 100 μl of trypsin/EDTA solution per well.
10. Incubate for 3–5 min at 20–25 °C.
11. Add 400 μl of complete DMEM and carefully resuspend the cells.
12. Divide the sample into 2 × 250 μl aliquots. Directly freeze (−80 °C) one of the samples for intracellular infectivity determination (*see* Subheading 3.3.1).
13. Spin the second 250 μl aliquot 2000–2500 × *g* for 5 min in a microfuge.
14. Remove supernatant and freeze (−80 °C) the cell pellet for HCV RNA load determination [13]. The impact of antiviral on

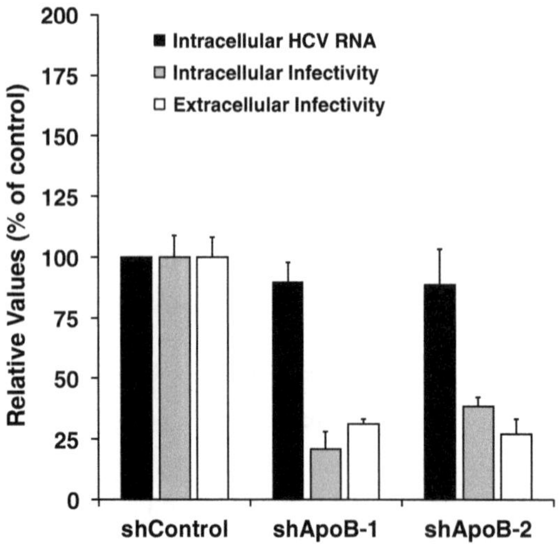

Fig. 3 Impact of silencing of a host gene involved in viral assembly in viral RNA load and progeny virus production in a persistent infection model. Persistently infected cultures were transduced with lentiviral vectors expressing shRNAs against apolipoprotein B (apoB), which has been shown essential to produce infectious virions in HCV [11]. At day 6 post-transduction, cells were plated after verification of host protein silencing, as described in Subheading 3.6, **step 5**, and samples were collected 24 h later and analyzed as described in Subheading 3.6. The data indicate that silencing did not impact HCV RNA replication as determined by reverse-transcription and real-time PCR, whereas infectivity titers (Subheading 3.3) were reduced both in intracellular and extracellular virions, indicating a defect in infectious virus assembly in cells with low apoB expression [11]

infectious intra- and extracellular virus production or HCV RNA replication may be determined by infectivity titration (Subheading 3.3) or by reverse-transcription and real-time PCR (Fig. 3).

4 Notes

1. Huh-7 are genetically unstable, and Huh-7 cell populations may therefore be genetically heterogeneous [19]. Depending on the culture conditions and passages, some Huh-7 cell populations may display reduced susceptibility to HCV infection, due to the presence of cell subpopulations which are poorly susceptible to infection. Depending on the laboratory of origin, Huh-7 cells may display differential susceptibility to infection [20].
2. Other primary antibodies may be used for detection of viral antigens.
3. Production and handling of recombinant hepatitis C virus requires BSL2+ and above biological containment laboratories

in Europe. Make sure your laboratory complies with biosafety regulations in your country.

4. Huh-7 cells may also be used for titration experiments. In the proposed experimental setup, infection foci will appear larger in clone 2 cells (10–30 cells/focus) than in Huh-7 cells (2–10 cells/focus).
5. Four to six wells are sufficient to titrate JFH-1 virus stocks using the described methodology as the maximum titers reach 1–5 × 10^4 FFU/ml.
6. It is highly recommended to repeat the experiment with different dilutions if the infectivity titer is not calculated out of 2–3 wells or if the endpoint dilution has not been reached.
7. At the peak of infection, cell duplication time is notably reduced. This phenomenon subsides in the following passages, where persistently infected Huh-7 cells display a slightly longer doubling time as compared with their uninfected counterparts.
8. It is recommended to maintain the persistently infected cultures for limited passages (6–8 weeks after inoculation approximately), as it has been previously documented that both virus and cell populations undergo genetic drift [5] that may reduce the reproducibility of the experiments.
9. The supernatant contains the antiviral drug candidate and this may interfere with the titration assay. It is possible that, at low dilutions, the apparent titer is reduced. If this interferes critically with the titration assay, virus may be separated from the compound using 100 kD cutoff microfiltration devices, provided the molecular weight of the compound is smaller than that cutoff.
10. Typically, HCV RNA titer is between 10^5 and 10^6 copies/μg of total RNA. Intracellular infectivity titers depend on the cell density, incubation period, and supernatant volume. Maximum infectivity titers in this system range approximately between 10^3 and 10^4 FFU/ml (Fig. 1b).

Acknowledgment

This protocol is based on work initially developed at Dr. Francis V. Chisari's laboratory at The Scripps Research Institute (La Jolla, CA) in collaboration with Dr. Takaji Wakita's group at the Tokyo Metropolitan Institute of Medical Science (Tokyo, Japan) and was originally generated and improved with the essential contribution of Jin Zhong (Pasteur Institute in Shanghai), Sharookh Kapadia (Genentech), Guofeng Cheng (Gilead), Susan Uprichard (University of Illinois, Chicago), and specially Stefan Wieland (University Hospital of Basel). We would like to thank Gema Calvo and

Martina Friesland (CNB-CSIC) for excellent technical assistance and Dr. Urtzi Garaigorta for critically reviewing the manuscript. This work was supported by grant SAF2014-52400-R from the Spanish Ministry of Economy.

References

1. Thrift AP, El-Serag HB, Kanwal F (2017) Global epidemiology and burden of HCV infection and HCV-related disease. Nat Rev Gastroenterol Hepatol 14:122–123
2. Guidotti LG, Chisari FV (2006) Immunobiology and pathogenesis of viral hepatitis. Annu Rev Pathol 1:23–61
3. Rehermann B (2013) Pathogenesis of chronic viral hepatitis: differential roles of T cells and NK cells. Nat Med 19:859–868
4. Zeuzem S, Herrmann E (2002) Dynamics of hepatitis C virus infection. Ann Hepatol 1:56–63
5. Zhong J, Gastaminza P, Chung J, Stamataki Z, Isogawa M, Cheng G et al (2006) Persistent hepatitis C virus infection in vitro: coevolution of virus and host. J Virol 80:11082–11093
6. Blight KJ, McKeating JA, Rice CM (2002) Highly permissive cell lines for subgenomic and genomic hepatitis C virus RNA replication. J Virol 76:13001–13014
7. Walters K-A, Syder AJ, Lederer SL, Diamond DL, Paeper B, Rice CM et al (2009) Genomic analysis reveals a potential role for cell cycle perturbation in HCV-mediated apoptosis of cultured hepatocytes. PLoS Pathog 5: e1000269
8. Mishima K, Sakamoto N, Sekine-Osajima Y, Nakagawa M, Itsui Y, Azuma S et al (2010) Cell culture and in vivo analyses of cytopathic hepatitis C virus mutants. Virology 405:361–369
9. Nakabayashi H, Taketa K, Miyano K, Yamane T, Sato J (1982) Growth of human hepatoma cells lines with differentiated functions in chemically defined medium. Cancer Res 42:3858–3863
10. Wakita T, Pietschmann T, Kato T, Date T, Miyamoto M, Zhao Z et al (2005) Production of infectious hepatitis C virus in tissue culture from a cloned viral genome. Nat Med 11:791–796
11. Gastaminza P, Cheng G, Wieland S, Zhong J, Liao W, Chisari FV (2008) Cellular determinants of hepatitis C virus assembly, maturation, degradation, and secretion. J Virol 82:2120–2129
12. Gastaminza P, Whitten-Bauer C, Chisari FV (2010) Unbiased probing of the entire hepatitis C virus life cycle identifies clinical compounds that target multiple aspects of the infection. Proc Natl Acad Sci U S A 107:291–296
13. Zhong J, Gastaminza P, Cheng G, Kapadia S, Kato T, Burton DR et al (2005) Robust hepatitis C virus infection in vitro. Proc Natl Acad Sci U S A 102:9294–9299
14. Pedersen IM, Cheng G, Wieland S, Volinia S, Croce CM, Chisari FV et al (2007) Interferon modulation of cellular microRNAs as an antiviral mechanism. Nature 449:919–922
15. Law M, Maruyama T, Lewis J, Giang E, Tarr AW, Stamataki Z et al (2008) Broadly neutralizing antibodies protect against hepatitis C virus quasispecies challenge. Nat Med 14:25–27
16. Wakita T (2009) Isolation of JFH-1 strain and development of an HCV infection system. Methods Mol Biol 510:305–327
17. Gastaminza P, Kapadia SB, Chisari FV (2006) Differential biophysical properties of infectious intracellular and secreted hepatitis C virus particles. J Virol 80:11074–11081
18. Friesland M, Mingorance L, Chung J, Chisari FV, Gastaminza P (2013) Sigma-1 receptor regulates early steps of viral RNA replication at the onset of hepatitis C virus infection. J Virol 87:6377–6390
19. Wild AT, Gandhi N, Chettiar ST, Aziz K, Gajula RP, Williams RD et al (2013) Concurrent versus sequential sorafenib therapy in combination with radiation for hepatocellular carcinoma. PLoS One 8:e65726
20. Sainz B Jr, Barretto N, Uprichard SL (2009) Hepatitis C virus infection in phenotypically distinct Huh7 cell lines. PLoS One 4:e6561
21. Pawlotsky JM, Feld JJ, Zeuzem S, Hoofnagle JH (2015) From non-A, non-B hepatitis to hepatitis C virus cure. J Hepatol 62:S87–S99
22. Migliaccio G, Tomassini JE, Carroll SS, Tomei L, Altamura S, Bhat B et al (2003) Characterization of resistance to non-obligate chain-terminating ribonucleoside analogs that inhibit hepatitis C virus replication in vitro. J Biol Chem 278:49164–49170

Chapter 11

Expression Cloning of Host Factors Required for the HCV Replication Cycle

Sandra Ciesek and Thomas von Hahn

Abstract

Chronic infection with the hepatitis C virus (HCV) is a common cause of chronic liver disease, cirrhosis, and liver cancer. Like most viruses, HCV depends of a number of factors encoded by its host cell to execute its replication cycle and create. Expression cloning is one of several possible approaches that have been employed to identify novel host factors essential for the HCV replication cycle. It involves generation of a cDNA library from a cell type with a desired trait (such as ability to bind E2 or being susceptible to HCV infection), expression of that library in a different cell type missing the trait of interest, and selection for cell clones that have acquired that trait through expression of a specific cDNA. This chapter describes an expression cloning approach similar to the one that was used to identify the tight junction component claudin-1 as an essential HCV cell entry factor.

Key words HCV, Host factor, Claudin-1, cDNA library, Selection, Cyclic packaging rescue, V1

1 Introduction

Like all viruses, hepatitis C virus (HCV) is an obligate cellular parasite that depends on the infected host cell to execute its replication cycle and generate progeny virions that are released into the blood and can then infect other cells either within the same or—following blood-borne transmission—in another host. Host cell components that are required for the execution of the viral replication cycle are referred to as essential host factors. By now, a number of essential HCV host factors have been identified through different approaches: the HCV receptor scavenger receptor class B type I (SR-BI) was identified through biochemical characterization of cellular binding partners of a soluble fragment of the HCV envelope protein E2 (soluble E2 [sE2]) in a CD81-negative cell line [1]. Cyclophilin A (CypA), an essential host factor for replication of the HCV genome, was discovered based on the observation that pharmacological CypA inhibitors also inhibit HCV replication [2, 3]. Identification of apolipoprotein E (ApoE) as a factor

Mansun Law (ed.), *Hepatitis C Virus Protocols*, Methods in Molecular Biology, vol. 1911,
https://doi.org/10.1007/978-1-4939-8976-8_11,

essential for assembly of HCV particles followed a more tortuous course: early indirect evidence included an association of serum HCV with ApoE containing very low-density lipoproteins [4] and the association of genetic variants of ApoE with the clinical course of hepatitis C [5–7]. Yet, it was later when more advanced cell culture systems for HCV had become available that its role was confirmed by more direct evidence including inhibition of HCV infection by antibodies against ApoE, inhibition of HCV assembly by knockdown of ApoE [8], and the demonstration that human ApoE was required to fully unlock HCV assembly in murine cells [9].

Expression cloning refers to an approach where a library of DNA clones is screened for ones encoding a specific trait of interest. In the HCV field, expression cloning approaches were employed very successfully to identify essential host factors involved in HCV cell entry. CD81, the first molecule to be identified as a HCV receptor, was discovered by screening a cDNA library, generated from a human T-cell lymphoma cell line that strongly bound sE2, in a mouse fibroblast cell line that did not bind sE2, for a cDNA conferring sE2 binding upon the latter cell line [10]. The cDNA thus isolated was found to encode CD81. Two other host cell factors essential for HCV entry, the tight junction components claudin-1 (CLDN1) and occludin (OCLN), were identified by screening cDNA libraries, generated from HCV-permissive cell lines, in cell lines that were resistant to HCV cell entry despite expressing all entry factors known at the time and selecting for cDNAs enabling HCV entry in the previously resistant cell line [11, 12]. In both cases, cyclic packaging rescue (CPR) was employed as a means to allow multiple rounds of selection and enrichment of specific cDNAs in order to overcome non-specific background.

This chapter describes materials and methods of cDNA expression cloning based on the screen that identified CLDN1 as a host factor essential for HCV cell entry [11]. These can and have been used with adaptions to identify other host factors, but clearly, every expression cloning project intended to identify novel essential HCV host factors will need to be tailored to the biological question at hand. It is highly unlikely that two such projects will ever be fully alike. Nonetheless, we hope that the methods described here may prove helpful in the planning and execution of future expression cloning approaches.

2 Materials

2.1 Cell Culture Reagents

1. Standard cell culture plasticware including 96- and 6-well plates; 10 and 15 cm dishes.

2. Sterile 2 mL syringes and 0.45 μm pore size syringe filters.
3. Sterile 500 mL/0.45 μm pore size bottle-top filter units.
4. Growth media: Dulbecco's Modified Eagle Medium supplemented with either 3% or 10% fetal bovine serum (*see* **Note 1**).
5. Opti-MEM serum-free media.
6. 0.5 g/L trypsin and 0.2 g/L EDTA.
7. Poly-L-lysine (PLL) hydrobromide 0.1 mg/mL (Sigma) in sterile water.
8. FuGENE-HD transfection reagent (Roche).
9. Polybrene (Sigma).
10. Phosphate-buffered saline (PBS).
11. PBS with 2% paraformaldehyde.
12. Beta-galactosidase staining buffer containing PBS with X-gal 1 mg/mL, $MgCl_2$ 2 mM, K-Ferricyanide 5 mM, K-Ferrocyanide 5 mM.

2.2 Cell Lines

1. Huh-7.5 human hepatoma cells.
2. HEK-293T human embryonic kidney cells.
3. TZM: a derivative of human HeLa cells containing a lacZ reporter integrated in the genome under the control of an HIV-1 long terminal repeat [13].

2.3 DNA Constructs

1. Vectors CSGW, CSPW, CSZW (*see* **Note 2**).
2. Vector V1 (*see* **Note 3**).
3. pcDNA3.1.
4. pcDNA-HCV/H77E1E2 encoding HCV isolate H77 glycoproteins E1 and E2 [11].
5. pcDNA-VSVG encoding the G-protein of vesicular stomatitis virus [11].
6. HIV gag-pol encoding HIV gag-pol [11].

2.4 Materials for cDNA Library Construction

1. SMART cDNA Library Construction Kit (Clontech Laboratories). Please refer to user manual for complete list of kit contents as well as additional materials required that are not contained in the kit.
2. TRIzol (Invitrogen).
3. Chloroform.
4. Oligotex Direct mRNA Midi/Maxi Kit (Invitrogen).
5. QIAquick PCR Purification Kit (Qiagen).
6. MinElute PCR Purification Kit (Qiagen).
7. SeaPrep agarose (Lonza).

3 Methods

3.1 Generation of a cDNA Library from Huh-7.5 Cells (See Note 4)

3.1.1 Preparation of Total RNA (See Also Invitrogen TRIzol Manual)

1. Pellet the cells (by centrifugation) and discard the supernatant. The amount of cells/tissue strongly depends on the cell or tissue type you are using. For example, we used 2.5 g of liver tissue (in liquid nitrogen) or around 5–8 confluent 15 cm dishes for liver-derived cells.
2. Add 1 mL TRIzol reagent per 50–100 mg of sample to the pellet. Do not wash cells before addition of TRIzol reagent to avoid mRNA degradation. Pipet the mixture up and down several times to homogenize.
3. Incubate for 5 min to permit complete dissociation of the nucleoprotein complex.
4. Add 0.2 mL of chloroform per 1 mL of TRIzol reagent used for lysis, then securely cap the tube, and incubate for 2–3 min.
5. Centrifuge the sample for 15 min at 12,000 × *g* at 4 °C. The mixture separates into a lower red phenol-chloroform phase, an interphase, and a colorless upper aqueous phase.
6. Transfer the aqueous phase containing the RNA to a new tube by angling the tube at 45° and pipetting the solution out. Avoid transferring any of the interphase or organic layer into the pipette when removing the aqueous phase.
7. Add 0.5 mL of isopropanol to the aqueous phase, per 1 mL of TRIzol reagent used for lysis.
8. Incubate for 10 min. Centrifuge for 10 min at 12,000 × *g* at 4 °C. Total RNA precipitate forms a white gel-like pellet at the bottom of the tube.
9. Discard the supernatant with a micropipettor.
10. Resuspend the pellet in 1 mL of 75% ethanol per 1 mL of TRIzol reagent used for lysis.
11. Vortex the sample briefly; then centrifuge for 5 min at 7500 × *g* at 4 °C.
12. Discard the supernatant with a micropipettor.
13. Vacuum or air-dry the RNA pellet for 5–10 min. Do not let the RNA pellet dry out, to ensure total solubilization of the RNA.
14. Resuspend the pellet in 20–50 μL of RNase-free water, 0.1 mM EDTA, or 0.5% SDS solution by pipetting up and down.
15. Incubate in a water bath or heat block set at 55–60 °C for 10–15 min. Quantify and store the RNA at −80 °C (*see* **Note 5**).

3.1.2 Isolation of Poly-A mRNA

1. Mix 3 mg of total RNA (solved in 650 μL H_2O) with 650 μL OBB buffer and 175 μL Oligotex.
2. Incubate the sample for 3 min at 70 °C in a water bath or heating block.
3. Remove sample from the water bath/heating block, and place at room temperature for 10 min.
4. Pellet the Oligotex/mRNA complex by centrifugation for 2 min at maximum speed (14,000–18,000 × *g*), and carefully remove the supernatant by pipetting.
5. Resuspend the Oligotex/mRNA pellet in 600 μL buffer OW2 by vortexing or pipetting, and pipet onto a large spin column placed in a 1.5 mL microcentrifuge tube.
6. Centrifuge for 1 min at maximum speed.
7. Wash 2 times with washing buffer (OW2) at maximum speed.
8. Elute in 2× 50 μL OEB.
9. Quantify. If concentration is less than 100 ng/μL, precipitate RNA overnight and resuspend in smaller volume of water.

3.1.3 First and Second Strand Synthesis

First- and second-strand synthesis should be performed as described in the Clontech SMART cDNA Library Construction Kit (*see* **Note 6**).

1. Following second-strand synthesis, clean up PCR product with QIAquick PCR Purification Kit.
2. Elute in 2× 30 μL EB buffer.

3.1.4 Digestion of cDNA for Cloning into V1 Vector

1. Digest DNA from Subheading 3.1.3 with SfiI: 10 μg of DNA in 100 μL reaction with 10 μL SfiI (20,000 U/mL) at 50 °C overnight in the thermocycler.
2. Also digest the vector (10 μg DNA with 10 μL SfiI at 50 °C overnight).
3. Run on 0.7% agarose gel in 9 lanes.
4. Remove center six lanes prior to exposing to UV light (to keep this DNA undamaged).
5. Notch gel to mark locations of linear vector and insert bands (discard this DNA).
6. Under normal light, replace center section, and use marks to cut approximate bands of vector (and insert to be used as control later).
7. Purify the gel by using with QIAquick PCR Purification Kit.
8. Elute in 50 μL EB buffer provided in the Purification Kit.
9. Redigest the DNA with 4 μL SfiI in 100 mL total at 50 °C for 4–8 h.
10. Clean up with QIAquick PCR Purification Kit and elute in 50 μL EB buffer.

3.1.5 cDNA Fractionation

1. Fractionate cDNA as described in the Clontech User manual. Prepare 20 tubes and wash the provided column 4 times with ten buffer. Then add 100 μL DNA and collect single drops of 35 μL per tube.
2. Quantify each drop. In total 200–400 ng cDNA are required.
3. Clean up with Qiagen's MinElute PCR Purification Kit.
4. Elute in 2× 10 μL 0.5× E1B provided in the kit.

3.1.6 Ligation

1. Combine 6 μL (~2 μg) prepared vector, 20 μL cDNA, 9 μL water, 4 μL 10x ligase buffer, and 1 μL ligase.
2. Keep at 16 °C overnight in thermocycler.
3. Clean up on MinElute Qiagen QIAquick Spin column.
4. Elute in 2 × 0.25× 10 μL EB buffer.

3.1.7 Quality Check of Generated Library

To ensure that the generated library has a high percentage of insert-containing vectors of the different sizes, a test transformation should be performed prior to embarking upon large-scale transformation as follows:

1. Add 100 μL competent cells (DH5 alpha) to 0.5 μL of the ligation product.
2. Incubate the mixture on ice for 20–30 min.
3. Heat shock the mixture at 42 °C for 30 s.
4. Put the mixture back on ice for 2–5 min.
5. Add 1 mL SOC medium.
6. Mix at 37 °C for 1 h.
7. Plate the transformation onto a 10 cm LB agar plate containing ampicillin.
8. Incubate the plate at 37 °C overnight.
9. Pick 20 colonies, expand, and isolate DNA.
10. Digest with BamH1 and XhoI.
11. Run on an agarose gel (Fig. 1) (*see* **Note** 7).

3.1.8 Large-Scale Transformation

For more details see Clontech user manual.

1. To each 500 mL bottle, add 450 mL 2× LB medium and 1.35 g SeaPrep agarose.
2. Add a stir bar.
3. Autoclave bottles.
4. Cool to 37 °C in water bath.
5. Add 450 μL 50 mg/mL carbenicillin.
6. Mix cDNA ligation with 400 μL competent cells.

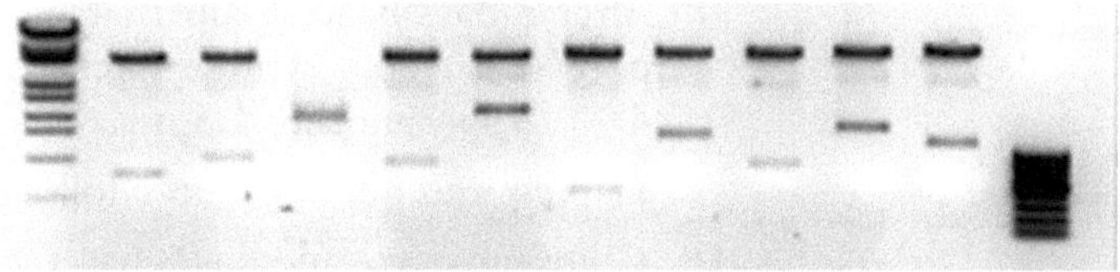

Fig. 1 Test digest prior to large-scale transformation of cDNA library. A fraction of a cDNA library generated in this case from human liver tissue and cloned into the V1 vector was transformed into competent bacteria. Twenty clones were picked, expanded, miniprepped, BamHI/XhoI digested, and run on an agarose gel. The lower bands of variable size correspond to the cDNA inserts. The larger bands of identical size correspond to the V1 vector. Size markers in lanes 1 and 12. Only lane 4 shows a defective clone

7. Electroporate in 40 μL aliquots, resuspend each in 2× 500 μL SOC medium.
8. Recover at 30 °C for a minimum of 2 h.
9. Add 1.7 mL cells to each SeaPrep bottle (on a stir plate).
10. Chill in ice bath for 1 h.
11. Incubate for 48 h at 30 °C.
12. Mix and heat the bottles to 37 °C.
13. Spin down bacteria at room temperature.
14. Discard supernatant.
15. Maxiprep cDNA using a commercially available maxiprep kit (e.g., Qiagen).

3.2 Screening of cDNA Library in 293T Cells

3.2.1 Titer Determination

1. Apply PLL to 6-well plate just covering well bottom.
2. Aspirate PLL.
3. Rinse once with water.
4. Air-dry under cell culture cabinet.
5. Plate 293T cells at 8 × 10^5 cells/well.
6. The next day transfect a total of 1.5 μg of DNA per well using FuGENE-HD following the manufacturer's instructions. To generate lentiviral pseudoparticles carrying library cDNAs enveloped with VSV-G (V1/library-VSVGpp), transfect 0.6 μg of V1-library plus 0.6 μg HIV gag-pol plus 0.3 μg VSV-G. To generate lentiviral pseudoparticles carrying a GFP reporter gene enveloped with VSV-G (V1/GFP-VSVGpp), use V1/GFP in lieu of V1-library.
7. After 6 h change media to DMEM/FCS3%, 1.5 mL/well.

8. After 2 days harvest supernatants and filter through 0.45 μm pore size sterile syringe filters. Store filtered supernatant at 4 °C. Apply another 1.5 mL/well DMEM/FCS3%.
9. Harvest and filter again the following day.
10. Combine day 2 and day 3 harvest. Store at −80 °C.
11. Plate 293T and TZM cells in plates on a 6-well plate (2×10^5 cells/well); plate TZM cells in wells on a 96-well plate (5×10^3 cells/well).
12. The following day thaw infectious supernatants (V1/library-VSVGpp and V1/GFP-VSVGpp).
13. Use both pseudoviruses (V1/library-VSVGpp and V1/GFP-VSVGpp) to transduce TZM cells in the 96-well plate: 48 wells each starting with undiluted virus in tenfold serial dilutions with 8 replicate wells per dilution. Dilutions should be made in DMEM/FCS3%. Polybrene at a final concentration of 4 μg/mL is included in all pseudovirus transductions. Pseudovirus containing supernatants is left on target cells for 6 h. Then they are aspirated and wells are replenished with DMEM/FCS3%.
14. Use V1/GFP-VSVGpp to transduce 293T and TZM cells in the 6-well plate with undiluted virus as well as 1:10 and 1:100 dilution with 1 well per dilution (*see* **Note 8**).
15. After 48 h supernatants from the 96-well plate are aspirated and beta-galactosidase staining is performed:
    (a) Wash once with PBS.
    (b) Apply PBS/paraformaldehyde (PFA) 2% and leave on ice for 15 min.
    (c) Wash 3× with PBS.
    (d) Put on 100 μL stain per well; incubate at RT for 2–4 h.
    (e) Wash 3× with PBS.
    (f) Determine TCID50 of V1/library-VSVGpp and V1/GFP-VSVGpp on TZM cells (*see* **Note 9**).
16. In parallel, supernatants from the wells in the 6-well plate are aspirated, and FACS detection of GFP is performed:
    (a) Trypsinize cells.
    (b) Pellet cells (300 × *g*/5 min).
    (c) Resuspend at in PBS/PFA0.5%.
    (d) Run through flow cytometer to determine percentage of GFP-positive cells.
    (e) Determine titer, i.e., transducing units (TUs) per mL, of V1/GFP-VSVGpp on TZM and on 293T cells (*see* **Note 10**).

3.2.2 Delivery of Library

1. Scale up procedure described above (Subheading 3.2.1) to generate at least 5 × 10^6 TU of V1/library-VSVGpp (*see* **Note 11**). A small amount of V1/GFP-VSVH should be generated in parallel and titer monitored as titer of lentiviral particles produced may vary from batch to batch.
2. Plate 2 × 10^6 cells of the target cell line (e.g., 293T) at low density (confluency <10%) aiming for just below the number of TUs to be delivered the following day.
3. Use V1/library-VSVGpp to transduce target cells aiming for a multiplicity of infection of between 1 and 5 TUs per target cells (*see* **Note 12**).

3.2.3 Selection for Specific cDNAs

1. Based on the procedure described above (Subheading 3.2.1), generate enough CSPW-HCVpp or CSZW-HCVpp to transduce 5x10^6 HCV-permissive cells (e.g., Huh-7.5) (*see* **Note 13**).
2. Forty-eight hours after transduction with the cDNA library (i.e., V1/library-VSVGpp), use CSPW- or CSZW-HCVpp to transduce the target cells aiming for a MOI of 1 (*see* **Note 14**).
3. After another 48 h, apply antibiotic selection. Cells not transduced with CSPW/CSZW should be exposed to antibiotic in parallel to control for efficiency of selection.

3.2.4 Cyclic Packaging Rescue (CPR)

1. After all untransduced cells have been eliminated by antibiotic selection, surviving transduced cells are pooled and further expanded.
2. A fraction of the cells is challenged with CSGW-HCVpp and CSGW-VSVGpp as well as CSGW lentiviral particles without a viral envelope protein (CSGW-NoEnvpp) and assayed by flow cytometry to gauge the permissiveness to HCVpp compared to controls (NoEnv and VSVG).
3. The remaining cells are either challenged again with HCVpp encoding resistance to a different antibiotic or subjected to CPR (*see* **Note 15**).
4. To perform CPR transduce surviving 293T cells expressing the post-selection cDNA library as described above (Subheading 3.2.1), but use only HIV gag-pol and VSV-G at a 2:1 (weight/weight) ratio.
5. Harvest V1/library-VSVGpp containing supernatant on days 2 and 3. Filter and pool as described above.
6. Use V1/library-VSVGpp to transduce a naïve batch or target cells (e.g., 293T) (Subheading 3.2.2).

3.2.5 Isolation and Sequence Identification of Hits

1. When testing of selected cells (Subheading 3.2.4, **step 2**) indicates that enhancement of HCV permissiveness has been achieved, cDNAs in surviving target cells should be amplified and sent for sequencing.
2. Isolate genomic DNA from target cells using any commercially available kit (*see* **Note 16**).
3. Use standard PCR to amplify cDNAs contained and send for sequencing.

4 Notes

1. Cell cultures are normally kept in DMEM/10% FCS. Only during pseudoparticle production and transduction they are kept in DMEM/3% FCS.
2. These vectors each encode a lentiviral genome containing a GFP (CSGW), puromycin resistance (CSPW), or zeocin resistance (CSZW) reporter under the control of a spleen focus forming virus promoter. The vectors contain a deletion in the 3′ LTR U3 region resulting in inactivation of the viral promoter upon reverse transcription. CSGW and CSPW have been described by Demaison and colleagues [14]. Construction of CSZW by modification of CSGW has also been described [11].
3. V1, used here as a lentiviral expression vector for the cDNA library, generates packaging competent transcripts including the cloned cDNAs from integrated proviruses, thus enabling their repackaging into new lentiviral particles and transfer into new cells—one precondition for CPR. It was generated in the laboratory of Paul Bieniasz and encodes a lentiviral genome from which several genes have been removed. A cloning site for insertion of cDNAs has been placed in lieu or the Nef gene. Tat, Rev, Vpu, and other elements remain intact so that packaging competent transcripts are generated from integrated V1 proviruses. V1-GFP is a variant containing GFP in lieu of a cDNA. More details on V1 have been described by the Bieniasz group [15, 16]. Other cloning vectors (such as Clontech's λTriplEx2) can be used instead of V1 with appropriate adjustment of the protocol. Most importantly, the oligonucleotides for first- and second-strand synthesis must match the vector employed.
4. For cDNA library generation from Huh-7.5 cells, we used several commercially available kits to perform parts of the overall protocol. These include Invitrogen TRIzol reagent for RNA preparation, Oligotex Direct mRNA Midi/Maxi Kit for mRNA isolation, and Clontech SMART cDNA Library

Construction Kit with modifications as described by Evans et al. [11] for synthesis of the actual cDNA library.

5. For library generation you will need around 5 mg total RNA at this point.
6. Alternatively, other RT-PCR systems (e.g., SuperScript first-strand synthesis system for RT-PCR, Invitrogen) and high-fidelity PCR systems (e.g., Expand Long Template PCR System, Roche) can be employed for first- and second-strand synthesis, respectively. Primers for first- and second-strand synthesis have to be adjusted depending on whether V1, Clontech's λTriplEx2, or yet another vector is used to clone the cDNA library into. In any case, at least 10 μg DNA should be available after second-strand PCR in total.
7. In our experience, library complexity should be at least 5×10^6 clones; over 70% of clones should contain inserts with a broad size distribution from 200 to 2000 bp. Of note, host factors encoded by very large cDNAs are more difficult to detect by this type of screen.
8. Since the V1 vector does not encode a reporter gene, pseudoviruses carrying the library cannot be titered directly. Instead, the TZM indicator cell line that has an integrated lacZ gene that is transcribed via the HIV-1 LTR can be used: infection of a TZM cell by a V1 containing pseudotype results in Tat expression from the V1 provirus, lacZ activation, and thus visualization of entry. This can be used to determine titer in a tissue culture infectious dose 50% (TCID50) assay. The effective titer may differ between cell lines and between pseudoviruses. To account for this, V1/library-VSVGpp and V1/GFP-VSVGpp are titered in parallel on TZM cells. In parallel V1/GFP-VSVGpp is titered using FACS detection of GFP expression on the cell line to which the library is to be delivered, i.e., 293T. Based on this information, the titer of the pseudoparticles carrying the library into the target cell population can be approximated.
9. Tissue culture infectious dose 50% (TCID50) is determined by counting the wells containing at least one positive cell per dilution and applying the method described by Reed and Muench [17].
10. Titer should be determined based on a well where between 2 and 20% of cells are GFP positive. Titer can be calculated based on percentage of GFP-positive cells, cell number at time of transduction, dilution of supernatant, and mL of supernatant used to inoculate well.
11. A higher or lower number of TUs may be required when the unknown cDNA of interest is expressed in the donor cell line with particularly low or high abundance, respectively.

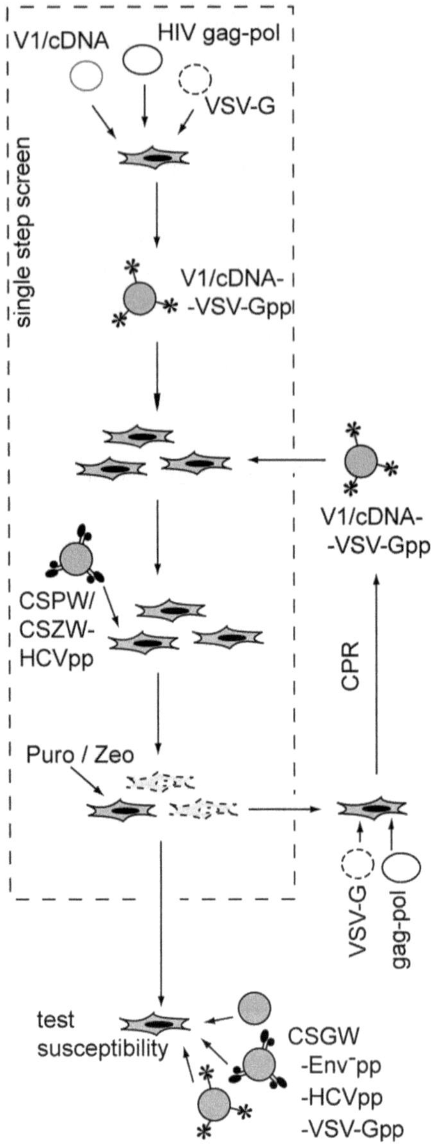

Fig. 2 Flow diagram of CPR screening. See text for details

12. If performing cyclic packaging rescue (CPR; Fig. 2), one may choose to use a lower MOI (0.5–1) in the last cycle, when planning to sequence surviving clones.
13. Since the actual titer of HCVpp on a formerly HCV-resistant cell line that has been rendered HCV-permissive through expression of a cDNA encoding an HCV entry factor is unknown, it is reasonable to use a naturally HCV-permissive cell line to measure titer of HCVpp preparations. To be able to approximate HCVpp titer quickly, one may choose to generate a small amount of CSGW-HCVpp in parallel to the antibiotic

resistance gene carrying (CSPW or CSZW) HCVpp that will be used to challenge target cells expressing the cDNA library.

14. When using HCVpp achieving a MOI of 1 may not be possible due to the rather low inherent titer of HCVpp compared to VSVGpp. In this case a lower MOI must be used, but efforts should be undertaken to optimize HCVpp production and come as close as possible to a MOI of 1.
15. CPR requires that a repackaging competent vector such as V1 is used and that the target cell line in which the cDNA library is expressed is capable of generating lentiviral particles with reasonable efficiency. In our experience, many cell lines generate lentiviral particles to some extent, but few reach the level of efficiency seen in 293T. This should be tested and factored into the design of the screening strategy before embarking on a CPR-based expression cloning project.
16. Genomic DNA can be isolated from the pooled target cells surviving the last round of selection. More ideally, one may choose to not pool cell clones surviving the final round of selection, but rather subclone and generate genomic DNA from individual clones.

Acknowledgments

The authors would like to acknowledge Theodora Hatziioannou and Paul Bieniasz of Aaron Diamond AIDS Research Center (NY, USA) who conceived, designed, generated, and provided the V1 vector enabling a CPR-based cloning strategy. The authors would also like to acknowledge the contribution of Charles Rice (The Rockefeller University, NY, USA) and Matthew Evans (Mount Sinai School of Medicine, NY, USA) who made very major contributions to devising the HCV-customized strategy described here.

References

1. Scarselli E, Ansuini H, Cerino R, Roccasecca RM, Acali S, Filocamo G et al (2002) The human scavenger receptor class B type I is a novel candidate receptor for the hepatitis C virus. EMBO J 21:5017–5025
2. Watashi K, Hijikata M, Hosaka M, Yamaji M, Shimotohno K (2003) Cyclosporin a suppresses replication of hepatitis C virus genome in cultured hepatocytes. Hepatology 38:1282–1288
3. Nakagawa M, Sakamoto N, Enomoto N, Tanabe Y, Kanazawa N, Koyama T et al (2004) Specific inhibition of hepatitis C virus replication by cyclosporin A. Biochem Biophys Res Commun 313:42–47
4. Andre P, Komurian-Pradel F, Deforges S, Perret M, Berland JL, Sodoyer M et al (2002) Characterization of low- and very-low-density hepatitis C virus RNA-containing particles. J Virol 76:6919–6928
5. Wozniak MA, Itzhaki RF, Faragher EB, James MW, Ryder SD, Irving WL et al (2002) Apolipoprotein E-epsilon 4 protects against severe liver disease caused by hepatitis C virus. Hepatology 36:456–463

6. Toniutto P, Fabris C, Fumo E, Apollonio L, Caldato M, Mariuzzi L et al (2004) Carriage of the apolipoprotein E-epsilon4 allele and histologic outcome of recurrent hepatitis C after antiviral treatment. Am J Clin Pathol 122:428–433
7. Price DA, Bassendine MF, Norris SM, Golding C, Toms GL, Schmid ML et al (2006) Apolipoprotein epsilon3 allele is associated with persistent hepatitis C virus infection. Gut 55:715–718
8. Chang KS, Jiang J, Cai Z, Luo G (2007) Human apolipoprotein e is required for infectivity and production of hepatitis C virus in cell culture. J Virol 81:13783–13793
9. Long G, Hiet MS, Windisch MP, Lee JY, Lohmann V, Bartenschlager R (2011) Mouse hepatic cells support assembly of infectious hepatitis C virus particles. Gastroenterology 141:1057–1066
10. Pileri P, Uematsu Y, Campagnoli S, Galli G, Falugi F, Petracca R et al (1998) Binding of hepatitis C virus to CD81. Science 282:938–941
11. Evans MJ, von Hahn T, Tscherne DM, Syder AJ, Panis M, Wolk B et al (2007) Claudin-1 is a hepatitis C virus co-receptor required for a late step in entry. Nature 446:801–805
12. Ploss A, Evans MJ, Gaysinskaya VA, Panis M, You H, de Jong YP et al (2009) Human occludin is a hepatitis C virus entry factor required for infection of mouse cells. Nature 457:882–886
13. Derdeyn CA, Decker JM, Sfakianos JN, Wu X, O'Brien WA, Ratner L et al (2000) Sensitivity of human immunodeficiency virus type 1 to the fusion inhibitor T-20 is modulated by coreceptor specificity defined by the V3 loop of gp120. J Virol 74:8358–8367
14. Demaison C, Parsley K, Brouns G, Scherr M, Battmer K, Kinnon C et al (2002) High-level transduction and gene expression in hematopoietic repopulating cells using a human immunodeficiency [correction of immunodeficiency] virus type 1-based lentiviral vector containing an internal spleen focus forming virus promoter. Hum Gene Ther 13:803–813
15. Cowan S, Hatziioannou T, Cunningham T, Muesing MA, Gottlinger HG, Bieniasz PD (2002) Cellular inhibitors with Fv1-like activity restrict human and simian immunodeficiency virus tropism. Proc Natl Acad Sci U S A 99:11914–11919
16. Zennou V, Bieniasz PD (2006) Comparative analysis of the antiretroviral activity of APOBEC3G and APOBEC3F from primates. Virology 349:31–40
17. Reed LJ, Muench H (1938) A simple method of estimating fifty per cent endpoints. Am J Hyg 27:493–497

Chapter 12

Application of Deep Mutational Scanning in Hepatitis C Virus

Nicholas C. Wu and Hangfei Qi

Abstract

Mutagenesis is one of the key techniques in virus research. The recent development of deep mutational scanning allows the assessment of replication fitness effects of a large number of viral mutants in a high-throughput manner. Here, we describe a protocol for studying hepatitis C virus (HCV) using deep mutational scanning, which includes the methodologies for mutant library construction, passaging, sequencing, and data analysis.

Key words HCV, Deep mutational scanning, Mutagenesis, Next-generation sequencing, Viral fitness

1 Introduction

Forward genetics and reverse genetics are commonly used approaches in virology research. Forward genetics aims to identify genotypes responsible for a particular phenotypic observation, whereas reverse genetics aims to determine the phenotypic consequences of a given genetic modification. While these approaches are useful and have facilitated many discoveries, they are relatively low-throughput. With the recent advance in next-generation sequencing, a high-throughput genetic approach that integrates forward and reverse genetics has been developed. High-throughput genetic approach employs random mutagenesis to generate a large number of mutants, and monitors the replication fitness of individual mutants under a specified condition using next-generation sequencing [1, 2]. Random mutagenesis can be performed by a transposon-based approach [3–5], error-prone PCR [6, 7], or random oligo synthesis [8–11]. Transposon-based approach generates insertional mutants, whereas error-prone PCR and random oligo synthesis generate amino acid substitutions. Therefore, both error-prone PCR and random oligo synthesis allow fitness profiling at single-residue resolution. Of note, a

Mansun Law (ed.), *Hepatitis C Virus Protocols*, Methods in Molecular Biology, vol. 1911,
https://doi.org/10.1007/978-1-4939-8976-8_12,

major difference between error-prone PCR and random oligo synthesis is that amino acid substitutions generated from error-prone PCR are mostly limited to 6 to 7 per codon because of genetic code arrangement. In contrast, random oligo synthesis can randomize a given codon to all possible amino acids. High-throughput genetics with single-residue resolution is often called deep mutational scanning. Deep mutational scanning has been applied to study different viruses [6–8, 11–13], including fitness profiling and drug resistance profiling of NS5A protein in hepatitis C virus (HCV) [8]. Here, we discuss the protocol of deep mutational scanning in HCV. The protocol described here is largely based on our deep mutational scanning study on the domain IA of NS5A [8].

2 Materials

1. Huh-7.5.1 cell line is cultured in Dulbecco's modified Eagle's medium (DMEM) supplemented with 10% of fetal bovine serum (FBS), 0.1 mM nonessential amino acids, 10 mM HEPES, penicillin (100 units/ml), streptomycin (100 μg/ml), and 2 mM L-glutamine at 37 °C with 5% CO_2.
2. A plasmid of pFNX-HCV that carries the viral genome was synthesized based on the chimeric sequence of J6/JFH1 virus [14].
3. KOD DNA polymerase (EMD Millipore).
4. T4 ligase (Life Technologies).
5. BsaI, BamHI, and RsrII (New England Biolabs).
6. MegaX DH10B T1R cells (Life Technologies).
7. T7 Ribomax Express kit (Promega).
8. DNase I (New England Biolabs).
9. BioRad electroporator.
10. QIAamp Viral RNA Mini Kit (QIAGEN).
11. Superscript III (Life Technologies).
12. Illumina HiSeq 2000/2500 platform.
13. Opti-MEM.

3 Methods

Carry out all procedures at room temperature unless otherwise specified.

3.1 Generation of a Plasmid Mutant Library

1. The area to be mutated is subdivided into small regions, with each composed of at most 18 amino acids.

2. Oligos spanning through each region are synthesized commercially so that each codon to be mutated is substituted with three continuous random nucleotides "N_1N_2K" (*see* **Note 1**). N_1 and N_2 represent random incorporation of A/T/G/C, and K represents random incorporation of T/G to minimize stop codon mutation. The number of amino acid substitution in each oligo can be easily manipulated with this approach. In this particular case [8], we conducted single amino acid mutagenesis. Therefore, each oligo only contained one randomized codon. BsaI recognition site is present on each end of the synthesized oligo (*see* **Note 2**). In addition, a constant region is present at the 3′ end of the oligo.
3. These oligos, each carrying a randomized codon, are mixed at equal molar ratio and duplexed using a one-cycle PCR using KOD DNA polymerase (EMD Millipore) with 1.5 mM $MgSO_4$, 0.2 mM of each dNTP (dATP, dCTP, dGTP, and dTTP), 600 nM of the oligo mix, and 600 nM of the reverse primer. The reverse primer annealed to the constant region and the BsaI site at the 3′ end of the oligos. The thermocycler was set as follows: 2.5 min at 95 °C, 30 s at 5 °C, and 2 min at 68 °C.
4. The DNA fragments are then digested with BsaI enzyme to generate the insert to make the plasmid mutant library.
5. To avoid amplifying long stretch of DNA fragments in PCR, a 2 kb fragment containing the target mutant regions is excised from the WT plasmid and circulated using T4 ligase.
6. The circulated DNA fragment is then used as a template to amplify the regions that flank (both 5′-end and 3′-end) the desired mutation region by PCR using primers containing a BsaI recognition sites on both ends. The PCR was performed using KOD DNA polymerase with 1.5 mM $MgSO_4$, 0.2 mM of each dNTP (dATP, dCTP, dGTP, and dTTP), 10 ng of the circulated DNA fragment, and 600 nM of forward and reverse primer. The thermocycler was set as follows: 2 min at 95 °C, then 20 three-step cycles of 20 s at 95 °C, 15 s at 58 °C, 1 min at 68 °C, and a 2-min final extension at 68 °C.
7. The PCR amplicons are digested with BsaI enzyme to produce the "sticky ends" matching the inserts generated with BsaI-digested duplexed oligos.
8. The BsaI-digested insert and flanking regions are ligated using T4 DNA ligase.
9. The ligated product is subcloned into the pFNX-HCV vector via BamHI and RsrII digestions that generate compatible ends and transformed into MegaX DH10B T1R cells.

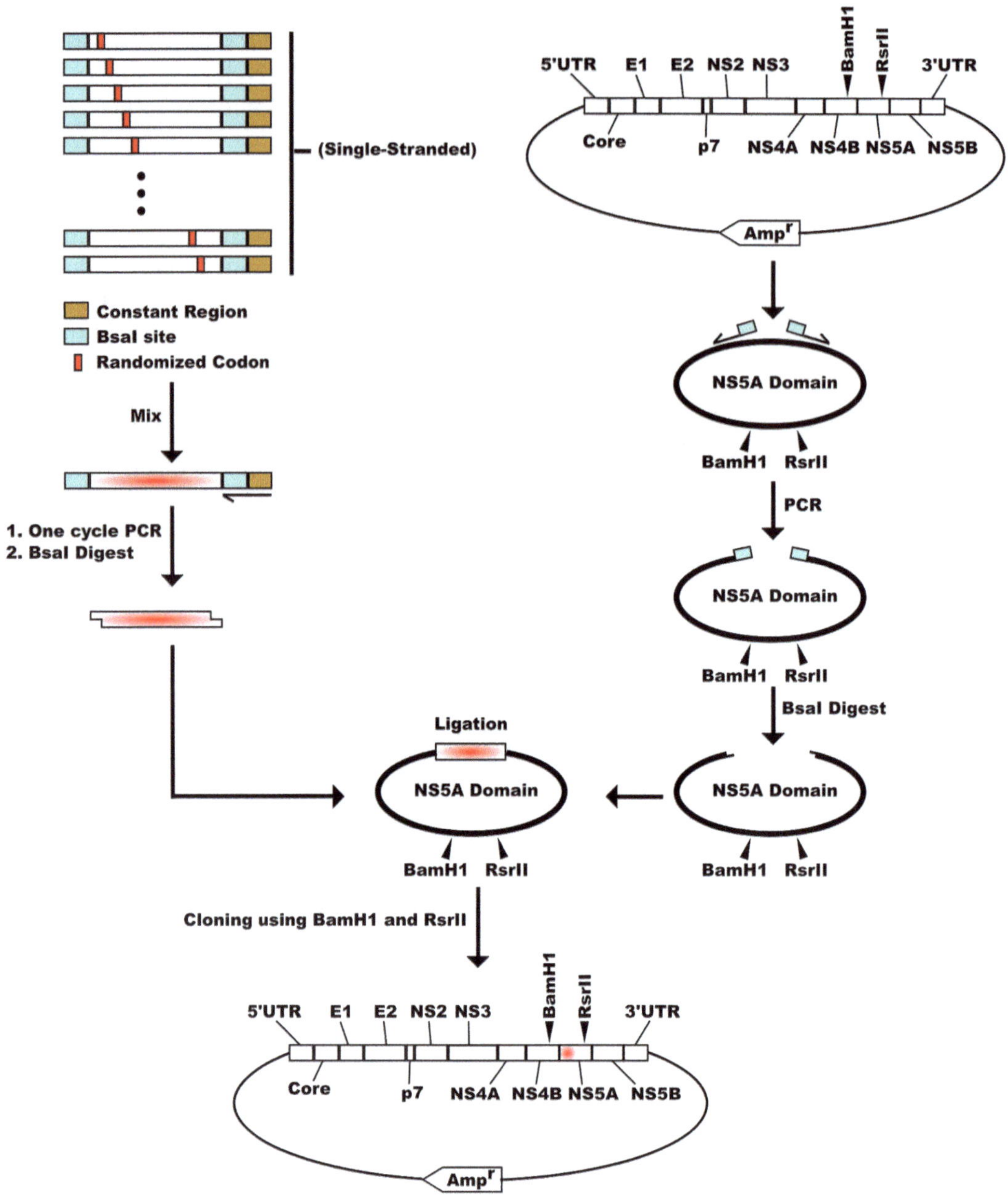

Fig. 1 Schematic diagram for generating a plasmid mutant library

10. Approximately 50,000 colonies are collected for each library to achieve an average coverage of at least 100-fold for each amino acid substitution.
11. The overall flow for this cloning process is shown in Fig. 1.

3.2 Passaging of the Virus Mutant Library

1. A total of 16 μg of the pFNX library DNA is used as the template for in vitro transcription using the T7 Ribomax Express Kit and is then treated with DNase I.
2. Four million Huh-7.5.1 cells are trypsinized and washed twice with ice-cold Opti-MEM transfection media and resuspended in Opti-MEM at 1×10^7 cells per ml.
3. 10 μg of in vitro transcribed RNA is mixed with 400 μl resuspended Huh-7.5.1 cells in 0.4 cm electroporation cuvettes.
4. Electroporation is performed by a BioRad electroporator with 270 V, 100 Ω, and 960 μF.
5. Cells are resuspended in 40 ml of complete DMEM and plated in T-75 culture flasks.
6. Medium is replaced at 18 h post-transfection.
7. Supernatant that contains the virus mutant library is harvested at 96 h post-transfection, and the virus titer is determined in focus-forming units per milliliter [3].
8. The mutant virus library is used to infect four million of naïve Huh-7.5.1 cells at a multiplicity of infection (MOI) of 0.2 (*see* **Note 3**).
9. The supernatant is collected at 144 h post-infection and used to infect naïve cells for the second round of selection.
10. **Steps 8** and **9** can be repeated if additional rounds of selection are desired.

3.3 Next-Generation Sequencing of the Mutant Libraries

1. The supernatant of each virus pool is spun at 1500 × *g* for 10 min to remove cell-associated RNA.
2. Total RNA is isolated from 1.4 ml of supernatant using QIAamp Viral RNA Mini Kit for viral RNA purification.
3. A total amount of 2 μg of extracted RNA is reverse transcribed into cDNA with Superscript III according to manufacturer's instruction.
4. The mutated regions from cDNA are PCR amplified with primers containing part of the Illumina TruSeq adapter region.
5. A second PCR is performed to add the rest of the TruSeq adapter region as well as barcode index sequence for multiplex sequencing.
6. The resultant PCR product is analyzed by Illumina HiSeq 2000/2500 platform.

3.4 Data Analysis

1. Forward and reverse reads from the same paired-end read are compared.
2. Paired-end reads with forward and reverse reads that are not reverse-complementary are discarded.

3. Each paired-end read is compared to the wild-type sequence using pairwise alignment to identify the amino acid substitutions.
4. The relative frequency of a given amino acid substitution, s, in a given sample N ($f_{s,N}$) is calculated as follows:

$$f_{s,N} = \frac{\text{Reads}_{s,N}}{\sum \text{Reads}_N}$$

where $\text{Reads}_{s,N}$ indicates the number of sequence reads for the amino acid substitution s in sample N and ΣReads_N represents the total reads in the sample N (*see* **Note 4**).

5. Similarly, the relative frequency of wild type (WT), wt, in a given sample N ($f_{wt,N}$) is calculated as follows:

$$f_{wt,N} = \frac{\text{Reads}_{wt,N}}{\sum \text{Reads}_N}$$

where $\text{Reads}_{wt,N}$ indicates the number of sequence reads for the WT in sample N and ΣReads_N represents the total reads in the sample N.

6. The relative fitness score of a given amino acid substitution (W_s) was determined as the antilogarithm of the slope of the regression:

$$\ln\left(\frac{f_{s,N}}{f_{wt,N}}\right) = \ln\left(\frac{f_{s,0}}{f_{wt,0}}\right) + \text{N} \ln W_s$$

where $\ln\left(\frac{f_{s,0}}{f_{wt,0}}\right)$ is the logarithm of the frequency of a given amino acid substitution s in the plasmid mutant library ($N = 0$) and $\ln\left(\frac{f_{s,N}}{f_{wt,N}}\right)$ represents the logarithm of the frequency of a given amino acid substitution s in virus mutant library ($N = 1$), round 1 passaged virus mutant library ($N = 2$), and round 2 passaged virus mutant library ($N = 3$). In this calculation, a $W_s = 1$ means the same fitness as the WT virus, whereas a $W_s < 1$ represents decreased fitness compared WT (attenuate, deleterious, or lethal), and a $W_s > 1$ is a mutation that shows higher fitness than WT in the selection (*see* **Note 5**).

4 Notes

1. We ordered the oligos from Integrated DNA Technologies in the 96-well format ("25 nmole DNA Plate Oligo") with normalized yield.

2. BsaI is a type IIS restriction enzyme, which recognizes asymmetric DNA sequence and cleaves outside of its recognition sequence.
3. In the study of profiling the drug resistance of the NS5A inhibitor, we applied 20 nM of Daclatasvir (NS5A inhibitor) for two rounds of infection during the selection. Any change of drug sensitivity (either decrease or increase) that is caused by mutations on the protein can be determined, and the positions that harbor these mutations are presumably the essential residues that mediate the compound-protein interactions. Presumably, any type of selection pressure can be employed to the screen for the interested phenotype, e.g., IFN to study immune modulation of the proteins.
4. For sequencing data with low coverage, a pseudocount of 1 is added to the read count $\mathrm{Reads}_{s,N}$ to avoid dividing by zero in the subsequent steps.
5. To validate the screen results, we constructed individual mutations that span the range of all phenotypes and span a range of functional and structural motifs along with a negative control virus that is defective in RNA polymerase activity (NS5B_GNN) and a positive control WT virus. Replication of these viruses was individually determined and compared to the screen data.

Acknowledgment

NCW was supported by the Croucher Foundation Fellowship.

References

1. Qi H, Wu NC, Du Y, Wu TT, Sun R (2015) High-resolution genetic profile of viral genomes: why it matters. Curr Opin Virol 14:62–70
2. Fowler DM, Fields S (2014) Deep mutational scanning: a new style of protein science. Nat Methods 11:801–807
3. Arumugaswami V, Remenyi R, Kanagavel V, Sue EY, Ngoc Ho T, Liu C et al (2008) High-resolution functional profiling of hepatitis C virus genome. PLoS Pathog 4:e1000182
4. Remenyi R, Qi H, Su SY, Chen Z, Wu NC, Arumugaswami V et al (2014) A comprehensive functional map of the hepatitis C virus genome provides a resource for probing viral proteins. MBio 5:e01469–e01414
5. Qi H, Chu V, Wu NC, Chen Z, Truong S, Brar G et al (2017) Systematic identification of anti-interferon function on hepatitis C virus genome reveals p7 as an immune evasion protein. Proc Natl Acad Sci U S A 114:2018–2023
6. Wu NC, Young AP, Al-Mawsawi LQ, Olson CA, Feng J, Qi H et al (2014) High-throughput profiling of influenza A virus hemagglutinin gene at single-nucleotide resolution. Sci Rep 4:4942
7. Wu NC, Olson CA, Du Y, Le S, Tran K, Remenyi R et al (2015) Functional constraint profiling of a viral protein reveals discordance of evolutionary conservation and functionality. PLoS Genet 11:e1005310
8. Qi H, Olson CA, Wu NC, Ke R, Loverdo C, Chu V et al (2014) A quantitative high-resolution genetic profile rapidly identifies sequence determinants of hepatitis C viral fitness and drug sensitivity. PLoS Pathog 10:e1004064

9. Olson CA, Wu NC, Sun R (2014) A comprehensive biophysical description of pairwise epistasis throughout an entire protein domain. Curr Biol 24:2643–2651
10. Wu NC, Grande G, Turner HL, Ward AB, Xie J, Lerner RA et al (2017) In vitro evolution of an influenza broadly neutralizing antibody is modulated by hemagglutinin receptor specificity. Nat Commun 8:15371
11. Doud MB, Bloom JD (2016) Accurate measurement of the effects of all amino-acid mutations on influenza hemagglutinin. Viruses 8: E155
12. Al-Mawsawi LQ, Wu NC, Olson CA, Shi VC, Qi H, Zheng X et al (2014) High-throughput profiling of point mutations across the HIV-1 genome. Retrovirology 11:124
13. Haddox HK, Dingens AS, Bloom JD (2016) Experimental estimation of the effects of all amino-acid mutations to HIV's envelope protein on viral replication in cell culture. PLoS Pathog 12:e1006114
14. Lindenbach BD, Evans MJ, Syder AJ, Wolk B, Tellinghuisen TL, Liu CC et al (2005) Complete replication of hepatitis C virus in cell culture. Science 309:623–626

Chapter 13

Studying the Hepatitis C Virus-Induced Epigenetic Signature After Cure with Direct-Acting Antivirals

Shira Perez and Meital Gal-Tanamy

Abstract

Hepatitis C virus (HCV) is the leading cause of hepatocellular carcinoma (HCC). While direct-acting antiviral (DAA) therapy efficiently eradicates HCV infection, epidemiological studies show that sustained virological response (SVR) following anti-HCV treatment reduces, but does not eliminate, the risk for HCC. We have recently demonstrated that HCV infection induces genome-wide epigenetic changes that reprogram host gene expression and persist as "epigenetic signature" following virus eradication by DAAs. We suggest that this epigenetic signature underlie the residual risk for HCC post-SVR. Here, we provide a methodology to study the HCV-induced epigenetic signature. We describe a ChIP-seq protocol to evaluate changes in epigenome profile following HCV infection, its cure with DAA, and after treatment with epigenetic modifier inhibitor. We also describe evaluation of changes in the gene expression profile using RNA-seq. The integration between detected alterations in epigenetic marks and gene expression allows for identification of biological processes that are involved in HCV-driven oncogenesis before and after cure.

Key words Hepatocellular carcinoma (HCC), Epigenetics signature, Direct-acting antivirals (DAAs), Epigenetic drugs

1 Introduction

The marked progress in epigenetic research over the past decade has reformed our understanding of the role of epigenetics in the development of various diseases, including cancer. Epigenetics is defined as the heritable elements of the genome that are not borne in the DNA sequence [1]. Epigenetics is dictated by a series of DNA methylations and posttranslational histone modifications (PTMs) that determine DNA packaging and, thus, the accessibility of the chromatin. Together, these form the "epigenetic code" that plays fundamental roles in the chromatin-mediated control of gene expression [2]. Transformation of normal cells into cancer cells involves aberrant epigenetic alterations, leading to reorganization of the epigenetic landscape of the cells, gene expression profile reprograming, and consequential cell dysregulation [2]. Since

Mansun Law (ed.), *Hepatitis C Virus Protocols*, Methods in Molecular Biology, vol. 1911,
https://doi.org/10.1007/978-1-4939-8976-8_13,

epigenetic codes, unlike somatic mutations, are reversible, understanding the mechanisms driving the formation and maintenance of these codes can provide fertile ground for development of epigenetic therapy [2]. Inhibitors of epigenetic modulatory enzymes are currently in clinical trials for treatment of a variety of cancers [3].

Epigenetic changes are dynamic and can be altered in response to environmental cues [4]. The changed epigenetic landscape of the cells may be imprinted in subsequent cell divisions and maintained by an epigenetic memory mechanism that is not well understood, leaving an "epigenetic signature" [5]. It has been suggested that viruses exploit the dynamic nature of the epigenetic landscape to impose epigenetic changes that reprogram gene expression profile essential for their life cycle [6, 7]. Hepatitis C virus (HCV) infection is the leading cause of chronic hepatitis, which often results in liver fibrosis, cirrhosis, and hepatocellular carcinoma (HCC). Recently, several combinations of interferon-free, direct-acting antiviral (DAA) regimes have dramatically improved HCV therapy, yielding SVR rates of >95%, with limited toxicity [8]. Still, HCC remains a prevailing complication in SVR populations [9, 10]. The development of HCC post-SVR was associated with a specific gene expression signature in HCV-infected patients that persisted even after antiviral cure [10, 11]. Recently, we have demonstrated that, although HCV is an exclusively cytoplasmic virus, infection with HCV leads to global changes in host gene expression and functions through global alteration of the epigenetic landscape of the host cell chromatin. Interestingly, these epigenetic and transcriptional changes were maintained as epigenetic signature even following cure with DAAs. Moreover, the HCV-induced epigenetic signature was reverted by epigenetic inhibitors. We suggest that this epigenetic signature may provide a mechanism for the residual risk for HCC development after HCV eradication by DAAs.

In this chapter, we provide a detailed strategy for exploring HCV-induced epigenetic signatures and their reversal by epigenetic drugs. The epigenetic landscape is explored by mapping the positioning of histone modifications that play critical roles in determining gene expression [2]. In general, histone modifications affect the inter- or intra-nucleosomal DNA-histone interactions and, thus, the formation of higher-order chromatin [12]. There are at least eight distinct types of modifications found on histones, with the covalent acetylation, methylation, and phosphorylation modifications being the most studied. Each type of modification is executed by a group of specific histone-modifying enzymes [12, 13]. Numerous lysine and arginine residues in the N-terminal tails of histones H3 and H4 are subject to acetylation, which is regulated by the opposing action of histone acetyltransferases (HATs) and histone deacetylases (HDACs) [14], and to methylation, which is regulated by histone methyltransferase enzymes (HMTs). As an example, we describe the profiling of the histone H3 lysine 9 acetylation

(H3K9Ac) that is associated with open chromatin and active gene expression and has been associated with liver-specific epigenetics [15, 16]. This modification results in a change in the electrostatic charge of histones, which alters the structural properties of histone and its binding to the DNA [14].

We describe a protocol for chromatin immunoprecipitation followed by sequencing (ChIP-seq) that is a standard method for genome-wide mapping of the position of specific modifications on chromatin-bound histones [17]. This process involves the covalently cross-linking of histones to the DNA. The cross-linked chromatin is purified and fragmented by enzymatic digestion or mechanical force. Chromatin fragments are then precipitated using a specific antibody that targets the specific histone modification being explored. The precipitated DNA is purified and used as a template for next-generation sequencing (NGS) [17]. The sequenced DNA fragments are then mapped to the reference genome to determine the genomic locations of the histone modification [18].

The ChIP-seq protocol provided here is adjusted for genome-wide profiling of the epigenetic alterations induced following HCV infection. This protocol describes ChIP-seq that is specific for histone modification H3K9Ac but can be modified to explore any histone modification of interest, by using an appropriate specific antibody. To characterize the epigenetic signature induced by HCV, the profile of H3K9Ac is explored in HCV-infected cells and compared to that of control non-infected cells. In addition, the signature of HCV-cured cells, where HCV was eradicated by DAAs (at least 1 month following HCV eradication), is compared to that of control cells treated with DAAs. Then, reversal of this epigenetic signature by the HAT inhibitor C646 [19] is assessed following treatment of HCV-cured cells with the inhibitor versus similarly treated control non-infected cells. To evaluate the effect of these treatments on gene expression, we also performed RNA-seq on the samples described above. All data is integrated and used to identify gene signature that their epigenetic and gene expression profiles are altered following HCV infection; these alterations remain persistent following HCV eradication and are reverted following epigenetic inhibitor treatment. The identified gene signature can be used for pathway analysis, to determine the biological processes that are involved. The involvement of a specific gene or a biological process of interest can then be future explored. The complete workflow is demonstrated in Fig. 1.

One technical limitation for the protocol we describe here is that it requires large number of cells. In addition, it requires a cell system supporting robust HCV infection to induce the epigenetic changes and that can be passaged for at least 1 month following virus eradication, which presents a further challenge. Therefore, this protocol cannot be implemented with primary hepatocytes

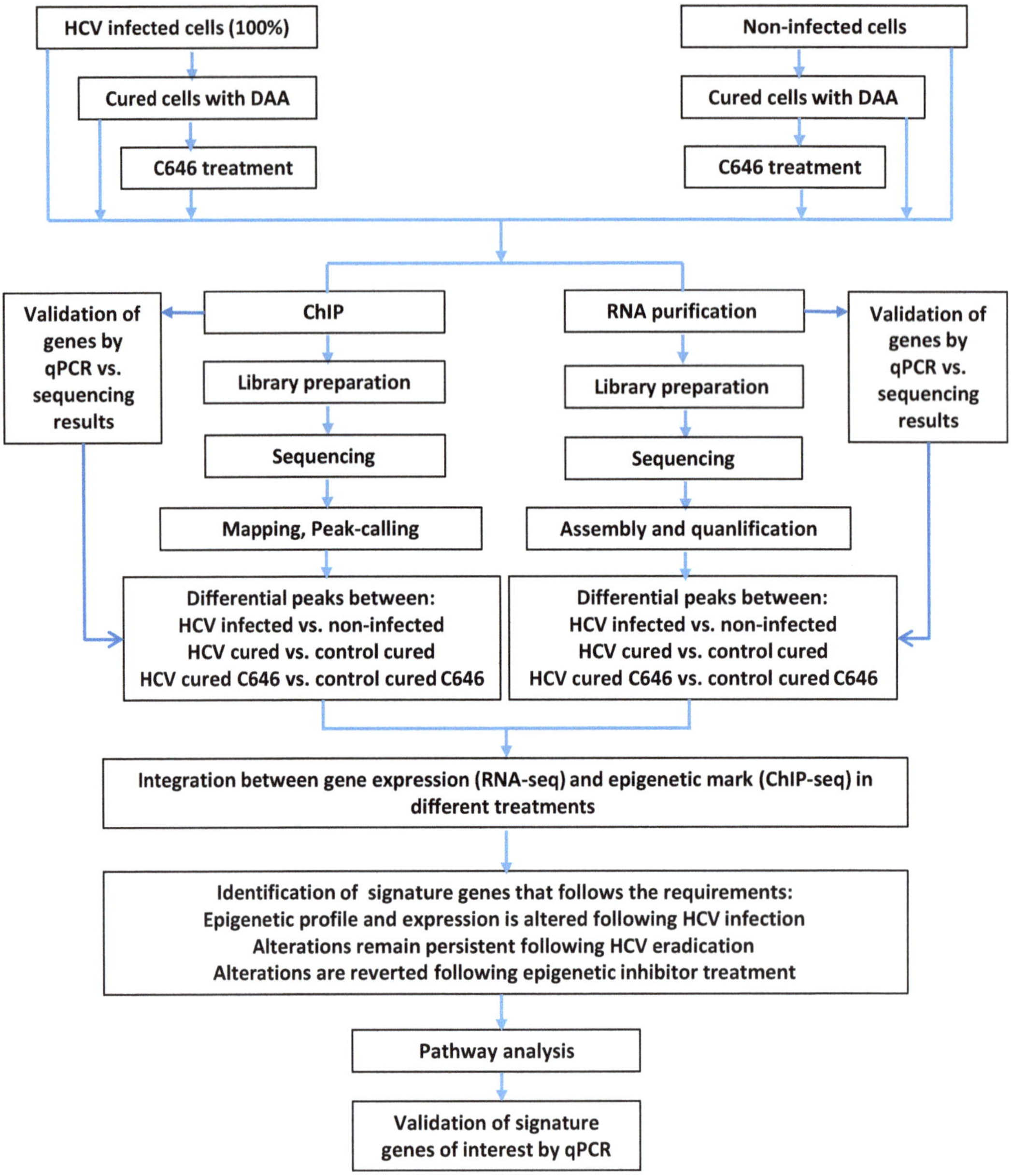

Fig. 1 Scheme of workflow for studying HCV-induced epigenetic signature after cure by direct-acting antivirals

that can only be infected for a short period of time and maintained in culture for only 2–3 weeks. Currently, Huh7.5 are the only cell culture model suitable for this protocol, as they support efficient HCV infection and can be maintained for weeks in culture. Although Huh7.5 are cancer cells, they have been extensively used in studies exploring the HCV life cycle, virus-host interactions, and viral pathogenesis, which have led to important landmarks in HCV research [20–31]. Huh7.5 cells can also be cultured in 2% human serum, to induce cell differentiation and produce hepatocyte-like cells, which produce high titers of infectious HCV that are stable for at least 65 days [32]. Moreover, gene expression profiles of these hepatocyte-like cells significantly overlap with those of primary human hepatocytes [33]. Exploring the epigenetic profile of liver biopsies requires significant medications to this protocol and other kits specialized to human tissues.

2 Materials

When working with chemicals, always wear a suitable lab coat, disposable gloves, and protective goggles. For more information, consult the appropriate safety data sheets, available from the product supplier.

Prepare all solutions using ultrapure water and analytical grade reagents. Prepare and store all ChIP reagents at room temperature (unless indicated otherwise) and RNA-seq and library preparation on ice (unless indicated otherwise). Diligently follow all waste disposal regulations when disposing waste materials such as formaldehyde.

2.1 Preparation of Cells and Virus

Stock of HCVcc HJ3-5 virus (a genotype 1a/2a chimeric virus containing structural genes from genotype 1a and nonstructural regions from genotype 2a JFH-1) is produced in Huh7.5 cells and is stored at −80 °C. Viral titer is determined by FFU assay in Huh7.5 cells, as described previously [34]; other viruses such as JHF-1 can be used.

2.2 Preparation of DAAs

We describe here curing HCV-infected cells with the DAA Viekirax/Exviera that contain dasabuvir (NS5B nonnucleoside polymerase inhibitor), ombitasvir (a NS5A inhibitor), paritaprevir (NS3/4A protease inhibitor), and ritonavir (pharmacokinetic enhancer) with the final concentrations of dasabuvir. Dilute Viekirax/Exviera in DMSO. Final concentration of Viekirax/Exviera to all experimental procedures was determined following XTT assay that evaluates the effective non-cytotoxic concentration of the drugs. Accordingly, for all assays use the following final concentrations of drugs: 0.4 μM dasabuvir, 3.66×10^{-2} μM ritonavir, 5.17×10^{2} μM paritaprevir, and 7.38×10^{-3} μM ombitasvir diluted in DMEM. Keep drug in aliquots at −80 °C.

This protocol can be implemented to explore the effect of other DAAs of interest.

2.3 Dissolving and Diluting Epigenetic Drug C646

Dissolve 10 mg C646 in 2.245 ml DMSO to prepare stock solution of 10 mM. Keep drug in aliquots at 80 °C. Final concentration of C646 to all experimental procedures was determined following XTT assay that evaluates the effective non-cytotoxic concentration of the drug. Accordingly, use a final concentration of 10 μM and prepare fresh prior to use.

2.4 ChIP

ChIP procedure was done using Magna Chip G kit (Millipore); the following reagents are included unless otherwise indicated.

2.4.1 Reagents

1. Glycine 2.5 M.
2. Phosphate-buffered saline (1× PBS).
3. Complete Lysis Buffer: 1% SDS, 10 mM EDTA, 50 mM Tris–HCl pH 8.0. Add 2.5 μl of Protease inhibitor cocktail II to each 0.5 ml of cell Lysis Buffer (*see* **Note 5**).
4. Complete Nuclear Lysis buffer: add 2.5 μl of Protease Inhibitor Cocktail II to each 0.5 ml of Nuclear Lysis Buffer (*see* **Note 5**).
5. Low-salt wash buffer: 0.1% SDS, 1% Triton X-100, 2 mM EDTA, 20 mM Tris–HCl pH 8.0, 150 mM NaCl.
6. High-salt wash buffer: 0.1% SDS, 1% Triton X-100, 2 mM EDTA, 20 mM Tris–HCl pH 8.0, 500 mM NaCl.
7. LiCl buffer: 0.25 M LiCl, 1% NP-40, 1% SDC, 1 mM EDTA, 10 mM Tris–HCl pH 8.0.
8. TE buffer: 20 mM Tris–HCl pH 8.0, 1 mM EDTA pH 8.0.
9. Antibody of interest for chromatin immunoprecipitation. For example, Anti-Trimethyl-Histone H3 Lys9 purified rabbit IgG (Millipore), not included.
10. 37% Formaldehyde. Not included (*see* **Note 3**).
11. Magnetic Protein G beads (Millipore).

2.4.2 Special Equipment

1. Magnetic Separator (Magna Grip Rack 8 Well).
2. Rotating wheel/platform.
3. Sonicator.
4. Thermomixer.
5. Cell scraper.

2.5 Library Preparation for ChIP-Seq

Library preparation procedure was done using NEBNext Ultra TM DNA Library Prep Kit (NEB) for Illumina.

2.5.1 Reagents Not Included in the Kit

1. Nuclease-free Water.
2. Prepare fresh 80% Ethanol—use high-quality Ethanol in nuclease-free water.
3. TE buffer X 0.1, pH 8.0.
4. AMPure XP Beads (Beckman Coulter).

2.5.2 Special Equipment

1. DNA LoBind Tubes (Eppendorf).
2. Thermal Cycler.
3. NEBNext Singleplex or Multiplex Oligos for Illumina (NEB) or customer supplied oligos.
4. Magnetic rack/stand PCR Machine.

2.6 RNA Purification

For total RNA purification from the cells, use the RNeasy Mini Kit (Qiagen).

2.6.1 Reagents Not Included in the Kit

1. Nuclease-free Water.
2. Prepare fresh 70% Ethanol—use high-quality Ethanol in nuclease-free water.
3. PBS × 1.
4. Sterile RNase-free pipet tips.

2.6.2 Special Equipment

1. 20-Gauge needle (0.9 mm diameter).
2. RNase-free syringe.

2.7 RNA-Seq

For preparation of RNA-seq libraries, use the NEBNext Ultra RNA Library Prep Kit for Illumina (NEB).

2.7.1 Reagents Not Included in the Kit

1. Nuclease-free Water.
2. Prepare fresh 80% Ethanol—use high-quality Ethanol in nuclease-free water.
3. NEBNext Singleplex or Multiplex Oligos for Illumina (NEB) or customer supplied oligos.
4. AMPure XP Beads (Beckman Coulter).
5. NEBNext Poly (A) mRNA Magnetic Isolation Module or NEBNext rRNA Depletion Kit (NEB).

2.7.2 Special Equipment

1. Magnetic rack/stand PCR Machine.
2. Thermal Cycler.

3 Methods

The protocol described here is an example for exploring the profile of PTM H3K9Ac and HAT inhibitor C646 as an epigenetic drug, but it can be adjusted to explore any PTM or epigenetic inhibitor of interest. Carry out all procedures at room temperature unless specified otherwise.

3.1 Preparation of Cells for the Experiment

For all experiments with infected cells:

3.1.1 Preparation of HCV-Infected Cells

1. Infect Huh7.5 cells with HJ3-5 chimeric virus at an MOI of 0.1.
2. Passage the cells for 2 weeks until approximately 100% of the cells are HCV positive, as we previously described [35].
3. Verify HCV infection by immunostaining using HCV-positive serum of infected subject and anti-human 488 Alexa fluor as secondary antibody or anti-core antibody. Visualize HCV infection under fluorescent microscope.

3.1.2 Preparation of HCV-Cured Cells

To explore the reversibility of HCV-induced changes in epigenetic profile following HCV eradication:

1. Infect Huh7.5 cells with HJ3-5 chimeric virus until approximately 100% of the cells are HCV positive as described above.
2. Treat HCV infected and non-infected cells with an effective non-cytotoxic concentration of the anti-HCV Viekirax/Exviera DAAs diluted in DMEM as described above. Treat the cells for 7 days and in every passage prepare fresh drug.
3. Following 7 days stop DAAs treatment. Continue to passage the cells and evaluate cure as described below.
4. During treatment and every passage until cells are harvested up to 2 months posttreatment, use immunostaining to detect HCV infection as described above to ensure that the cell are cured from HCV and no recurrence of infection is observed.
5. Evaluate HCV cure also by performing qRT-PCR with specific primers for HCV mRNA (HCV 3′ UTR forward: GCCATGGCGTTAGTATGAGTGT; HCV 3′ UTR reverse CCCTATCAGGCAGTACCACAA) to evaluate HCV levels before and 1 month after HCV eradication (termed here as "cured cells"), and compare to treated non-infected cells.

3.1.3 Preparation of HCV-Cured Cells Treated with Epigenetic Modifying Drug

1. Treat HCV Huh7.5 cured cells, and control cells with 10 μM of HAT inhibitor C646 for 7 days. In every passage prepare fresh drug.
2. Posttreatment, replace the cell medium to fresh medium not containing drug, and continue to grow for 1 day before harvesting the cells.

3.2 Chromatin Immunoprecipitation (ChIP)

The ChIP procedure includes five steps: (1) cross-linking the protein to the DNA, (2) chromatin isolation, (3) chromatin fragmentation, (4) immunoprecipitation of the protein of interest using specific antibodies, and (5) reverse cross-links of protein/DNA complexes and DNA purification.

Take 154 mm dishes for ChIP assay. Perform ChIP on fresh plates, or perform the cross-linking steps (Subheading 3.2.1 Cross-linking and lysis **steps 1–6**), and freeze immediately at −80 °C. Prior to use thaw frozen plates on ice (*see* **Note 4**).

3.2.1 Cross-Linking and Lysis

Prior to starting this section:
(a) Grow the cells (HCV-infected, HCV-cured cells, HCV-cured cell treated with C646 and control non-infected cells treated as the infected cell, as described in Fig. 1) to ~80–90% confluency in a 154 mm culture dish. (b) Cool PBS in advance, and obtain on ice for **step 3** and for **step 6** and for the washes of each 154 mm culture dish. (c) Warm nuclear lysis buffer to RT to ensure SDS is in solution before proceeding with cell lysis. (d) Thaw protease inhibitor cocktail II at room temperature for use in **steps 3** and **14** (*see* **Note 5**).

1. Add 550 μl of 37% formaldehyde to 20 ml of growth media to cross-link. Gently swirl dish to mix. Final concentration of formaldehyde is 1% (*see* **Note 3**).
2. Incubate at room temperature for 5 min. Agitation of cells is not necessary.
3. Add 15 μl of protease inhibitor cocktail II to 3 ml of PBS for each dish, and leave on ice (*see* **Note 5**).
4. Add 2 ml of glycine to each dish to quench unreacted formaldehyde. Swirl to mix, and incubate at room temperature for 5 min, and then place dishes on ice.
5. Aspirate medium, removing as much medium as possible; be careful not to disturb the cells.
6. Add 10 ml of cold PBS to wash cells and repeat wash once again.
7. Add 3 ml cold PBS containing Protease Inhibitor Cocktail II to each dish (prepared in **step 3**).
8. Scrape cells from each dish into a separate 5 ml tube.
9. Spin at 800 × *g* at 4 °C for 5 min to pellet the cells.
10. Remove supernatant. Cell pellet can be frozen at −80 °C at this step.
11. Resuspend cell pellet in complete cell Lysis Buffer. Prepare fresh.
12. Incubate on ice for 15 min; vortex the cell suspension briefly every 5 min.
13. Spin the cell suspension at 800 × *g* at 4 °C for 5 min.
14. During spin, prepare fresh complete nuclear lysis buffer, remove supernatant, and resuspend cell pellet in 0.5 ml of complete nuclear lysis buffer.

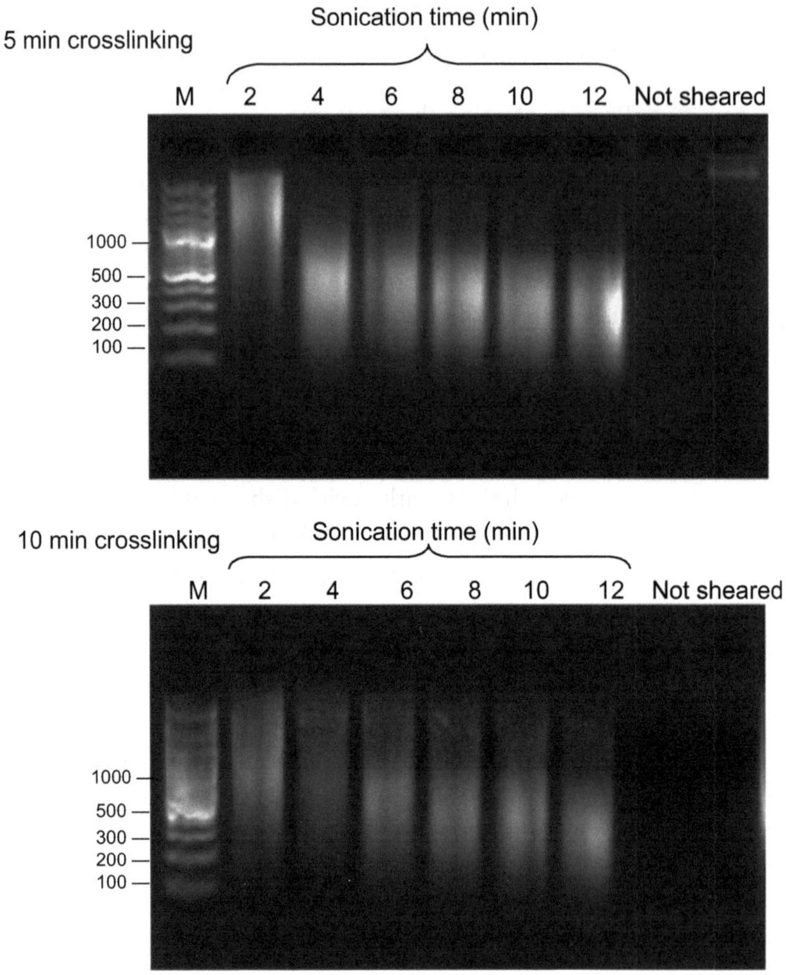

Fig. 2 Calibration of cross-linking and sonication of DNA for ChIP assay. Sonication of sheared chromatin following formaldehyde cross-linking was performed. Different cross-linking times (5 and 10 min) and sonication times were tested using the Covaris 220. 20 μl of sheared chromatin was then electrophoresed through a 2% agarose gel and stained with GelRed. At 5-min cross-linking and 6-min sonication conditions, the majority of the DNA was sheared to fragments between 200 bp and 1000 bp that are optimal for ChIP. M-100 bp plus DNA ladder (Thermo Fisher Scientific)

3.2.2 Sonication to Shear DNA

Prior to starting this section, (a) optimal conditions of cross-linking time with formaldehyde and sonication time need to be determined to shear cross-linked DNA to ~200–1000 base pairs in length as shown in Fig. 2; (b) if cell lysate from Subheading 3.2.1, **step 10**, was previously frozen, thaw on ice.

1. Sonicate.
   (a) Keep cell lysate ice-cold. Sonication produces heat, which can denature the chromatin.
2. Spin at a minimum of 10,000 × *g*, but not exceeding 15,000 × *g*, at 4 °C for 10 min to remove insoluble material.

3. Remove supernatant to fresh microfuge tubes in 50 μl aliquots. Each 50 μl aliquot contains 2×10^6 cells which is sufficient for one immunoprecipitation.
   (a) Sheared cross-linked chromatin can be stored at −80 °C for up to 3 months.

3.2.3 Immunoprecipitation (IP) of Cross-Linked Protein/DNA

Prior to starting this section, (a) thaw protease inhibitor cocktail II at room temperature for use in **step 2**; (b) as a negative control, use IgG of the same species as the antibody of interest or use beads only.

1. For each ChIP prepare a microfuge tube containing 50 μl of sheared cross-linked chromatin (Subheading 3.2.2, **step 3**), and put on ice. If chromatin has been previously frozen, thaw on ice.
2. Add 450 μl of dilution buffer containing 2.25 μl protease inhibitor cocktail II into each tube containing 50 μl of the fragmented chromatin (*see* **Note 5**).
3. Remove 5 μl (1%) of the supernatant as "input" from each tube, and save at 4 °C until Subheading 3.2.4, **step 1**.
4. Add 1–10 μg of the immunoprecipitation antibody per tube and 20 μl of fully suspended protein G magnetic beads (*see* **Note 1**).
5. Incubate overnight at 4 °C with rotation.
6. Pellet protein G magnetic beads with the magnetic separator, and remove the supernatant completely.
7. Wash the protein G bead-antibody/chromatin complex by resuspending the beads in each of the 0.5 ml cold buffers in the order listed below, and incubate for 8 min at 4 °C on a rotating platform followed by magnetic clearance and careful removal of the supernatant fraction:
   (a) Low-salt immune complex wash buffer, one wash.
   (b) High-salt immune complex wash buffer, one wash.
   (c) LiCl immune complex wash buffer, one wash.
   (d) TE buffer, one wash.

3.2.4 Elution of Protein/DNA Complexes and Reverse CrossLinks of Protein/DNA Complexes to Free DNA

1. Thaw proteinase K and warm the ChIP elution buffer to room temperature. Prepare 100 μl ChIP elution buffer and 1 μl proteinase K for each IP tube as well as input tube (*see* Subheading 3.2.3, **step 3**).
2. Incubate at 62 °C for 2 h with shaking.
3. Incubate at 95 °C for 10 min.
4. Cool the samples down to room temperature followed by short spin.

5. Separate beads using a magnet and remove supernatant to a new tube.

3.2.5 DNA Purification

For DNA purification use any kit which enables bind-wash-elute procedure. Binding buffer is added directly to the sample from Subheading 3.2.4, and the mixture is applied to a spin column. Nucleic acids adsorb to the silica membrane in the high-salt conditions provided by the buffer. Impurities are washed away, and pure DNA is eluted with a small volume of low-salt buffer provided or water, ready to use immediately in all subsequent applications or stored frozen at −20 °C.

(a) Heat the elution buffer to 60 °C before applying it to the column, and incubate for 2 min before centrifugation.

3.2.6 Validation of ChIP by qPCR

Prior to library preparation, validate the enrichment of PTM H3K9Ac following ChIP by performing qRT-PCR for genes representing positive and negative markers for each PTM. For example, ALB is a positive marker (ALB F: GGCAGCCAATGAAATACAAAGAT; ALB R:CAGAAGAGAAAAGCTAGGACAAACG), and PGR is a negative marker (PGR F:GGCTGCGACGGCAATTTA; PGR R: TCACAGCATGCACGAGTTTG) for PTM H3K9Ac in hepatocytes.

3.3 Library Preparation for ChIP-Seq

Prepare libraries using the NEBNext® Ultra TM DNA Library Prep Kit for Illumina®, and sequence using Illumina HiSeq 2500 platform with single-end reads of 50 bp according to the manufacturer's instructions. The expected average fragment size with indexes and adaptors of the ChIP library is demonstrated in Fig. 3 as measured by the Agilent 2200 TapeStation (*see* **Note 6**).

3.4 RNA Purification

Purify total RNA from the cells (HCV-infected, HCV-cured cells, HCV-cured cell treated with C646, and control non-infected cells treated as the infected cell, as described in Fig. 1) using a dedicated kit according to the manufacturer's instructions.

1. It is recommended to purify RNA fresh. If it is impossible, make sure that RNA pellet is snapped frozen in liquid nitrogen *immediately* after you collect it or that culture dishes are washed twice with PBS, frozen *immediately* at −80 °C, and thawed on ice prior to use.
2. Use pipets that are dedicated to RNA procedures.
3. Cell lysates can be homogenized using a syringe and needle. Pass the lysate through a 20-gauge (0.9 mm) needle attached to a sterile plastic syringe at least 5–10 times or until a homogeneous lysate is achieved.
4. DNase digestion during RNA purification is recommended.

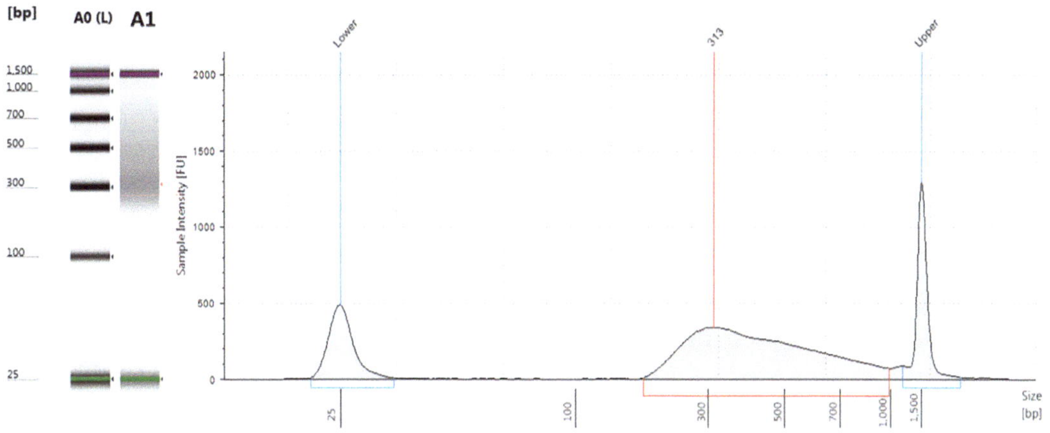

Peak Table

Size [bp]	Calibrated Conc. [ng/µl]	Assigned Conc. [ng/µl]	Peak Molarity [nmol/l]	% Integrated Area	Peak Comment	Observations
25	5.80	-	357	-		Lower Marker
313	20.4	-	101	100.00		
1,500	6.50	6.50	6.67	-		Upper Marker

Fig. 3 Measurement of ChIP library using Agilent 2200 TapeStation system. The expected average fragment size for the ChIP library with indexes and adaptors is 300–1100 bp as demonstrated by a single peak and by the Gel Image using Agilent 2200 TapeStation system

3.5 RNA-Seq Protocol

1. Treat 1 µg of total RNA with the NEBNext poly(A) mRNA Magnetic Isolation Module (NEB) according to manufacturer's instructions (*see* **Note 1**).
2. Prepare RNA-seq libraries by using the NEBNext Ultra RNA Library Prep Kit for Illumina (NEB) according to manufacturer's instructions (*see* **Note 6**).
3. Sequence by Illumina HiSeq 2500 platform with single-end reads of 50 bp according to the manufacturer's instructions.

3.6 Bioinformatic Analysis

The ChIP-seq and RNA-seq data should be analyzed by specialized bioinformatic tools that are not specified here. The bioinformatic analysis should provide lists of genes that were identified as differentially expressed (by RNA-seq), or the epigenetic marker at their genomic location was altered by identifying differential peaks (for ChIP-seq) for the different treatments compared to their corresponding controls: (1) HCV-infected versus control non-infected cells, (2) HCV-cured cells versus DAAs-treated control non-infected cells, and (3) HCV-cured cell treated with C646 versus DAAs-treated control non-infected cells treated with C646 (Fig. 1). In general the bioinformatic analysis should include the following steps:

3.6.1 ChIP-Seq Bioinformatic Analysis

1. Trim short sequence tags and discard sequences shorter than 35 nt.

2. Align the sequences to the GRCh37 (hg19) reference genome or newer version using aligner, and remove duplicate alignments.
3. Generate peaks by a peak caller with an input sample as a background.
4. Align reads.
5. Perform differential analysis.

3.6.2 RNA-Seq Bioinformatic Analysis

1. Trim short sequence tags and discard sequences shorter than 35 nt.
2. Align the sequences to the GRCh37 (hg19) reference genome or newer version using aligner, and remove duplicate alignments.
3. Count the aligned reads.
4. Perform differential analysis.

3.6.3 Integrate ChIP-Seq and RNA-Seq Data

1. Integrate the list of genes generated from ChIP-seq and RNA-seq data to identify gene signature that (a) their epigenetic profile and expression are altered following HCV infection and (b) these alterations remain persistent following HCV eradication and (c) are reverted following epigenetic inhibitor treatment.
2. Perform pathway analysis for the identified gene signature. Identify significant biological process that is involved in HCV-induced epigenetic signature. Choose pathway analysis tool that can integrate both ChIP-seq and RNA-seq data such as the Thompson Reuters MetaCore analysis (https://portal.genego.com/cgi/data_manager.cgi).
3. Identify genes of interest for validation.

3.7 Validation of Gene Signature by qPCR and ChIP-qPCR

1. Design oligonucleotides using the Primer3 PCR primer design tool for the gene of interest according to the RNA-seq and ChIP-seq data. For validation of ChIP-seq data, design primers that target the peak of the tested genes according to the ChIP-seq data.
4. Take equal amounts of isolated RNA from treated or control cells.
5. Prepare cDNA using the high-capacity cDNA reverse transcriptase kit (Applied Biosystems).
6. Perform qRT-PCR using Power SYBR Green Master Mix (Life Technologies) according to the manufacturer's instructions. Use the following qRT-PCR thermal program: 10 min incubation at 95° followed by 40 cycles of 95° for 10 s, 58° for 10 s, and 72° for 20 s.

7. For validation of RNA-seq, calculate differential expression using the equation $2^{(-\Delta\Delta Ct)}$, with the GAPDH as the endogenous controls. For validation of ChIP-seq data, normalize values relative to ChIP performed with irrelevant control antibody or beads only.

4 Notes

1. Agencourt AMPure XP Magnetic Particle Solution should be handled properly:
   (a) Store at 4 °C upon arrival, for up to 12 months.
   (b) Mix the reagent well before use, as magnetic beads will settle on the bottom of the tube over time. It should appear homogenous and consistent in color for IP.
   (c) *Do not freeze* beads at any step.
2. Products that contain DMSO will remain frozen below 18.4 °C.
3. Do not use formaldehyde past the expiration date suggested by the manufacturer. Use high-quality formaldehyde.
4. Make sure to keep samples ice-cold for ChIP assay up to the reverse cross-links of protein/DNA complexes to free DNA step and for any RNA-related procedure.
5. Add protease inhibitors to PBS or lysis buffer just prior to use. Protease inhibitors have a half-life of approximately 30 min in aqueous solutions.
6. For measuring double-stranded DNA or RNA concentrations, or after library preparation step for ChIP-seq or RNA-seq prior to sequencing, use the Qubit dsDNA HS (high-sensitivity) assay kit suitable for concentrations from 10 pg/μl to 100 ng/μl. Dedicated 500 μl thin-walled PCR tubes are required but not included in the kit.

Acknowledgment

The authors acknowledge grant support from the Leona M. and Harry B. Helmsley Charitable Trust Grant #2012PG-ISL013 and Israel Cancer Association #20160120. The authors thank Prof. Izhak Haviv (Bar-Ilan University), Prof. Assam El-Osta (Monash University), and Dr. Antony Kaspi (Monash University) for their insight and helpful discussions.

References

1. Gronbaek K, Hother C, Jones PA (2007) Epigenetic changes in cancer. APMIS 115:1039–1059
2. Esteller M (2007) Cancer epigenomics: DNA methylomes and histone-modification maps. Nat Rev Genet 8:286–298
3. Ribich S, Harvey D, Copeland RA (2017) Drug discovery and chemical biology of cancer epigenetics. Cell Chem Biol 24:1120–1147
4. Gramling S, Rogers C, Liu G, Reisman D (2011) Pharmacologic reversal of epigenetic silencing of the anticancer protein BRM: a novel targeted treatment strategy. Oncogene 30:3289–3294
5. Halley-Stott RP, Gurdon JB (2013) Epigenetic memory in the context of nuclear reprogramming and cancer. Brief Funct Genomics 12:164–173
6. Lilley CE, Chaurushiya MS, Weitzman MD (2010) Chromatin at the intersection of viral infection and DNA damage. Biochim Biophys Acta 1799:319–327
7. Knipe DM, Lieberman PM, Jung JU, McBride AA, Morris KV, Ott M et al (2013) Snapshots: chromatin control of viral infection. Virology 435:141–156
8. D'Ambrosio R, Della Corte C, Colombo M (2015) Hepatocellular carcinoma in patients with a sustained response to anti-hepatitis C therapy. Int J Mol Sci 16:19698–19712
9. Kanwal F, Kramer J, Asch SM, Chayanupatkul M, Cao Y, El-Serag HB (2017) Risk of hepatocellular cancer in HCV patients treated with direct-acting antiviral agents. Gastroenterology 153(996–1005):e1001
10. Ono A, Goossens N, Finn RS, Schmidt WN, Thung SN, Im GY et al (2017) Persisting risk of hepatocellular carcinoma after hepatitis C virus cure monitored by a liver transcriptome signature. Hepatology 66:1344–1346
11. Nakagawa S, Wei L, Song WM, Higashi T, Ghoshal S, Kim RS et al (2016) Molecular liver cancer prevention in cirrhosis by organ transcriptome analysis and lysophosphatidic acid pathway inhibition. Cancer Cell 30:879–890
12. Iwasaki W, Miya Y, Horikoshi N, Osakabe A, Taguchi H, Tachiwana H et al (2013) Contribution of histone N-terminal tails to the structure and stability of nucleosomes. FEBS Open Bio 3:363–369
13. Rothbart SB, Strahl BD (2014) Interpreting the language of histone and DNA modifications. Biochim Biophys Acta 1839:627–643
14. Bannister AJ, Kouzarides T (2011) Regulation of chromatin by histone modifications. Cell Res 21:381–395
15. Faure AJ, Schmidt D, Watt S, Schwalie PC, Wilson MD, Xu H et al (2012) Cohesin regulates tissue-specific expression by stabilizing highly occupied cis-regulatory modules. Genome Res 22:2163–2175
16. Polak P, Karlic R, Koren A, Thurman R, Sandstrom R, Lawrence MS et al (2015) Cell-of-origin chromatin organization shapes the mutational landscape of cancer. Nature 518:360–364
17. Furey TS (2012) ChIP-seq and beyond: new and improved methodologies to detect and characterize protein-DNA interactions. Nat Rev Genet 13:840–852
18. Pundhir S, Poirazi P, Gorodkin J (2015) Emerging applications of read profiles towards the functional annotation of the genome. Front Genet 6:188
19. Bowers EM, Yan G, Mukherjee C, Orry A, Wang L, Holbert MA et al (2010) Virtual ligand screening of the p300/CBP histone acetyltransferase: identification of a selective small molecule inhibitor. Chem Biol 17:471–482
20. Rusyn I, Lemon SM (2014) Mechanisms of HCV-induced liver cancer: what did we learn from in vitro and animal studies? Cancer Lett 345:210–215
21. Li Q, Zhang YY, Chiu S, Hu Z, Lan KH, Cha H et al (2014) Integrative functional genomics of hepatitis C virus infection identifies host dependencies in complete viral replication cycle. PLoS Pathog 10:e1004163
22. Iqbal J, McRae S, Banaudha K, Mai T, Waris G (2013) Mechanism of hepatitis C virus (HCV)-induced osteopontin and its role in epithelial to mesenchymal transition of hepatocytes. J Biol Chem 288:36994–37009
23. Chen Y, Chen J, Wang H, Shi J, Wu K, Liu S et al (2013) HCV-induced miR-21 contributes to evasion of host immune system by targeting MyD88 and IRAK1. PLoS Pathog 9:e1003248
24. Lupberger J, Zeisel MB, Xiao F, Thumann C, Fofana I, Zona L et al (2011) EGFR and EphA2 are host factors for hepatitis C virus entry and possible targets for antiviral therapy. Nat Med 17:589–595
25. Zhang Q, Wei L, Yang H, Yang W, Yang Q, Zhang Z et al (2016) Bromodomain containing protein represses the Ras/Raf/MEK/ERK pathway to attenuate human hepatoma cell

proliferation during HCV infection. Cancer Lett 371:107–116
26. Luna JM, Scheel TK, Danino T, Shaw KS, Mele A, Fak JJ et al (2015) Hepatitis C virus RNA functionally sequesters miR-122. Cell 160:1099–1110
27. Yamane D, McGivern DR, Wauthier E, Yi M, Madden VJ, Welsch C et al (2014) Regulation of the hepatitis C virus RNA replicase by endogenous lipid peroxidation. Nat Med 20:927–935
28. Neufeldt CJ, Joyce MA, Levin A, Steenbergen RH, Pang D, Shields J et al (2013) Hepatitis C virus-induced cytoplasmic organelles use the nuclear transport machinery to establish an environment conducive to virus replication. PLoS Pathog 9:e1003744
29. Zona L, Lupberger J, Sidahmed-Adrar N, Thumann C, Harris HJ, Barnes A et al (2013) HRas signal transduction promotes hepatitis C virus cell entry by triggering assembly of the host tetraspanin receptor complex. Cell Host Microbe 13:302–313
30. Li Q, Sodroski C, Lowey B, Schweitzer CJ, Cha H, Zhang F et al (2016) Hepatitis C virus depends on E-cadherin as an entry factor and regulates its expression in epithelial-to-mesenchymal transition. Proc Natl Acad Sci U S A 113:7620–7625
31. Li Q, Pene V, Krishnamurthy S, Cha H, Liang TJ (2013) Hepatitis C virus infection activates an innate pathway involving IKK-alpha in lipogenesis and viral assembly. Nat Med 19:722–729
32. Steenbergen RH, Joyce MA, Thomas BS, Jones D, Law J, Russell R et al (2013) Human serum leads to differentiation of human hepatoma cells, restoration of very-low-density lipoprotein secretion, and a 1000-fold increase in HCV Japanese fulminant hepatitis type 1 titers. Hepatology 58:1907–1917
33. El-Shamy A, Eng FJ, Doyle EH, Klepper AL, Sun X, Sangiovanni A et al (2015) A cell culture system for distinguishing hepatitis C viruses with and without liver cancer-related mutations in the viral core gene. J Hepatol 63:1323–1333
34. Yi M, Ma Y, Yates J, Lemon SM (2007) Compensatory mutations in E1, p7, NS2, and NS3 enhance yields of cell culture-infectious intergenotypic chimeric hepatitis C virus. J Virol 81:629–638
35. Gal-Tanamy M, Zemel R, Bachmatov L, Jangra RK, Shapira A, Villanueva RA et al (2010) Inhibition of protease-inhibitor-resistant hepatitis C virus replicons and infectious virus by intracellular intrabodies. Antivir Res 88:95–106

Chapter 14

Measuring Hepatitis C Virus Envelopment by Using a Proteinase K Protection Assay

Allison E. Roder and Stacy M. Horner

Abstract

The infectious virion of hepatitis C virus (HCV) is made up of the viral nucleocapsid surrounded by an envelope that contains an ER-derived membrane bilayer, cellular lipids, and the viral E1 and E2 glycoproteins. Because the infectious HCV particle contains both protein and lipid layers, selective disruption of these layers and analysis for the presence or absence of resulting virion components can be used to study the virion assembly process. This chapter describes an experimental method to measure HCV virion envelopment, which can reveal the mechanisms of how specific viral protein–protein interactions and host factors contribute to the process of HCV envelopment.

Key words Hepatitis C virus, Envelopment, Proteinase K, Immunoblot, Electroporation, Protection assay

1 Introduction

The hepatitis C virus (HCV) virion is composed of an inner RNA/ Core protein nucleocapsid, surrounded by an envelope, which contains a lipid membrane bilayer and the viral E1 and E2 glycoproteins (*see* Fig. 1). The HCV envelope serves many important functions during infection: organizing the viral glycoproteins to aide in viral entry, protecting the viral RNA from degradation, shielding the viral RNA from sensing by the intracellular innate immune system, and encompassing viral and cellular components that contribute to the effective release of the viral particle (Reviewed in [1]). Studies on the formation of the HCV virion envelope can reveal several key viral and host regulators of HCV assembly and envelopment. Indeed, a previous study used a method like the one described below to reveal critical protein–protein interactions necessary for successful HCV envelopment [2].

For enveloped viruses, including HCV, the viral envelope is required for infectivity. For HCV, this process of virion envelopment takes place in endoplasmic reticulum (ER) invaginations near

Mansun Law (ed.), *Hepatitis C Virus Protocols*, Methods in Molecular Biology, vol. 1911,
https://doi.org/10.1007/978-1-4939-8976-8_14,

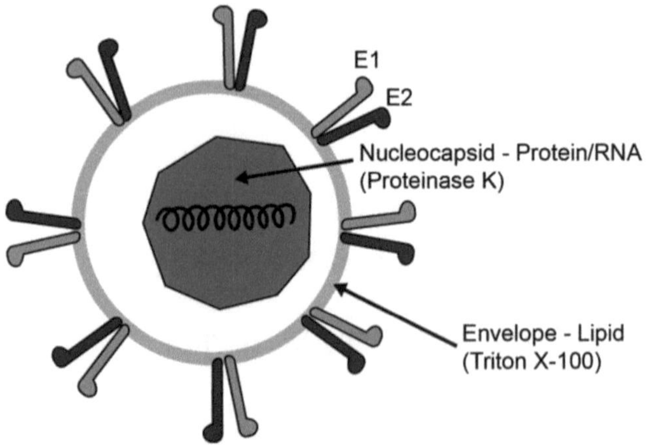

Fig. 1 The HCV virion. The HCV virion contains a central Core/RNA complex known as the nucleocapsid, surrounded by lipids and an ER-derived membrane, referred to as the envelope. This envelope also contains the HCV E1 and E2 glycoproteins. Each of these main virion layers can be antagonized (as noted) for studies of HCV virion formation, as described in the text

cytosolic lipid droplets [3, 4]. In the process of HCV assembly, the HCV Core protein aggregates at lipid droplets where it binds to viral RNA and then shuttles to the ER. At the ER, the Core/RNA complex forms a nucleocapsid which buds into the ER lumen to acquire the envelope. This process of HCV envelopment occurs through a series of coordinated viral protein–protein interactions, organized by the viral NS2 protein. Specifically, NS2 directly interacts with the E2 viral glycoprotein, as well as with the p7, NS3, and NS5A proteins to bring the nonstructural HCV proteins into close proximity to the structural proteins, including E2, E1, and Core [5–12]. These NS2-driven interactions likely play a role in the transfer of viral RNA from replication complexes to assembly/envelopment complexes, although the exact mechanisms for how these processes occur are not fully understood.

The processes of HCV assembly and envelopment are thought to occur in one step, as nucleocapsids of the RNA and the Core protein form at the same time as envelopment of this nucleocapsid [2, 13]. However, because mutations in the viral RNA that cause envelopment defects result in accumulation of Core on cytosolic lipid droplets and do not preclude the formation of intact nucleocapsids (Core/RNA complexes), it is possible that in the case of defective viral envelopment, viral assembly and nucleocapsid formation occur as an identifiable step of the HCV lifecycle [2]. Ultimately, as the HCV lifecycle progresses, the virion buds into the ER lumen acquiring the viral membrane glycoproteins, E1 and E2, which become glycosylated in the ER [14]. As the HCV virion leaves the cell through the secretory pathway, lipids, such as cholesterol esters, cholesterol, sphingomyelin, and phosphatidylcholine, as well as host apolipoproteins such as ApoE, are incorporated into the virion resulting in the characteristic low density of the HCV particle [15–17].

Given that infectious HCV particles consist of an outer membrane lipid layer (the envelope), a protein layer (the Core protein), and RNA (the viral genome), antagonizing any one of these layers in a protection assay, using detergent, proteinase, or RNase, respectively, can reveal which step(s) of virion formation have occurred. Methods to specifically measure viral assembly and the formation of nucleocapsids based on rate-zonal centrifugation of infected cell lysates through a sucrose gradient followed by protection assays have been described previously [2]. Here, we will describe a method to measure HCV envelopment.

In this method, lysates from cells transfected with HCV RNA are treated with proteinase K. Incomplete or unsuccessful envelopment would result in exposed Core protein that would be degraded by proteinase K resulting in a loss of Core signal by immunoblot, whereas the presence of an envelope would protect Core from this proteinase K degradation. As a positive control for this assay, lysates are treated with the detergent Triton X-100 to remove the envelope before proteinase K treatment. In addition, HCV RNA with a deletion of the E1 and E2 genes (ΔE1/E2) can also be used as a control for the envelopment assay, as these glycoproteins are necessary for acquisition of the membrane bilayer that forms the envelope [2]. As an alternative to immunoblotting, Core protein levels can also be measured by ELISA [2].

2 Materials

2.1 Electroporation of Infectious HCV RNA into Cells

1. Huh7.5 cells [18] (or other human cells that replicate HCV to high titers) plated in 15 cm^2 tissue culture plates, ~1 confluent plate per experimental condition.
2. 1× PBS: Dilute 10× in dH_2O and filter sterilize or purchase sterile 1× PBS.
3. Complete DMEM (cDMEM): Dulbecco's modified Eagle's medium, 10% fetal bovine serum, 2.5 mM HEPES.
4. 0.05% Trypsin-EDTA.
5. Tabletop centrifuge with swinging bucket rotor to accommodate 15 mL or 50 mL conical tubes (such as Thermo Scientific Legend XTR).
6. Cytomix electroporation buffer: 120 mM potassium chloride, 10 mM potassium phosphate buffer (pH 7.6), 5 mM magnesium chloride, 25 mM HEPES, 0.15 mM calcium chloride, 2 mM EGTA (pH 7.6). Combine in dH_2O; titrate to final pH

of 7.6 with potassium hydroxide. Filter sterilize and aliquot into 1.5 mL microfuge tubes. Store at −20 °C until use.

7. Electroporator (such as BioRad GenePulser XL).
8. Cuvettes (4 mm).

2.2 Plating Cells and Harvesting Lysates

1. Proteinase K buffer: 50 mM Tris pH 8, 10 mM $CaCl_2$, 1 mM DTT in dH_2O, filter sterilize and store at 4 °C.

2.3 Testing HCV Envelopment

1. Dry ice/100% ethanol bath: Fill ice bucket or Styrofoam container with dry ice. Add 100% ethanol slowly to cover dry ice.
2. 10% Triton X-100: Dilute 100% Triton X-100 to a 10% solution in dH_2O.
3. Proteinase K (Thermo-Fisher): 20 mg/mL stock, dilute 1:16 in RNase-free dH_2O to obtain 1.25 mg/mL working concentration.
4. Phenylmethylsulfonyl fluoride (PMSF): Reconstitute to 5 mg/mL in 100% ethanol, aliquot in dark tubes, desiccate, and store at 4 °C in dark container or container covered in foil.
5. 6× sodium dodecyl sulfate (SDS) loading dye: For 10 mL, 1.2 g SDS, 6 mg bromophenol blue sodium salt, 4.7 mL glycerol, 0.6 mL 1 M Tris pH 6.8, fill to 10 mL with dH_2O, heat gently at 37 °C, and shake to dissolve; add 0.93 g DTT, dissolve, and store at −20 °C.
6. HCV Core antibody (Thermo-Fisher).

3 Methods

This method requires in vitro transcribed HCV RNA. All materials and methods for generation of HCV RNA from a HCV cDNA clone can be found in Springer Protocols Second Edition, Chapter 11, Section 2.2 [19].

3.1 Electroporation of Infectious HCV RNA into Cells (See Note 1)

1. Aspirate media from plates, wash adherent cells with 1× PBS, trypsinize with 3 mL trypsin per plate at 37 °C for 5–7 min to detach cells, resuspend cells with 7 mL cDMEM (10 mL total volume), and move to 15 mL conical tube.
2. Centrifuge to pellet cells, 5 min at 1000 × *g* at 4 °C using a tabletop centrifuge with swinging bucket rotor.
3. Aspirate media from tubes and resuspend cell pellets in 40 mL 1× PBS to wash. Pipet well to thoroughly resuspend cells and break up clumps (*see* **Note 2**). Remove 10 μL cell suspension for counting.
4. Centrifuge cells to pellet, 5 min at 1000 × *g* at 4 °C in tabletop centrifuge with swinging bucket rotor.

5. Count cells.
6. Aspirate PBS from cell pellet and resuspend cells in cytomix buffer to a final concentration of 1×10^7 cells/mL.
7. Aliquot 400 μL of cells (4×10^6 cells) into a 4 mm cuvette. Add 5 μg in vitro transcribed HCV RNA and tap side of cuvette to mix.
8. Electroporate RNA into cells using the settings 250 V and 950 μF on the electroporator.
9. Allow cells to rest on ice for 10 min post-electroporation.
10. Prepare one 50 mL conical tube with 20 mL of cDMEM per electroporation condition.
11. Take 1 mL of media from the prepared conical, and add to the corresponding cuvette. Pipet up and down using a P1000 pipettor to thoroughly suspend cells. Move media and cells back to the conical.
12. Repeat **step 11** several times to transfer all the electroporated cells to the conical.
13. Mix media and cells in conical well using a 10 mL serological pipet (*see* **Note 3**).

3.2 Plating Cells and Harvesting Lysates

1. Plate 2.5 mL of each electroporation suspension ($\sim 5 \times 10^5$ cells) into one well of a six-well cell culture plate. Incubate for 4 h in a 37 °C, 5% CO_2 cell culture incubator to allow cells to settle and attach to the plate (*see* **Note 4**).
2. After at least 4 h (before 24 h), wash plates three times with 1× PBS and one time with cDMEM to remove any dead cells or cell debris from the plates. Add fresh cDMEM to plates and return to the incubator.
3. At 48 h post-electroporation, aspirate media and wash one time with PBS. Add 200 μL of ice-cold proteinase K buffer to each well of the experiment.
4. Use cell scraper to scrape back and forth and in circular motion to lift cells from dish. Pipet cells up and down using a P200 pipettor to break up any clumps, and transfer into a 1.5 mL microcentrifuge tube (*see* **Note 5**).
5. Keep suspended cells on ice, or freeze at −80 °C until ready to proceed with experiment (*see* **Note 6**).

3.3 Testing HCV Envelopment

1. Prepare a bath of dry ice and 100% ethanol for quick freezing.
2. Thaw the cells suspended in proteinase K buffer at 37 °C (*see* **Note 7**) and vortex to mix.
3. Freeze cells in the dry ice/ethanol bath.

Table 1
Treatments for testing HCV envelopment

	Triton X-100	Proteinase K (1.25 mg/mL)	PMSF (5 mg/mL)
Tube 1	None	None	None
Tube 2	5 μL	2 μL	2 μL
Tube 3	None	2 μL	2 μL

4. Repeat this freeze/thaw cycle (**steps 2** and **3**) five times total (*see* **Note 8**).
5. Divide sample into three microfuge tubes with 50 μL of cell suspension in each (*see* **Note 9**). Label the tubes one to three for each experimental condition.
6. Set the #1 tube aside—this will be the untreated control lysate.
7. Add 5 μL 10% Triton X-100 to tube #2 (only #2, do not add to #3, *see* **Note 10**).
8. Vortex and allow to sit on ice for 5 min.
9. Add 2 μL proteinase K (1.25 mg/mL) to tubes #2 and #3 (the final concentration of proteinase K is ~50 μg/mL).
10. Vortex to mix and incubate on ice for 30 min.
11. Add 2 μL PMSF stock to tubes #2 and #3 for each experimental condition.
12. Vortex to mix and incubate on ice for 10 min (*see* **Note 11**).
13. Treatment conditions for tubes one to three are consolidated in Table 1.
14. Pre-warm 6× SDS buffer in 37 °C water bath for at least 5 min (*see* **Note 12**).
15. Add 12 μL 6× SDS buffer to each tube (#1, #2, and #3). Pipet gently to mix (*see* **Note 13**).
16. Following addition of SDS loading buffer, samples can be frozen at −80 °C or used immediately for SDS-PAGE analysis. Load 25 μL of tube #1 and 50 μL of #2 and #3 onto SDS-PAGE gel (*see* **Note 14**).
17. Immunoblot for HCV Core protein using Core-specific antibody (*see* Fig. 2 **Notes 15** and **16**).

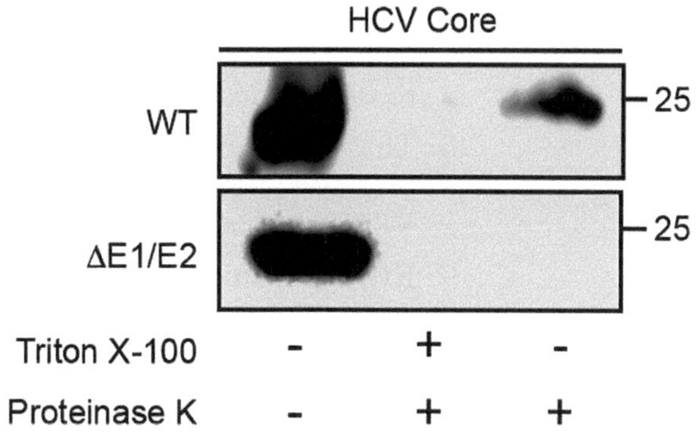

Fig. 2 Immunoblot analysis for HCV Core protein of Huh7.5 cell lysates at 48 h post-electroporation with in vitro transcribed HCV RNA [wild-type (WT) or with a deletion in the E1/E2 genes (ΔE1/E2)] subjected to the indicated treatments in a proteinase K protection assay

4 Notes

1. Any materials that touch the HCV RNA or media containing HCV RNA after electroporation, up until lysis step (Subheading 3.3, **step 5**), should be considered infections and decontaminated appropriately before disposal into biohazard waste.
2. Add 5 mL of PBS first with a 5 mL serological pipet, and use this to thoroughly resuspend the cells. The cell clumps are easier to break with a smaller volume pipet. Once mixed, add the remaining 35 mL of PBS, and pipet a few times to mix with 50 mL serological pipet.
3. In vitro transcribed HCV RNA can also be transfected into cells by using the TransIT-mRNA transfection kit from Mirus Bio or a similar RNA transfection reagent.
4. It is normal for some cell death to occur following electroporation. While some cells will fail to adhere to the plate, this will not affect the outcome of the experiment as a sufficient number of cells will attach.
5. Corning cell scrapers with a 1.8 cm blade work well for scraping cells from the individual wells of a six-well dish.
6. Cell suspensions should not be frozen for longer than 1 week prior to the start of the experiment. If working from frozen cell suspensions, be sure to thaw suspensions and pipet and/or vortex to thoroughly resuspend cells before starting freeze/thaw cycles. Once lysed, proceed immediately to proteinase K treatment.

7. Thawing of lysates should be done at a 37 °C incubator, bead bath, or in gloved hands. Thawing in a water bath if not recommended as the virions will not be protected.
8. This will be extremely cold; take care not to contact with skin or gloves; wear eye protection. Freezing of suspensions takes ~30 s; thawing of suspensions takes ~2 min.
9. Extra remaining lysate will be present. This can be used for other SDS-PAGE analysis.
10. The detergent Triton X-100 will degrade any present envelope, allowing for proteinase K treatment to degrade Core protein. Upon immunoblotting, Core protein should be absent from these samples.
11. PMSF is quite unstable, and maximum protease inhibitor activity is essential for proper completion of this protocol. Therefore, it is recommended to store PMSF in dark tubes, desiccate, and wrap outer container in foil for extra protection from light. Replace PMSF with a new stock every 4–6 months to ensure viability.
12. SDS does not dissolve well at this concentration. Heating for 5–10 min at 37 °C should dissolve SDS enough for efficient pipetting. Do not vortex after heating; gently rock tube back and forth to dissolve SDS.
13. Aggressive agitation will create many bubbles in the SDS buffer and should be avoided.
14. The untreated sample (tube #1) will contain a much higher quantity of Core protein than the other treated samples. Loading half of this sample will reduce the likelihood of signal blowout in this lane when detected by immunoblot.
15. The protected Core signal (lane 3) on the immunoblot sometimes varies by experiment, as the proteinase K and PMSF are often lose activity after storage. To increase the protected Core signal, decrease the concentration or length of proteinase K treatment or use fresh PMSF.
16. Ratios of protected Core (lane 3) to untreated Core (lane 1) can be calculated to identify viral mutants with possible defects in virion envelopment.

Acknowledgments

Research on hepatitis C virus in the Horner laboratory is supported by funds from the National Institutes of Health (NIH) (R01AI125416 and R21AI124100), a Duke School of Medicine Whitehead Scholarship, and the Burroughs Wellcome Fund. AER has received support from NIH 1F31AI131477-01A1 and T32CA009111.

References

1. Vieyres G, Dubuisson J, Pietschmann T (2014) Incorporation of hepatitis C virus E1 and E2 glycoproteins: the keystones on a peculiar virion. Viruses 6:1149–1187
2. Gentzsch J, Brohm C, Steinmann E, Friesland M, Menzel N, Vieyres G et al (2013) Hepatitis c virus p7 is critical for capsid assembly and envelopment. PLoS Pathog 9: e1003355
3. Miyanari Y, Atsuzawa K, Usuda N, Watashi K, Hishiki T, Zayas M et al (2007) The lipid droplet is an important organelle for hepatitis C virus production. Nat Cell Biol 9:1089–1097
4. Counihan NA, Rawlinson SM, Lindenbach BD (2011) Trafficking of hepatitis C virus core protein during virus particle assembly. PLoS Pathog 7:e1002302
5. Steinmann E, Penin F, Kallis S, Patel AH, Bartenschlager R, Pietschmann T (2007) Hepatitis C virus p7 protein is crucial for assembly and release of infectious virions. PLoS Pathog 3:e103
6. Jones CT, Murray CL, Eastman DK, Tassello J, Rice CM (2007) Hepatitis C virus p7 and NS2 proteins are essential for production of infectious virus. J Virol 81:8374–8383
7. Popescu CI, Callens N, Trinel D, Roingeard P, Moradpour D, Descamps V et al (2011) NS2 protein of hepatitis C virus interacts with structural and non-structural proteins towards virus assembly. PLoS Pathog 7:e1001278
8. Boson B, Granio O, Bartenschlager R, Cosset FL (2011) A concerted action of hepatitis C virus p7 and nonstructural protein 2 regulates core localization at the endoplasmic reticulum and virus assembly. PLoS Pathog 7:e1002144
9. Stapleford KA, Lindenbach BD (2011) Hepatitis C virus NS2 coordinates virus particle assembly through physical interactions with the E1-E2 glycoprotein and NS3-NS4A enzyme complexes. J Virol 85:1706–1717
10. Ma Y, Anantpadma M, Timpe JM, Shanmugam S, Singh SM, Lemon SM et al (2011) Hepatitis C virus NS2 protein serves as a scaffold for virus assembly by interacting with both structural and nonstructural proteins. J Virol 85:86–97
11. Jirasko V, Montserret R, Appel N, Janvier A, Eustachi L, Brohm C et al (2008) Structural and functional characterization of nonstructural protein 2 for its role in hepatitis C virus assembly. J Biol Chem 283:28546–28562
12. Phan T (2009) Hepatitis C virus NS2 protein contributes to virus particle assembly via opposing epistatic interactions with the E1-E2 glycoprotein and NS3-NS4A enzyme complexes. J Virol 83:8379–8395
13. Steinmann E, Doerrbecker J, Friesland M, Riebesehl N, Ginkel C, Hillung J et al (2013) Characterization of hepatitis C virus intra- and intergenotypic chimeras reveals a role of the glycoproteins in virus envelopment. J Virol 87:13297–13306
14. Vieyres G, Thomas X, Descamps V, Duverlie G, Patel AH, Dubuisson J (2010) Characterization of the envelope glycoproteins associated with infectious hepatitis C virus. J Virol 84:10159–10168
15. Catanese MT, Uryu K, Kopp M, Edwards TJ, Andrus L, Rice WJ et al (2013) Ultrastructural analysis of hepatitis C virus particles. Proc Natl Acad Sci U S A 110:9505–9510
16. Merz A, Long G, Hiet MS, Brugger B, Chlanda P, Andre P et al (2011) Biochemical and morphological properties of hepatitis C virus particles and determination of their lipidome. J Biol Chem 286:3018–3032
17. Chang T-HH (2007) Human Apolipoprotein E is required for infectivity and production of hepatitis C virus in cell culture. J Virol 81:13783–13793
18. Sumpter R Jr, Loo YM, Foy E, Li K, Yoneyama M, Fujita T et al (2005) Regulating intracellular antiviral defense and permissiveness to hepatitis C virus RNA replication through a cellular RNA helicase, RIG-I. J Virol 79:2689–2699
19. Lohmann V (2009) HCV replicons: overview and basic protocols. Methods Mol Biol 510:145–163

Chapter 15

Membrane Fusion Assays for Studying Entry Hepatitis C Virus into Cells

Solène Denolly, François-Loïc Cosset, and Natalia Freitas

Abstract

The membrane fusion properties of HCV envelope glycoproteins can be evaluated using several assays. Fusion assays generally require contacts between glycoproteins expressed on a donor membrane, such as those from a cell or a viral particle, and an acceptor membrane that may or may not express cognate viral receptor, such as those from an indicator cell or a liposome. In this chapter, we describe three well-established methods in the field that use either cell surface expression of glycoproteins, HCV pseudoparticles (HCVpp), or cell culture-grown HCV (HCVcc) particles for donor membrane and cells or liposomes as acceptor membrane in which specific components can be included to monitor and quantify fusion. We provide details of cell-cell fusion assay, virus-liposome fusion assay, and finally virus-plasma membrane fusion assay. We also describe inhibitors that can block HCV envelope membrane fusion.

Key words Fusion, Hepatitis C, Liposomes, Inhibitors, Plasma membrane, Cell-cell fusion

1 Introduction

HCV entry is a complex process that is not fully understood. Entry begins by attachment of viral particles to several receptors including HSPGs and SR-BI. This allows the clustering of particles with CD81 and their lateral diffusion and interactions with Claudin-1 and other co-factors [1], which is followed by their endocytosis and internalization. The last step of entry is the fusion of viral membrane with that of the host cell. This last step allows the release of the genome in the cytosol and the initiation of virus replication. This process includes conformational changes of the fusion proteins at low endosomal pH, which induce insertion of a fusion peptide into the host membrane, lipid mixing of external membrane leaflets, and finally full fusion of the two membranes.

As this fusion step is critical for the infection, it has been studied since many years. It was first studied by using some peptides derived from HCV glycoproteins, allowing determining putative fusion peptides [2]. Cell-cell fusion assays were then developed [3]

Mansun Law (ed.), *Hepatitis C Virus Protocols*, Methods in Molecular Biology, vol. 1911,
https://doi.org/10.1007/978-1-4939-8976-8_15,

as for other viruses, based on the expression of glycoproteins at the cell surface of donor cells and a reporter gene expressed only after fusion with indicator cells. Although HCV does not induce formation of syncytia in this assay, the measurement of reporter gene allowed determining that fusion occurs at optimal pH of *ca.* 5 and requires both E1 and E2 glycoproteins. The development of HCV pseudoparticles (HCVpp) [4] allowed the study of fusion of HCV glycoproteins at the surface of particles instead of cell surface. This was determined using liposome fusion assays [5] that confirmed the requirement of both glycoproteins for inducing membrane fusion at pH 5.5 but also allowed the discovery of specific determinants in either glycoprotein that govern fusion [6]. The study of fusion of fully infectious HCV particles, HCVcc, was developed later and is based on the fusion of viral particles at the plasma membrane [7], as previously achieved for other viruses or with liposomes. Again, these assays confirmed the dependency of acid pH and for both E1E2 glycoproteins in cell entry. These three fusion assays were also used to find inhibitors of membrane fusion that block cell entry and infection.

According to findings obtained with these assays, membrane fusion of HCV particles might occur in the endosome compartment and requires its acidic environment. Interestingly, the optimal pH values required for membrane fusion is dependent of the assay used, indicating that this crucial step for HCV entry is complex and may require additional factors. As the study of membrane fusion *in cellulo* is technically difficult to achieve and there is still no assay for studying fusion of HCV particles in endosomes of hepatoma cells, it is clear that the assays described below, individually, are not perfect. However, their combined study could allow better understanding of this critical step and the discovery of new HCV drugs. Here we describe these three assays as well as previously described fusion inhibitors.

2 Materials

2.1 Cell-Cell Fusion Assay

2.1.1 Cell Lines

1. Donor cells: HEK-293T cells.
2. Indicator cells: Typically cells that are permissive to HCV replication such as Huh-7 and Huh7.5 cells transiently or stably expressing HIV-1 transactivator Tat protein. Huh7-Lunet cells expressing high levels of human CD81 can also be used.

2.1.2 Cell Culture

1. Common cell culture solutions and equipment.
2. Additional solutions: Nonenzymatic cell dissociation reagent Versene 1:5000 (1×) solution.
3. Recommended cell culture media: Dulbecco's modified Eagle's medium (DMEM) supplemented with 10% fetal bovine serum and 1% Pen-Strep antibiotics solution.

2.1.3 Transfection Reagents

1. JetPEI (Polyplus-transfection Inc.) for transfection of donor HEK-293T cells.
2. GeneJammer (Agilent) for transitory transfection of indicator cells.

2.1.4 Plasmid DNA for Transfection

1. Luciferase reporter system, which consists in a pLTR-luc plasmid containing a fragment of HIV-1 long terminal repeat (LTR) driving luciferase expression in donor cells upon transactivation by HIV-1 Tat [6], as provided in acceptor cells by plasmid LXSN-Tat [6], upon cell-cell fusion.
2. Constructs expressing HCV glycoproteins consist of, e.g., plasmid phCMV-H77-wt expressing wild-type E1E2 from genotype 1a strain H77, or alternative/mutated HCV glycoproteins, under control of the CMV promoter [4]. Optionally, plasmid pGFP encoding green fluorescent protein (GFP) can be used to access transfection efficiency.
3. Controls plasmids such as plasmid phCMV, also called "empty vector" that does not contain any DNA sequences encoding for viral glycoproteins and is used as negative control to establish the background levels of cell-cell fusion assay.
4. Constructs such as plasmids phCMV-HA and phCMV-NA encoding the fowl plaque virus hemagglutinin and neuraminidase, respectively, or phCMV-VSVg expressing the vesicular stomatitis virus glycoprotein, which are used as positive controls of the cell-cell fusion assay. Both HA/NA and VSVg glycoproteins display maximum activity at pH 5.

2.1.5 Buffers

1. 1× DPBS.
2. 1× fusion buffer (135 mM NaCl, 15 mM sodium citrate, 10 mM MES, and 5 mM HEPES). First prepare a 10× concentrated fusion buffer by weighing 15.8 g of NaCl (M.W. 58.44 g/mol), 8.8 g of sodium citrate (M.W. 294.1 g/mol), 3.9 g of morpholineethanesulfonic acid (MES; M.W. 195.2 g/mol), and 2.3 g of HEPES (M.W. 238.1 g/mol) and dissolving all components in 200 ml of molecular biology grade water. To prepare 1× fusion buffer, simply dilute 5 ml of the 10× concentrated fusion buffer in 40 ml of water. Measure the pH using a calibrated pH meter, and adjust the pH to the desired value. Use a 1N NaOH solution to adjust pH values above 5 and 37% HCl solution for fusion buffer pH 5 or below. After obtaining the desired pH, add water to a final volume of 50 ml, and sterilize the solution by filtration using 0.20 μm filter units.
3. 5× passive lysis buffer (Promega): Prepare a 1× working solution by diluting the 5× concentrated stock in molecular biology grade water, and pre-chill the 1× lysis buffer at 4 °C.

4. Firefly luciferase reagent: First prepare 1 l of firefly buffer (25 mM Tricine buffer, pH 7.8, 5 mM $MgSO_4$, 0.5 mM EDTA, pH 8.0, 5 mM DTT, and 0.5 mM ATP). Make 100 ml aliquots, keep one vial to reconstitute firefly luciferase substrate mix, and freeze the remaining tubes. In order to prepare firefly luciferase substrate mix, resuspend 50 mg of D-Luciferin sodium salt (Synchem OHG) and 25 mg of coenzyme A sodium salt hydrate (Sigma) in 100 ml of firefly buffer. After reconstitution, this solution should be protected from light but can undergo several freeze/thaw cycles.

2.1.6 Special Equipment

1. Multimode plate reader like Mithras LB 940 or equivalent.

2.2 Virus-Liposome Fusion Assay

2.2.1 HCV Particles

1. HCVpp, β-lactamase HIV-based HCV pseudoparticles or HCVcc particles (for a detailed protocol of HCVpp and HCVcc production, *see* Chapters 2 and 7). A protocol for production of β-lactamase HIV-based HCV pseudoparticles can be found in [5].
2. Control particles devoid of envelope (negative control).
3. Pseudoparticles harboring alternative envelope virus glycoproteins used as positive or negative controls for acid pH sensitivity, such as HApp or MLVpp, respectively.

2.2.2 Reagents and Materials

1. Phosphatidylcholine from egg yolk (PC, 99% pure; Sigma).
2. Cholesterol (chol, 99% pure; Sigma).
3. Sphingomyelin (SM, 99% pure; Sigma).
4. Triton X-100.
5. Octadecyl rhodamine B chloride (R_{18}; Molecular Probes).
6. Beta-lactamase substrate CCF2-FA; free acid (Invitrogen).
7. Polycarbonate filters.

2.2.3 Equipment

1. SLM Aminco 8000 spectrofluorometer or equivalent.

2.3 Virus-Plasma Membrane Fusion Assay

2.3.1 Cell Lines

1. Target cells: Huh7.5 or derivatives (Huh7-Lunet/hCD81).

2.3.2 Cell Culture

1. Classical medium for target cell culture: for example, DMEM supplemented with 100 U/ml of penicillin, 100 μg/ml of streptomycin, and 10% fetal bovine serum.
2. Incubator for cell culture at 37 °C, 5% CO_2.
3. Six-well plates.

2.3.3 HCV Particles

1. HCVcc particles containing a luciferase reporter genome (*see* Chapter 7 for production).

2.3.4 Reagents

1. Inhibitor of endosomal acidification: Bafilomycin A1 (Sigma-Aldrich, CAS 88899-55-2) or concanamycin A (Sigma-Aldrich 80,890-47-7). Dissolve the powder in DMSO and store the solutions at −20 °C. A stock solution could be made at 25 μM and 5 μM for bafilomycin A1 and concanamycin A, respectively.
2. Ice-cold DPBS.
3. Pre-warm citric buffer (pH 7 or pH 5): 15 mM citric acid, 150 mM NaCl. For 100 ml, weigh 288.2 mg of citric acid and 876.6 mg of NaCl, and dissolve all components in 100 ml of molecular biology grade water. Measure pH using a calibrated pH meter and adjust pH to the desired value (either 7 or 5). Use 1N of NaOH or HCl to adjust pH (*see* **Note 1**). Just before the assay, pre-warm buffer at 37 °C.
4. Passive lysis buffer: *see* Subheading 2.1.5 for the recipe.
5. Luciferase reagent: *see* Subheading 2.1.5 for the recipe.
6. Multimode plate reader like Mithras LB 940 or equivalent.

2.4 Fusion Inhibitors

A list of HCV fusion inhibitors described in previous studies is provided in Table 1. They can be used as controls in fusion experiments.

3 Methods

3.1 Cell-Cell Fusion Assay

Cell-cell fusion assays typically consist in transient overexpression of viral glycoproteins in one cell type (donor cells) and co-culture (co-cultivation) with a second cell type that expresses viral receptors (indicator cells). Viral glycoproteins mediated cell-cell fusion is quantified by measuring the transcriptional activation of a reporter gene such as luciferase under the control of the HIV-1 3′ long terminal repeat (LTR) in donor cells by the HIV-1 transactivator Tat produced in indicator cells. Fusion of the two cell types results in cytoplasmic content mixing between donor cells and indicator cells, allowing the expression of luciferase, which is measured upon cell lysis (*see* Fig. 1). This method allows investigating the direct role of the viral envelope proteins in membrane fusion independently of virus RNA replication and virion production. This is particularly advantageous when identifying fusion determinants and/or characterizing E1E2 point mutations that may significantly affect other steps of HCV life cycle and/or impair viral particle production. Nevertheless, this assay relies on efficient viral glycoprotein expression and transport to the cell surface, which should be properly monitored for cell-cell fusion normalization.

Table 1
HCV fusion inhibitors

Drug/Compound	Application	Results	Concentration range	Mode of action	References
Arbidol (ARB)	HCVpp-HCVcc/liposome fusion	Dose-dependent inhibition of HCVpp- and HCVcc-mediated lipid fusion	0.1–6 μg/ml (11.3 μM)	Membrane association	[8, 9]
Monoclonal antibody CBH-5 directed to an epitope in E2	HCVcc/liposome fusion	Dose-dependent inhibition of HCVcc-mediated lipid fusion	1–25 μg/ml	Blocks E2 binding to CD81 [10]	[8]
Amphipathic DNA polymers	HCVcc/liposome fusion	Inhibition of the rate and maximum extent of lipid fusion	1–10 nM	Target HCV postbinding step by an unknown mechanism	[11]
Peptide fusion inhibitor from E2 stem domain (E27)	Cell-cell fusion	Dose- dependent inhibition of E1E2-mediated membrane fusion	EC_{50} of 734.2 ± 9,3 nM	Interfere with E1E2 heterodimerization	[12]
Flunarizine	Plasma membrane fusion	Reduction of HCV infection	10 μM	Targets the potential fusion peptide within E1	[13]
Concanamycin A (ConA), bafilomycin A (BafA1)	Plasma membrane fusion assay	Block HCVpp and HCVcc fusion at the endosomes	25–50 nM BafA1 5 nM ConA	Inhibitors of vacuolar ATPases that disturb endosome acidification	[13]
Chloroquine, ammonium chloride	Plasma membrane fusion assay	Block HCVpp fusion at the endosomes [14]	EC_{50} of 50 μM chloroquine and 10 mM NH_4Cl	Weak bases that neutralize the pH of endosomes	[13]

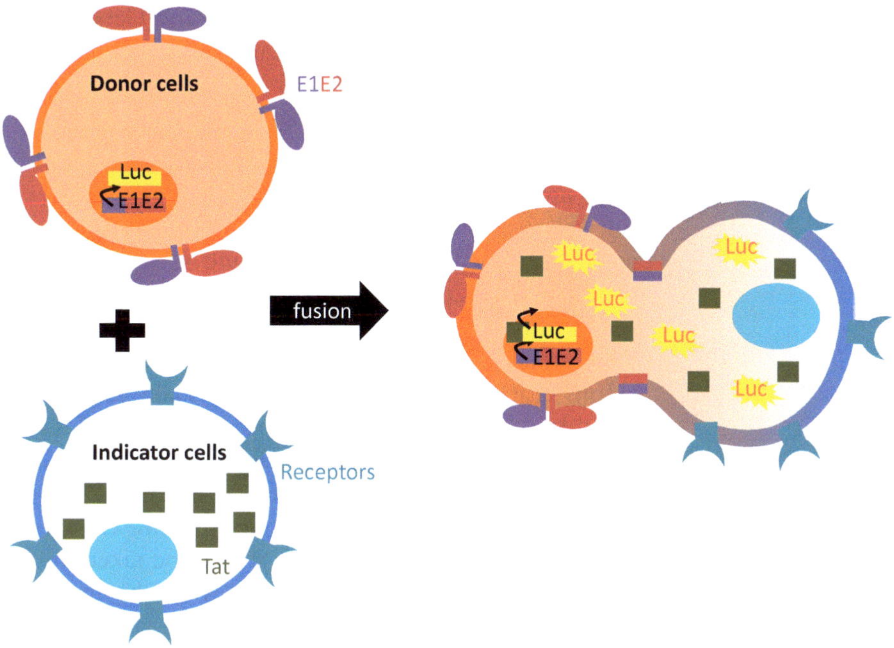

Fig. 1 Schematic representation of cell-cell fusion assay. *See* Subheading 3 for further details

Importantly, this assay can be easily employed to screen for anti-HCV compounds that specifically inhibit HCV entry.

In the following protocol, Huh7-Tat indicator cells are generated by transient transfection of Huh-7 cells with the plasmid LXSN-Tat. Alternatively, stably transfected Huh-7 cell lines expressing Tat can also be used.

1. Day 1—Cells seeding for transient transfection. Approximately 24 h before transfection seed 3 × 10^5 HEK-293T cells and 5 × 10^5 Huh7 cells per well in a six-well plate (*see* **Notes 2** and **3**).
2. Day 2—Transient transfection of donor and indicator cells. Twenty-four hours after cell seeding, replace the culture media by adding 1 ml of fresh pre-warmed media, and return cells to CO_2 incubator while preparing transfection mixes. Co-transfect HEK-293T donor cells using JetPEI transfection reagent following the manufacturer's instructions with 3 μg of phCMV-H77-wt, empty vector or positive control plasmids, and 50 ng of LTR-luc per well. Routinely, a JetPEI to DNA ratio of 2:1 is used but further optimization of transfection conditions is advisable to obtain maximum transfection efficiencies (*see* **Note 4**). Six to eight hours after transfection, replace the culture media by adding 1–2 ml of fresh pre-warmed media, and incubate cells for 24 h.

In parallel, transfect Huh7 cells using GeneJammer transfection reagent following the manufacturer's instructions with 2 μg of LXSN-Tat per well, and incubate cells for 24 h without replacing the culture media (*see* **Notes 5** and **6**).

3. Day 3—Co-cultures of donor and indicator cells. Following a 24-h incubation, dissociate transfected HEK-293T cells from the wells using Versene dissociation reagent, and suspend them in DMEM medium. Count cells and seed them in a 12-well plate at a density of 1×10^5 cells/well. Swirl the plates to spread the cells evenly and return them to the incubator. Let them attach, and repeat cell dissociation with the indicator cells (Huh7-Tat). Cell attachment takes usually less than 2 h but should be checked by microscopy. Count Huh7-Tat cells and adjust cell density to 2×10^5 cells/ml. Aspirate culture media from HEK-293T cells and seed 2×10^5 Huh7-Tat cells/well (1 ml) on the top of the donor cells. HEK-293T cells detach very easily; therefore, extra care should be taken during media replacement. Return the co-cultures to the incubator for a period of 24 h (*see* **Note 7**).
4. Day 4—pH shock. To initiate fusion, remove the media from the co-cultures, and rinse the wells carefully once with 1× fusion buffer adjusted to the desired pH, at room temperature. Replace the 1× fusion buffer by fresh 1× fusion buffer and incubate for 3–5 min. After pH shock, remove 1× fusion buffer and replace it by 1 ml of pre-warmed media. Repeat media exchange one more time and return the co-cultures to the incubator. Incubate the co-cultures for a period of 24 h.
5. Day 5—Luciferase quantification. Twenty-four hours after pH shock, remove the media from the co-cultures, and wash once with 1× DPBS. Add 175 μl of 1× Passive Lysis buffer pre-chilled at 4 °C per well and incubate for 30 min with agitation at 4 °C. Transfer the cell lysates to 1.5 ml Eppendorf tubes, and centrifuge the samples at maximum speed for 5 min at 4 °C to remove cell debris. Transfer the supernatants to new tubes, and freeze the cell lysates at −80 °C or proceed with luciferase quantification assay.
6. For luciferase quantification assay, aliquot 30 μl of each supernatant into a 96-well black plate.
7. Detect luciferase luminescence using Mithras LB 940 luminescence reader with integrated injectors (Berthold Technologies, Germany) after automatic addition of firefly luciferase reagent (substrate).

3.2 Virus-Liposome Fusion Assay

Hemifusion/fusion lipid mixing assays have been invaluable tools in characterizing the biochemical properties of viral fusion steps, including the pH and the lipid species dependencies.

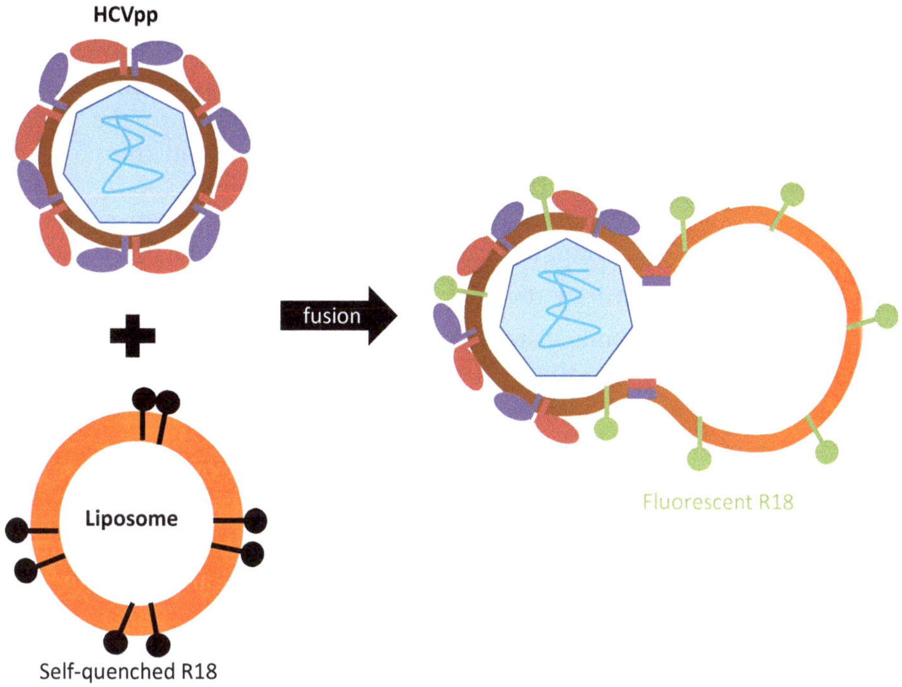

Fig. 2 Schematic representation of virus-liposome fusion assay. *See* Subheading 3 for further details

Sphingomyelin, together with cholesterol, was shown to enhance HCVcc fusion using a liposome fusion assay [8]. Lipid mixing kinetics between the virus and the liposomes is monitored by measuring changes in fluorescence upon dilution of fluorescent lipid dyes such as R_{18}, incorporated into either the liposomes or the viral particles (*see* Fig. 2).

Discrimination between hemifusion and complete lipid fusion with internal content mixing between viral particles and liposomes can be achieved by labeling the liposomes with CCF2 and incorporating β-lactamase (BlaM) into the core of HIV-1-based HCVpp. CCF2 is a fluorogenic substrate of BlaM composed of two moieties, a coumarin derivative and fluorescein, that upon contents mixing between pseudoparticles and liposomes is cleaved by BlaM resulting in a shift of the fluorescence spectrum of CCF2. BlaM is expressed as a chimeric protein encoding BlaM fused to the *N*-terminus of the HIV-1 viral protein R (Vpr).

HCVpp with incorporated BlaM-Vpr may also be used in entry/fusion assays *in cellulo*. However, this method has not been yet tested or published for HCV despite having been well described for HIV [15] and Ebola virus [16] (*see* **Note 8**).

3.2.1 PC, PC/chol, and PC/chol/SM Liposomes Preparation and R_{18} Labeling

1. Dissolve the different components of liposomes (different lipids) individually in chloroform. To prepare PC, PC/chol (70:30 molar ratio), and PC/chol/SM (65:30:5 molar ratio)

containing liposomes, mix the dissolved lipids in chloroform in the required molar proportion.

2. Add R_{18} in ethanol at self-quenching concentrations, equivalent to approximately 5 mol% relative to the phospholipid content of the liposome. Incubate the mixture for 1 h in the dark at room temperature, and allow the lipids to dry by vacuum evaporation for 30 min. Alternatively to liposome labeling, R_{18} can be incorporated into HCVpp.
3. Hydrate the lipid film in PBS 1× pH 7.4.
4. Conduct five freeze/thaw cycles (liquid nitrogen and 37 °C, respectively). The vesicle suspension is then sized to 100 nm diameter vesicles by extrusion through polycarbonate filters.

3.2.2 HCVpp/Liposome Lipid Mixing

1. Pipette different volumes (5–40 μl) of HCVpp or HCVcc in PBS (pH 7.4) to a 37 °C –thermostable cuvette.
2. Add R_{18}-labeled liposomes (final lipid concentration of 15 μM), and incubate the mixtures at 37 °C for 10 min to allow temperature equilibration.
3. Decrease the pH to 5 by adding an appropriate volume of diluted HCl to the cuvette, and record the fusion kinetics on an SLM Aminco 8000 spectrofluorometer over a 30-min period, with an excitation wavelength at 560 nm and an emission wavelength at 590 nm.
4. Add 0.1% Triton X-100 (final concentration, vol/vol), and measure the maximal R_{18} quenching.

3.2.3 CCF2-FA Liposome Preparation

1. Resuspend the lipid film into 100 μM CCF2-FA solution in 25 mM HEPES, 150 mM NaCl pH 7.5.
2. Remove unincorporated CCF2 by gel filtration using PD-10 columns.
3. Mix CCF2-loaded liposomes (150 μM phospholipid and 25 μM CCF2 final) with BlaM HIV core-based HCVpp.
4. After temperature equilibration at 37 °C, decrease the pH to 5 by adding an appropriate volume of diluted HCl to the cuvette, and record the fusion kinetics on an SLM Aminco 8000 spectrofluorometer over a 30-min time period, with an excitation wavelength at 409 nm and an emission wavelength between 420 and 560 nm.

3.3 Virus-Plasma Membrane Fusion Assay

This assay allows the study of fusion of viral particles *in cellulo* which has two main advantages: the use of HCVcc particles and the use of cells as target, rather than liposomes. The use of inhibitors of acidification of endosomes prevents viral entry and membrane fusion by the normal route through acidified endosomes, which is technically difficult to investigate. It was previously

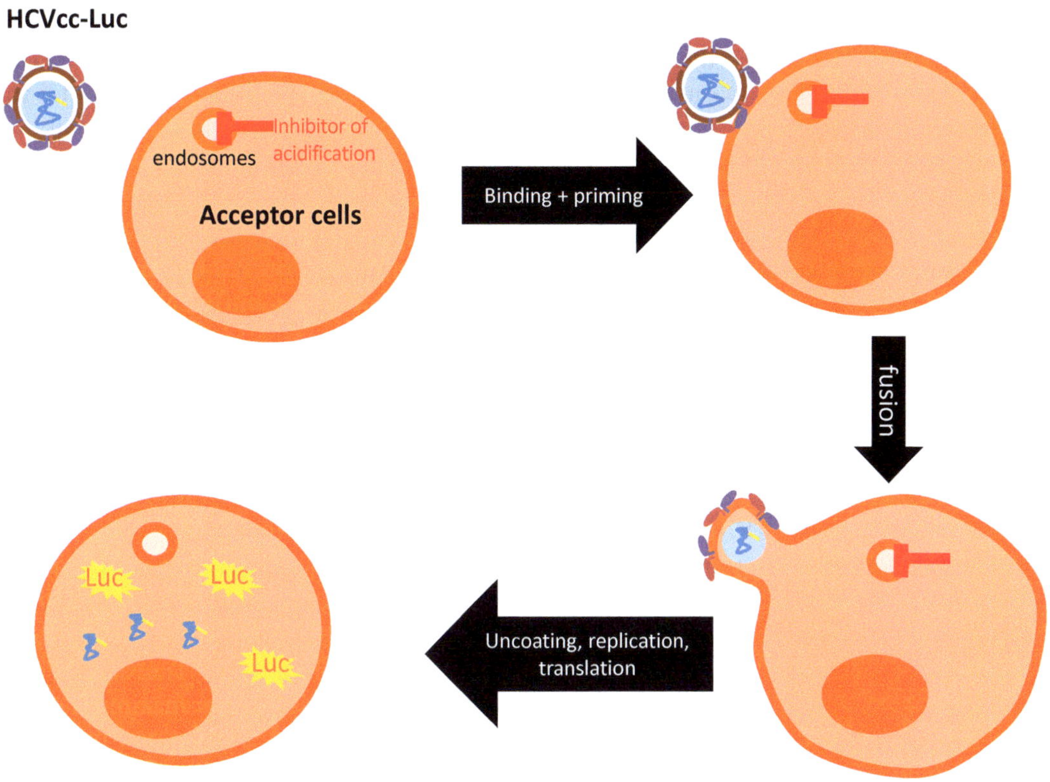

Fig. 3 Schematic representation of virus-plasma membrane fusion assay. *See* Subheading 3 for further details

demonstrated that viral particles have to be primed by receptor interactions and that fusion can only occur after 1 h incubation of viral particles with cells (*see* Fig. 3) as compared to other viruses that can directly fuse after binding to the cells. In contrast to the other above-described assays, this assay requires the uncoating of the viral genome and subsequent viral genome translation and replication, meaning that this assay is not directly related to fusion; thus, results obtained with this assay should be confirmed with other assays to ensure that fusion is the step that is really involved.

1. *Day 1.* Seed target cells in a six-well plate at 7–8 × 10^5 cells/well (*see* **Note 9**).
2. *Day 2.* Treat target cells with bafilomycin A1 (25 nM) or concanamycin A (5 nM) for 1 h at 37 °C (*see* **Note 10**). For that add 1.5 ml of cell culture medium + 1.5 μl of stock solution of inhibitor per well.
3. Infect target cells in the presence of bafilomycin A1 or concanamycin A for 2 h at 4 °C (*see* **Note 11**). For that remove medium and replace it with ice-cold medium containing viral particles + inhibitors at the final concentration of 25 nM for bafilomycin A1 and 5 nM for concanamycin A1.

4. Wash cells with cold DPBS twice (*see* **Note 12**).
5. Incubate cells with medium containing bafilomycin A1 or concanamycin A for 1 h at 37 °C (*see* **Note 13**). For that add 1.5 ml of cell culture medium + 1.5 μl of stock solution of inhibitor per well.
6. Incubate cells for 5 min at 37 °C with warm citric acid buffer (*see* **Note 14**). For that remove medium, wash with cold DPBS, add 1.5 ml/well of warm citric acid buffer, and incubate cells 5 min at 37 °C. After that remove buffer and wash with medium.
7. Add fresh medium in the presence of bafilomycin A1 or concanamycin A for 3 h at 37 °C. For that add 1.5 ml of cell culture medium + 1.5 μl of stock solution of inhibitor per well.
8. Change medium and add 1.5 ml cell culture medium per well.
9. Incubate cells at 37 °C for 24 h (*see* **Note 15**).
10. *Day* 3. Harvest cells and measure luciferase at 24 h (*see* **Note 15**). Remove media from the cells and wash once with DPBS. Add 350 μl of 1× passive lysis buffer at 4 °C per well and incubate for 30 min with agitation at 4 °C. Transfer the cell lysates to a tube and centrifuge the samples at maximum speed for 5 min at 4 °C. Transfer the supernatant to a new tube, and aliquot 30 μl of each supernatant into a 96-well black plate. Freeze the cell lysates at −80 °C or proceed with luciferase quantification assay.
11. Detect luciferase luminescence using a luminescence reader with integrated injectors after automatic addition of firefly luciferase reagent.

3.4 Fusion Inhibitors

Several drugs/compounds have been shown to modulate HCV infection by affecting viral membrane fusion. These molecules can be applied in combination with the different fusion assays described above to further characterize viral membrane fusion. Furthermore, fusion assays are powerful tools that can be used for screening and characterizing new antivirals that target virus entry. Fusion inhibitors are a diverse group of molecules that display different mechanisms of action but ultimately interfere with the required conformational changes of fusion peptides during virus fusion (Table 1). In this section, we provide examples of fusion inhibitors previously used in combination with the three fusion methods described.

3.4.1 Arbidol

The broad-spectrum antiviral arbidol has been shown to inhibit efficiently HCV membrane fusion in liposome fusion assays, either using HCVpp or cell culture HCVcc [8, 9]. The extent of fusion inhibition was shown to be directly proportional to arbidol concentration, with some variations among genotypes. Fusion of HCVpp coated with E1E2 of genotype 1a was totally abolished at 1 μg/ml, while 6 μg/ml were required to block fusion of genotype 1b. For genotype 2a, HCVpp fusion was still observed using 6 μg/ml of arbidol, which is equivalent to 11.3 μM [9]. The concentration of arbidol able to inhibit fusion of genotype 2a HCVcc by 50% (IC50) was determined to be 2 μg/ml [8].

1. Dissolve ARB first in 100% ethanol and then dilute ARB to a final concentration of 100 mM in water.
2. Preincubate ARB in the desired concentration for 2 min with either HCVpp or 15 μM of R_{18}-labeled liposomes at 37 °C in buffer at neutral pH.
3. Then, add the R_{18}-labeled liposomes or HCVpp to the mixture and start fusion by decreasing the pH to 5.0.
4. Monitor lipid mixing as described before.

3.4.2 Human Monoclonal Antibody CBH-5

CBH-5 against an epitope within E2, with neutralizing activity against HCVpp harboring diverse HCV genotypes [10], was shown to exert a dose-dependent inhibition of HCVcc fusion using liposome fusion assays [13]. It was determined that 25 μg/ml of CBH-5 completely abrogates E1E2-mediated fusion and 5 μg/ml was estimated to inhibit 50% of the fusion activity.

1. Incubate HCVcc with increasing concentrations of CBH-5 in cell suspension buffer (0.85% (w/v) NaCl, 10 mM Tricine-NaOH, pH 7.4) for 15 min at pH 7.4.
2. Then, add the liposomes, induce fusion, and monitor lipid mixing.

3.4.3 Amphipathic DNA Polymers (APs)

Phosphorothioate oligonucleotides (PS-ONs) inhibit HCV infection of different genotypes without affecting replication or attachment. HCVpp fusion with liposomes was significantly inhibited when 10 nM of PS-ONs was incubated with HCVpp before the addition of R_{18}-labeled liposomes [12].

3.4.4 Fusion Peptide Inhibitors

Recently, a peptide library screening that covered the entire E1E2 amino acid sequence led to the identification of a peptide-based fusion inhibitor derived from the E2 stem domain, amino acids 671–705, with an anti-E1E2 fusion EC_{50} value of 734 ± 9 nM in cell-cell fusion assays [12].

1. Add the peptide fusion inhibitor at a final concentration of 5 μM to the co-cultures, after pH shock.

3.4.5 Flunarizine

In the plasma membrane assay, HCV can only access cells by exogenous acidification since endosomal acidification is prevented by continuous presence of bafilomycin A or concanamycin A as described. Alternatively, chloroquine or ammonium chloride can be used instead of bafilomycin A or concanamycin A.

Using this setting, it was observed that flunarizine reduces HCV infection when added after HCVcc inoculation and before the pH shock [13].

4 Notes

1. Buffer at pH 7 is used as a negative control (to set the background level) since no fusion occurs at this pH.
2. Prior to transfection, cells can be detached with 0,05% Trypsin-EDTA (1X) solution, although after transfection it is highly recommended to detach cells using a nonenzymatic cell dissociation reagent such as Versene 1:5000 (1×) solution.
3. Calculate the number of wells and plates required taken into account that each pH condition should be tested at least in duplicate and extra cells might be required to seed co-cultures at different ratios.
4. When performing multiple transfections with the same plasmid combinations, it is recommended to prepare a master mix of transfection complexes.
5. GFP reporter vector can be added to the transfection mixes, 5% of the total amount in μg of plasmid DNA used for transfection, for determination of transfection efficiency.
6. The described luciferase reporter system works well in HEK-293T and Huh7 cells. However, if using different cell lines, it is recommended to check the activity of the HIV-1 3′ LTR promoter/Tat inducible system before performing cell-cell fusion assays. This can be achieved by simply co-transfecting both donor and indicator cells with 50 ng of LTR-luc and 2 μg of LXSN-Tat and measuring luciferase activity after cell lysis.
7. Different ratios of donor to indicator cells should be tested in order to maximize contacts between adjacent donor and indicator cells. For cell-cell fusion optimization, we recommend to compare luciferase activity between ratios of donor to indicator cells of 1:2, 1:5, and 5:1.
8. After infection with HCVpp containing BlaM-Vpr at 37 °C for 1–3 h, cells are washed in CO_2-independent medium and loaded with CCF2-AM dye at room temperature for 1 h. After two additional washes, the BlaM reaction is allowed to develop for 7 h at room temperature in CO_2-independent medium supplemented with 10% FBS and 2.5 mM probenecid, a nonspecific inhibitor of anion transport, without antibiotics. Finally, cells are washed twice with PBS and fixed in a 1.2%

paraformaldehyde solution prior to analysis. The change in emission fluorescence of CCF2 after cleavage by BlaM-Vpr is monitored by flow cytometer (520 nm before cleavage and 447 nm after upon excitation at 409 nm).

9. Density of cells should be adjusted depending on cell types and time of measurement of luciferase.
10. This pre-treatment allows inhibition of acidification of endosomes.
11. This step allows binding of particles. Concentrations of particles should be high and could be adjusted depending on the strain used. It is essential to put the same amounts of particles for all the conditions tested, especially when mutants are compared.
12. Temperature of DPBS is important to prevent internalization of particles. This step is used to remove unbound particles.
13. This step allows the priming of glycoproteins. Indeed, unlike other viruses, e.g., VSV or Semliki Forest virus, HCV requires additional step(s) to activate the pH-dependent fusion of glycoproteins [7]. This step could be skipped to have an additional negative control.
14. This step triggers fusion. Incubation with citric acid buffer at pH 7 is used as a negative control. Temperature of buffer is important for the quality of the assay, and buffer at 37 °C should be used to allow fusion of cells.
15. The analysis was made at 24 h postinfection in [7] or 48 h postinfection in [13]. Since luciferase is a very sensitive way to detect infection, the result of this assay should be detected as soon as possible after infection to prevent the possibility of replication, assembly, secretion, or second round of infection. High amounts of particles could help to reduce the time of analysis.

Acknowledgment

Our work is supported by the French "Agence Nationale de la Recherche sur le SIDA et les hépatites virales" (ANRS), the European Research Council (ERC-2008-AdG-233130-HEP-CENT), and the LabEx ECOFECT (ANR-11-LABX-0048).

References

1. Douam F, Lavillette D, Cosset FL (2015) The mechanism of HCV entry into host cells. Prog Mol Biol Transl Sci 129:63–107
2. Perez-Berna AJ, Moreno MR, Guillen J, Bernabeu A, Villalain J (2006) The membrane-active regions of the hepatitis C

virus E1 and E2 envelope glycoproteins. Biochemistry 45:3755–3768

3. Takikawa S, Ishii K, Aizaki H, Suzuki T, Asakura H, Matsuura Y et al (2000) Cell fusion activity of hepatitis C virus envelope proteins. J Virol 74:5066–5074
4. Bartosch B, Dubuisson J, Cosset FL (2003) Infectious hepatitis C virus pseudo-particles containing functional E1-E2 envelope protein complexes. J Exp Med 197:633–642
5. Lavillette D, Bartosch B, Nourrisson D, Verney G, Cosset FL, Penin F et al (2006) Hepatitis C virus glycoproteins mediate low pH-dependent membrane fusion with liposomes. J Biol Chem 281:3909–3917
6. Lavillette D, Pecheur EI, Donot P, Fresquet J, Molle J, Corbau R et al (2007) Characterization of fusion determinants points to the involvement of three discrete regions of both E1 and E2 glycoproteins in the membrane fusion process of hepatitis C virus. J Virol 81:8752–8765
7. Tscherne DM, Jones CT, Evans MJ, Lindenbach BD, McKeating JA, Rice CM (2006) Time- and temperature-dependent activation of hepatitis C virus for low-pH-triggered entry. J Virol 80:1734–1741
8. Haid S, Pietschmann T, Pecheur EI (2009) Low pH-dependent hepatitis C virus membrane fusion depends on E2 integrity, target lipid composition, and density of virus particles. J Biol Chem 284:17657–17667
9. Pecheur EI, Lavillette D, Alcaras F, Molle J, Boriskin YS, Roberts M et al (2007) Biochemical mechanism of hepatitis C virus inhibition by the broad-spectrum antiviral arbidol. Biochemistry 46:6050–6059
10. Owsianka AM, Tarr AW, Keck ZY, Li TK, Witteveldt J, Adair R et al (2008) Broadly neutralizing human monoclonal antibodies to the hepatitis C virus E2 glycoprotein. J Gen Virol 89:653–659
11. Matsumura T, Hu Z, Kato T, Dreux M, Zhang YY, Imamura M et al (2009) Amphipathic DNA polymers inhibit hepatitis C virus infection by blocking viral entry. Gastroenterology 137:673–681
12. Chi X, Niu Y, Cheng M, Liu X, Feng Y, Zheng F et al (2016) Identification of a Potent and Broad-Spectrum Hepatitis C Virus Fusion Inhibitory Peptide from the E2 Stem Domain. Sci Rep 6:25224
13. Perin PM, Haid S, Brown RJ, Doerrbecker J, Schulze K, Zeilinger C et al (2016) Flunarizine prevents hepatitis C virus membrane fusion in a genotype-dependent manner by targeting the potential fusion peptide within E1. Hepatology 63:49–62
14. Ashfaq UA, Javed T, Rehman S, Nawaz Z, Riazuddin S (2011) Lysosomotropic agents as HCV entry inhibitors. Virol J 8:163
15. Cavrois M, De Noronha C, Greene WC (2002) A sensitive and specific enzyme-based assay detecting HIV-1 virion fusion in primary T lymphocytes. Nat Biotechnol 20:1151–1154
16. Yonezawa A, Cavrois M, Greene WC (2005) Studies of ebola virus glycoprotein-mediated entry and fusion by using pseudotyped human immunodeficiency virus type 1 virions: involvement of cytoskeletal proteins and enhancement by tumor necrosis factor alpha. J Virol 79:918–926

Chapter 16

Functional Characterization of Apolipoproteins in the HCV Life Cycle

Luhua Qiao and Guangxiang (George) Luo

Abstract

Apolipoprotein E (apoE) plays dual functions in the HCV life cycle by promoting HCV infection and virion assembly and production. ApoE is a structural component on the HCV envelope. It mediates HCV cell attachment through specific interactions with the cell surface receptors such as syndecan-1 (SDC-1) and SDC-2 heparan sulfate proteoglycans (HSPGs). It also interacts with NS5A and E2, resulting in an enhancement of HCV morphogenesis. It can bind HCV extracellularly and promotes HCV infection. It is critical for HCV cell-to-cell transmission and may also play a role in HCV persistence by interfering with the action of HCV-neutralizing antibodies. Other apolipoproteins particularly apoB and apoC1 were also found on the HCV envelope, but their roles in the HCV life cycle remain unclear. In the last decade, a number of genomic, immunological, structural, and cell biology methodologies have been developed and used for determining the importance of apoE in the HCV life cycle. These methods and protocols will continue to be valuable to further understand the importance and the underlying molecular mechanism of various apolipoproteins in HCV infection and pathogenesis.

Key words Apolipoprotein, ApoE, ApoB, ApoC1, HCV, Attachment, Infectivity, Morphogenesis

1 Introduction

It has long been known that HCV is associated with lipoproteins in the plasma of hepatitis C patients [1–6]. Apolipoproteins B (apoB), C1 (apoC1), and E (apoE) were all detected in the low-density fractions of HCV RNA-containing particles, which could also be captured by apoB-, apoC1-, and apoE-specific antibodies [1, 4, 7–11]. Delipidation of the low-density HCV particles resulted in nucleocapsid-like structures [1, 2, 4, 8, 11, 12]. These observations suggest that HCV is assembled as lipo-viro-particles (LVPs) with lipoproteins [1, 2]. However, the importance of apolipoproteins in the HCV life cycle has not been experimentally examined until HCV propagation systems in cell culture were successfully developed in 2005 [13–16]. It was speculated that HCV production depends on the pathways of very-low-density lipoprotein (VLDL)

Mansun Law (ed.), *Hepatitis C Virus Protocols*, Methods in Molecular Biology, vol. 1911,
https://doi.org/10.1007/978-1-4939-8976-8_16, © Springer Science+Business Media, LLC, part of Springer Nature 2019

assembly and secretion [17, 18]. However, our studies have demonstrated that microsomal triglyceride transfer protein (MTP) inhibitors and knockdown of apoB expression by specific siRNAs potently reduced VLDL secretion but had no significant effects on HCV production [8], suggesting that HCV production is independent of VLDL. Our findings were confirmed by others using an engineered HCV-permissive HepG2 cell line, which is competent for VLDL assembly and secretion. Similarly, the later study did not find any correlation between HCV production and VLDL secretion in HepG2 cells [19]. These findings suggest that HCV is assembled and/or maturated with apolipoproteins incorporation independent of VLDL assembly and secretion.

Several lines of substantial evidence obtained from our previous studies demonstrate that apoE plays important roles in both HCV infection and virion assembly [7, 8, 20–24]. First of all, infectious HCV particles are enriched in apoE, and their infectivity was closely correlated with the amounts of apoE incorporated into HCV [7]. ApoE is a structural component on the HCV envelope [8], which could be visualized under immunogold electron microscope [25–27]. More importantly, apoE-specific monoclonal antibodies could potently inhibit HCV infection via a specific blockade of its attachment to the surface of hepatocytes [8, 22, 23] as well as prevent HCV cell-to-cell transmission [28]. Synthetic peptides derived from the receptor-binding domain of apoE could block HCV cell attachment and entry [22, 29]. Interestingly, apoE mediates HCV cell attachment through binding to cell surface attachment receptors such as syndecan-1 (SDC-1) and SDC-2 heparan sulfate proteoglycans (HSPGs) [24, 28]. Additionally, our studies have found that apoE promotes HCV virion assembly and production via a specific interaction with NS5A [8, 20, 21], although others argued for possible interactions between apoE and HCV E2 [30]. Strikingly, apoE is able to evade immune response by interfering with antibody-mediated neutralization of HCV [31, 32]. These interesting characteristics and properties of apoE highlight the importance of apolipoproteins in the HCV life cycle and pathogenesis in vivo.

In the past, a number of state-of-the-art technologies and methodologies have been developed and utilized for the characterization of apolipoproteins in the HCV infectious cycle, including genomics, reverse genetics, immunological, structural, and cell biological approaches. In this chapter, we summarize the methods and protocols used for the determination of the importance and underlying mechanism of action of apolipoproteins in HCV infection and pathogenesis.

2 Materials

2.1 Cell Culture

1. HCV-permissive cell lines such as Huh-7.5, Huh-7.5.1, primary human hepatocytes, or hESC-derived human hepatocyte-like cells [33].
2. Cell culture medium: Dulbecco's modified Eagle's medium (DMEM) supplemented with 10% fetal bovine serum (FBS), 1× nonessential amino acids, and penicillin-streptomycin (100 U/ml).
3. Genotype 2a (JFH1) HCV virus: adapted in Huh-7.5 cells [34].
4. 37 °C cell culture incubator with 5% CO_2.
5. 1× phosphate-buffered saline (PBS) and 0.25% trypsin-EDTA.
6. Trypan blue, cell counting slide, and TC20 automated cell counter (Bio-Rad).
7. Cell culture flasks and plates.

2.2 Small Interfering RNA (siRNA) Transfection

1. OPTI-MEM (Invitrogen).
2. RNAiMax (Invitrogen).
3. ApoB-, apoC1-, and apoE-specific siRNA and a non-specific control siRNA (#2) were purchased from Dharmacon. The siRNA-targeting sequences for apoB and apoE mRNAs are 5′-AGACAGAGCCGGAGCCCGA-3′ and 5-′-AGACAGAGCCGGAG CCCGA-3′, respectively.
4. 1.5 ml, 15 ml, and 50 ml centrifuge tubes.

2.3 Construction of ApoE-Expressing Plasmid DNA and DNA Transfection

1. ApoE3 expression vector pCMV6XL5/ApoE.
2. pCMV6XL5 control vector.
3. QuikChange Lightning site-directed mutagenesis kit (Agilent).
4. Synthetic oligonucleotide primers, mApoE/Up (5 ′-GCAAG CGGTGGAGACG GAACCCGAACCGGAGCTGCGCCAG CAG-3′) and mApoE/Bt (5′-CTGCTGGC GCAGCTCCGG TTCGGGTTCCGTC TCCACCGCTTGC-3′).
5. Lipofectamine 3000 Transfection Reagent (Invitrogen).

2.4 ApoE-Derived Peptides and HSPG-Blocking Peptide

1. ApoE-derived peptides include E3-21 (TEELRVRLASHLR KLRKRLLR), E3-21C (CEELRVRLASHLRKLRKRLLR), and hEP-1 (EELRVRLASHLRKLRKRLLRDA DDLQKR LAVYEEQAQQIRLQAEAFQARLKSWFEPLVEDM).
2. ApoE peptide mutants include E3-21m (TEELRVRLASHL-RELRERLLR) and E3-21Cm (CEELRVRLASHLRELRER LLR).

3. HSPG-binding peptide 6a-P (KSVRGKGKGQKRKRKKSR YK) and its mutant form 6a-Pm (KSVRGKGKGQERERKES RYK).

2.5 ApoE Antibody Neutralization of HCV Infectivity

1. HCV E2-specific monoclonal antibody CBH2 and CBH5 [35].
2. ApoE-specific monoclonal antibody 23 (mAb23).
3. ApoE monoclonal antibody WuE4 (ATCC).
4. Isotype-matched normal mouse and human IgGs.

2.6 Construction of Apolipoprotein Gene-Knockout Cell Lines Using LentiCRISPRv2 System

1. LentiCRISPRv2 vector (Addgene).
2. sgRNAs targeting specific apolipoprotein genes.
3. T4 polynucleotide kinase and T4 DNA ligase (NEB).
4. BsmBI restriction enzyme (NEB).
5. Stbl3 competent cells (TransGen Biotech).
6. DNA isolation and purification kits (Qiagen).
7. HEK293T cells.
8. Plasmids: pVSVg and psPAX2 (Addgene).
9. Lipofectamine 2000 and lipofectamine 3000 reagents (Invitrogen).
10. 0.45 μm low-protein-binding membrane unit (Millipore).
11. Puromycin (Invitrogen).
12. DNeasy Blood & Tissue Kits (Qiagen).
13. 5× Taq master mix (NEB).
14. PCR thermocycler (Bio-Rad).
15. QIAquick Gel Extraction Kit (Qiagen).
16. Polybrene (Invitrogen).

2.7 qRT-PCR

1. Trizol reagent (Invitrogen).
2. StepOnePlus real-time PCR system (Applied Biosystems).
3. HCV primers HCV-F (5′-AGCCATGGCGTTAGTATGA G TGTC-3′) and HCV-R (5′-ACAAGGCCTTTCGCAACCC AA-3′) as well as the HCV-specific probe (5′-AAACCCACTC TATGCCCGGCCATTT-3′) containing 6-carboxyfluorescein at the 5′ end and 6-carboxytetramethylrhodamine at the 3′ end were synthesized by Integrated DNA Technologies.
4. GAPDH Control Primer-Probe Set (Thermo Fisher Scientific).
5. One-Step Real-Time RT-PCR Master Mix (GE Healthcare).

2.8 Western Blot

1. HCV NS3 (mAb15) and NS5A (9E10) monoclonal antibodies [13, 14].

2. β-actin monoclonal antibody (AC15, Sigma-Aldrich).
3. Horseradish peroxidase (HRP)-conjugated goat anti-mouse IgG (sc2005) secondary antibody (Santa Cruz).
4. RIPA buffer: 10 mM Tris–Cl (pH 8.0), 1 mM EDTA, 1% Triton X-100, 0.1% sodium deoxycholate, 0.1% SDS, 140 mM NaCl, 1 mM PMSF.
5. ECL Western Blotting Substrate (Bio-Rad).
6. Gel Doc MP gel documentation system (Bio-Rad).

2.9 Titration of Infectious HCV Titers

1. 96-well cell culture plates.
2. DMEM containing 1% methylcellulose.
3. 10% neutralized formalin.
4. NS3-specific monoclonal antibody (mAb15).

3 Methods

Cell culture models based on the human) hepatocellular carcinoma (HepG2, Huh-7 and its variants such as Huh-7.5 or Huh-7.5.1) cell lines have been developed for HCV infection and/or replication. To determine the importance of apolipoproteins in HCV life cycle, specific apolipoprotein genes can be downregulated using siRNAs or gene-editing technologies (e.g., CRISPR/Case9) or upregulated using recombinant DNA expression vectors and/or adenovirus and lentivirus. The effects of down- and upregulation of apolipoproteins on HCV infection, replication, and/or production can be quantified by various methods including Western blot, qRT-PCR, and infectious HCV titers.

3.1 Cell Culture

1. Huh-7.5 cells are maintained in high-glucose DMEM supplemented with 10% FBS, 1 mM nonessential amino acids, and penicillin-streptomycin at 37 °C cell incubator with 5% CO_2.
2. Cells are split by trypsinization with 1 ml of 0.25% trypsin-EDTA for a 75cm^2 flask. Trypsin-EDTA is immediately removed after covering cell monolayer.
3. After incubation at 37 °C incubator for a few minutes, cells are resuspended in 10 ml fresh DMEM.
4. The numbers of cells will depend on the size of cell culture plates, for instance, 1.5×10^5 cells per well in a 12-well cell culture plate (*see* **Note 1**).

3.2 Silencing of Apolipoprotein Gene Expression Using Synthetic siRNAs

1. Add 2.5 μl of RNAiMax to a 1.5 ml Eppendorf (EP) tube containing 250 μl of OPTI-MEM.
2. Add 3.125 μl of siRNAs (20 μM) to a 1.5 ml EP tube containing 250 μl of OPTI-MEM.

3. Combine RNAiMax and siRNAs in tubes 1 and 2 with incubation at room temperature for 15 min.
4. Transfer 100 μl RNAiMax/siRNA mix to a tube containing 400 μl OPTI-MEM (fivefold dilution), and repeat this step again for another fivefold dilution, resulting in final concentrations of siRNAs of 50 nM, 10 nM, and 2 nM.
5. Trypsinize and resuspend Huh-7.5 cells in DMEM medium containing 10% FBS.
6. Pellet down cells by centrifugation at 800 × *g* for 5 min.
7. Resuspend cells in OPTI-MEM.
8. Mix 10 μl of cells with 10 μl trypan blue, and then count the number of cells by transferring 10 μl of trypan blue-stained cells into the chamber of cell counting slide using TC20 automated cell counter.
9. Dilute cells to 2.5×10^5 cells per ml using OPTI-MEM.
10. Add 400 μl of the above-prepared siRNA/RNAiMax mix to each well containing 600 μl of cell suspension to a 12-well plate.
11. At 6 h post-transfection, replace siRNA/RNAiMax OPTI-MEM with 1 ml of DMEM medium containing 10% FBS.

3.3 Construction of Stable Cell Lines with Specific Apolipoprotein Gene Knockout Using CRISPR/Cas9 System

1. Two oligonucleotide primers are used for construction of sgRNA cDNA. For instance, a pair of oligonucleotide primers apoE-F (5′-CACCGGCTTTTGGGATTA CC GCGC-3′) and apoE-R (5′-AAACGCGCAGGTAATCCCAAAAGCC-3′) is used for construction of apoE-specific sgRNA. Two primers are initially treated with T4 polynucleotide kinase. The reaction mixture (50 μl) consists of 5 μl of T4 PNK reaction buffer (10×), 5 μl forward and reverse primers, 2 μl T4 PNK, and 33 μl H_2O. Incubate at 37 °C for 30 min.
2. Two oligonucleotide primers are annealed to form double-stranded DNA fragment in T4 DNA ligation buffer.
3. Digest lentiCRISPRv2 (Addgene) using BsmBI restriction enzyme.
4. Digested lentiCRISPRv2 DNA is purified by electrophoresis in a 1% agarose gel followed by DNA extraction using a Qiagen kit.
5. The double-stranded oligonucleotides are cloned into BsmBI-digested lentiCRISPRv2 by using T4 DNA ligase. The ligation mixture is transformed into Stbl3 competent cells (TransGen Biotech).
6. Plasmid DNAs are purified using a DNA isolation kit (Qiagen) and confirmed by sequence analysis.
7. 3.0×10^6 HEK293T cells are seeded in a 6-well plate on the day before DNA transfection.

8. 0.58 μg of pVSVg, 1.25 μg of psPAX2, and 0.83 μg of lenti-CRISPRv2 vector expressing Cas9 and sgRNAs and 5 μl of Lipofectamine 2000 are mixed together in 1 ml of OPTI-MEM with an incubation of 15 min at room temperature.
9. HEK293T cells are washed with 1× PBS prior to transfer with the above DNA and lipid mixture.
10. After 6 h post-transfection, the DNA/lipid-containing OPTI-MEM is replaced with fresh DMEM containing 10% FBS.
11. At 72 h post-transfection, the supernatant is collected and filtered through a 0.45-μm low-protein-binding membrane unit (Millipore). The resulting recombinant lentiviruses are aliquoted and stored in a −80 °C freezer.
12. Huh-7.5 cells in 100-mm cell culture dishes are transduced with about 5 to 10 μL lentiviruses expressing sgRNAs targeting apolipoprotein genes. Cell clones are selected in the presence of 2 μg/ml of puromycin for 2–3 weeks. Individual cell clones are picked up for expansion.
13. Genomic DNAs are extracted from each cell clone using DNeasy Blood & Tissue Kits and used for PCR amplification of each specific gene with gene-specific primers. For instance, apoE gene is amplified by PCR using the primer pairs 5′-GCTTTCCAAGTGATTAAACCGAC-3′ and 5′-ACCTGGGGAGGTATAGCCG-3′. PCR products are subject to DNA sequence verification.
14. The knockout of specific gene is further validated by Western blotting.

3.4 DNA Construction and Transfection

1. pCMV6XL5/mApoE3 is constructed by introducing five silent nucleotide mutations into the siRNA-targeting region of apoE3 gene (*see* **Note 2**), using a QuikChange Lightning site-directed mutagenesis kit (Agilent) and two synthetic oligonucleotide primers, mApoE/Up (5′-GCAAGCGGTGGAG ACGGAACCCGAACCGGAGCTGCGCCAGCAG-3′) and mApoE/Bt (5′-CTGCTGGCGCAGCTCCGGTTC GGGTTCCGTCTCCACCGC TTGC-3′).
2. The DNA construct is confirmed by sequence analysis and purified with a DNA isolation kit from Qiagen. The vector DNA pCMV6XL5 is used as a control.
3. ApoE siRNA transfection is described in Subheading 3.2.
4. At 12 h after siRNA transfection, change the medium to OPTI-MEM.
5. Prepare DNA/lipid mixture by adding 250 μl of OPTI-MEM and 2.5 μl Lipofectamine 3000 to a 1.5 ml EP tube, followed by adding 250 μl OPTI-MEM containing 4 μL P3000 reagent and 2 μg of pCMV6XL5/mApoE3 or control DNA

pCMV6XL5. The DNA and lipid mixture is incubated at room temperature for 15 min.

6. Transfer the DNA/lipid mix to each well. After 6 h incubation, replace the OPTI-MEM with fresh DMEM containing 10% FBS.
7. At 48 h post-transfection, the cells are infected with HCV at a multiplicity of infection (m.o.i.) of 10.
8. At 24 h post-infection (for single-cycle HCV growth), the levels of HCV proteins and RNAs are determined by Western blot and qRT-PCR, respectively.

3.5 HCV Attachment Assay

1. Huh-7.5 cells are incubated with HCV at an m.o.i. of 10 on ice for 2 h.
2. Unbound HCV is removed by washing with PBS for three times.
3. Total RNAs are extracted with Trizol reagent from the HCV-bound cells.
4. The levels of HCV vRNA are determined by a real-time qRT-PCR method.

3.6 HCV Infection

1. Huh-7.5 cells are infected with HCV at an m.o.i. of approximately 1 (for multiple-cycle HCV growth) or 10 (for single-cycle HCV growth) for 2 h at 37 °C (*see* **Note 3**).
2. Wash the HCV-infected cell with 1× PBS three times.
3. Add fresh DMEM containing 10% FBS, and incubate for another 24 h (single-cycle HCV growth) or 72 h (multiple-cycle HCV growth).
4. Lyse cells with 70 μl RIPA buffer for the determination of HCV proteins by Western blotting or extract total cellular RNAs with 500 μl Trizol (for 12-well plate) for quantification of HCV RNAs by qRT-PCR.
5. Collect the cell culture supernatants for quantification of apolipoprotein levels by Western blot or titration of infectious HCV titers by a limiting dilution assay in conjunction with IHC or IFA using a NS3-specific monoclonal antibody [13].

3.7 Blockade of HCV Cell Attachment and Infection by Apolipoprotein-Specific Antibodies, ApoE-Derived Peptides, or a HSPG-Binding Peptide

1. Huh-7.5 cells in 12-well plates are infected with HCV in the presence of increasing amounts of HCV E2-specific or apolipoprotein-specific antibodies, apoE-derived peptides (apoE-21, apoE-21c, hEP), or an HSPG-binding peptide 6a-P as described in Subheading 2.4.
2. For HCV attachment assay, Huh-7.5 cells are incubated with HCV on ice for 2 h in the presence of various amounts of E2 and apolipoprotein-specific antibodies or apoE-derived and HSPG-binding peptides.

3. After 2 h incubation on ice (attachment), unbound HCV is removed by washing with PBS three times. Total cellular RNAs are extracted with Trizol reagent. The levels of HCV vRNA are determined by qRT-PCR.
4. At 2 h post-infection, HCV-infected cells are washed with PBS and incubated with fresh DMEM containing 10% FBS for 24 h (single-cycle HCV growth) or 72 h (multiple-cycle HCV growth).
5. The levels of HCV proteins and RNA in the HCV-infected cells are determined by Western blot and qRT-PCR, respectively.
6. The cell culture supernatants are collected for the determination of infectious HCV titers by a limiting dilution assay.

3.8 qRT-PCR

1. Determine the concentrations of total RNAs by spectrometry (*see* **Note 4**).
2. Dilute total RNAs to 20 ng/μl using diethyl pyrocarbonate (DEPC)-treated H_2O (*see* **Note 5**).
3. The levels of positive-strand HCV RNA are determined by one-step RT-PCR in a 20 μl reaction volume containing 10 μl of 2× One-Step Real-Time RT-PCR Master Mixes, 0.24 μl primers (100 μM), 0.4 μl TaqMan probe (100 μM), 5 μl of total RNA, 1 μl GAPDH control primers and probe, and 2.92 μl H_2O.
4. The qRT-PCR is carried out with Applied Biosystems StepOnePlus Real-Time PCR System using the following program: stage 1, 95 °C for 10 min, and stage 2 (40 cycle) 95 °C 30 s, 55 °C 30 s.

3.9 Western Blot

1. Proteins in cell lysates or cell culture supernatants are separated by electrophoresis in 10% polyacrylamide-SDS gel.
2. Proteins are transferred onto PDF membrane using transfer-blot cell (Bio-Rad).
3. The membrane is blocked by incubation with 5% dry milk.
4. Antibodies specific to cellular proteins (apolipoproteins and β-actin) and viral proteins (e.g., NS3 and NS5A) are used to react with specific proteins.
5. Specific viral and cellular proteins are stained with horseradish peroxidase-conjugated secondary antibodies, visualized with a chemiluminescence substrate (ECL kits) and quantified with an imaging system. Alternatively, fluorescent secondary antibodies can be used for quantification of proteins by Li-Cor Odyssey Imaging System.

3.10 Titration of Infectious HCV Titers

1. HCV in the culture supernatants is tenfold serially diluted.
2. 50 μl of diluted HCV is used to infect Huh-7.5 cells in each well of 96-well plates (triplicate).
3. After 8 h incubation at 37 °C, HCV-infected cells are washed with PBS and incubated with 100 μl of fresh DMEM containing 1% methylcellulose.
4. At 48 h p.i., Huh-7.5 cells are fixed with 10% neutralized formalin (*see* **Note 6**).
5. Focus-forming units (FFU) are stained by immunofluorescence (IFA) or immunohistochemistry (IHC) assay using an NS3-specific monoclonal antibody, peroxidase-conjugated secondary anti-mouse antibody, and 3, 3′ diaminobenzidine (DBA) substrate (Vector Laboratories).
6. HCV titers are determined by the average number of NS3-positive focus-forming units (FFU/ml) in triplicate assays.

4 Notes

1. The number and density of cells used for HCV infection are critical to the experimental success. Cell numbers should be counted, and cell density should be kept the same between different samples. In general, 2×10^5 cells are seeded to each well in a 12-well plate for single-cycle HCV growth, and 1×10^5 cells are recommended if multiple-cycle HCV growth is used.
2. For ectopic expression of siRNA-targeted cellular genes, mutations should be introduced into the siRNA-targeting regions of exogenous gene in order to confer resistance to siRNA-mediated silencing. Thus, gene-specific siRNA would only silence endogenous but not ectopic expression of target genes [24].
3. For HCV infection experiments, an m.o.i. of 10 is used for single-cycle HCV growth, and an m.o.i. of <1 is used for multiple-cycle HCV growth experiments.
4. The amounts of RNA used for quantification by qRT-PCR should be kept constant between different samples for accurate comparison.
5. For RNA isolation and quantification, all solutions should be prepared with diethyl pyrocarbonate (DEPC)-treated water. RNA samples should be resuspended in DEPC-treated water.
6. The DMEM containing 1% methylcellulose in Subheading 3.10 should be aspirated, and cells are then washed with 1× PBS three times prior to IHC or IFA staining.

Acknowledgment

This work was supported by NIH grants AI097318 and AI091953 and partially by the University of Alabama at Birmingham (UAB) Center for AIDS Research (CFAR), an NIH-funded program (P30 AI027767).

References

1. Andre P, Komurian-Pradel F, Deforges S, Perret M, Berland JL, Sodoyer M et al (2002) Characterization of low- and very-low-density hepatitis C virus RNA-containing particles. J Virol 76:6919–6928
2. Andre P, Perlemuter G, Budkowska A, Brechot C, Lotteau V (2005) Hepatitis C virus particles and lipoprotein metabolism. Semin Liver Dis 25:93–104
3. Hijikata M, Shimizu YK, Kato H, Iwamoto A, Shih JW, Alter HJ et al (1993) Equilibrium centrifugation studies of hepatitis C virus: evidence for circulating immune complexes. J Virol 67:1953–1958
4. Nielsen SU, Bassendine MF, Burt AD, Martin C, Pumeechockchai W, Toms GL (2006) Association between hepatitis C virus and very-low-density lipoprotein (VLDL)/LDL analyzed in iodixanol density gradients. J Virol 80:2418–2428
5. Thomssen R, Bonk S, Propfe C, Heermann KH, Kochel HG, Uy A (1992) Association of hepatitis C virus in human sera with beta-lipoprotein. Med Microbiol Immunol 181:293–300
6. Thomssen R, Bonk S, Thiele A (1993) Density heterogeneities of hepatitis C virus in human sera due to the binding of beta-lipoproteins and immunoglobulins. Med Microbiol Immunol 182:329–334
7. Chang KS, Jiang J, Cai Z, Luo G (2007) Human apolipoprotein e is required for infectivity and production of hepatitis C virus in cell culture. J Virol 81:13783–13793
8. Jiang J, Luo G (2009) Apolipoprotein E but not B is required for the formation of infectious hepatitis C virus particles. J Virol 83:12680–12691
9. Meunier JC, Engle RE, Faulk K, Zhao M, Bartosch B, Alter H et al (2005) Evidence for cross-genotype neutralization of hepatitis C virus pseudo-particles and enhancement of infectivity by apolipoprotein C1. Proc Natl Acad Sci U S A 102:4560–4565
10. Meunier JC, Russell RS, Engle RE, Faulk KN, Purcell RH, Emerson SU (2008) Apolipoprotein c1 association with hepatitis C virus. J Virol 82:9647–9656
11. Nielsen SU, Bassendine MF, Martin C, Lowther D, Purcell PJ, King BJ et al (2008) Characterization of hepatitis C RNA-containing particles from human liver by density and size. J Gen Virol 89:2507–2517
12. Maillard P, Huby T, Andreo U, Moreau M, Chapman J, Budkowska A (2006) The interaction of natural hepatitis C virus with human scavenger receptor SR-BI/Cla1 is mediated by ApoB-containing lipoproteins. FASEB J 20:735–737
13. Cai Z, Zhang C, Chang KS, Jiang J, Ahn BC, Wakita T et al (2005) Robust production of infectious hepatitis C virus (HCV) from stably HCV cDNA-transfected human hepatoma cells. J Virol 79:13963–13973
14. Lindenbach BD, Evans MJ, Syder AJ, Wolk B, Tellinghuisen TL, Liu CC et al (2005) Complete replication of hepatitis C virus in cell culture. Science 309:623–626
15. Wakita T, Pietschmann T, Kato T, Date T, Miyamoto M, Zhao Z et al (2005) Production of infectious hepatitis C virus in tissue culture from a cloned viral genome. Nat Med 11:791–796
16. Zhong J, Gastaminza P, Cheng G, Kapadia S, Kato T, Burton DR et al (2005) Robust hepatitis C virus infection in vitro. Proc Natl Acad Sci U S A 102:9294–9299
17. Gastaminza P, Cheng G, Wieland S, Zhong J, Liao W, Chisari FV (2008) Cellular determinants of hepatitis C virus assembly, maturation, degradation, and secretion. J Virol 82:2120–2129
18. Huang H, Sun F, Owen DM, Li W, Chen Y, Gale M Jr et al (2007) Hepatitis C virus production by human hepatocytes dependent on assembly and secretion of very low-density lipoproteins. Proc Natl Acad Sci U S A 104:5848–5853
19. Jammart B, Michelet M, Pecheur EI, Parent R, Bartosch B, Zoulim F et al (2013) Very-low-density lipoprotein (VLDL)-producing and hepatitis C virus-replicating HepG2 cells

secrete no more lipoviroparticles than VLDL-deficient Huh7.5 cells. J Virol 87:5065–5080
20. Benga WJ, Krieger SE, Dimitrova M, Zeisel MB, Parnot M, Lupberger J et al (2010) Apolipoprotein E interacts with hepatitis C virus nonstructural protein 5A and determines assembly of infectious particles. Hepatology 51:43–53
21. Cun W, Jiang J, Luo G (2010) The C-terminal alpha-helix domain of apolipoprotein E is required for interaction with nonstructural protein 5A and assembly of hepatitis C virus. J Virol 84:11532–11541
22. Jiang J, Cun W, Wu X, Shi Q, Tang H, Luo G (2012) Hepatitis C virus attachment mediated by apolipoprotein E binding to cell surface heparan sulfate. J Virol 86:7256–7267
23. Jiang J, Wu X, Tang H, Luo G (2013) Apolipoprotein E mediates attachment of clinical hepatitis C virus to hepatocytes by binding to cell surface heparan sulfate proteoglycan receptors. PLoS One 8:e67982
24. Shi Q, Jiang J, Luo G (2013) Syndecan-1 serves as the major receptor for attachment of hepatitis C virus to the surfaces of hepatocytes. J Virol 87:6866–6875
25. Catanese MT, Uryu K, Kopp M, Edwards TJ, Andrus L, Rice WJ et al (2013) Ultrastructural analysis of hepatitis C virus particles. Proc Natl Acad Sci U S A 110:9505–9510
26. Gastaminza P, Dryden KA, Boyd B, Wood MR, Law M, Yeager M et al (2010) Ultrastructural and biophysical characterization of hepatitis C virus particles produced in cell culture. J Virol 84:10999–11009
27. Merz A, Long G, Hiet MS, Brugger B, Chlanda P, Andre P et al (2011) Biochemical and morphological properties of hepatitis C virus particles and determination of their lipidome. J Biol Chem 286:3018–3032
28. Fan H, Qiao L, Kang KD, Fan J, Wei W, Luo G (2017) Attachment and postattachment receptors important for hepatitis C virus infection and cell-to-cell transmission. J Virol 91 (13):e00280–e00217 In press
29. Liu S, McCormick KD, Zhao W, Zhao T, Fan D, Wang T (2012) Human apolipoprotein E peptides inhibit hepatitis C virus entry by blocking virus binding. Hepatology 56:484–491
30. Lee JY, Acosta EG, Stoeck IK, Long G, Hiet MS, Mueller B et al (2014) Apolipoprotein E likely contributes to a maturation step of infectious hepatitis C virus particles and interacts with viral envelope glycoproteins. J Virol 88:12422–12437
31. Bankwitz D, Doepke M, Hueging K, Weller R, Bruening J, Behrendt P et al (2017) Maturation of secreted HCV particles by incorporation of secreted ApoE protects from antibodies by enhancing infectivity. J Hepatol 67:480–489
32. Fauvelle C, Felmlee DJ, Crouchet E, Lee J, Heydmann L, Lefevre M et al (2016) Apolipoprotein E mediates evasion from hepatitis C virus neutralizing antibodies. Gastroenterology 150:206–217 e204
33. Wu X, Robotham JM, Lee E, Dalton S, Kneteman NM, Gilbert DM et al (2012) Productive hepatitis C virus infection of stem cell-derived hepatocytes reveals a critical transition to viral permissiveness during differentiation. PLoS Pathog 8:e1002617
34. Jiang J, Luo G (2012) Cell culture-adaptive mutations promote viral protein-protein interactions and morphogenesis of infectious hepatitis C virus. J Virol 86:8987–8997
35. Keck ZY, Xia J, Cai Z, Li TK, Owsianka AM, Patel AH et al (2007) Immunogenic and functional organization of hepatitis C virus (HCV) glycoprotein E2 on infectious HCV virions. J Virol 81:1043–1047

Chapter 17

Investigating Hepatitis C Virus Infection Using Super-Resolution Microscopy

Pedro Matos Pereira, Caron Jacobs, and Joe Grove

Abstract

Super-resolution microscopy (SRM) can provide a window on the nanoscale events of virus replication. Here we describe a protocol for imaging hepatitis C virus-infected cells using localization SRM. We provide details on sample preparation, immunostaining, data collection, and super-resolution image reconstruction. We have made all efforts to generalize the protocol to make it accessible to all budding super-resolution microscopists.

Key words Microscopy, Super-resolution, Imaging, Virus, Hepatitis

1 Introduction

Since the mid-1940s, visualization of virus particles and infected cells by electron microscopy (EM) has been a mainstay of virology. While EM provides an unrivalled resolution, it also has severe limitations, such as the inability to observe living samples and difficulties in identifying specific molecular components. In contrast, fluorescence microscopy is minimally invasive, making it ideal for live imaging, and offers a variety of means for specific molecular labelling (e.g., antibodies or fluorescent protein tags). Nevertheless, the intrinsic diffraction limit of conventional fluorescence microscopy prevents structures smaller than ~300 nm from being resolved, therefore limiting its use for studying the molecular events of virus replication.

However, recent advances in optical physics, data analysis, and photochemistry have brought about a new era of super-resolution microscopy. SRM is able to overcome the diffraction limit of conventional fluorescence microscopy to provide ultrastructural information of virus particles and infected cells that was previously only accessible by EM. As such, SRM is likely to supersede EM to become an essential component of the virologist's tool kit

Mansun Law (ed.), *Hepatitis C Virus Protocols*, Methods in Molecular Biology, vol. 1911,
https://doi.org/10.1007/978-1-4939-8976-8_17,

[1]. The various SRM techniques differ wildly in their technical and analytical approaches, the details of which can be found in these review articles [2, 3]. In this chapter, we outline the use of one SRM technique, localization microscopy, to image hepatitis C virus (HCV)-infected cells.

In conventional fluorescence microscopy, a sample is decorated with a specific label (e.g., a fluorescent dye), the individual molecules of which are usually extremely crowded; consider the many thousand copies of HCV proteins that associate with intracellular membranes in infected cells. When illuminated, these fluorescent molecules emit light simultaneously, and, due to the intrinsic diffraction of light, their signals spread and merge such that the fine detail of their molecular distribution is lost. This limits the resolution of conventional fluorescence microscopy to 200–350 nm.

In localization SRM techniques, such as STORM and PALM [4, 5], photochemical tricks are used to induce fluorescent dyes and proteins to photoswitch, such that they stochastically turn on and off (commonly referred to as "blinking"). Under these conditions, closely apposed molecules are unlikely to emit light simultaneously, and the signals from individual fluorescent molecules can be captured in isolation. Mathematical fitting of these individual signals allows the true location of fluorescent molecules to be estimated with a high precision. Prolonged imaging allows thousands or millions of fluorescent molecules to be located and mapped, resulting in a high-resolution reconstruction of the molecular distribution of fluorescent label across a sample (Fig. 1).

Localization SRM does not require a particularly specialized (or expensive) microscope [6], and there is a wealth of open-source analysis software packages available for image reconstruction [7], making it accessible even to novice microscopists. Here we provide a detailed description of SRM of HCV-infected cells, paying particular attention to the measures that need to be taken to ensure optimal sample preparation. We provide a step-by-step guide to acquiring high-quality data and performing image reconstruction using a state-of-the-art SRM analysis package [8]. The protocol outlined here uses immunostaining of fixed samples; however, it could be adapted for live-cell SRM of fluorescently tagged virus proteins and/or virions.

2 Materials

1. #1.5 25 mm circular coverslips.
2. Sterile double-deionized H_2O.
3. Absolute ethanol.
4. Absolute methanol.

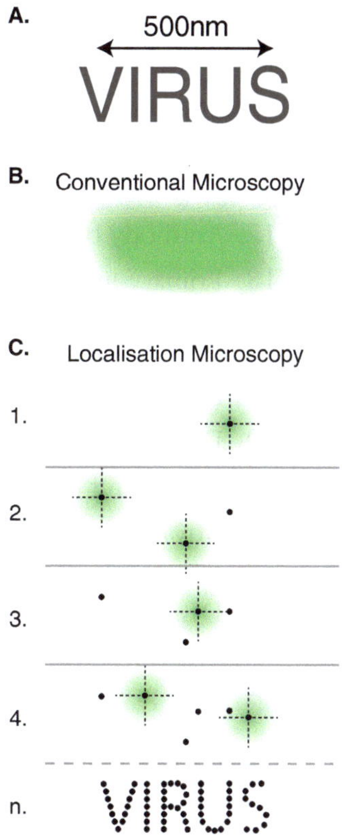

Fig. 1 The principles of localization SRM. (**a**) Consider a nanoscale object decorated with fluorescent dye. (**b**) When viewed by conventional microscopy, the signals from closely apposed dye molecules merge, and their nanoscale distribution is lost. (**c**) In localization microscopy the dye molecules are induced to "blink" on and off stochastically, such that the signals from individual molecules can be viewed in isolation. Mathematical analysis of these signals allows the true location of the dye molecules to be estimated with high precision. Through prolonged imaging it is possible to locate and map sufficient numbers of molecules such that the ultrastructure of the nanoscale object can be reconstructed

5. Acetone.
6. 3-Aminopropyltriethoxysilane (APES) (Sigma-Aldrich).
7. Wash-N-Dry coverslip holder (Sigma-Aldrich).
8. Fine-point tweezers.
9. 19G syringe needle.
10. 6-well culture plate.
11. Cell media: DMEM, supplemented with 10% fetal calf serum (FCS), nonessential amino acids, penicillin, and streptomycin.
12. Infection media: DMEM, supplemented with 3% FCS, nonessential amino acids, penicillin, and streptomycin.

13. Cells: Huh-7.5 cells (Apath, LLC).
14. Virus: J6/JFH HCVcc stock [9].
15. Phosphate-buffered saline (PBS): 137 mM NaCl, 2.7 mM KCl, 1.8 mM KH_2PO_4, 10 mM Na_2HPO_4, pH 7.4.
16. Blocking buffer: PBS + 1% bovine serum albumin (BSA).
17. Antibody buffer: PBS + 0.5% BSA.
18. Primary antibody: 9E10 anti-NS5A (Charles M. Rice, Rockefeller University, New York).
19. Secondary antibody: anti-mouse IgG Alexa Fluor 647 (Thermo Fisher Scientific).
20. DAPI (4′,6-Diamidino-2-phenylindole dihydrochloride).
21. Post-fixing buffer: PBS + 1% EM grade formaldehyde.
22. Attofluor chamber (Thermo Fisher Scientific).
23. Imaging buffer: PBS + 3% OxyFluor (Oxyrase Inc.), 100 μM DL-lactate, 50 mM β-Mercaptoethylamine hydrochloride (MEA), pH adjusted to 8.0–8.5 with NaOH.
24. This procedure can be performed on many intermediate or advanced imaging systems. In this case the work was performed on a Zeiss Elyra PS.1 fitted with a 100× TIRF objective and an EMCCD camera (iXon Ultra 897, Andor). Imaging was performed using 405 nm (50 mW) and 642 nm (150 mW) diode solid-state lasers, with band-pass 420–480/long-pass 655 and long-pass 655 emission filters.
25. ImageJ/Fiji open-source image analysis software with the NanoJ super-resolution radial fluctuations (SRRF) plug-in installed (*see* Subheading 3.6) [8, 10, 11].

3 Methods

3.1 Cleaning Coverslips

SRM is particularly susceptible to background sources of fluorescence; therefore, coverslip cleaning is essential for optimal imaging. In our experience it is preferable to prepare fresh coverslips on the day of each experiment.

1. Place a #1.5 25 mm coverslip into each well of a 6-well plate for cleaning.
2. Perform two 5-min washes in 2 ml ddH_2O; agitate gently (*see* **Note 1**).
3. Perform two 5-min washes in 2 ml 100% ethanol; agitate gently.
4. Perform two 5-min washes in 2 ml 100% methanol; agitate gently (*see* **Note 2**).
5. Replace methanol with 2 ml ddH_2O.

3.2 Coating Coverslips

Cleaned glass coverslips are typically coated with a suitable substrate to promote cell adhesion. Treatment with APES covalently coats the glass with a layer of positively charged free amine groups; this provides superior adhesion and can also be used to immobilize virus particles or purified proteins.

1. In a fume hood, prepare two 100 ml glass beakers, one containing 80 ml 100% acetone and the other 80 ml 2% APES diluted in acetone.
2. Carefully transfer the cleaned coverslips into a Wash-N-Dry coverslip rack (*see* **Notes 3** and **4**).
3. Using tweezers, dunk the coverslip rack into the beaker of acetone, and then transfer to the beaker containing 2% APES, ensuring complete immersion of the coverslips; incubate for 2 min.
4. Transfer the coverslip rack back into the beaker of acetone to wash off excess APES. Transfer the coverslip rack into a beaker containing fresh ddH_2O, stand for 30 s, then transfer into another beaker of ddH_2O. This step will ensure the removal of excess acetone. Discard the acetone and APES (*see* **Note 5**).

3.3 Sample Preparation

The precise nature of the sample(s) will depend on the particular experiment; nonetheless, careful sample preparation is critical to successful SRM. This simple protocol generates HCV-infected cell samples for imaging, without any particular experimental question in mind. SRM is a powerful but time-consuming approach; we would recommend using it to address focused questions, with a small number of experimental parameters. More extensive studies with multiple parameters and/or treatments are better suited to lower-resolution imaging techniques; SRM should be used to provide fine details.

1. Once cleaned, place one prepared coverslip into each well of a 6-well plate, seed 1.2×10^5 Huh-7.5 cells per well in 4 ml media, and allow the cells to adhere and grow (*see* **Note 6**).
2. Inoculate the samples with J6/JFH HCVcc at a low multiplicity of infection in 2 ml DMEM + 3% FCS; allow the infection to proceed for 72 h (*see* **Note 7**).
3. To fix, remove the media from each well, and rinse with 1 ml cold PBS. Remove the PBS, and add 4 ml of ice-cold methanol to each well, and incubate at room temperature (RT) for 15 min (*see* **Note 8**). Following proper fixation, virus particles will be inactivated, and the samples can be handled on the bench.

3.4 Immunostaining

Immunostaining is the most commonly used labelling strategy, but labelling can also be achieved through genetically encoded tags such as fluorescent proteins. In either case, an optimal label will be highly specific to the target and provide a strong fluorescent signal. Careful consideration should also be given to the possible pitfalls of any given labelling method. For example, some fluorescent proteins have a tendency to multimerize causing aberrant clustering of the target protein [12], whereas use of primary and secondary antibodies creates a physical separation between the target protein and the fluorescent signal. Discussions on the various properties of different labelling methods can be found here [13–16].

1. Rinse each well twice with 4 ml PBS to wash away methanol and rehydrate the samples (*see* **Notes 9** and **10**).
2. Retrieve the coverslips into a fresh 6-well plate, containing 2 ml PBS per well.
3. Block the samples with 2 ml PBS + 1% BSA for >90 min (*see* **Note 11**).
4. Centrifuge primary antibody (anti-NS5A 9E10) at full speed in a microcentrifuge for 5 min to pellet any large aggregates of antibody. Dilute the antibody to 100 ng/ml in PBS + 0.5% BSA. Add 1 ml of diluted antibody per well, and incubate for 30 min at RT with gentle agitation.
5. Aspirate and discard the primary antibody, and wash the sample in 2 ml PBS at RT for 5 min with gentle agitation.
6. Repeat **step 5**, twice.
7. Centrifuge secondary antibody (anti-mouse IgG Alexa Fluor 647) (*see* **Note 12**) as above. Dilute to 2 μg/ml in PBS + 0.5% BSA; pass the prepared solution through a 0.2 μm syringe filter prior to adding 1 ml to each well (*see* **Note 13**). Incubate for 120 min at RT with gentle agitation.
8. Perform three PBS washes as described in **step 5**.
9. Incubate samples with 1 ml of 2 μg/ml DAPI diluted in PBS for 10 min to counterstain nuclear DNA.
10. Postfix the samples in PBS + 1% formaldehyde for 10 min at RT with gentle agitation (*see* **Note 14**).
11. Perform two PBS washes as described in **step 5**.
12. Store the samples in 2 ml PBS at 4 °C until ready to use; they will be stable for 3–7 days.

3.5 Imaging

The imaging presented here was performed on a commercial SRM system; however the procedure and subsequent image reconstruction are capable of producing high-quality super-resolution images on less specialized microscopes. This method of SRM can be performed on any wide-field microscope with the following specifications: (1) a high magnification/high numerical aperture objective (e.g., 60–100 × oil objectives); (2) a sensitive, low-noise camera (e.g., most EMCCD or sCMOS cameras); and (3) a strong fluorescent light source, ideally a laser, although a powerful LED source is sufficient for SRM, albeit with reduced resolution [8]. Systems matching these minimal requirements are common, making the approach outlined here accessible to most, if not all, researchers.

1. Transfer a coverslip with prepared sample into an Attofluor chamber. Ensure the coverslip is sitting faceup within the central lip, and carefully tighten the chamber to seal. Add 1.5 ml of imaging buffer to the chamber (*see* **Note 15**).
2. Place a fresh clean 25 mm coverslip on top of the buffer in the chamber, taking care to avoid the formation of air bubbles. Press down gently on the coverslip to create a loose seal, and remove excess buffer with a pipette (*see* **Note 16**).
3. Turn on the microscope 30–60 min before acquisition (*see* **Notes 17** and **18**).
4. Place a small drop of oil on the 100× objective.
5. Place your sample securely on the microscope sample holder, and carefully find focus, using the DAPI reference channel to observe the sample.
6. Select the 642 nm laser for Alexa Fluor 647 excitation and the appropriate emission filter.
7. Find an appropriate region of interest by searching the sample at low laser power.
8. Engage the microscope focus lock (*see* **Note 19**), and leave the sample to rest for 5 min (*see* **Note 20**).
9. Set the microscope to acquire widefield images using epifluorescence illumination.
10. Take diffraction-limited reference images of the field of view using the DAPI and Alexa Fluor 647 channels. This will provide a comparison for the super-resolution image.
11. Acquire 60,000 frames at 30 frames per second (33 ms exposure time) at 100% 642 nm laser power (approximately 2 kW/cm^2). The combination of imaging buffer and high-intensity illumination will induce the Alexa Fluor 647 dye molecules to photoswitch; the sample will appear to start "twinkling" as the individual dyes blink on and off stochastically. Examples of typical fluorescent blinking can be seen in these video tutorials [17, 18]. It is this property that allows mathematical reconstruction of the super-resolution image (*see* **Notes 21–24**).
12. Repeat **steps 9–13** until an appropriate number of fields have been acquired.

3.6 Image Reconstruction

Localization SRM uses complex data analysis algorithms to extract super-resolved image reconstructions from conventional microscopy data. There is a bewildering variety of software packages for performing this task, each with their own relative flaws and strengths, as outlined in a recent comprehensive comparison [7]. Here, we outline the use of super-resolution radial fluctuations analysis (SRRF, pronounced "surf"); this is available as a plug-in for the commonly used image analysis platform ImageJ/Fiji [10, 11]. SRRF is capable of improving resolution across a range of sample types, microscope specifications, and data qualities [8]. For these reasons, it is a good place to start, particularly for novice users (*see* **Note 25**).

1. Install SRRF in ImageJ/Fiji (*see* **Note 26**).
2. Ensure your dataset is in an appropriate format (i.e., .tif, .czi, . nd2, or .nji), and open the dataset in ImageJ/Fiji (*see* **Note 27**).
3. Estimate sample drift in the data (in ImageJ/Fiji select Plugins -> NanoJ-SRRF -> Estimate drift) using the default values for time averaging and maximum expected drift. The drift estimate function will generate a drift table that must be saved and then applied in the next step.
4. Perform image reconstruction analysis: In ImageJ/Fiji, select Plugins -> NanoJ-SRRF -> SRRF analysis, set each of the analysis parameters to the default values stated in the dialogue box, and select "Do Drift-Correction." To run the analysis, click "OK," and, when prompted, select the relevant drift correction table (generated in the previous step) (*see* **Note 28**).
5. Once the analysis is completed, SRRF will provide a super-resolved reconstructed image as a 32bit .tif file, which can be saved and processed as desired. The default reconstruction setting uses a magnification factor of 5; in the examples provided here, the reconstructed image has a pixel size of 20 nm (i.e., five times smaller than the 100 nm pixels in the raw data) (*see* **Note 29**).

Example SRRF reconstructions of HCV-infected cells stained with anti-NS5A are shown in Fig. 2 alongside conventional diffraction-limited images as a comparison. HCV induces extensive remodeling of the endoplasmic reticulum to produce a structure known as the membranous web, where genome replication and virus assembly are thought to occur [19, 20]. The ultrastructure of the membranous web has previously been studied using EM, as conventional microscopy does not provide a sufficiently high resolution. The SRRF reconstructions in Fig. 2 provide fine details of this convoluted membrane structure, which are not apparent in the conventional images. Vesicular structures ranging from 200 to 400 nm in diameter are clearly visible; these likely represent double-membrane vesicles and/or lipid droplets [20–22].

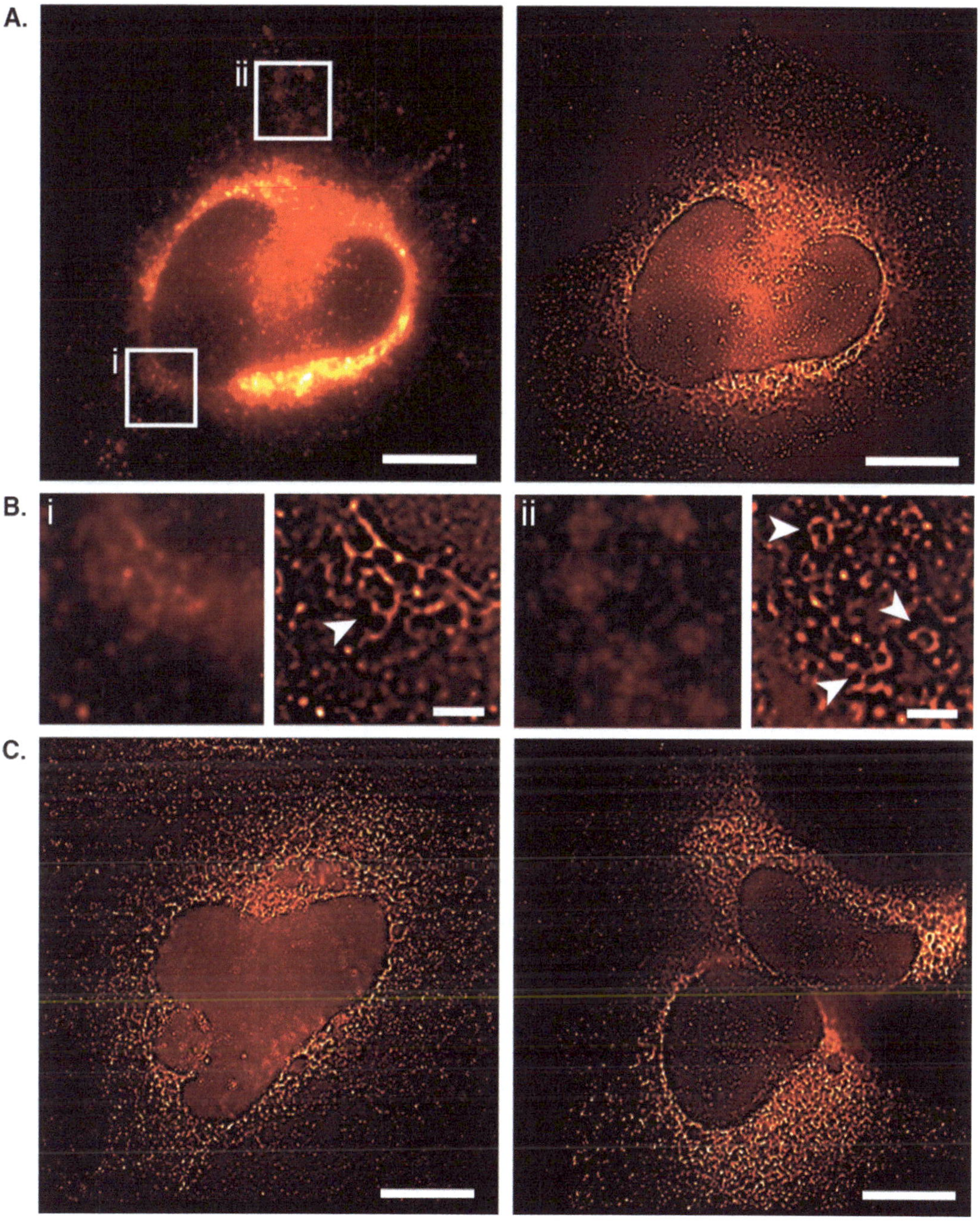

Fig. 2 Super-resolution reconstructions of HCV-infected Huh-7.5 cells. Samples and images were generated as described in the text. The "red hot" lookup table has been applied to the images and brightness and contrast settings altered slightly for display purposes. (**a**) Conventional diffraction-limited (left) and super-resolution (right) images of HCV-infected cells stained with anti-NS5A. (**b**) Enlarged views of the inset areas (i and ii) shown in A, conventional (left) and super-resolution (right) images are shown for each. The super-resolution images clearly show looping and vesicular structures (white arrowheads) that likely represent the HCV-induced membranous web. These structures are barely discernable in the conventional microscopy images. (**c**) Two additional representative super-resolution reconstructions of HCV-infected cells. Scale bars 10 μm (**a** and **c**) and 2 μm (**b**). The images shown here achieved a resolution of 75–100 nm; conventional microscopy typically achieves 200–300 nm

The protocol presented here provides an example of how SRM can be used to visualize the basic biology of HCV-infected cells. More advanced variations on this procedure may include multicolor SRM of viral and/or host components; while this presents technical challenges, particularly in properly aligning the two SRM channels, there are a variety of technical articles to provide guidance [23–25]. Another advantage to SRRF is its ability to extract super-resolution reconstructions from live-cell microscopy images. To attempt this, viral and/or host components will need to be tagged with photoswitching fluorescent proteins (such as mEOS) or dye-accepting peptides [12, 26]. Users wishing to perform live-cell SRM should consult the original SRRF publication [8] and the associated online resources at https://bitbucket.org/rhenriqueslab/nanoj-srrf.

4 Notes

1. At this stage, the coverslip is prone to floating on top of the water. If this occurs, use tweezers to gently resubmerge.
2. More thorough cleaning methods can be employed if necessary. For alternative methods, see these articles [27, 28].
3. Fine-pointed tweezers and a 19G syringe, with the tip bent into a slight hook, are ideal for manipulating the coverslips.
4. Other coverslip holders are also suitable; however, this particular brand easily fits within a 100 ml beaker. It is also sufficient to manually hold each coverslip with fine-pointed tweezers during the coating procedure.
5. As an alternative, poly-L-lysine (PLL) coating also promotes cell adhesion.
6. A low/medium density of cells is ideal for high-resolution microscopy.
7. It is ideal to have a mixture of both infected and uninfected cells; this can assist the discrimination of signal over background.
8. The manner of fixation is an important parameter and may need to be optimized, taking consideration of the characteristics of each method. Methanol fixation dehydrates the sample leading to denaturing and precipitation of proteins and some lipid extraction, resulting in simultaneous fixation and permeabilization of the sample. However methanol fixation is not appropriate for imaging of fluorescent proteins. Aldehyde-based fixatives (e.g., 4% EM grade formaldehyde in PBS) covalently cross-link juxtaposed proteins resulting in a fixed meshwork. The addition of low-percentage glutaraldehyde (<1%) may be necessary to immobilize membrane-associated

proteins. Also detergent treatment (e.g., 0.1% Triton X-100 in PBS) will be necessary to permeabilize aldehyde-fixed cells prior to antibody staining of intracellular compartments. More detailed discussion of fixation methods can be found here [29, 30].

9. In all subsequent steps, the samples should be treated with care; aspirate and apply buffer from the edge of the well avoiding contact with the coverslip.
10. During immunostaining buffers should be passed through a 0.2 or 0.45 μm filter prior to use. This is not to ensure sterility but to minimize the introduction of dust and precipitates.
11. The blocking step is critical to producing a high-quality sample. It serves to minimize the non-specific binding of antibodies and may need to be optimized. BSA blocking is normally sufficient, and concentrations of up to 5% are commonly used. Blocking for extended periods or overnight may also help.
12. Alexa Fluor 647 is an excellent and commonly used dye for localization SRM. However, if performing dual labelling, it should be noted that Alexa Fluor 647 and related far-red dyes such as Cy5 exhibit spurious fluorescence in the near-red (i.e., ~550–600 nm) and, therefore, should not be combined with red dyes [31]. In this case, a green dye such as Alexa Fluor 488 would be a more appropriate partner.
13. Syringe filtration is an additional precaution to eliminate secondary antibody aggregates, which can be particularly detrimental to sample quality.
14. This optional postfix step can preserve the sample and minimize gradual dissociation of antibodies.
15. The buffer is an optimized photoswitching buffer for inducing blinking in fluorescent dyes [32]. It contains an oxygen scavenging system (Oxyrase) and a reducing agent (MEA) to promote most commonly used dyes to photoswitch but may not be suitable for some fluorescent proteins.
16. Placing a coverslip on top of the buffer minimizes exposure to fresh oxygen in air and ensures optimal imaging conditions for hours. However, avoid a perfect airtight seal on top of the chamber as this risks generating a slight vacuum, which can crack the sample coverslip.
17. High-quality SRM reconstructions require tens of thousands of microscopy images to be acquired. One of the biggest challenges is the lateral drift (*x*–*y* plane) of the sample during acquisition. If possible, the microscope should be turned on 60 min before the start of the experiment, and the sample should be left on the stage for 15 min before imaging to allow warm-up and equilibration. Some algorithms (including

SRRF, used here) can calculate and correct lateral drift using the raw imaging data. However, this requires structures that remain visible throughout the entire image sequence, which is not always the case. As an additional precaution, a low density of fiducial markers can be added to the sample to assist accurate drift calculation and correction. Fiducials are bright fluorescent markers that remain visible for the duration of the experiment; they are also necessary when aligning channels during multi-color SRM [17, 32, 33].

18. Many commonly used microscopes have an additional magnification lens that can be introduced into the light path to provide modest additional magnification; if available, this should be used. This example employs a 1.6 times magnifying lens, which, when combined with a 100× objective, yields an image pixel size of 100 nm and results in superior data for image reconstruction.
19. Drifting focus, like lateral movement of the sample, is detrimental to data quality over long acquisitions. Many commercial systems are fitted with a focus-lock system that will maintain focus for many hours.
20. Resting the sample for a few minutes will allow the stage to settle and further minimize lateral drift during acquisition.
21. In epifluorescence illumination, the excitation light passes through the entire depth of the sample; consequently the resulting image includes some out-of-focus emissions from above and below the focal plane; this will moderately decrease the resolution achieved (by corrupting the precision by which single molecules can be localized). It is possible to decrease fluorophore excitation outside of the focal plane using alternative illumination modalities such as highly inclined thin illumination (HILO) or total internal reflection fluorescence (TIRF). For instance, TIRF allows selective illumination to the basal part of the cell (within 100–200 nm of the coverslip) [25, 34].
22. For true localization microscopy, the signals from individual dye molecules should be clearly discernable from adjacent signals. However, for densely labelled structures, it can be difficult to achieve sparsely distributed signals. Therefore, in this protocol, we use a versatile reconstruction algorithm (SRRF) that can extract super-resolution images from suboptimal data, albeit with reduced resolution.
23. The duration of the acquisition, imaging speed, and laser intensity can be empirically tailored for the user's experimental and technical conditions [34, 35]. However, for a classical localization microscopy dataset, we recommend a high number of frames (50,000–100,000), at maximum illumination power and maximum frame rate.

24. The rate of blinking by Alexa Fluor 647 may reduce during the acquisition, resulting in very sparse signals. If this occurs, low-level illumination with the 405 nm laser (for instance, 0.01–1%) can reactivate the dye molecules to increase the density of signals.
25. Alternatively, ThunderSTORM is an excellent reconstruction software that works particularly well on sparsely separated dye signals [36].
26. The original SRRF software package is provided in the supplementary materials of Gustafsson et al. [8]. Up-to-date versions and user guides can be found here: https://bitbucket.org/rhenriqueslab/nanoj-srrf.
27. Given the high-intensity illumination used to acquire the data, the first few thousand frames of the image series are often saturated by the high signal. This saturated data is not suitable for reconstruction; either remove these frames from the image stack or direct SRRF to crop them during analysis.
28. SRRF does not perform typical localization microscopy but uses sophisticated image analysis to estimate the distribution of dye molecules across a sample. It is a particularly versatile reconstruction method as it can extract super-resolved images from difficult datasets with high densities of fluorescent signals. SRRF exploits both the spatial and temporal characteristics of the emitted fluorescence across an image series to extract information about the underlying fluorophore distribution. An in-depth description of the SRRF method and parameters can be found here [8], and various online resources can be found here: https://bitbucket.org/rhenriqueslab/nanoj-srrf. If necessary the reconstruction process can be optimized to produce higher-quality images. If unsure about which parameters to choose for a given dataset, crop a small representative part of the data (including different fluorophore densities and/or structures of interest), and run the Parameter Sweep macro, available for download with a user guide from https://bitbucket.org/rhenriqueslab/nanoj-srrf/downloads/.
29. The time taken to analyze the image will depend on the size of the data set and the specifications of the analysis computer. SRRF exploits the GPU processors found in computer graphic cards, as such SRRF will run more quickly on computers with powerful graphic engines. On a moderately powerful desktop computer, a large, 512 by 512 pixel, image series is likely to take tens of minutes to analyze.

Acknowledgments

We would like to thank Ricardo Henriques and Lucas Walker for advice and technical support. JG is supported by a Sir Henry Dale Fellowship jointly funded by the Wellcome Trust and the Royal Society (107653/Z/15/Z). PMP is supported by a Biotechnology and Biological Sciences Research Council grant (BB/M022374/1). CJ is a Commonwealth Scholar, funded by the UK government. The imaging systems used in this work were funded by the Medical Research Council (MR/K015826/1).

References

1. Grove J (2014) Super-resolution microscopy: a virus' eye view of the cell. Viruses 6:1365–1378
2. Sydor AM, Czymmek KJ, Puchner EM, Mennella V (2015) Super-resolution microscopy: from single molecules to supramolecular assemblies. Trends Cell Biol 25:730–748
3. Galbraith CG, Galbraith JA (2011) Super-resolution microscopy at a glance. J Cell Sci 124:1607–1611
4. Bates M, Huang B, Dempsey GT, Zhuang X (2007) Multicolor super-resolution imaging with photo-switchable fluorescent probes. Science 317:1749–1753
5. Betzig E, Patterson GH, Sougrat R et al (2006) Imaging intracellular fluorescent proteins at nanometer resolution. Science 313:1642–1645
6. Holm T, Klein T, Löschberger A et al (2014) A blueprint for cost-efficient localization microscopy. ChemPhysChem 15:651–654
7. Sage D, Kirshner H, Pengo T et al (2015) Quantitative evaluation of software packages for single-molecule localization microscopy. Nat Methods 12:717–724
8. Gustafsson N, Culley S, Ashdown G et al (2016) Fast live-cell conventional fluorophore nanoscopy with ImageJ through super-resolution radial fluctuations. Nat Commun 7:12471
9. Lindenbach BD, Evans MJ, Syder AJ et al (2005) Complete replication of hepatitis C virus in cell culture. Science 309:623–626
10. Schneider CA, Rasband WS, Eliceiri KW (2012) NIH Image to ImageJ: 25 years of image analysis. Nat Methods 9:671–675
11. Schindelin J, Arganda-Carreras I, Frise E et al (2012) Fiji: an open-source platform for biological-image analysis. Nat Methods 9:676–682
12. Zhang M, Chang H, Zhang Y et al (2012) Rational design of true monomeric and bright photoactivatable fluorescent proteins. Nat Methods 9:727–729
13. Uno S-N, Tiwari DK, Kamiya M et al (2015) A guide to use photocontrollable fluorescent proteins and synthetic smart fluorophores for nanoscopy. Microscopy 64:263–277
14. Dempsey GT, Vaughan JC, Chen KH et al (2011) Evaluation of fluorophores for optimal performance in localization-based super-resolution imaging. Nat Methods 8:1027–1036
15. Durisic N, Cuervo LL, Lakadamyali M (2014) Quantitative super-resolution microscopy: pitfalls and strategies for image analysis. Curr Opin Chem Biol 20C:22–28
16. Ḍurisic N, Laparra-Cuervo L, Sandoval-Álvarez Á et al (2014) Single-molecule evaluation of fluorescent protein photoactivation efficiency using an in vivo nanotemplate. Nat Methods 11:156–162
17. Metcalf DJ, Edwards R, Kumarswami N, Knight AE (2013) Test samples for optimizing STORM super-resolution microscopy. J Vis Exp 79. https://doi.org/10.3791/50579
18. Ji C, Lou X (2016) Single-molecule Super-resolution Imaging of Phosphatidylinositol 4,5-bisphosphate in the Plasma Membrane with Novel Fluorescent Probes. J Vis Exp 116. https://doi.org/10.3791/54466
19. Paul D, Madan V, Bartenschlager R (2014) Hepatitis C Virus RNA Replication and Assembly: Living on the Fat of the Land. Cell Host Microbe 16:569–579
20. Wang H, Tai AW (2016) Mechanisms of Cellular Membrane Reorganization to Support Hepatitis C Virus Replication. Viruses 8:e142
21. Romero-Brey I, Merz A, Chiramel A et al (2012) Three-dimensional architecture and biogenesis of membrane structures associated

with hepatitis C virus replication. PLoS Pathog 8:e1003056

22. Camus G, Herker E, Modi AA et al (2013) Diacylglycerol acyltransferase-1 localizes hepatitis C virus NS5A protein to lipid droplets and enhances NS5A interaction with the viral capsid core. J Biol Chem 288:9915–9923
23. Tam J, Cordier GA, Borbely JS et al (2014) Cross-talk-free multi-color STORM imaging using a single fluorophore. PLoS One 9: e101772
24. Valley CC, Liu S, Lidke DS, Lidke KA (2015) Sequential Superresolution Imaging of Multiple Targets Using a Single Fluorophore. PLoS One 10:e0123941
25. Shroff H, Galbraith CG, Galbraith JA et al (2007) Dual-color superresolution imaging of genetically expressed probes within individual adhesion complexes. Proc Natl Acad Sci U S A 104:20308–20313
26. Lelek M, Di Nunzio F, Henriques R et al (2012) Superresolution imaging of HIV in infected cells with FlAsH-PALM. Proc Natl Acad Sci U S A 109:8564–8569
27. Pereira PM, Almada P, Henriques R (2015) High-content 3D multicolor super-resolution localization microscopy. Methods Cell Biol 125:95–117
28. Chandradoss SD, Haagsma AC, Lee YK et al (2014) Surface passivation for single-molecule protein studies. J Vis Exp 86. https://doi.org/10.3791/50549
29. Whelan DR, Bell TDM (2015) Image artifacts in Single Molecule Localization Microscopy: why optimization of sample preparation protocols matters. Sci Rep 5:7924
30. Tanaka KAK, Suzuki KGN, Shirai YM et al (2010) Membrane molecules mobile even after chemical fixation. Nat Methods 7:865–866
31. Stone MB, Veatch SL (2014) Far-red organic fluorophores contain a fluorescent impurity. ChemPhysChem 15:2240–2246
32. Nahidiazar L, Agronskaia AV, Broertjes J et al (2016) Optimizing imaging conditions for demanding multi-color super resolution localization microscopy. PLoS One 11:e0158884
33. Barna L, Dudok B, Miczán V et al (2016) Correlated confocal and super-resolution imaging by VividSTORM. Nat Protoc 11:163–183
34. van de Linde S, Sauer M (2014) How to switch a fluorophore: from undesired blinking to controlled photoswitching. Chem Soc Rev 43:1076–1087
35. Fox-Roberts P, Marsh R, Pfisterer K et al (2017) Local dimensionality determines imaging speed in localization microscopy. Nat Commun 8:13558
36. Ovesny M, Křížek P, Borkovec J et al (2014) ThunderSTORM: a comprehensive ImageJ plug-in for PALM and STORM data analysis and super-resolution imaging. Bioinformatics 30:2389–2390

Chapter 18

Live Cell Imaging of Hepatitis C Virus Trafficking in Hepatocytes

Yasmine Baktash and Glenn Randall

Abstract

Standard fixed cell confocal microscopy is inherently limited in visualizing dynamic processes involving two- and three-dimensional movement. To overcome these limitations, live cell imaging approaches have been developed to study hepatitis C virus (HCV) entry, replicase protein trafficking, virion assembly, and egress. These studies have relied on fluorescent labeling of viral proteins by epitope tag insertion, genome labeling via nucleophilic dyes, or using lipophilic dyes to label the virion envelope. In this method review, we describe two approaches to study HCV virion trafficking in live cells. Lipophilic labeling of the envelope allows for study of the early events (through virion uncoating/fusion) in the HCV lifecycle. Tetracysteine (TC) tag insertion into the capsid protein permits study of virion assembly and capsid trafficking via binding of a fluorogenic biarsenical dye.

Key words Hepatitis C virus, Fluorescent labeling, Live cell imaging, Viral entry, DiD, Tetracysteine tag, TC-core

1 Introduction

HCV, a member of the *Flaviviridae*, is a small, single-stranded positive-sense virus of roughly 9600 nucleotides. Initial investigations into HCV protein dynamics focused on NS5A, in part because it is the only HCV gene identified that tolerates large insertions, such as GFP fusions. NS5A-GFP was shown to partially localize to replication compartments. Live cell imaging demonstrated two dynamic groups—larger, less motile foci, and smaller, motile puncta [1, 2]; such movement could be reduced by NS5A inhibitors [3].

Advances in cell culture systems and infectious viral strains have allowed for elucidation of all aspects of the viral lifecycle [4–7]. Live cell imaging of virion trafficking was hindered by the inability to generate infectious HCV containing fluorescent protein fusions with its structural genes. This hurdle has been cleared using two distinct approaches: lipophilic dyes to study HCV entry and small

Mansun Law (ed.), *Hepatitis C Virus Protocols*, Methods in Molecular Biology, vol. 1911,
https://doi.org/10.1007/978-1-4939-8976-8_18,

tetracysteine tags in the capsid to study HCV assembly and egress [8–10].

HCV is enveloped by a highly lipidated bilayer. For two other enveloped viruses, influenza A virus and dengue virus, protocols were developed wherein DiD, a membrane-permeable lipophilic dye, is used to label the envelope [11–14]. The fluorogenic DiD (1,1′-dioctadecyl-3,3,3′,3′-tetramethylindodicarbocyanine) readily intercalates viral envelopes without compromising infectivity. These fluorescently labeled virions were then used to image viral entry. The limitation of this approach is that viruses, such as HCV, with poor specific infectivity (i.e., 1 in ~2000 HCV virions produced in cell culture is infectious) cannot be studied with confidence because the vast majority of the particles visualized will not lead to a productive infection.

We overcame this limitation by taking advantage of the observation that infectious HCV is highly lipidated and has a different buoyant density than noninfectious HCV [15]. In our protocol, several rounds of purification ensure specific labeling of highly infectious virus with a specific infectivity of ~10 [8]. Characterization of the DiD-HCV particles showed that >95% of them colocalize with the capsid and E2, DiD-HCV could be immunodepleted with anti-E2 antibodies, and that DiD-HCV uncoating required the HCV receptor CD81 and acidic pH, as does bona fide HCV entry [8]. The study used these DiD-labeled particles to examine the requirements of HCV entry, using live cell imaging to investigate DiD-HCV association with its receptors, actin, its internalization pathway, and subsequent endosomal trafficking. Sainz et al. used DiD-HCV to study the roles and requirements of cellular proteins in regulating HCV entry [16]. Although this imaging protocol is highly reproducible, the specificity of DiD labeling and specific infectivity of the labeled inoculum should be confirmed for each preparation.

We and others adopted the TC tag system for labeling viral core proteins in an infected cell [9, 10]. TC motifs (CCPGCC) allow for binding of a small fluorescein derivative, FlAsH (fluorescein arsenical hairpin binder) or ReAsH (resorufin arsenical hairpin binder) [17]. The TC-core protein was first used to study the dynamics of core recruitment to lipid droplets [10]. Our group used a virtually identical TC-core HCV to characterize viral egress. The trafficking of TC-core with fluorescently labeled secretory components showed that HCV is released from the cell via various components of the secretory pathway, including the Golgi, recycling endosomes, and VAMP1 vesicles [9]. Importantly, processive TC-core movements require NS2, which is essential for virion assembly. Thus, there is confidence that motile TC-core is indeed an assembled nucleocapsid. The TC tag was also inserted into the nonstructural NS5A protein, looking at the requirements for its trafficking and localization [1]. Finally, the TC-core protein can be used to

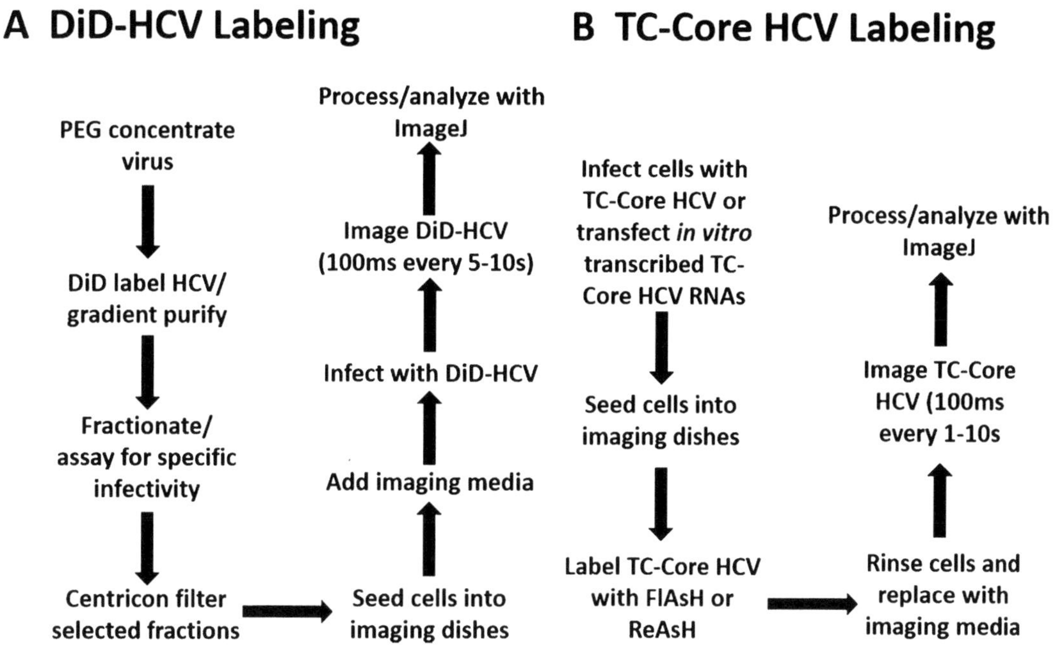

Fig. 1 Workflow for the labeling and imaging of fluorescently labeled HCV. (**a**) DiD-labeled HCV: HCV is first PEG-concentrated, then labeled with the lipophilic dye, DiD. Following iodixanol gradient purification, each fraction is assayed for specific infectivity. The fraction with the best (lowest) specific infectivity is then concentrated/purified via Centricon filtration. One day prior to imaging, cells are plated onto imaging dishes. Before infecting, imaging media is added and then cells are incubated with DiD-HCV on ice. A series of images are then acquired every 5–10 s, with an exposure time of roughly 100 ms for DiD. Resulting images are then processed and analyzed with ImageJ. (**b**) TC-Core Labeled HCV: Cells are either infected with TC-Core virus or transfected with in-vitro transcribed TC-Core HCV RNA. The day before imaging, cells are seeded onto imaging dishes. Just before imaging, TC-Core is labeled with either FlAsH or ReAsH, then washed and replaced with imaging media. TC-core labeled HCV is then imaged every 1–10 s (at 100 ms exposure). Following acquisition, images are processed and analyzed with ImageJ plugins

assay cell–cell spread, visualizing the transfer of TC core puncta from producer to recipient cells [18].

Here we provide methods to visualize HCV virions via both DiD and the TC tag. The procedures will detail virus concentration, DiD-labeling and purification for enhanced infectivity, TC-core HCV staining, imaging setup, and basic guidelines for live cell imaging and analysis of these fluorescent viruses (Fig. 1).

2 Materials

HCV stocks are generated in Huh-7.5 cells or related derivatives by standard protocols [19]. Ideally, viral stocks should have a starting TCID50/mL (preconcentration) of at least 7×10^5–3×10^6.

2.1 Virus Concentration: 40% PEG 8000

1. Double distilled water (ddH_2O).
2. Phosphate-buffered saline (PBS, 10×): 137 mM NaCl, 2.7 mM KCl, 8 mM Na_2HPO_4, 2 mM KH_2PO_4.
3. PEG 8000.
4. Dark-colored bottle (or bottle covered in foil).
5. Stir bar.
6. Magnetic stir plate.
7. Sterile filter unit: bottle top filter, bottle.
8. 100 mL graduated cylinder.
9. Aluminum foil.
10. High-speed centrifuge tubes with sealing cap.
11. High speed centrifuge.
12. 50 mL conical tubes.
13. 10% DMEM: DMEM, 10% fetal bovine serum, 1% nonessential amino acids, 1% penicillin–streptomycin.

2.2 DiD-HCV Labeling

1. 1.5 mL centrifuge tubes.
2. DiD lipophilic tracer or related lipophilic dyes, e.g., DiI, DiO (ThermoFisher Scientific).
3. Shaker/rotator.
4. OptiPrep (60%, w/v, solution of iodixanol in water; Sigma).
5. Sterile ddH_2O.
6. Gradient maker: 20 mL total volume (10 mL per side).
7. Open top polyallomer ultracentrifuge tubes (14 × 89 mm).
8. Ultracentrifuge, SW 40Ti rotor and swinging buckets.
9. Ring stand and clamp.
10. Sterile needle.
11. 24-well plate.
12. Amicon Ultra centrifugal filters (0.5 mL, Ultracel 100 k).
13. 10% DMEM: *See* Subheading 2.1.

2.3 Imaging Setup

1. 35 mm imaging dishes, 17 mm thickness (must have interlocking lids to prevent spillage).
2. Type 1 Collagen solution, 1%.
3. Sterile water.
4. 10% DMEM: *see* above.
5. Sterile PBS (for cell culture).

6. Imaging media: DMEM-F12 (no phenol red), 10% fetal bovine serum, 1% nonessential amino acids, 1% penicillin–streptomycin, 50 μM HEPES.
7. Parafilm.

2.4 TC-Core HCV Labeling

1. Collagen-treated imaging dishes (*see* Subheading 2.3).
2. Cells infected with TC-core HCV (*see* **Note 1**).
3. Opti-MEM.
4. 1,2-Ethanediol (EDT): 100 mM, in DMSO (*see* **Note 2**).
5. FlAsH or ReAsH Biarsenical dye (ThermoFisher Scientific): 1.25 μM final concentration, diluted in Opti-MEM (*see* **Note 3**).
6. 1×BAL Wash Buffer: 100xBAL buffer (ThermoFisher Scientific), Opti-MEM, 500 μM EDT.
7. Imaging media: (*see* Subheading 2.3).
8. Parafilm.

2.5 Imaging

1. Olympus DSU Spinning Disc Confocal microscope with 100× NA 1.45 oil objective, Hamamatsu back-thinned EM-CCD high speed/sensitivity camera.
2. Heated stage with imaging dish insert.
3. Slidebook software (Intelligent Imaging Innovations, Inc.).
4. Microscope immersion oil (for objective).
5. 70% Ethanol in a spray bottle.
6. Cling wrap.

2.6 Image Analysis

1. ImageJ (NIH).
2. RGB profiler plugin (Christophe Laummonerie).
3. Colocalization highlighter (ImageJ).
4. JACoP plugin [20].
5. Multiple kymograph plugin (ImageJ).
6. Manual tracking plugin (ImageJ).

3 Methods

All work with infectious virus (with the exception of live cell imaging) is done in BSL2 cell culture hoods. Materials coming into contact with infectious virus should undergo proper decontamination procedures as detailed in institutional biosafety committee (IBC) protocols. Live cell imaging is done with an interlocking imaging dish (*see* Subheading 2.3), then wrapped with Parafilm. The prepared imaging dish is then transported to imaging facilities in a sealed Styrofoam container to prevent infectious virus spills. Imaging can be performed in a BSL2 microscopy facility in the USA. A bottle of 70% ethanol should also be kept at the imaging facility in case of accidental spills.

3.1 PEG Concentration

1. Combine 40 mL ddH_2O and 10 mL 10× PBS in a dark-colored bottle. Add 40 g PEG 8000. Shake to combine, then add stir bar. Mix on magnetic stir plate until powder has dissolved (30–45 min) (*see* **Note 4**). Refrigerate until no bubbles remain (*see* **Note 5**).
2. Using 5 mL ddH_2O, prewet the filter. Pour the PEG solution into the graduated cylinder, being careful to avoid bubbles. Note the volume of PEG, then pour into the filter unit. Add enough ddH_2O to the graduated cylinder to bring the total volume up to 95 mL (The final volume will be 100 mL). Swirl the graduated cylinder to dissolve any residual PEG solution, then pour into the filter unit. Cover filter unit with foil and apply vacuum to filter the PEG (*see* **Note 6**).
3. Once filtered, wrap the bottle of 40% PEG 8000 in foil and store at 4 °C (*see* **Note 7**).
4. In high speed centrifuge tubes, combine virus stock and 40% PEG 8000 at a 4:1 ratio (bringing the final PEG concentration to 8%). Cap and invert 8–10 times to mix. Cover in aluminum foil and incubate at 4 °C overnight.
5. Centrifuge tubes at 9000 × g for 20 min, 4 °C (*see* **Note 8**).
6. Pipette off all but roughly 15 mL of the supernatant (*see* **Note 9**). Resuspend the pellet and transfer to a 50 mL conical tube.
7. Spin down again, 9000 × g for 10 min, 4 °C.
8. Pipette off all supernatant. Resuspend in serum free DMEM at 1/100th of initial concentration. Wrap in foil and store at 4 °C (*see* **Note 10**).

3.2 DiD Labeling

1. Prechill the rotor and buckets.
2. In a 1.5 mL centrifuge tube, combine 1 mL of concentrated virus with 5 μL DiD dye. Incubate on a rotator or shaker, protected from light, for 90 min at 4 °C.
3. Using a gradient maker, prepare a 10–60% OptiPrep gradient in the ultracentrifuge tubes—1 tube per 1 mL of labeled, concentrated virus. Leave roughly 2–3 cm space at the top of the tube. Chill until use.
4. Load 1 mL of labeled virus onto the top the ultracentrifuge tube, being careful not to disturb the gradient. Spin tubes in the ultracentrifuge at 142,000 × g for 16.5 h at 4 °C (*see* **Note 11**).

5. After spinning, secure the tube in a ring stand. Puncture the bottom of the tube with a needle and separate into 1 mL fractions using a 24-well plate (see **Note 12**).

3.3 Infectivity and Labeling Specificity of DiD

1. Remove two 50 μL aliquots from each fraction. One aliquot is used to determine density, then TRIzol-LS extracted for RNA. These RNA extracts are assayed for RNA levels via qPCR. The second aliquot is titered for infectious virus (*see* **Note 13**). Determine the specific infectivity of each fraction, measured as the ratio of RNA level to infectious virus particles (*see* **Notes 14** and **15**).
2. After selecting the fraction with lowest (best) specific infectivity, use an Amicon filter to remove the OptiPrep solution: Add 0.5 mL of the selected fraction to each Amicon filter. Spin at 14,000 × *g* for 20 min at 4 °C. Invert the filter into a new tube and spin at 2000 × *g* for 2 min at 4 °C. Pool the concentrated virus and resuspend in 1 mL 10% DMEM (or alternate concentration of FBS).

3.4 Imaging Setup

3.4.1 DiD Imaging Setup

1. Prepare collagen imaging dishes (in sterile tissue culture hood): coat dishes with 500 μL of collagen solution for 3–4 h at 37 °C. Aspirate excess collagen from plate and rinse twice with sterile water. Sterilize imaging dishes via overnight exposure to UV light in the tissue culture hood.
2. The day before imaging, seed cells in 10% DMEM so that they are roughly 70% confluent at the time of imaging (*see* **Note 16**). Just prior to imaging, wash with PBS and replace with 500 μL of cold imaging media (*see* **Note 17**).
3. Add 100 μL DiD labeled virus to imaging dish and incubate on ice for 1 h, protected from light (*see* **Note 18**). Replace media with 500 μL fresh imaging media (optional). Wrap the edge of the dish in Parafilm. Cells should be immediately transported to the imaging facility for visualization.

3.4.2 TC-Core HCV Labeling and Imaging Setup

1. Prepare collagen-treated dishes and cell monolayers as in Subheading 3.4.1.
2. Prewarm Opti-MEM, 1× BAL wash buffer, and imaging media. For staining, wash cells with Opti-MEM. Aspirate, then replace with 400 μL FlAsH or ReAsH dye labeling solution. Incubate 30 min at 37 °C (*see* **Note 19**).
3. Remove dye and wash cells for 5 min at 37 °C with the 1× BAL wash buffer. Repeat three more times. Rinse cells for 5 min at 37 °C with Opti-MEM.
4. Replace with imaging media (*see* **Note 20**). Wrap the edge of the dish in Parafilm and incubate cells at 37 °C until ready to image. Cells should be visualized within 2 h of staining.

3.5 Live Cell Imaging

1. *See* **Note 21** for safety considerations.
2. Preheat stage. If uncovered, set to 39° (so that center of the dish is 37 °C). If enclosed/incubator type set-up, then 37 °C is sufficient.
3. Image acquisition is dependent on the microscope used. Our setup uses Slidebook software for image acquisition, with intensification set to 255 (maximum). Live cell imaging is carried out by acquiring sequential static images of labeled virus and cellular markers of interest.

3.5.1 DiD Detection

DiD is detected with a Cy5 filter set (excitation: 635/20; emission: 685/40); DiO is detected with a EGFP filter set (excitation: 480/25; emission: 525/40); DiI is detected with the Cy3 filter set (excitation: 555/28; emission 617/73). DiD images are usually taken every 5–10 s. At 100–200 ms exposure times, 100 frame time courses are readily achievable (*see* **Note 22**).

3.5.2 TC-Core Detection

FlAsH is detected with an EGFP filter set (excitation: 480/25; emission: 525/40). ReAsH can be visualized with a DSReD filter set (excitation: 565/25; emission 620/60). Biarsenical dyes photobleach quickly during live cell imaging. For confocal imaging, 100–200 ms every 1–10 s maximizes data capture while minimizing photobleaching. Nonspecific labeling of organelles (e.g., mitochondria) can often occur. This can be alleviated by increasing EDT concentration or using additional wash steps.

3.6 Image Analysis

1. Images/videos are processed using ImageJ. Channels can be separated, then processed via the "subtract background" tool, followed by the "smooth" function. Channels can then be merged and contrast adjusted (especially for DiD/TC-core puncta, as they tend to be less bright than cellular markers).
2. Colocalization: For individual frames, colocalization between the HCV particle and other cellular/viral proteins can be determined using several plugins.

   RGB profiler plugin can be used to generate an intensity profile of each channel along a selected line.

   Colocalization highlighter plugin: For simple colocalization analysis, split channels can be analyzed for colocalization of particles given a set threshold.

   JACoP: A more sophisticated plugin, this allows analysis via various methods, including Pearson's and Manders coefficients.
3. Montages/kymographs: Montages can be created from either single or multiple channels via the "Make Montage" command. Kymographs are produced with the "Multiple kymograph" plugin.

4. Calculating distance: Following generation of a kymograph, the "Manual tracking" plugin can be used to calculate the distance from origin traveled as well as velocity (distance vs time).

4 Notes

1. TC-Core HCV is produced by inserting a tetracysteine tag (CCPGCC) into the HCV core protein after amino acid 3 [9]. In vitro transcribed TC-Core HCV RNAs are electroporated into cells and allowed to propagate for 48 h following electroporation, after which they can be seeded into imaging dishes. Imaging is performed at roughly 72 h post-electroporation. Alternatively, stocks of HCV TC-core virus can be prepared beforehand and used to infect cells; TC-core imaging can be performed using these infected cells.
2. Keep at room temperature with desiccator. Stock is usable for 2–3 weeks.
3. Prepare in tissue culture hood. Prepare only what is needed for staining. Optionally, the solution can be spiked with 10 μM EDT.
4. After 15–20 min, check to make sure that no clumps of PEG remain. If so, shake until clumps have broken up.
5. To maximize titer, it is best to chill reagents before adding them to HCV. Refrigeration during this step is not required, but saves time at the end of the process.
6. The PEG solution is quite thick. In order to speed up the process, make sure all connections are as tight as possible.
7. Stored at 4 °C and protected from light, PEG solution is stable for up to 4 days. However, we find it best to prepare the solution the same day as concentrating the virus.
8. It is essential not to disturb the pellet at this point. It will look white, somewhat fluffy, and run the length of the tube (often to the lip of the bottle).
9. Do not aspirate the supernatant, as it will clog the tube.
10. Pellet can be resuspended in alternate media compositions (serum-free) depending on the desired experiment. One can also resuspend in less media for a higher concentration, depending on the initial titer of virus (For DiD labeling, the TCID50/mL of the concentrated virus should be roughly 3×10^8). At this stage, the pellet is very compact and difficult to resuspend. We recommend storing the concentrated virus in the conical tube so as not to lose any of the pellet. Additionally, a resting period of 30–60 min will allow the all the bubbles to

dissipate in the resuspension so as to utilize all of the concentrated virus for the next stage.

11. Subsequent steps with DiD labeled virus are done in low light so as to preserve fluorescence.
12. Initial high density fractions may clog the puncture site. Wider gauge needles may help with this problem. Be aware that larger holes will lead to uncontrolled fractionation at lower densities. We prefer to use a 25 gauge needle (5/8 in. in length).
13. Protocols for qPCR and titering via limiting dilution are standard in the field [21, 22].
14. The specific infectivity of the virus should ideally be somewhere in the range of 10 (particles or RNA copies per infectious virus).
15. Additionally, a squash may be performed to ensure specific labeling of the virus, wherein DiD labeled particles are immunostained for components of the virion (e.g., core, E2) [8].
16. Often, cells have been previously transfected or transduced with a fluorescently tagged protein of interest to study cotrafficking. Alternatively, there are many commercially available organelle-specific dyes. These are labeled according to the manufacturer's instructions.
17. The imaging media can be made in bulk, but it is best to spike with HEPES the day of imaging. The cold imaging media is necessary to prevent endocytosis of the virus during the ice incubation phase.
18. Calculate the volume and/or resuspension of Amicon-purified virus based on titer of the fraction. An multiplicity of infection (MOI) of 0.5–1 will provide enough DiD particles for successful imaging. Higher MOIs enable a larger *n* value per experiment.
19. To minimize reagent requirements, use a silicone well insert. The insert is placed in the well after washing with Opti-MEM and before adding the dye. If using these inserts, incubate at 37 °C with 50 μL dye solution for 5 min. Replace the dye solution 3×, for a total of 20 min incubation.
20. If using the well inserts, add 500 μL warm imaging media around the insert to maintain temperature.
21. Always use gloves when handling the dishes at the imaging facility. However, make sure that such gloves do not come in contact with common use equipment (It is often useful to place cling wrap on the keyboard to minimize glove changing).

    Following imaging, gather up used gloves, cling wrap, etc. to dispose of in the proper biohazard container. Wipe down everything with ethanol.

22. The HCV entry process occurs within 2 h of the temperature shift to 37 °C. The majority of DiD-labeled HCV particles internalize within 30 min of this shift. Roughly 50% of particles have undergone uncoating by 85 min post-temperature shift [8].

Acknowledgments

We thank Kelly Coller Metzinger, The University of Chicago Light Microscopy Facility and its director Vytas Bindokas for helping develop this protocol. This work was funded by NIAID (AI080703). Y.B. was funded by NIH training grant T32 GM007183.

References

1. Eyre NS, Fiches GN, Aloia AL, Helbig KJ, McCartney EM, McErlean CS et al (2014) Dynamic imaging of the hepatitis C virus NS5A protein during a productive infection. J Virol 88:3636–3652
2. Wolk B, Buchele B, Moradpour D, Rice CM (2008) A dynamic view of hepatitis C virus replication complexes. J Virol 82:10519–10531
3. Chukkapalli V, Berger KL, Kelly SM, Thomas M, Deiters A, Randall G (2015) Daclatasvir inhibits hepatitis C virus NS5A motility and hyper-accumulation of phosphoinositides. Virology 476:168–179
4. Lindenbach BD, Evans MJ, Syder AJ, Wolk B, Tellinghuisen TL, Liu CC et al (2005) Complete replication of hepatitis C virus in cell culture. Science 309:623–626
5. Lindenbach BD, Meuleman P, Ploss A, Vanwolleghem T, Syder AJ, McKeating JA et al (2006) Cell culture-grown hepatitis C virus is infectious in vivo and can be recultured in vitro. Proc Natl Acad Sci U S A 103:3805–3809
6. Wakita T, Pietschmann T, Kato T, Date T, Miyamoto M, Zhao Z et al (2005) Production of infectious hepatitis C virus in tissue culture from a cloned viral genome. Nat Med 11:791–796
7. Zhong J, Gastaminza P, Cheng G, Kapadia S, Kato T, Burton DR et al (2005) Robust hepatitis C virus infection in vitro. Proc Natl Acad Sci U S A 102:9294–9299
8. Coller KE, Berger KL, Heaton NS, Cooper JD, Yoon R, Randall G (2009) RNA interference and single particle tracking analysis of hepatitis C virus endocytosis. PLoS Pathog 5:e1000702
9. Coller KE, Heaton NS, Berger KL, Cooper JD, Saunders JL, Randall G (2012) Molecular determinants and dynamics of hepatitis C virus secretion. PLoS Pathog 8:e1002466
10. Counihan NA, Rawlinson SM, Lindenbach BD (2011) Trafficking of hepatitis C virus core protein during virus particle assembly. PLoS Pathog 7:e1002302
11. Brandenburg B, Lee LY, Lakadamyali M, Rust MJ, Zhuang X, Hogle JM (2007) Imaging poliovirus entry in live cells. PLoS Biol 5:e183
12. Lakadamyali M, Rust MJ, Babcock HP, Zhuang X (2003) Visualizing infection of individual influenza viruses. Proc Natl Acad Sci U S A 100:9280–9285
13. van der Schaar HM, Rust MJ, Chen C, van der Ende-Metselaar H, Wilschut J, Zhuang X et al (2008) Dissecting the cell entry pathway of dengue virus by single-particle tracking in living cells. PLoS Pathog 4:e1000244
14. van der Schaar HM, Rust MJ, Waarts BL, van der Ende-Metselaar H, Kuhn RJ, Wilschut J et al (2007) Characterization of the early events in dengue virus cell entry by biochemical assays and single-virus tracking. J Virol 81:12019–12028
15. Hijikata M, Shimizu YK, Kato H, Iwamoto A, Shih JW, Alter HJ et al (1993) Equilibrium centrifugation studies of hepatitis C virus: evidence for circulating immune complexes. J Virol 67:1953–1958
16. Sainz B Jr, Barretto N, Martin DN, Hiraga N, Imamura M, Hussain S et al (2012) Identification of the Niemann-pick C1-like 1 cholesterol absorption receptor as a new hepatitis C virus entry factor. Nat Med 18:281–285

17. Hoffmann C, Gaietta G, Zurn A, Adams SR, Terrillon S, Ellisman MH et al (2010) Fluorescent labeling of tetracysteine-tagged proteins in intact cells. Nat Protoc 5:1666–1677
18. Liu Z, He JJ (2013) Cell-cell contact-mediated hepatitis C virus (HCV) transfer, productive infection, and replication and their requirement for HCV receptors. J Virol 87:8545–8558
19. Berger KL, Cooper JD, Heaton NS, Yoon R, Oakland TE, Jordan TX et al (2009) Roles for endocytic trafficking and phosphatidylinositol 4-kinase III alpha in hepatitis C virus replication. Proc Natl Acad Sci U S A 106:7577–7582
20. Bolte S, Cordelieres FP (2006) A guided tour into subcellular colocalization analysis in light microscopy. J Microsc 224:213–232
21. Randall G, Chen L, Panis M, Fischer AK, Lindenbach BD, Sun J et al (2006) Silencing of USP18 potentiates the antiviral activity of interferon against hepatitis C virus infection. Gastroenterology 131:1584–1591
22. Randall G, Panis M, Cooper JD, Tellinghuisen TL, Sukhodolets KE, Pfeffer S et al (2007) Cellular cofactors affecting hepatitis C virus infection and replication. Proc Natl Acad Sci U S A 104:12884–12889

Chapter 19

Cloning and Analysis of Authentic Patient-Derived HCV E1/E2 Glycoproteins

Richard A. Urbanowicz, Jonathan K. Ball, and Alexander W. Tarr

Abstract

Experimental characterization of the properties of authentic viruses circulating in infected individuals presents a problem when investigating RNA viruses with error-prone polymerases. The hepatitis C virus provides an extreme example of RNA virus genetic variability, as the nucleotide composition of HCV genomes can vary by more than 30% between strains. The envelope glycoproteins E1 and E2 in particular are able to tolerate a particularly high level of variation. They are under continual selection pressure from the host antibody response during chronic infection and can tolerate adaptive mutations, leading to great diversity in a single host. The diversity of E1/E2 in circulating viruses has hindered investigations of their function and development of a vaccine that will generate antibodies able to potently neutralize entry of genetically distinct strains.

Here we describe methods used in our laboratory to overcome the limitations of investigating the properties of the envelope glycoproteins representing only small numbers of HCV variants. Using a high-fidelity, limiting dilution ("endpoint") PCR approach to amplify single E1/E2 cDNA templates, which can then generate recombinant model viral particles using retrovirus packaging/reporter constructs. These retroviral pseudoparticles (pseudotypes) facilitate investigation of the properties of authentic E1/E2 glycoproteins in a single-round infection assay. We also describe optimized methods for generation of infectious pseudoparticles from patient-isolated E1/E2 and methods for performing neutralization assays with both anti-virus and anti-host antibodies.

Key words Hepatitis C virus, Pseudoparticle, Pseudotype, Vaccine, Neutralizing antibodies, Virus entry, Therapy

1 Introduction

Hepatitis C virus (HCV) possesses a positive-sense RNA genome. Replication of the genome is catalyzed by an error-prone, virus-encoded RNA-dependent RNA polymerase (RdRp). HCV replication results in high viral turnover, with more than 10^{11} particles produced per day [1] that, together with error-prone replication, results in generation of a quasispecies of genetically diverse variants in a single host. In addition to intrahost diversity, HCV exhibits great genetic diversity between host populations. HCV is currently

Mansun Law (ed.), *Hepatitis C Virus Protocols*, Methods in Molecular Biology, vol. 1911,
https://doi.org/10.1007/978-1-4939-8976-8_19,

classified into six major genotypes, with a seventh genotype described more recently [2]. These genotypes are geographically distributed and vary in their pathogenesis and sensitivity to therapy.

The tropism of HCV is restricted to human cells. Hepatocytes are the main target for viral infection. Susceptibility to entry of HCV is defined by expression of a specific set of host receptor proteins (Reviewed in [3]). The viral proteins responsible for attachment and entry of HCV virions into host hepatocytes are the glycoproteins E1 and E2 [4–6]. These proteins are the major target of virus-neutralizing antibodies. As such, analysis of these glycoproteins is important for vaccine discovery and drug targeting (Reviewed in [7]). Many studies of glycoprotein structure and function have focused on a limited number of well-characterized E1/E2 from cloned viruses that are potentially unrepresentative of naturally occurring variants [8]. To examine the importance of genetic diversity we developed methods to enable recovery of authentic functional E1 and E2 proteins from patient samples for expression on the surface of retroviral pseudoparticles. These samples have contributed to identification of common entry pathways of genetically diverse HCV strains and analysis of the differences in the glycoprotein phenotype during disease progression.

Recovery and expression of accurate representations of the HCV E1 and E2 genes poses a significant technical challenge. Genetic diversity is observed at each end of the coding region, from the signal peptide of E1 to the C terminal region of E2. This has implications for robust PCR amplification of these genes, requiring sensitive, specific PCR protocols. PCR introduces error during the polymerization of new DNA molecules, and it is important to discriminate between authentic genetic variation and experimentally induced mutations. In addition, high efficiency recovery of HCV RNA from clinical samples is difficult, requiring sensitive extraction and purification. At each stage, particular attention must be paid to preventing contamination.

Here, building on our previously published approaches [9], we describe optimized methods for recovery of authentic, functional E1 and E2 genes from HCV-infected blood samples, while minimizing introduction of experimentally induced error. We also describe optimized methods for characterizing the phenotype of these glycoproteins using a retroviral pseudoparticle cell entry assay.

2 Materials

1. Patient-derived plasma or serum samples.
2. pcDNA3.1 V5-HIS-DTOPO cloning vectors (ThermoFisher Scientific).
3. Viral RNA Isolation Kit (Qiagen).

4. Thermoscript cDNA Synthesis Kit (ThermoFisher Scientific).
5. Custom oligonucleotide primers (MWG Biotech).
6. NEB LongAmp Taq DNA polymerase (New England Biolabs).
7. HotStarTaq (Qiagen).
8. *E. coli* strain NEB10 (New England Biolabs).
9. GenElute HP Plasmid DNA Midiprep kit (Sigma).
10. Mammalian cell lines: HEK293T (ATCC CRL-3216), HuH7 (ATCC VR-2092).
11. Primaria-coated 10 cm dishes (Corning).
12. Dulbecco's modified Eagle medium (DMEM).
13. Fetal bovine serum (FBS).
14. OptiMEM medium.
15. Nonessential amino acids.
16. Sterile PBS.
17. Trypsin-EDTA 0.05% (Gibco).
18. Polyethylenimine, PEI (Polysciences, Inc).
19. 96-well white plate (Corning).
20. Luciferase Assay system (Promega).
21. Antibodies to HCV E1, E2, and MLV gag-pol [4, 10].
22. Lysogeny broth (LB).
23. LB Agar plates containing 100 μg/mL ampicillin.
24. 2× nonreducing Laemmli loading buffer.
25. 10% Mini-TGX polyacrylamide gels (Bio-Rad).
26. Lysis buffer (20 mM Tris–HCl, pH 7.4; 20 mM iodoacetamide; 1 mM EDTA; 150 mM NaCl; 1% Igepal C630).
27. PageRuler protein molecular weight marker (Thermo Fisher Scientific).

3 Methods

The methods are organized into three sections.

1. PCR amplification and cloning of authentic HCV E1/E2-encoding cDNA isolated from HCV-infected sera.
2. Generation and functional analysis of HCV pseudoparticles (HCVpp) representing these authentic patient-derived HCV isolates.
3. Analysis of incorporation of E1/E2 glycoproteins into retroviral pseudoparticles by western blotting.

3.1 Isolation of HCV E1E2-Encoding Sequences from Different Genotypes and Cloning into a Plasmid Vector for Expression in HEK293T Cells

3.1.1 Extraction of Hepatitis C Virus RNA for PCR Amplification

1. Viral RNA is isolated from patient sera obtained from clinic using BD Vacutainer serum separation tubes (Becton Dickinson, inverting the tube five times to mix with the clot activator. Incubate for no longer than 1 h then centrifuge at 2000 × *g* for 10 min to separate out the cells and clotting factors. Aspirate the serum from the tube and aliquot. Centrifuge blood samples at 2000 × *g* for 10 min, recovering the serum fraction for long-term storage at −80 °C). HCV genotype is determined clinically as part of the diagnostic protocol (*see* **Note 1**).
2. PCR amplification of the E1 and E2 coding region requires extracted viral RNA. Our comparison of different methods has found that the Qiagen Viral RNA Extraction Kit provides greatest recovery of RNA for amplification by reverse-transcription polymerase chain reaction (RT-PCR), without the need for phenol–chloroform RNA extraction. Viral RNA is isolated from 140 μL of serum at room temperature.
3. Firstly, prepare the AVL lysis buffer with the addition of carrier RNA, and heating to 80 °C before cooling to room temperature. Prepared AVL buffer is stored at 4 °C and resuspended before use.
4. Pulse-vortex the serum sample for 10 s in a Class I laminar flow cabinet, then add to 560 μL of AVL lysis buffer in a 1.5 mL microcentrifuge tube. The mixture is vortexed for 15 s and incubated at room temperature for 10 min, before pulse centrifuging the sample to recover all of the solution. To this, add 560 μL of molecular grade 100% ethanol and mix by vortexing for 15 s. It is essential to thoroughly mix the sample, then briefly pulse-centrifuge the tubes to recover the entire sample.
5. Placing a QIAamp spin column into a 2 mL collection tube, carefully pipette 630 μL of sample onto the membrane then centrifuge the tube for 1 min at 6000 × *g*. The flow-through is then discarded and the remaining sample added to the column, repeating the centrifugation step.
6. Placing the column into a clean collection tube, add 500 μL AW1 buffer. Closing the tube, centrifuge the assembly for 1 min at 6000 × *g*. The flow-through is again discarded and 500 μL AW2 wash buffer added. Centrifugation is performed at 20,000 × *g* for 3 min. To ensure complete removal of wash buffer, the column is placed once again into a clean collection tube and centrifuged for 1 min at 20,000 × *g*.
7. Remove the lid of a sterile 1.5 mL microcentrifuge tube and insert the column. To this assembly add 60 μL of elution buffer AVE, centrifuging at 6000 × *g* to recover purified RNA. Smaller volumes of elution buffer can increase the concentration of recovered RNA, but generally result in lower yield.
8. RNA recovered in this way is stored at −80 °C in small aliquots, and is stable for prolonged periods (>1 year).

Table 1
Primers used for antisense E1/E2 cDNA synthesis from positive strand HCV RNA templates

Genotype 1a	GGGATGCTGCATTGAGTA
Genotype 1b	CCGGCCACGGACGCCGCATTG
Genotype 2	RGACCATTGGMRCTAGCAGC
Genotype 3	TGCGCTGAGGGCGTTCAG
Genotype 4	CACCAGCGGGTGAAGCAGCATTGA
Genotype 5	GCCAAGCGAAGCAAATAACGAGCGAACCCCAGAAAA
Genotype 6	GCAGGGCCAGGATTAGCAGGAGGAGCGGCCA

3.1.2 cDNA Synthesis from HCV RNA

1. cDNA covering the entire E1/E2 coding regions is generated with an antisense primer specific for HCV RNA. Primers are designed to be complementary to the conserved sites within the p7/NS2 coding regions. The synthesized negative-strand cDNA is used as template for nested PCR amplification of the E1 and E2 genes.
2. cDNA is synthesized from 8 μL of purified RNA, using genotype-dependent primers (Table 1). Mix the RNA template with 15 pmol of primer and 2 μL of a 10 mM stock of dNTPs, in a final volume of 12 μL. Heat the samples to 65 °C for 5 min and then rapidly cool the tubes on ice.
3. To the primer–template mixture add 4 μL of 5× reaction buffer, 1 μL of 100 mM DTT, 40 units of RNaseOUT, 1 μL of Thermoscript reverse transcriptase (Invitrogen), and 1 μL of RNase-free water. The samples are centrifuged briefly to draw the sample to the bottom of the tube, then mixed with a pipette tip.
4. Incubate the mixture at 50 °C for 1 h.
5. Heat the reaction to 85 °C for 5 min to inactivate the polymerase. To degrade the template RNA and leave single-stranded cDNA product, add 2 units of RNaseH and incubate the sample at 37 °C for 20 min.
6. cDNA generated in this way is stable at 4 °C for weeks to months, but can be stored at −20 °C for prolonged storage. This single-stranded DNA is used as template during PCR amplification of the E1 and E2 genes for cloning and expression.

3.1.3 Limiting-Dilution Polymerase Chain Reaction of Full-Length E1E2 Genes

1. Thaw the template cDNA on ice. Two nested rounds of PCR are used for PCR amplification of E1/E2.
2. Optimization of the PCR primers for E1/E2 amplification of specific genotypes required analysis of publicly available

sequence repositories, as well as analyses of sequence variability in the core-NS2 region of HCV strains. In the first round of amplification, use outer PCR primers located in the flanking regions in the Core gene and in p7/NS2. The second round nested PCR amplification uses inner primers located such that encoded ORF following amplification includes amino acids 170–746 (numbering referenced to GenBank accession AF011751).

3. It is important to determine the titer of the cDNA used as a template for PCR amplification, and perform PCR amplification for cloning on limiting amounts of template [11–13]. cDNA samples with low copy number will provide an artificially low estimate of the diversity in a HCV population. To estimate the amount of template required as input into the first round of amplification, prepare a preliminary experiment with serially diluted cDNA samples, performing multiple identical PCR reactions at each dilution. Starting at a dilution of 1/8 (1 μL of cDNA and 7 μL of nuclease-free water), make twofold dilutions of cDNA to a final dilution of 1/128.

4. Performing five replicates at each dilution, amplify each reaction with two nested rounds of PCR, using genotype-specific primers and LongAmp *Taq* polymerase. Add 0.5 units of polymerase for amplification of E1 and E2 in a 25 μL reaction containing 5 pmol each of an appropriate sense primer and genotype-specific antisense primer (Table 2), 200 μM concentration of dATP, dCTP, dGTP, and dTTP. Prepare the mix with the addition of 5 μL 5× LongAmp *Taq* reaction buffer (*see* **Note 2**).

5. Amplification cycle parameters are optimized for PCR templates with potential secondary structure. The thermal cycle for both rounds of is:

   1× 94 °C for 30 s;

   35× [94 °C for 20 s; 53 °C for 20 s; 65 °C for 120 s];

   1× 65 °C for 5 min.

6. The second round of PCR amplification uses the inner sense primer and genotype-specific antisense primers, in reactions prepared in the same manner as the first round. One microliter of first-round product is used as template in this reaction, with the same cycling parameters as the first round (*see* **Notes 3** and **4**).

7. Analyze E1/E2 PCR products by electrophoresing them through an ethidium bromide-stained 2% agarose gel for 40 min at 90 V. The size of the PCR product depends on the genotype of the template: genotype 1a/b, 1734 bp; genotype

Table 2
Primer sequences used for PCR amplification of the E1 and E2 regions of HCV genomes of the major HCV genotypes. The table describes nested outer (first round) and inner (second round) pairs of primers used in PCR reactions. The sense primer in both first and second round amplifications is common to all genotypes. Antisense primers are described to amplify each specific genotype. The CACC tag (underlined), which facilitates TOPO cloning, acts as a Kozak sequence together with the initial Met (ATG) for efficient mammalian translation. Each inner antisense primer introduces a stop codon at the end of the expression construct (TAA)

Sense primers			Nucleotide position
All genotypes	OUTER	GGACGGGGTAAACTA TGCAACAGG	818–841
	INNER	CACC**ATG**GGTTGCTCTTTTTC TATC	843–869
Antisense primers			
Genotype 1	INNER	**TTA**CGCCTCCGCTTGGGATA TGAGTAACATCAT	2550–2582
	OUTER(1a)	GGGATGCTGCATTGAGTA	2599–2616
	OUTER(1b)	CCGGCCACGGACGCCGCATTG	2603–2623
Genotype 2	INNER	**TTA**CGCTTCGGCTTGGCCCA	2575–2591
	OUTER	RGACCATTGGMRCTAGCAGC	2627–2646
Genotype 3	INNER	**TTA**TGCTTCCGCCTGWGAWA TC	2580–2598
	OUTER	TGCGCTGAGGGCGTTCAG	2623–2640
Genotype 4	INNER	**TTA**CGCCTCAACTTGAC TTACCATAAACATCAT	2549–2578
	OUTER	CACCAGCGGGTGAAGCAGCA TTGA	2601–2624
Genotype 5	INNER	**TTA**TGCTTCGGCC TGACAAACCAAG	2563–2580
	OUTER	GCCAAGCGAAGCAAA TAACGAGCGAACCCCAGAAAA	2638–2671
Genotype 6	INNER	**TTA**TGCCTCTACCTGGCCGA TGATCAACATGA	2563 2591
	OUTER	GCAGGGCCAGGA TTAGCAGGAGGAGCGGCCA	2718–2748

2, 1746 bp; genotype 3, 1752 bp; genotype 4, 1734 bp; genotype 5, 1737 bp; genotype 6, 1749 bp (Fig. 1a).

8. The number of cDNA copies can be estimated using the formula of Reed and Muench [12], using the frequency of successful reactions at a dilution factor that results in both positive and negative PCR reactions:

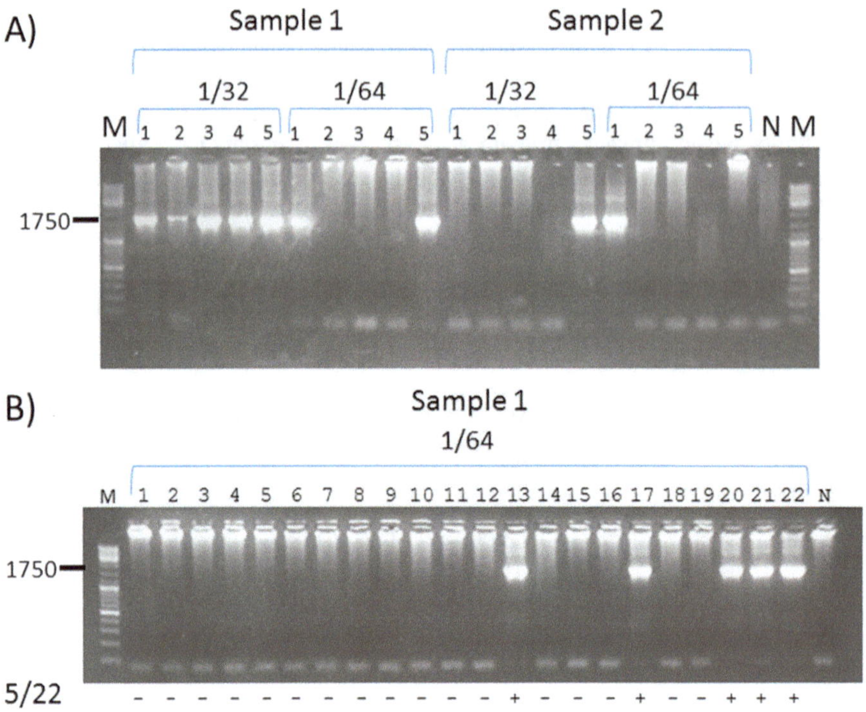

Fig. 1 An example of PCR amplification of E1/E2 genes using end-point diluted cDNA synthesized with genotype-specific primers. (**a**) Serial dilutions (1/8, 1/16, 1/32, 1/64. 1/128) of template cDNA are performed with five replicates. PCR reactions are shown representing two genotype 1a templates diluted 1/32 or 1/64. This serial dilution identifies the dilution at which limiting amounts of template are available. (**b**) 22 PCR reactions at the limiting dilution are prepared. In this case, 5/22 reactions are positive, indicating that each of these reactions are likely to possess a single cDNA template molecule. In this example, cDNA template from a genotype 1a-infected individual was amplified to generate a product of 1742 bp

$$Titer = \frac{-\mathrm{Ln}([No.of\ \ negative\ PCR\ reactions]/[No.of\ \ replicates])}{[Dilution\ factor]^{-1}}$$

9. This titer is determined from calculating the distribution of positive reactions at any single dilution point from a Poisson distribution. This is only possible for dilutions where there are positive and negative reactions.
10. Having identified a dilution endpoint at which individual PCR reactions possess a single template, prepare multiple reactions at this dilution (Fig. 1b). A positive–negative ratio of approximately 3:10 indicates the presence of single-molecules of cDNA in the positive reactions [12].
11. The PCR products are purified using a PCR Purification Kit (Qiagen) to remove unincorporated primers and nucleotides, then quantified using a NanoDrop spectrophotometer (NanoDrop Technologies).

Table 3
Primer sequences used for sequencing

E1IS	TGGGATATGATGATGAACTGG
E1OA	TCATTGCAGTTCAGGGCAGTCCTGTTGATG
T7	TAATACGACTCACTATAGGG
BGH	TAGAAGGCACAGTCGAGG

3.1.4 Cloning of Amplified E1E2 Genes into an Expression Vector for Pseudoparticle Genesis

1. The E1/E2 PCR products generated from endpoint-diluted cDNA templates are each sequenced directly using internal primers E1IS and E1OA (Table 3). As such this sequence accurately represents the template molecule. Identification of mixed sequence chromatograms at this stage indicate that the template is not a single molecule, and should be discarded [12].
2. Clone the bulk PCR products into the mammalian expression vector pcDNA3.1D/V5-His-TOPO vector (Thermo Scientific, Fig. 2). Approximately 50 ng of PCR product is inserted into the vector under the action of the covalently attached topoisomerase enzyme. Mix the PCR product with 1 μL of plasmid, 1 μL of salt solution (supplied with the cloning kit), and add water to a final volume of 6 μL. Incubate at room temperature for 15 min (*see* **Note 5**).
3. The plasmid mixture is used to transform chemically competent cells. Add the entire ligation reaction to a 50 μL aliquot of NEB10 cells (New England Biolabs) and incubate the mixture on ice for 20 min. Following this incubation heat the cells to 42 °C for 30 s and return them to ice. Add 250 μL of SOC medium to the cells and incubate the mixture at 37 °C for 1 h. Competent cells that have taken up plasmid are then selected by spreading the culture onto a Luria–Bertani (LB) agar plate containing 100 $\mu g m L^{-1}$ ampicillin. Colonies are generated following incubation overnight at 37 °C.
4. Directly inoculate PCR reactions with cells from individual colonies to identify bacteria possessing cloned E1E2 genes (Fig. 3). Prepare PCR reactions with 5 pmol each of vector-specific primers T7 and BGH (Table 3), or the T7 primer combined with an E2-specific antisense primer (E1OA; Table 3). Each reaction is prepared to a final volume of 25 μL, containing 0.15 μL of HotStarTaq (Qiagen), 200 μM concentration of each dNTP and 2.5 μL of 10× reaction buffer. To these reactions inoculate bacterial cells as template into the reaction with a sterile, nuclease-free pipette tip. Amplify with the following thermal cycling parameters:

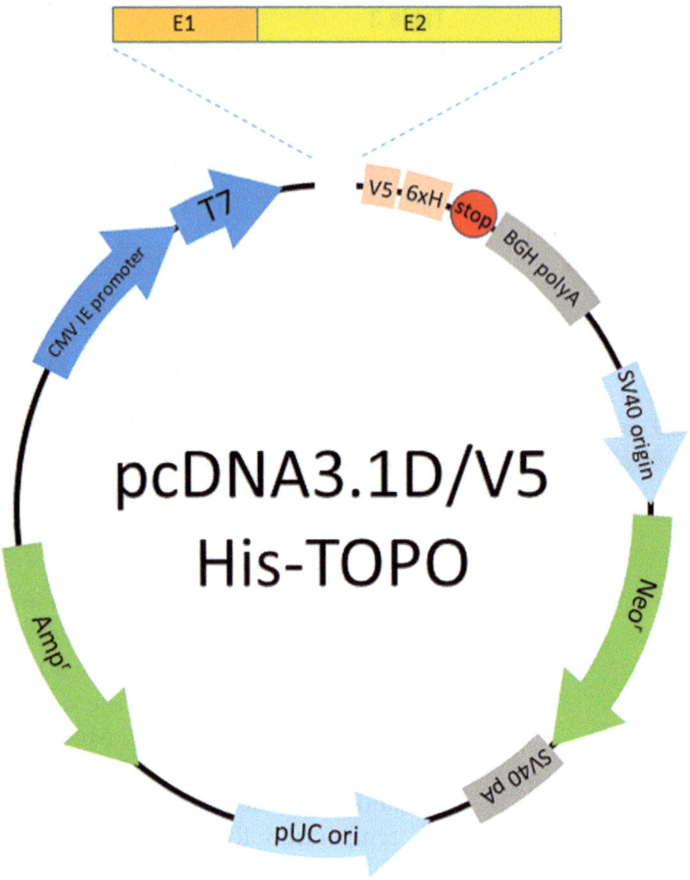

Fig. 2 Cartoon of plasmid pcDNA3.1D/V5-His-TOPO, used for construction of expression plasmid E1E2 clones for generating HCV pseudoparticles to characterize patient-derived E1/E2 genes. Expression of the HCV E1/E2 genes in mammalian cells is driven by the CMV immediate-early (IE) promoter. While the V5 and 6xHistidine tags are included in the vector, these are not incorporated into the protein construct due to the stop codon introduced by the antisense PCR primer

1× 95 °C, 15 min;

25× [94 °C, 15 s; 50 °C, 30 s; 72 °C, 2 min];

1× 72 °C, 7 min.

5. Following a single round of amplification, products are isolated on a 2%, ethidium bromide-stained agarose gel, a successful cloning reaction being visualized as a product at approximately 2 kb. Colonies identified as possessing a correctly oriented clone are then used to inoculate a 50 mL LB culture containing 100 $\mu g mL^{-1}$ ampicillin, shaking at 225 rpm overnight at 37 °C.

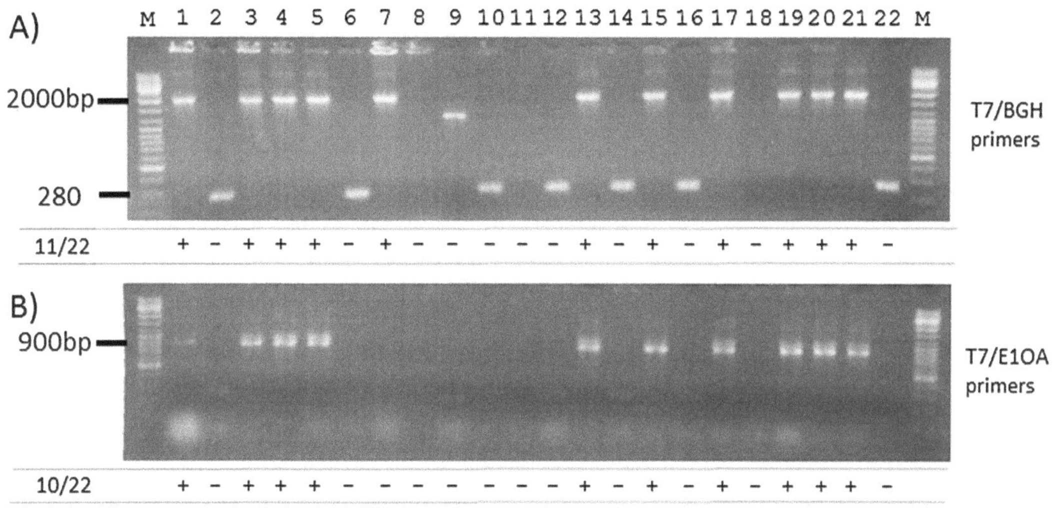

Fig. 3 Screening of transformed bacterial colonies for presence of cloned E1/E2 genes. In each sample, a single transformed bacterial colony was used directly as template in two PCR reactions using (**a**) vector-specific primers T7 and BGH, or (**b**) screening primers that discriminate the orientation of cloning of the PCR product. (**a**) The presence of a cloned E1/E2 gene is indicated by the presence of a PCR product of approximately 2000 bp. In the case illustrated above, 11 of the 22 selected colonies possess an insert corresponding to E1/E2. (**b**) Directional screening allows inserts that are ligated in a reverse orientation to be discounted. In this experiment 10 out of the 11 colonies positive in (**a**) are also positive for a band of approximately 900 bp. In lane 7, the sample is positive in panel (**a**) but negative in (**b**), indicating that the insert is cloned in reverse

6. Cultures are used to prepare a glycerol stock of bacteria possessing the plasmid for long-term storage, and a plasmid midiprep for generation of HCV pseudoparticles. Firstly, centrifuge 1.5 mL of culture at 13,000 × *g* for 2 min. The supernatant is discarded and 500 μL of fresh LB medium added. Vortex this sample to thoroughly resuspend the pellet and add 100 μL of sterile glycerol. This sample is mixed and stored at −80 °C as a long-term storage of each clone.
7. The remaining culture is used to prepare a DNA midiprep for HCVpp generation. The GenElute HP DNA Plasmid Midiprep kit (Sigma) provides a high-yield, endotoxin-free preparation of plasmid DNA suitable for transfection into human cells. Midipreps are performed according to the manufacturer's instructions.
8. Purified plasmid DNA is quantified using a NanoDrop spectrophotometer. The E1/E2-coding DNA is sequenced using Big Dye chemistry using primers T7, E1IS and BGH (Table 3). The cloned sequence must be cross-referenced against the sequence of the original bulk PCR product obtained from the single-molecule cDNA template. Sequence reads can be used to generate a contig using freely available assembly software

such as Bioedit (http://www.mbio.ncsu.edu/BioEdit/BioEdit.zip). Phylogenetic and molecular evolutionary analysis of samples is performed with the MEGA7 software [14].

3.2 Characterization HCV Glycoproteins Using HCV Pseudoparticles (HCVpp)

3.2.1 Transfection of HEK293T Cells to Generate HCVpp

1. Human Embryonic Kidney (HEK) 293T and HuH-7 human hepatoma cell lines can be purchased from ECACC. Both cell lines are grown in Dulbecco's modified Eagle's medium (DMEM), supplemented with 10% fetal bovine serum (FBS, heat-inactivated at 56 °C for 30 min), and 1× nonessential amino acids at 37 °C and 5% CO_2. HEK 293T cells are sub-cultured when reaching 80% confluence by removing the media, washing with sterile PBS and adding 0.05% Trypsin/EDTA. Cells are then incubated at 37 °C until they have detached from the cell culture surface and 10 mL of complete media added. Cells are pelleted by centrifugation (300 × *g* for 10 min) and resuspended in fresh, complete DMEM.
2. Approximately 16 h before transfection, seed 1.2×10^6 HEK293T cells into a 10 cm diameter Primaria coated dish (Corning) in 10 mL of DMEM supplemented with nonessential amino acids and heat-inactivated FBS (complete DMEM). Incubate overnight at 37 °C and 5% CO_2. The cell density after 16 h should be approximately 50%.
3. Prepare transfection mixtures for each pcDNA3.1D-E1/E2 clone by mixing 2 μg of phCMV MLV Gag/Pol packaging construct (plasmid phCMV-5349), 2 μg of luciferase encoding reporter plasmid (plasmid pTG126) and 2 μg of pcDNA3.1D-E1/E2 together in a 1.5 mL microcentrifuge tube, in a final volume of 300 μL OptiMEM. For most glycoprotein clones this ratio of 1:1:1 produces infectious particles. However, for some E1/E2 clones optimization of plasmid ratios is required (Subheading 3.2.2).
4. In a separate tube, mix 276 μL of OptiMEM with 24 μL cationic polymer transfection reagent (Polyethylenimine; Polysciences). PEI should be added directly into OptiMem, to minimize binding of PEI to plastic surfaces.
5. Add the 300 μL PEI–OptiMEM mix to the tube containing the plasmid mix to give a total of 600 μL A negative control is prepared (ΔE1/E2) by preparing plasmid mixtures omitting the glycoprotein-encoding plasmid.
6. Leave at room temperature for 1 h.
7. During this hour, the media is aspirated from the 10 cm dish containing HEK 293T cells, taking care not to disrupt the cell layer. Add 7 mL of OptiMEM by tilting the dish and pipetting into the bottom so as not to disturb the cells.

8. Add all 600 μL of the plasmid–PEI mix to the cells, dropwise across the top to ensure even spread, and then return the dish to an incubator at 37 °C and 5% CO_2 for 6 h.
9. Remove the OptiMEM and replace with 10 mL of complete DMEM media.
10. Leave for 72 h at 37 °C and 5% CO_2.
11. During this incubation, 24 h before harvesting the supernatants containing HCV pseudoparticles, seed 1.5×10^4 HuH7 cells per well into a 96-well white plate (Corning) in 100 μL of complete DMEM and leave overnight at 37 °C and 5% CO_2.
12. Using a sterile 10 mL syringe, aspirate the supernatant from the 10 cm transfection dish and attach a 0.45 μM filter. Filter the supernatant into a sterile 15 mL Falcon tube to remove cells from the suspension. This supernatant contains the pseudoparticles.
13. To recover intracellular E1/E2 proteins, lyse the transfected HEK293T cells with addition of 1 mL per dish of lysis buffer for 30 min on ice, centrifuge 13,000 × *g* for 5 min and collect the clarified supernatant. The clarified lysate may be stored at −20 °C.
14. If using pseudoparticles immediately, remove the media from the HuH7 cells seeded the previous day, and add 100 μL of the pseudoparticle-containing media to the wells, using at least triplicate infections for each sample (*see* **Note 6**).
15. Leave for 4 h at 37 °C and 5% CO_2. Following infection aspirate the pseudoparticle-containing media.
16. Add 250 μL of complete DMEM and leave for 72 h.
17. Discard the media from each well, gently tapping the plate on a clean tissue to remove residual liquid from each well. Add 50 μL of Promega Cell Lysis Buffer (diluted from 5× stock solution with distilled water; provided with Luciferase Reporter System; Promega) to each well.
18. Place on a laboratory rocker at room temperature for 15 min at 10 revolutions/min.
19. Luminescence is then detected using a luminometer equipped with direct injection, such as the multimode BMG Labtech FluoStar Omega plate reader. For this machine, select top optic reading and set PMT gain to 3600. Program the machine to inject 50 μL of Promega Luciferase Substrate, shake the plate for 2 s and read luminescence immediately afterward for 1 s.
20. Infectivity of E1/E2 clones can differ greatly between genetically related variants in this model (Fig. 4). Optimized HCVpp production may be necessary to generate infectious particles.

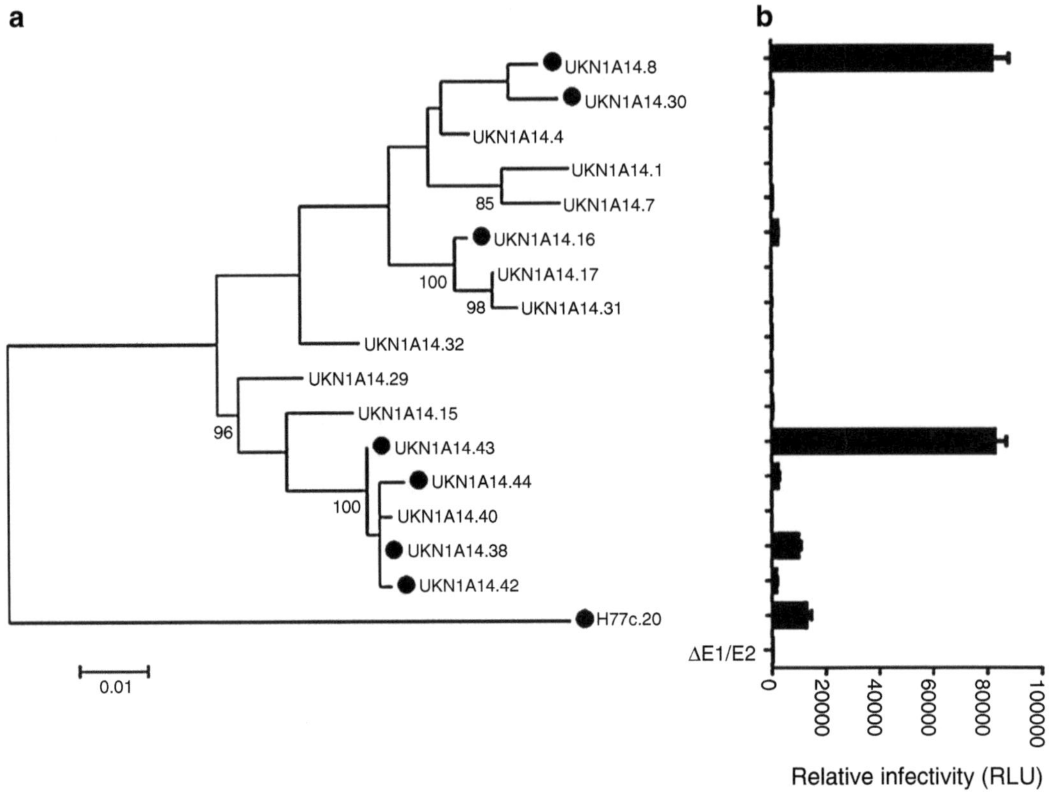

Fig. 4 Infectivity of HCV pseudoparticles bearing genetically related HCV E1/E2 genes generated from PCR products from a single infected individual. (**a**) Phylogenetic analysis of amplified, cloned E1/E2 was performed with MEGA, version 7, using a Clustal W alignment of sequences, followed by maximum likelihood reconstruction of the phylogenetic relationships between sequences (WAG model). Bootstrap analysis of phylogeny using 1000 replicates was performed and values >80 are highlighted). Clones identified as infectious in this model are indicated with (●). (**b**) Luciferase reporter readout for infectivity of HCVpp for each E1/E2 sequence in HuH7 cells. Reproduced from ref. 16 under CC BY 4.0 license

3.2.2 Preparing Matrix Transfections to Generate Functional HCVpp

1. Follow **steps 1** and **2** in Subheading 3.2.1.
2. If the plasmid ratio of 1:1:1 does not produce functional pseudoparticles it may be possible to generate functional particles by changing the ratio of packaging construct to E1/E2 while fixing the amount of luciferase-encoding reporter plasmid (*see* **Note** 7).
3. As above, mix the plasmids together in a 1.5 mL microcentrifuge tube, topped up to 300 μL with OptiMEM.
4. Follow **steps 4–20** in Subheading 3.2.1. An example infectivity plot of UKN1A14.38 (GenBank: AY734971) [6, 15, 16] is shown in Fig. 5.

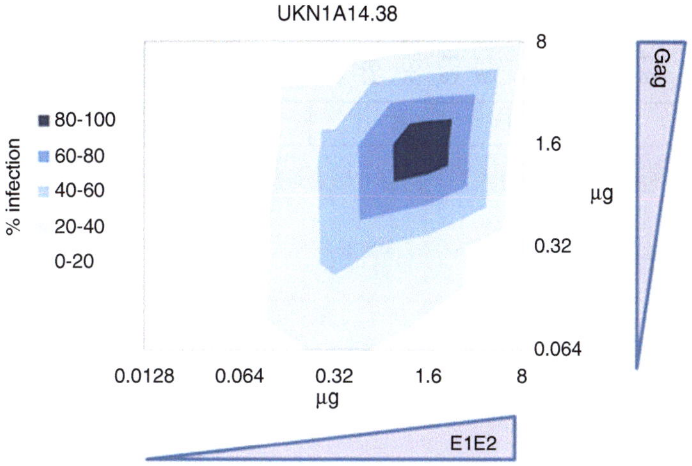

Fig. 5 HCVpp infectivity with pseudoparticles representing patient clone UKN1A14.38. Pseudoparticles were made with a matrix of varying quantities of the plasmids encoding packaging vector and E1/E2 glycoproteins. Relative infectivity was normalized to the greatest signal observed (~10,000 RLU) and the signal achieved with a negative control preparation (ΔE1/E2). The graph is presented as an *x–y* contour plot of Gag- and E1/E2-encoding plasmid concentrations. Greater infectivity is presented as darker blue, stratified into 20% increments, normalized to maximal infectivity. Reproduced from ref. 16 under CC BY 4.0 license

3.3 Analysis of E1/E2 Protein Expression and Incorporation into Pseudoparticles

3.3.1 Protein Analysis—Western Blot

1. Samples for western blotting are recovered from both the transfected HEK 293T cell lysates and pelleted HCVpp preparations. For each sample a 15 μL aliquot is mixed with an equal volume of 2× nonreducing Laemmli loading buffer, and loaded onto a polyacrylamide gel possessing a 10% resolving gel (Mini TGX-Bio-Rad). A molecular weight marker (PageRuler, ThermoFisher Scientific) is loaded into one lane of the gel to determine the apparent MW of expressed proteins. Running at a constant voltage of 150 V, proteins are separated after 90 min of electrophoresis.
2. Transfer the separated proteins from the polyacrylamide gel to Hybond N nitrocellulose membranes (GE Healthcare) using a Trans-Blot® Turbo semidry blotting apparatus (Bio-Rad) for 30 min at 1.3 A in transfer buffer (25 mM Tris–HCl pH 8.3, 192 mM glycine, 20% (v:v) methanol), following the manufacturer's instructions.
3. Add the membrane to a 5% milk solution in phosphate buffered saline–0.05% Tween (PBS-T), then washed once with 30 mL PBS-T and incubate for 1 h to block nonspecific protein binding sites.
4. Prepare a mixture of mouse anti-E2 antibodies AP33 (epitope: aa412-420 in E2) and ALP98 epitope: aa644-651 in E2) [10]

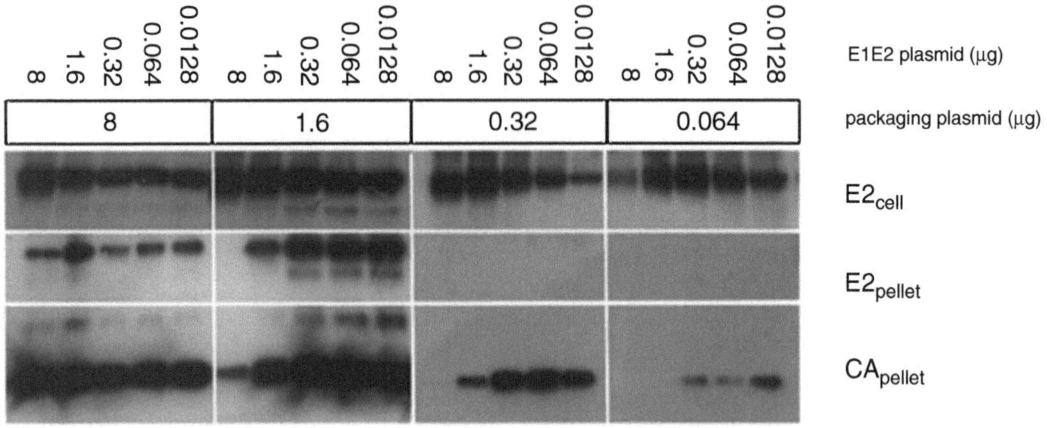

Fig. 6 Protein expression in cell lysates and particles when performing transfections with different amounts of E1/E2-encoding plasmid and MLV packaging plasmid. Western blotting of transfected cell lysates producing UKN1A14.38 HCVpp, using anti-E2 antibodies AP33 and ALP98 (upper panel), and pelleted pseudoparticles with AP33/ALP98 (middle panel), or anti-MLV capsid (CA; lower panel). Incorporation of E2 was observed when either 8 or 1.6 μg of packaging vector was used, while capsids were detected at all concentrations of the packaging vector. Greater concentrations of glycoprotein plasmid resulted in reduced expression of capsid. Reproduced from ref. 16 under CC BY 4.0 license

each at a concentration of 1 $\mu g mL^{-1}$ in PBS-T. Add the antibody mixture to the membrane for 1 h at room temperature, then wash the membrane a further three times with PBS-T.

5. Add a secondary antibody, rabbit polyclonal anti-mouse IgG, conjugated to horseradish peroxidase (Dako), at a dilution of 1/1000 in PBS-T, incubating at room temperature for 1 h. Wash the membrane a final three times with PBS-T.
6. Proteins are visualized using enhanced chemiluminescence (ECL Plus, GE healthcare). Add the ECL substrate to the membrane for 5 min, then dry the membrane and wrap in cling film. Detect luminescence by placing the membrane in contact with Kodak Biomax MR film, typically exposing the film for 1 min and developing with Kodak GBX developer. An example of a protein blot for of UKN1A14.38 (GenBank: AY734971) is shown in Fig. 6.

3.4 Modeling Neutralization of HCV Entry with Pseudoparticles

3.4.1 Neutralization of Entry of Pseudoparticles Representing Patient-Derived E1/E2 Genes with Antibodies Directed to E1/E2

1. After the pseudoparticles have been validated as infectious they can then be used for neutralization assays. If using an anti-virus antibody, the HCVpp and mAb are mixed together than added to the cells. If an anti-receptor mAb is being assessed, the antibody is added to the cells first before inoculating with HCV pseudoparticles.
2. Seed 1.5×10^4 HuH7 cells per well into a 96-well white plate (Corning) in 100 μL of complete DMEM and leave overnight for 16 h at 37 °C and 5% CO_2.

3. When using anti-virus antibodies, each neutralization is performed in triplicate. Each well contains a total volume of 100 μL (90 μL HCVpp and 10 μL of diluted mAb). Serially dilute the antibody preparation in a V-bottom microtiter plate to a concentration of 500 $\mu g m L^{-1}$–0.5 $\mu g m L^{-1}$. 30 μL of diluted antibody is required at each concentration for each HCVpp being tested. Add 270 μL of HCVpp preparation to each dilution of antibody, plus one additional well containing 30 μL PBS. The final concentrations of antibodies will be between 50 $\mu g m L^{-1}$ and 0.05 $\mu g m L^{-1}$.
4. Mix well and incubate at room temperature for 1 h.
5. Remove the media from the 96 well white plate containing HuH7 cells and add 100 μL of the pseudoparticle–mAb mixture, in triplicate, to the wells.
6. Leave for 4 h at 37 °C and 5% CO_2.
7. Aspirate the inoculum and replace with 250 μL of complete DMEM in each well and return to the incubator for 72 h.
8. Discard the media from each well and add 50 μL of Cell Lysis Buffer (made using the manufacturer's protocol; Promega) to each well.
9. Place the plate on a rocker at room temperature for 15 min.
10. Assess luminescence output as described in Subheading 3.2.1.
11. Neutralization curves are then calculated using the ΔE1/E2 HCVpp as 0% and the uninhibited control as 100% signal. An example of anti-virus mAb neutralization is shown in Fig. 7.

3.4.2 Neutralization of Entry of Pseudoparticles Using Anti-Receptor Antibodies

1. Seed 1.5×10^4 HuH7 cells per well into a 96-well white plate (Corning) in 100 μL complete DMEM and leave overnight at 37 °C and 5% CO_2.
2. Each neutralization experiment with anti-receptor antibodies is performed in triplicate. Dilute mAbs in DMEM to the required range of concentrations (0.01–50 $\mu g m L^{-1}$) allowing 50 μL of diluted antibody for each well.
3. Remove the media from the 96-well white plate seeded the previous day, and add 50 μL of the diluted anti-receptor mAb, in triplicate, to the wells. Fifty microliters of DMEM should be added to an extra set of wells as an uninhibited (positive) control.
4. Incubate the antibodies with target cells for 1 h at 37 °C and 5% CO_2.
5. Remove mAb and add 100 μL of HCVpp prepared as described in Subheading 3.2.1 to each well.
6. Incubate the virus with mAb-treated cells for 4 h at 37 °C and 5% CO_2.

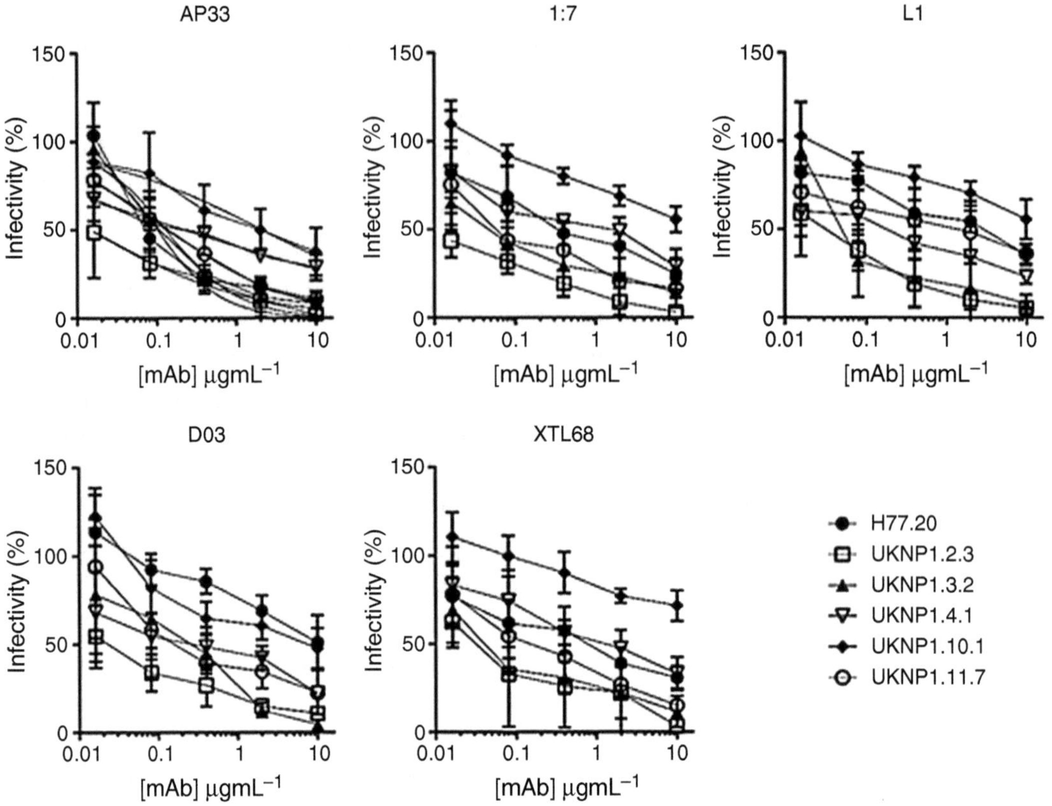

Fig. 7 Patient-derived E1/E2 genes display great differences in neutralization resistance. Selected genotype 1 clones exhibited both antibody-resistant and -sensitive phenotypes. Entry of HCVpp generated with genotype 1 E1/E2 proteins H77.20 (GenBank AF011751), UKNP1.2.3 (KU285154), UKNP1.3.2 (KU285157), UKNP1.4.1 (KU285161), UKNP1.10.1 (KU285172), and UKNP1.11.7 (KU285179) were neutralized by increasing concentrations of mAb AP33, mAb 1:7, mAb L1, nB D03, and mAb XTL68, and the resulting infectivity was plotted against the antibody concentration. Reproduced from ref. 8 under CC BY 4.0 license

7. Remove the virus inoculum and replace with 250 μL of complete media and leave for 72 h.
8. Discard the media from each well and add 50 μL of Cell Lysis Buffer to each well.
9. Place on a rocker at RT for 15 min. Measure luminescence as described in Subheading 3.2.1.
10. Neutralization curves can be calculated using the ΔE1/E2 HCVpp as 0% and the uninhibited control as 100%.

4 Notes

1. Serum that has been isolated from venous blood is an ideal source of HCV RNA. Collect blood in a BD Vacutainer Plus plastic serum tube (red top). Serum can be frozen at −80 °C.

Virus-containing samples stored this way are stable for >10 years. However, repeated freezing–thawing cycles can limit recovery of HCV RNA molecules required to reverse transcribe the E1/E2 coding regions.

2. Selection of an appropriate DNA polymerase is critical to successful PCR amplification of viral cDNA generated from HCV RNA, Having compared different commercially available polymerases the best results have been achieved with LongAmp *Taq* polymerase (NEB).
3. Contamination presents a significant problem when PCR-amplifying the E1/E2 genes at single molecule sensitivity. To prevent contamination, methods must be performed with the use of filter tips, nuclease-free reagents and appropriate decontamination of surfaces. In particular, amplified products or E1/E2-containing plasmids must never be handled in areas used for pre-PCR manipulation of samples, as contaminating aerosols are a significant problem.
4. During the second round of nested PCR the primers introduce a start codon at the 5′ end of the signal peptide of E1, as well as a stop codon in place of the first amino acid of p7, resulting in an E1/E2 expression construct that can be used to generate retroviral pseudoparticles. The positioning of these start and stop codons is important for producing functional constructs.
5. The inner sense primer also possesses a CACC sequence upstream of the initiation codon, facilitating insertion of the PCR product in a single orientation and minimizing the occurrence of incorrectly oriented clones when using Directional TOPO Cloning vectors (Thermo Fisher).
6. HCV pseudoparticle preparations can be used either immediately or stored at 4 °C for a maximum of 2 weeks. Long term storage can be performed at −80 °C, although loss of pseudoparticle infectivity can occur when stored for prolonged periods.
7. To empirically determine the optimal amounts of each plasmid for a specific E1/E2 clone, a series of transfections should be performed with a matrix of fivefold serial dilutions of the plasmid preparations. Suggested plasmid quantities are between 8 μg and 0.0128 μg. While not all E1/E2 sequences will yield infectious particles, infectivity can be recovered for some patient-isolated clones that initially are noninfectious when HCVpp are generated at the standard plasmid ratios.

Acknowledgments

We thank François-Loïc Cosset for packaging constructs phCMV5349 and reporter construct pTG126, J. Dubuisson for polyclonal anti-MLV core antibody, Jens Bukh for the molecularly cloned E1/E2 of strain H77, and Arvind Patel for antibodies AP33 and ALP98.

References

1. Ramratnam B, Bonhoeffer S, Binley J, Hurley A, Zhang L, Mittler JE et al (1999) Rapid production and clearance of HIV-1 and hepatitis C virus assessed by large volume plasma apheresis. Lancet 354:1782–1785
2. Smith DB, Bukh J, Kuiken C, Muerhoff AS, Rice CM, Stapleton JT et al (2014) Expanded classification of hepatitis C virus into 7 genotypes and 67 subtypes: updated criteria and genotype assignment web resource. Hepatology 59:318–327
3. Lindenbach BD, Rice CM (2013) The ins and outs of hepatitis C virus entry and assembly. Nat Rev Microbiol 11:688–700
4. Bartosch B, Dubuisson J, Cosset FL (2003) Infectious hepatitis C virus pseudo-particles containing functional E1-E2 envelope protein complexes. J Exp Med 197:633–642
5. Lindenbach BD, Evans MJ, Syder AJ, Wolk B, Tellinghuisen TL, Liu CC et al (2005) Complete replication of hepatitis C virus in cell culture. Science 309:623–626
6. Lavillette D, Tarr AW, Voisset C, Donot P, Bartosch B, Bain C et al (2005) Characterization of host-range and cell entry properties of the major genotypes and subtypes of hepatitis C virus. Hepatology 41:265–274
7. Tarr AW, Khera T, Hueging K, Sheldon J, Steinmann E, Pietschmann T et al (2015) Genetic diversity underlying the envelope glycoproteins of hepatitis C virus: structural and functional consequences and the implications for vaccine design. Viruses 7:3995–4046
8. Urbanowicz RA, McClure CP, Brown RJ, Tsoleridis T, Persson MA, Krey T et al (2016) A diverse panel of hepatitis C virus glycoproteins for use in vaccine research reveals extremes of monoclonal antibody neutralization resistance. J Virol 90:3288–3301
9. Tarr AW, Owsianka AM, Szwejk A, Ball JK, Patel AH (2007) Cloning, expression, and functional analysis of patient-derived hepatitis C virus glycoproteins. Methods Mol Biol 379:177–197
10. Owsianka A, Clayton RF, Loomis-Price LD, McKeating JA, Patel AH (2001) Functional analysis of hepatitis C virus E2 glycoproteins and virus-like particles reveals structural dissimilarities between different forms of E2. J Gen Virol 82:1877–1883
11. Simmonds P, Zhang LQ, Watson HG, Rebus S, Ferguson ED, Balfe P et al (1990) Hepatitis C quantification and sequencing in blood products, haemophiliacs, and drug users. Lancet 336:1469–1472
12. Simmonds P, Balfe P, Peutherer JF, Ludlam CA, Bishop JO, Brown AJ (1990) Human immunodeficiency virus-infected individuals contain provirus in small numbers of peripheral mononuclear cells and at low copy numbers. J Virol 64:864–872
13. Brown RJ, Hudson N, Wilson G, Rehman SU, Jabbari S, Hu K et al (2012) Hepatitis C virus envelope glycoprotein fitness defines virus population composition following transmission to a new host. J Virol 86:11956–11966
14. Kumar S, Stecher G, Tamura K (2016) MEGA7: molecular evolutionary genetics analysis version 7.0 for bigger datasets. Mol Biol Evol 33:1870–1874
15. Brown KS, Keogh MJ, Owsianka AM, Adair R, Patel AH, Arnold JN et al (2010) Specific interaction of hepatitis C virus glycoproteins with mannan binding lectin inhibits virus entry. Protein Cell 1:664–674
16. Urbanowicz RA, McClure CP, King B, Mason CP, Ball JK, Tarr AW (2016) Novel functional hepatitis C virus glycoprotein isolates identified using an optimised viral pseudotype entry assay. J Gen Virol 97:2265–2279

Chapter 20

Simultaneous Quantification of Hepatitis C Virus Envelope Glycoproteins E1 and E2 by Dual-Color Fluorescence Immunoblot Analysis

Fang Chen, Erick Giang, Radhika Gopal, and Mansun Law

Abstract

The hepatitis C virus (HCV) envelope glycoproteins, E1 and E2, are crucial for HCV assembly and entry, and are promising vaccine antigens. However, they are challenging to study because of technical difficulties in protein production and in quality control for protein folding and glycosylation. To study E1 and E2 in different experimental systems, e.g. infected cells, virus culture, virus-like particles, and clinical samples, a standardized method to accurately quantify the glycoproteins will be essential for most research projects. Here we outline a sensitive assay based on dual-color fluorescence immunoblot and the Odyssey imaging system to detect and quantify HCV E1 and E2 glycoproteins either using a purified E1E2 complex, or an engineered protein standard containing E1 and E2 at equal molar ratio. The method is capable of simultaneously detecting and quantifying as little as 7 ng of E1 and 5 ng of E2 in HCV pseudoparticles, and will be useful to quantify E1 and E2 from a wide variety of samples.

Key words E1, E2, Glycoprotein, Immunoblot, Protein standard, Antibody

1 Introduction

The HCV envelope glycoproteins E1 and E2 are targets of neutralizing antibody responses. They are known to form heterodimers on the viral surface [1–3], and play essential roles at different steps of the HCV life cycle, including viral entry, assembly, fusion and budding [4]. As the lead vaccine candidate antigens and the machinery responsible for a largely unknown viral fusion mechanism, there is significant interest in deciphering the structural and molecular details of E1 and E2. However, E1 and E2 are difficult to produce, consequently pure proteins are not readily available for experiments demanding high quality materials. E1 and E2 on virions, e.g., HCV pseudoparticles (HCVpp), HCV cell culture (HCVcc) or virus-like particles (VLPs), may not be easily detected or quantified depending on the source of materials [5–7]. Glycoproteins on virus particles expressed in cell culture or found in

Mansun Law (ed.), *Hepatitis C Virus Protocols*, Methods in Molecular Biology, vol. 1911,
https://doi.org/10.1007/978-1-4939-8976-8_20,

patient sera are highly heterogeneous and are associated with cellular components like apolipoproteins, which further complicate their characterization [8–10]. Thus, a sensitive assay to detect and quantify their presence in different conditions will be valuable for the studies of E1 and E2. Here we describe a strategy based on dual-color fluorescence immunoblot and the Odyssey imaging system to detect and quantify HCV envelope glycoproteins. By using a recombinant E1E2 complex or an engineered soluble E1E2 protein as a protein standard, the assay was able to simultaneously detect as little as 7 ng of E1 and 5 ng of E2 in HCVpp on the same immunoblot, and will be useful to quantify E1 and E2 in a wide variety of samples.

2 Materials

2.1 Preparation of E1E2 Standard

1. FreeStyle 293-F cells (Gibco).
2. FreeStyle 293 expression medium (Gibco).
3. pCMVtpa_sE1E2v4 construct for expression of soluble fused E1E2 lacking transmembrane domains [11].
4. pAdvantage plasmid (Promega).
5. Opti-MEM reduced serum medium (Invitrogen).
6. Polyethylenimine (PEI), linear, MW 2,500 (Polysciences). Dissolve in deionized water and filter-sterilize to make 1 mg/ml stock aliquots.
7. 5 mM stock solution of kifunensine (GlycoSyn).
8. Phosphate-buffered saline (PBS).
9. Strep-Tactin Sepharose, 50% suspension (IBA).
10. 2 L Erlenmeyer cell culture flasks (Corning).
11. 0.22 μm syringe filters (Corning).
12. Steritop bottle top filters, 0.22 μm (Millipore).
13. 37 °C, 8% CO_2 incubator shaker (Innova).
14. NanoDrop 2000c spectrophotometer (Thermo Fisher Scientific).
15. Vivaflow 200, 30,000 MWCO Hydrosart (Sartorius).
16. Avidin (Rockland). Make 1 mg/ml stock aliquots with deionized water.

2.2 Generation of HCVpp

1. 293T cells (ATCC CRL-3216).
2. 293T medium: Dulbecco's Modified Eagle Medium (DMEM, Gibco) supplemented with 1 mM sodium pyruvate (Gibco), 2 mM L-glutamine (Gibco), 100 units/ml of penicillin

(Gibco), 100 μg/ml of streptomycin (Gibco), and 5% heat-inactivated fetal bovine serum (FBS, Gibco).

3. phCMV_E1E2 constructs for expression of full-length E1E2 from different HCV genotypes/isolates [12].
4. pNL4-3.Luc.R-E$^-$ plasmid (NIH AIDS Reagent program).
5. 20% (wt/volume) sucrose (Sigma).
6. 150 mm × 25 mm tissue culture dishes (Corning).
7. 0.22 μm syringe filters (Corning).
8. 0.4 μm syringe filters (Corning).
9. Oak Ridge high-speed PPCO centrifuge tubes (Nalgene).
10. 37 °C, 5% CO_2 incubator (Thermo Fisher Scientific).
11. Avanti J-20i floor model centrifuge (Beckman Coulter).
12. Polyethylenimine (PEI), linear, MW 25,000 (Polysciences). Dissolve in deionized water and filter-sterilize to make 1 mg/ml stock aliquots.

2.3 Dual-Color Fluorescence Immunoblot

1. 4–20% Mini-PROTEAN TGX precast gels (Bio-Rad).
2. 10× Tris-Glycine SDS buffer (pH 8.3, Pierce).
3. 4× Laemmli sample buffer (Bio-Rad).
4. 1 M Dithiothreitol (DTT, Sigma).
5. Precision Plus Protein dual color standards (Bio-Rad).
6. Trans-Blot Turbo transfer system (Bio-Rad).
7. Blocking buffer: 5% nonfat dry milk (Bio-Rad) in 1× PBS–0.2% Tween 20.
8. Wash solution: 1× PBS–0.05% Tween 20.
9. Chiron E1E2 (Novartis Vaccines and Diagnostics).
10. Murine anti-E1 monoclonal antibody A4 or equivalent.
11. Human anti-E2 monoclonal antibody HCV1 or equivalent.
12. IRDye 680RD goat anti-mouse IgG (H + L) (LI-COR Biosciences) or equivalent.
13. IRDye 800CW goat anti-human IgG (H + L) (LI-COR Biosciences) or equivalent.
14. Odyssey Imaging system and Image Studio software (LI-COR Biosciences).

3 Methods

3.1 Preparation of E1E2 Protein Standard

The E1E2 standard will be used as a reference to estimate the amount of E1 and E2 present in various samples by immunoblotting. It can be a purified E1E2 complex or an engineered protein containing E1 and E2 at equal molar ration. Two standards were used in this study: Chiron E1E2, a purified discontinued full-length E1E2 vaccine candidate by Novartis Vaccines and Diagnostics (*see* **Note 1**), and sE1E2v4, a soluble strep-tagged fusion protein of E1 and E2 ectodomains. Here we describe a protocol to obtain sE1E2v4 using the mammalian transfection system.

1. *Day 1*: Seed 5×10^8 FreeStyle 293-F cells into a 2 L Erlenmeyer cell culture flask with 1 L of FreeStyle 293 expression medium and incubate overnight at 130 rpm, 37 °C and 8% CO_2.
2. *Day 2*: Check cell density and variability. Ensure that cells are in the range of 8.0×10^5–1.2×10^6 per ml and the viability of cells is over 90%.
3. Mix 0.6 mg of pCMVtpa_sE1E2v4 with 0.4 mg of pAdvantage in 10 ml of Opti-MEM. Filter-sterilize the mixture through a 0.22 μm syringe filter, then add Opti-MEM to a final volume of 44 ml.
4. Add 6 ml of 1 mg/ml PEI. Mix by pipetting and incubate for 20 min at room temperature (*see* **Note 2**).
5. Add the DNA-PEI complexes dropwise into the cell culture supplemented with 7.5 μM kifunensine and incubate for 72 h at 130 rpm, 37 °C and 8% CO_2.
6. *Day 5*: Transfer the cell culture into centrifugation tubes and spin at $4000 \times g$ for 15 min at 4 °C.
7. Filter the supernatant with a 0.22 μm Steritop bottle top filter.
8. Set up a peristaltic pump system with the Vivaflow 200 module (30,000 MWCO) according to manufacturer's protocol to begin removing free biotin from and concentrating filtered supernatant.
9. Concentrate the supernatant tenfold (e.g., 1 L to 0.1 L), add 1× PBS to 1 L and repeat concentration three times.
10. Add 80 μl of 1 mg/ml avidin to the concentrated supernatant, mix and incubate at room temperature for 30 min to inactivate residual free biotin.
11. Proceed with Strep-Tactin affinity purification according to IBA manufacturer's protocol using the Strep-Tactin Sepharose 50% suspension. In brief, after passing the supernatant through the Strep-Tactin Sepharose, wash the column five times with 1 Column Volume (CV) of Buffer W, then add six times 0.5 CVs of Buffer E and collect the eluate in 0.5 CV fractions (The Strep-tag fusion protein usually elutes in the second to fifth fractions).
12. Buffer exchange and concentrate eluted protein samples into 1× PBS.

13. Measure the protein concentration using a spectrophotometer (e.g., NanoDrop 2000c). Store the proteins at −20 °C in aliquots.

3.2 Production of HCVpp

HCVpp are widely used to study HCV viral entry and neutralization. This section describes a protocol for generation of HCVpp by transfection.

1. *Day 1*: Seed 293T cells on poly L-Lysine treated 150 mm × 25 mm dishes at a density of 8 × 10^6 cells per dish.
2. *Day 2*: Check cell density and variability. Ensure that cells are 60–80% confluent and evenly spread out across the dish. Change the medium before transfection (*see* **Note 3**).
3. Mix 80 μg of pNL4-3.Luc.R^-E^- and 20 μg of phCMV_E1E2 plasmid DNA in 10 ml of Opti-MEM for each dish. Filter-sterilize the mixture through a 0.22 μm syringe filter into a 15 ml conical tube.
4. Add 400 μl of 1 mg/ml PEI (MW 25,000). Mix by pipetting and incubate for 20 min at room temperature.
5. Add the DNA-PEI complexes dropwise onto the cell monolayer and gently agitating the dish to ensure even distribution.
6. Incubate for 6 h at 37 °C and 5% CO_2.
7. Remove the media and replace with 30 ml fresh 293T media. Incubate for 72 h at 37 °C and 5% CO_2.
8. *Day 5*: Harvest supernatant at 600 × *g* for 10 min at 4 °C to remove cellular debris.
9. Filter the clarified supernatant with a 0.4 μm syringe filter.
10. Transfer the 30 ml supernatant into Oak Ridge centrifuge tubes and centrifuge at 31,000 × *g* for 1 h. Carefully remove the supernatant.
11. Resuspend the virus pellet in 1 ml DMEM and overlay on 300 μl of 20% sucrose (*see* **Note 4**).
12. Centrifuge at 31,000 × *g* for 4 h at 4 °C. A tiny pellet can be visible at the bottom of the tube.
13. Gently remove all medium, resuspend the wet HCVpp pellet in 50 μl 1× DPBS and store at −80 °C before use.

3.3 Dual-Color Fluorescence Immunoblot

Although chemiluminescence-based immunoblot analysis is one of the most common techniques for protein detection and quantification, dual-color fluorescence detection greatly improves the sensitivity, quality and accuracy of immunoblot data, and allows for simultaneously detecting and quantifying two target proteins on the same blot.

Table 1
Representative E1- and E2-specific antibodies to linear epitopes on E1E2

		Epitope		
Antibody	**Species**	**Residues**	**Sequence (H77 isolate)**	**Source**
A4	Murine	E1 197-207	SSGLYHVTNDC	[1]
H-111	Human	E1 192-202	YQVRNSSGLYH	[13]
RM1-18	Rhesus Macaque	E1 316-329	HRMAWDMMMNWSPT	Law lab
HCV1	Human	E2 412-423	QLINTNGSWHIN	[14]
HC33.1	Human	E2 412-423	QLINTNGSWHIN	[15]
AP33	Murine	E2 412-423	QLINTNGSWHIN	[16]
19B3	Murine	E2 412-423	QLINTNGSWHIN	[11]
AP320	Murine	E2 464-471	DFAQGWGP	[16]
ALP98	Murine	E2 644-651	CNWTRGER	[16]
ALP1	Murine	E2 648-659	RGERCDLEDRDR	[16]

1. Make appropriate dilutions of E1E2 standard (sE1E2v4 or Chiron E1E2) and HCVpp in Laemmli samples buffer with reducing agents, e.g., 10 mM DTT.
2. Denature sample by heating at 95 °C for 5 min or 75 °C for 15 min before loading on a 4–20% Mini-PROTEAN TGX precast gel along with Precision Plus Protein dual color standards.
3. Perform SDS-PAGE at 250 V until dye of loading buffer reaches bottom of gel and the protein standards are well separated.
4. Transfer the proteins from the gel to an PVDF membrane using the Trans-Blot Turbo transfer system according to the manufacturer's instructions (*see* **Note 5**).
5. Block the membrane with blocking buffer and incubate for 1 h at room temperature with rocking.
6. Remove the blocking buffer.
7. Dilute E1- and E2-specific monoclonal antibodies (Table 1, *see* **Note 6**) at 1–5 μg/ml in blocking buffer. The two primary antibodies must be derived from different species. For example, mouse anti-E1 monoclonal antibody A4 and human anti-E2 monoclonal antibody HCV1 are generally recommended.
8. Incubate the membrane with 10 ml of the primary antibody solution on a rocking platform overnight at 4 °C (*see* **Note 7**).

9. Wash the membrane three times for 5 min each with wash buffer.
10. Dilute the fluorescent conjugated secondary antibodies at 1:10,000 in blocking buffer. The second antibodies need to be derived from the same host species as each of the primary antibodies and labeled with different dyes. For primary antibodies from mouse and human, the IRDye 680RD goat anti-mouse IgG (H + L) and IRDye 800CW goat anti-human IgG (H + L) are generally recommended.
11. Incubate the membrane with 10 ml of the secondary antibody dilution at room temperature for 1 h with gentle shaking. Protect the membrane from light during incubation.
12. Wash the membrane four times for 5 min each with wash buffer and once with 1× PBS to remove residual Tween 20.
13. Keep the membrane in 1× PBS buffer and protected from light until scanned and imaged with the Odyssey imaging system.
14. Analyze and quantify the proteins using the LI-COR Odyssey Image Studio software according to the manufacturer's instructions (*see* **Note 8**). Representative results are shown in Fig. 1 (*see* **Note 9**).
15. The Odyssey imaging system can be replaced by the standard chemiluminescence method although only a single protein can be measured in each blot. There is also exciting development of new fluorescence-based detection systems that can detect multiple colors (proteins) simultaneously on the same blot.

4 Notes

1. The recombinant E1E2 antigen (Chiron E1E2) used in a Phase 1 clinical trial was kindly provided by Novartis Vaccines and Diagnostics [17, 18]. The protein is highly pure and the non-covalent heterodimer appears to be trimeric in size-exclusion chromatography. In the presence of the denaturing detergent SDS, the complex is resolved into two distinct protein bands (E1 and E2) in SDS-PAGE without the need of a reducing agent.
2. Transfection mixture should be pipetted gently and intermittently mixed during the 20-min incubation by gently inverting the tube.
3. Do not scrape or pipet medium directly onto the cell monolayer. Tilt the dish at an angle and place pipette at a corner of the dish to add/remove medium gently. Poly-L-lysine coating promotes cell adhesion to the plastic.

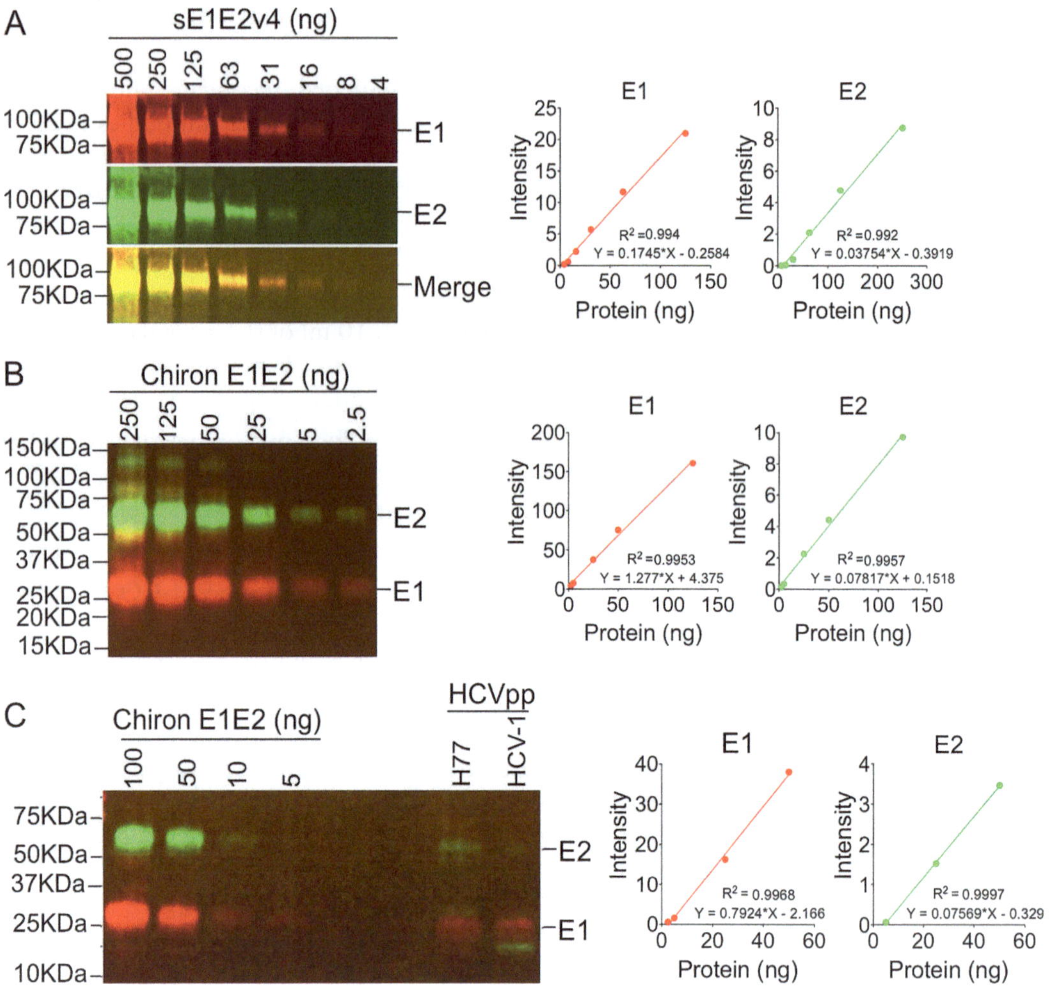

Fig. 1 Immunoblot analysis and quantification of E1 and E2 proteins. (**a**) and (**b**) Determination of the linear range of detection for E1 and E2 in the protein standards. Serial dilutions of sE1E2v4 (**a**) and Chiron E1E2 (**b**) were blotted and detected using murine anti-E1 antibody A4 and human anti-E2 antibody HCV1, followed by IRDye 680RD goat anti-mouse IgG (H + L) (red signal) and IRDye 800CW goat anti-human IgG (H + L) (green signal). Intensity plot of the protein bands in the linear range for each protein was shown in the right panel. (**c**) Quantification of E1 and E2 in HCVpp by using Chiron E1E2 as a standard. The E1 and E2 in HCVpp were quantified as 6.7 ng and 6.3 ng for HCVpp-H77, and 6.6 and 4.6 ng for HCVpp-HCV-1, respectively

4. DMEM with supplements is recommended for resuspension of pellet since they will help visualize the overlay for the overlay procedure. Slowly pipet the resuspended virus pellet solution along the wall of the tubes so it does not mix with sucrose to form a discrete layer on top of tubes.

5. The PVDF membrane should be soaked in methanol for 2–3 min followed by transfer buffer for 1–2 min before setting up the transfer apparatus. Forceps should be used to handle PVDF membrane. The transfer time of 7 min is optimized for transfer of 1 mini gel at a time.

6. Monoclonal antibodies to conserved linear epitopes on E1E2 are useful for immunoblot. However, there is limited choice of antibodies for E1. The murine antibody A4 has been shown to detect genotype 1 E1 but not HCV from other genotypes. The RM1-18 mAb from rhesus macaque binds to a conserved E1 epitope and cross-reacts with multiple HCV genotypes at moderate affinity. There is a need for an anti-E1 antibody that can recognize diverse HCV genotypes. For E2, several cross-genotype binding antibodies are available, including those recognize the conserved E2 antigenic site 412-423 (AS412). These antibodies, e.g., human antibody HCV1, and murine antibodies AP33 and 19B3, generally produce good results in immunoblot. Additional information on antibodies useful for immunoblots can be found in Triyatni et al. [19].
7. Overnight incubation of the blot with primary antibodies helps enhance signal for sample with low protein concentration. A 1-h incubation with primary antibody is usually satisfactory when the protein of interest is abundant in the sample.
8. Ensure that the shape around the band covers entire band and does not overlap or touch bands in the next lane or nonspecific bands close to band of interest. This can be achieved by enlarging the image to make sure the shape is drawn only around band of interest. For HCV glycoproteins, it may require inclusion of a larger area covering the different glycoforms of the proteins. If the Image Studio software isn't available to use in the analysis, other densitometric analysis software like ImageJ or similar can be substituted.
9. Figure 1a, b are protein standards with known concentrations of proteins while Fig. 1c has two HCVpp samples with unknown concentrations. In Fig. 1a, the ectodomains of E1 and E2 in sE1E2v4 are linked at equal ratio. In Fig. 1b, c, calculations were based on the assumption that E1 and E2 in Chiron E1E2 was present at equal stoichiometry in the E1E2 complex.

Acknowledgments

This work was supported by NIH grants AI079031, AI106005, AI123365, and AI123861.

References

1. Dubuisson J, Hsu HH, Cheung RC, Greenberg HB, Russell DG, Rice CM (1994) Formation and intracellular localization of hepatitis C virus envelope glycoprotein complexes expressed by recombinant vaccinia and Sindbis viruses. J Virol 68:6147–6160
2. Duvet S, Cocquerel L, Pillez A, Cacan R, Verbert A, Moradpour D et al (1998) Hepatitis

C virus glycoprotein complex localization in the endoplasmic reticulum involves a determinant for retention and not retrieval. J Biol Chem 273:32088–32095

3. Rouille Y, Helle F, Delgrange D, Roingeard P, Voisset C, Blanchard E et al (2006) Subcellular localization of hepatitis C virus structural proteins in a cell culture system that efficiently replicates the virus. J Virol 80:2832–2841
4. Lindenbach BD, Rice CM (2013) The ins and outs of hepatitis C virus entry and assembly. Nat Rev Microbiol 11:688–700
5. Vieyres G, Thomas X, Descamps V, Duverlie G, Patel AH, Dubuisson J (2010) Characterization of the envelope glycoproteins associated with infectious hepatitis C virus. J Virol 84:10159–10168
6. Clayton RF, Owsianka A, Aitken J, Graham S, Bhella D, Patel AH (2002) Analysis of antigenicity and topology of E2 glycoprotein present on recombinant hepatitis C virus-like particles. J Virol 76:7672–7682
7. Sandrin V, Boulanger P, Penin F, Granier C, Cosset FL, Bartosch B (2005) Assembly of functional hepatitis C virus glycoproteins on infectious pseudoparticles occurs intracellularly and requires concomitant incorporation of E1 and E2 glycoproteins. J Gen Virol 86:3189–3199
8. Piver E, Boyer A, Gaillard J, Bull A, Beaumont E, Roingeard P et al (2017) Ultrastructural organisation of HCV from the bloodstream of infected patients revealed by electron microscopy after specific immunocapture. Gut 66:1487–1495
9. Thomssen R, Bonk S, Propfe C, Heermann KH, Kochel HG, Uy A (1992) Association of hepatitis C virus in human sera with beta-lipoprotein. Med Microbiol Immunol 181:293–300
10. Nielsen SU, Bassendine MF, Burt AD, Martin C, Pumeechockchai W, Toms GL (2006) Association between hepatitis C virus and very-low-density lipoprotein (VLDL)/LDL analyzed in iodixanol density gradients. J Virol 80:2418–2428
11. Ruwona TB, Giang E, Nieusma T, Law M (2014) Fine mapping of murine antibody responses to immunization with a novel soluble form of hepatitis C virus envelope glycoprotein complex. J Virol 88:10459–10471
12. Giang E, Dorner M, Prentoe JC, Dreux M, Evans MJ, Bukh J et al (2012) Human broadly neutralizing antibodies to the envelope glycoprotein complex of hepatitis C virus. Proc Natl Acad Sci U S A 109:6205–6210
13. Keck ZY, Sung VM, Perkins S, Rowe J, Paul S, Liang TJ et al (2004) Human monoclonal antibody to hepatitis C virus E1 glycoprotein that blocks virus attachment and viral infectivity. J Virol 78:7257–7263
14. Broering TJ, Garrity KA, Boatright NK, Sloan SE, Sandor F, Thomas WD Jr et al (2009) Identification and characterization of broadly neutralizing human monoclonal antibodies directed against the E2 envelope glycoprotein of hepatitis C virus. J Virol 83:12473–12482
15. Li Y, Pierce BG, Wang Q, Keck ZY, Fuerst TR, Foung SK et al (2015) Structural basis for penetration of the glycan shield of hepatitis C virus E2 glycoprotein by a broadly neutralizing human antibody. J Biol Chem 290:10117–10125
16. Owsianka A, Clayton RF, Loomis-Price LD, McKeating JA, Patel AH (2001) Functional analysis of hepatitis C virus E2 glycoproteins and virus-like particles reveals structural dissimilarities between different forms of E2. J Gen Virol 82:1877–1883
17. Stamataki Z, Coates S, Evans MJ, Wininger M, Crawford K, Dong C et al (2007) Hepatitis C virus envelope glycoprotein immunization of rodents elicits cross-reactive neutralizing antibodies. Vaccine 25:7773–7784
18. Frey SE, Houghton M, Coates S, Abrignani S, Chien D, Rosa D et al (2010) Safety and immunogenicity of HCV E1E2 vaccine adjuvanted with MF59 administered to healthy adults. Vaccine 28:6367–6373
19. Triyatni M, Vergalla J, Davis AR, Hadlock KG, Foung SK, Liang TJ (2002) Structural features of envelope proteins on hepatitis C virus-like particles as determined by anti-envelope monoclonal antibodies and CD81 binding. Virology 298:124–132

Chapter 21

Overcoming Challenges of Hepatitis C Virus Envelope Glycoprotein Production in Mammalian Cells

Samantha A. Yost, Jillian Whidby, Abdul G. Khan, Yuanyuan Wang, and Joseph Marcotrigiano

Abstract

Posttranslational modifications (PTMs) are often required for proper folding and physiological function of proteins, including the envelope glycoproteins 1 and 2 (E1 and E2) of hepatitis C virus (HCV). Commonly used expression systems such as bacteria, yeast, and baculovirus produce soluble HCV E1 and E2 at very low yields, as the cellular environment and molecular machinery necessary for PTMs may be suboptimal or missing. Here, we describe an expression system for HCV E2 ectodomain (eE2) with 11 N-linked glycans and eight disulfide bonds, which combines lentivirus transduction of mammalian cells and a continuous growth, adherent cell bioreactor. It is environmentally friendly, as well as cost- and time-efficient compared to other methods of recombinant protein expression in mammalian systems with final yields of eE2 approaching 60 mg/L of cell culture supernatant. eE2 produced by this system is amenable to a variety of biophysical studies, including structural determination by X-ray crystallography. Considering the ease of use and flexibility, this method can be applied to express an array of difficult target proteins in a variety of mammalian cell lines.

Key words E2, Lentivirus, Protein expression, Purification, HCV, Hepatitis C virus, Glycoproteins, Posttranslational modification

1 Introduction

Historically, E1 and E2 have been difficult to produce in high enough quantities or sufficient quality necessary for thorough biochemical and biophysical study to facilitate vaccine design. Traditional production of recombinant proteins relies heavily on expression systems developed in bacteria or lower eukaryotes (yeast, baculovirus, etc.); however, proteins with PTMs (e.g., glycosylation, disulfide bonding, phosphorylation, myristoylation) are often misfolded or nonphysiological when produced this way as these systems fail to accurately generate these modifications. Expression of the HCV envelope glycoproteins in human cells should be superior to expression in prokaryotic or lower eukaryotic

Mansun Law (ed.), *Hepatitis C Virus Protocols*, Methods in Molecular Biology, vol. 1911,
https://doi.org/10.1007/978-1-4939-8976-8_21,

systems because the innate molecular biology is in place for efficient translation, folding, PTM, and subcellular targeting.

Production of milligram quantities of recombinant protein using mammalian cell expression systems has relied on two predominant techniques: (1) large-scale transient transfection, or (2) production of a constitutively expressing, stable cell line. Large-scale transient transfection is a relatively quick procedure involving introduction of DNA into 1×10^8–1×10^9 mammalian cells followed by harvesting of expressed protein a few days later [1, 2]. Gene expression using the transient transfection method only lasts for a few days to a week because the expression plasmid is not integrated into the cellular genomic DNA and is therefore lost after several rounds of cell division. Furthermore, large-scale transfection depends on reagents, such as calcium phosphate and polyethylenimine, which are prone to highly variable transfection efficiencies depending on the cell line. Adapting this method to high-throughput expression of multiple proteins requires a large, constant supply of naïve cells (greater than 1×10^{10} cells), DNA, and plastic consumables. The process is costly, as well as time-, labor-, and space-intensive.

Producing a cell line that stably expresses a protein of interest requires introducing the expression plasmid into cells followed by selection for cells with stable integration using a selection marker (e.g., a drug resistance gene). There are advantages to using this method. First, the introduction of the expression plasmid into cells can be done on a smaller scale (1×10^5–1×10^6 cells) using efficient transfection reagents (e.g., Lipofectamine or FuGENE), making it better suited for parallel, simultaneous processing of multiple expression constructs. Second, highly expressing stable cell lines are a sustainable resource, as additional protein can be produced without backtracking to the beginning of the transfection process (unlike transient transfection). The main limitation of stable cell line expression is the time required to isolate a highly expressing clone during the selection process, which can take several months to achieve [3, 4].

The method described here combines the speed and efficiency of lentivirus transduction with adherent cell bioreactors to enable large-scale production of protein in mammalian cells (Table 1). This approach significantly reduces plastic waste and cost, while improving the quantity and quality of challenging PTM-containing proteins that are important for human health. Viral and human proteins that have failed to express in any other system have yielded up to 60 mg of protein per liter of cell culture supernatant using this method. Importantly, high quality HCV E2 ectodomain (eE2) produced by this technique has been successfully used for structure determination, yielding new insights into HCV entry [5].

Lentiviral vectors have become an established tool for reliable and safe gene transduction in a wide variety of mammalian cell

Table 1
Outline of workflow for lentivirus infection of mammalian cells for expression of eE2

Week 1	• Construct preparation • Small scale expression testing
Week 2	• Lentivirus production • Infection
Week 3	• Monitor GFP expression • Expansion of transduced cells
Week 4	• Seed bioreactor • Monitor bioreactor stability • Initial protein production harvests
Week 5 onward	• Harvest supernatant containing expressed protein every other day • Continuous protein production for up to 3 months

types, including nondividing cells [6]. Immediately following infection, the viral RNA genome is reverse-transcribed into DNA, which is transported to the nucleus and integrated into the genome by a viral-encoded integrase enzyme. By separating the viral components across multiple plasmids, recombination of genetic material is prevented and an intact viral genome cannot be produced. This results in a "pseudo" lentivirus capable of only a single round of infection.

Lentivirus transduction results in a substantial increase in protein yield, owing to the high copy number of expression cassettes introduced to the cell and integration at sites of active transcription. The number of copies of the expression cassette introduced into a cell is correlated to the number of virus particles entering the cell, or multiplicity of infection (MOI). The protocol below is designed to ensure all cells are infected at extremely high MOIs (approximately 100–1000), eliminating the need for selection markers to identify clones that contain the expression cassette. By eliminating the selection step, stable clone production is shortened considerably from 2–3 months to less than 2 weeks. In short, the lentiviral transduction method is more rapid and efficient than previous approaches because it does not require additional selection reagents, eliminates the need to isolate clonal, stably expressing cell lines, and increases protein yield.

For optimal cell growth and density, the BelloCell-500 bioreactor (Cesco Bioengineering, Taiwan) is used, as it is simple to operate, cost effective, and provides ideal growth conditions via a programmable stage that controls feeding and aeration times. Relative to traditional roller bottles and multilayer flasks, the BelloCell system consumes significantly less disposable plastic waste, resulting in a cost savings of approximately 50%. The three-dimensional

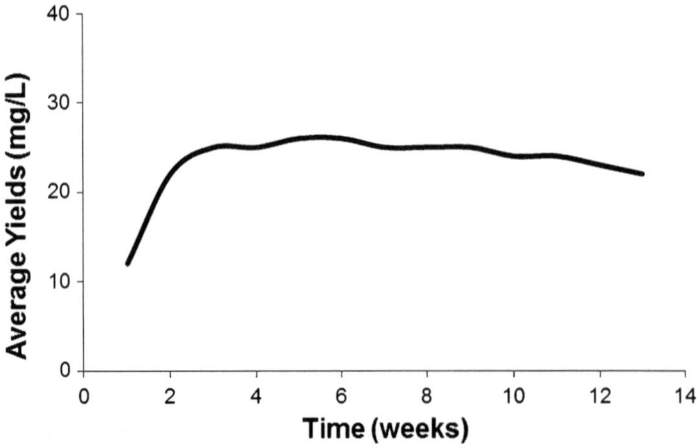

Fig. 1 eE2 Yield Over Time. Yields of an expressed eE2 construct remain stable at approximately 22 mg/L of cell supernatant from Week 2 through Week 13

attachment matrix in the bioreactor provides an ideal growth environment by evenly distributing temperature and facilitating gas exchange, allowing cells to grow to high density [7]. The programmable stage can accommodate up to four bioreactors at a time and fits inside a standard cell culture incubator. In our experience, HEK293T cells can be maintained for at least 3 months in this bioreactor with continuous media harvesting and replacement, if optimal sterile technique is used in combination with antibiotic/antimycotic supplementation (Fig. 1).

We have applied this method to a highly glycosylated protein containing numerous disulfide linkages: HCV eE2. Figure 2 illustrates an approximate 100-fold increase in expression of eE2 for lentiviral transduced HEK293T cells when compared to a transfected and stably selected cell clone. eE2 is secreted into the cell culture supernatant via a prolactin signal sequence as described in Subheading 3 below. Protein-A was chosen as a purification tag because it binds to a recombinant IgG column (GE Healthcare) with high affinity, yielding highly purified target protein. The protein-A tag lacks cysteine residues as to avoid potentially disrupting the disulfide bonding pattern of eE2 during folding. Furthermore, a 3X-FLAG tag (Sigma) can be fused to the protein-A tag in tandem, allowing for easy detection by Western blotting.

Originally described by Trono et al., two components of the lentiviral system described here are currently available through Addgene (Subheading 2.2) [6, 8, 9]. pMD2.G encodes vesicular stomatitis virus (VSV) glycoproteins for incorporation into the viral envelope and versatility of cell tropism and psPAX2 contains the standard HIV Gag/Pol cassette. The third plasmid, pJG, was reengineered to further improve virus production (Fig. 3). pJG includes a cytomegalovirus (CMV) promoter to drive expression, a Rev

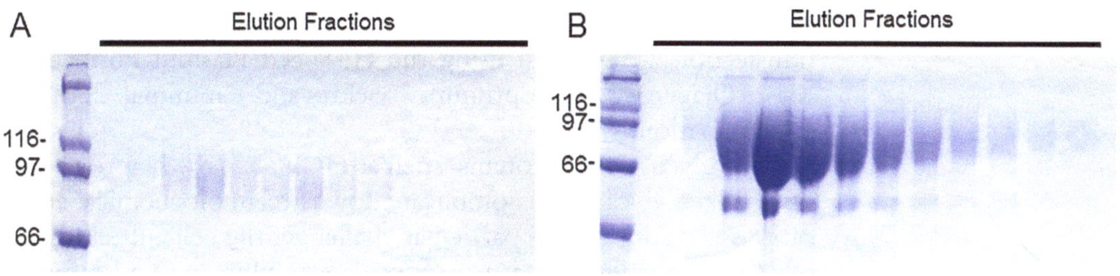

Fig. 2 Method Comparison of eE2 Yields from (**a**) a transfected and stably selected HEK293T cell line and (**b**) a lentiviral transduced HEK293T cell line after purification over IgG column (Subheading 3.6)

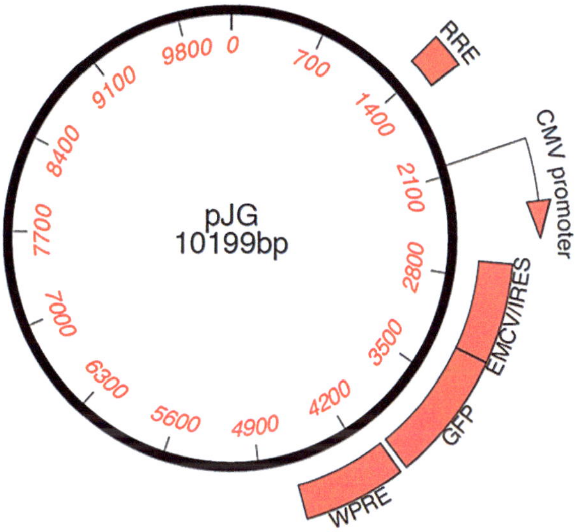

Fig. 3 Schematic of the pJG Plasmid, engineered to include a CMV promoter to enhance expression of the protein of interest, a Rev. Response Element (RRE), a Woodchuck Hepatitis promoter response element (WPRE), and a green fluorescent protein (GFP) gene controlled by an EMCV/IRES. The gene of interest (including chosen tags and protease cleavage sties) is subcloned into the pJG vector immediately following the CMV promoter

Response Element (RRE), a Woodchuck Hepatitis promoter response element (WPRE), and a green fluorescent protein (GFP) gene driven by an encephalomyocarditis virus internal ribosome entry site (EMCV/IRES) [6, 10]. GFP provides a reliable, visible indication of infection/transduction efficiency correlating to the expression of the protein of interest; it can also be used to identify and isolate highly expressing cells using fluorescence-activated cell sorting (FACS) for especially low-yield proteins [11]. The InFusion-HD (Clontech) cloning kit is recommended for use with pJG, since the large vector size reduces the efficiency of typical cloning reagents. InFusion-HD cloning is executed according to the manufacturer's instructions and the reaction product is transformed into HB101 cells. All three DNA vectors that are used for lentiviral transfection are purified using the Hi-Speed Plasmid Purification Kit (Qiagen) for optimum yield and minimal ethanol contamination.

For glycosylated proteins, such as HCV eE2, the heterogeneity of N-linked glycans can complicate downstream biochemical applications. To address this particular challenge, the cell culture media can be supplemented with mannosidase inhibitors (e.g., kifunensine) prevent the addition of complex and hybrid glycosylation during folding and processing in the secretory pathway. The resulting protein is more homogeneous and amenable to downstream applications [12, 13]. Alternatively, the recombinant lentivirus can be used to infect glycosylation-deficient cell lines, such as HEK293 N-acetylglucosaminyltransferase deficient (GnTI-) or CHO Lec cell lines, circumventing the need for mannosidase inhibitors [14]. Both of these techniques enable deglycosylation with Endoglycosidase H/F (EndoH, EndoF), leaving one N-acetylglucosamine attached per asparagine to help maintain protein solubility. In our laboratory, we routinely use HEK293 GnTI- cells to produce protein of high homogeneity that is easy to deglycosylate with EndoH treatment.

2 Materials

2.1 Cloning and Small Scale Testing

1. pJG plasmid.
2. InFusion-HD (Clontech).
3. Desired restriction enzymes.
4. *E. coli* strain capable of replicating large plasmids (e.g., HB101).
5. Hi-Speed Plasmid Purification Kit (Qiagen).
6. HEK29T cells (alternatively HEK293T GnTI1-, CHO Lec, etc. as described in Subheading 1).
7. DMEM supplemented with 10% fetal bovine serum (FBS) (Life Technologies).
8. FuGENE HD Transfection Reagent (Promega).
9. Inverted microscope with GFP visualization capabilities.
10. Cell culture incubator at 37 °C with 5% CO_2.

2.2 Lentivirus Production

1. HEK293T cells.
2. DMEM supplemented with 10% FBS (Life Technologies).
3. DMEM supplemented with 1% antibiotic/antimycotic (A/A) and 10% FBS (Life Technologies).

4. DMEM supplemented with 1% A/A and 20% FBS (Life Technologies).
5. Cell culture incubator at 37 °C with 5% CO_2.
6. Plasmids per transfection: 90 μg pJG vector with gene of interest, 60 μg psPAX2 (Addgene plasmid #12260), 30 μg pMD2.G (Addgene plasmid #12259).
7. 2 M $CaCl_2$.
8. 2× HEPES buffered saline (HEBS).
9. Microscope with GFP visualization capabilities.
10. Thinwall, ultra-clear ultracentrifuge Tubes (Beckman).
11. Ultracentrifuge fitted with SW28 rotor (Model LB-70M, Beckman).
12. Polybrene.
13. Vesphene.

2.3 Infection

1. HEK293T cells.
2. DMEM supplemented with 10% FBS and 1% A/A.
3. Harvested lentivirus suspension from previous step.
4. Cell culture incubator at 37 °C with 5% CO_2.

2.4 Bioreactor Seeding

1. HEK293T cells transduced with gene of interest from previous step.
2. DMEM supplemented with 10% FBS and 1% A/A.
3. Cell culture incubator at 37 °C with 5% CO_2.
4. BelloCell 500 bottle (CESCO Bioengineering).
5. BelloStage 3000 (CESCO Bioengineering).

2.5 Bioreactor Harvest

1. BelloCell 500 bottle seeded with transduced cells from previous step.
2. DMEM supplemented with 10% FBS and 1% A/A.
3. Glucose meter and test strips.
4. 500 mL Steritop-GP 0.22 μm fast flow filters (Millipore).

2.6 Purification

1. 5 mL HiTrap IgG Sepharose FF column (GE Healthcare).
2. Harvested cell culture supernatant containing protein of interest from previous step.
3. Equilibration buffer: 20 mM sodium phosphate pH 7.0.
4. 1 M NaN_3.
5. 1 M Tris pH 9.0.
6. Elution buffer: 0.1 M sodium citrate pH 3.0 and 20 mM KCl.

7. IgG column storage buffer: 50 mM Tris pH 7.6, 150 mM NaCl, 0.05% Tween, and 10–20% ethanol.
8. Dialysis buffer: 20 mM HEPES pH 7.5, 250 mM NaCl, and 5% glycerol.
9. Rhinovirus 3C protease or protease of choice for affinity tag removal.

3 Methods

3.1 Cloning and Small Scale Testing

1. Subclone the gene of interest into pJG plasmid downstream of the CMV promoter (*see* **Note 1**).
2. Amplify and purify pJG plasmid with insert in a suitable *E. coli* strain (e.g., HB101).
3. Confirm by restriction enzyme digestion and sequencing.
4. Seed 4.3×10^5 HEK293T cells in a 6-well plate and transfect the following day with 4 μg pJG using FuGENE HD transfection reagent as recommended by the manufacturer's protocol. The following day, check for GFP expression via microscopy to confirm transfection efficiency. Take a sample of the supernatant for western blot analysis.

3.2 Lentivirus Production

1. One day prior to the planned transfection, seed a single T-225 monolayer flask with 6.2×10^6 HEK293T cells in 30 mL of DMEM supplemented with 10% FBS. Cells are grown in a 37 °C incubator with 5% CO_2.
2. Combine the following in a 50 mL Falcon tube: 90 μg pJG containing the gene of interest, 60 μg psPAX2, 30 μg pMD2. G, 450 μL of 2 M $CaCl_2$, q.s. to 4.5 mL of ddH_2O. Mix gently and add 4.5 mL of 2X HEBS by bubbling with a serological pipette for 10 s. Incubate at room temperature for 2 min. Incubation time should be optimized for each batch of $CaCl_2$ and 2× HEBS.
3. Add mixture directly to the culture media in the T-225 flask seeded the previous day. Aspirate the media 6–8 h posttransfection and replace with 40 mL DMEM supplemented with 1% A/A and 10% FBS.
4. Two days posttransfection, check for GFP expression via microscopy to confirm transfection efficiency. Seed 2×10^4 fresh cells into a single well of a 96-well plate in a final volume of 100 μL (1×10^5 cells/mL).
5. Three days posttransfection, harvest the cell culture supernatant containing the produced lentivirus and centrifuge for 30 min at 4000 × *g* at 4 °C to pellet cellular debris. Transfer ~37 mL of clarified supernatant to a Beckman ultracentrifuge

tube. Pellet the virus at 28,000 rpm (150,000 × *g*) in an SW28 rotor for 1 h and 30 min at 4 °C. Decant supernatant into a waste container filled with disinfectant, e.g., 1% Vesphene solution). Allow pellet to air-dry for 5 min, then resuspend in 120 μL of DMEM containing 20% FBS, 1% A/A, and 8 μg/mL Polybrene by pipetting gently to homogenize the virus pellet (*see* **Note 2**).

3.3 Infection (Transduction)

1. Aspirate media from the prepared 96-well plate (Subheading 3.2, **step 4**). Add 60 μL of virus suspension and incubate overnight at 37 °C with 5% CO_2. The remaining 60 μL of virus suspension can be frozen and stored at −80 °C.
2. The following morning, add 100 μL of fresh DMEM supplemented with 10% FBS and 1% A/A to the infected well.
3. Three days posttransduction, remove the media and replace with 100–200 μL of fresh DMEM supplemented with 10% FBS and 1% A/A. At this time, check for GFP expression via microscopy to confirm transduction efficiency.
4. Cell expansion can begin as soon as confluence is reached. We recommend step-wise expansion from 96-well, to 12-well, to 6-well, to 75 cm^2 flask, and finally to a 175 cm^2 flask over the course of about 1 week. When the 175 cm^2 flask has reached confluence, the transduced cells can be expanded into four 175 cm^2 flasks, three of which can be used for bioreactor seeding and one of which can be used for liquid nitrogen storage.

3.4 Bioreactor Seeding

1. Add 170 mL of fresh DMEM supplemented with 10% FBS and 1% A/A to a new bioreactor bottle and allow 30 min of pre-incubation on the BelloStage 3000 to warm the bottle and equilibrate the matrix. Program the stage with an oscillation rate of 2.0 mm/s up and down, a top hold time of 20 s, and a bottom hold time of 0 s for seeding the bioreactor. Seeding protocol is followed according to the manufacturer's instructions.
2. Seed a single BelloCell 500 bioreactor with ~8–10 × 10^7 HEK293T cells, the equivalent of three confluent 175 cm^2 monolayer flasks (Subheading 3.3). Collect the cells in a final volume of 30 mL, and slowly eject from a sterile pipette onto the matrix drop-by-drop ensuring even distribution of cells over the entire matrix surface. Place the seeded bottle on the BelloStage 3000 and run the seeding program for 3–4 h.
3. Pause the BelloStage 3000 and reduce the oscillation rate to 1.0 mm/s, with a top hold time of 10 s and a bottom hold time of 1 min for cell growth.

3.5 Bioreactor Harvest

1. We recommend using a standard glucose meter and test strips to monitor the glucose concentration of the media each day after seeding for at least the first week. Daily monitoring will help ensure that the bioreactor reaches a stable cell population and will prevent sudden culture death.
2. Approximately 3 days after seeding, or when the glucose reading for the media falls below 1 g/L, aspirate the media from the bottom chamber of the BelloCell 500 bioreactor and replace with 500 mL of fresh DMEM supplemented with 10% FBS and 1% A/A.
3. For regular maintenance of the cells, harvest the media and replace every 2–3 days.
4. Centrifuge harvested cell culture supernatant at 5000 rpm (~7200 × *g*) for 20 min at 4 °C to pellet cell debris. We recommend using large volume conical tubes by Corning with appropriate centrifuge rotor adapters.
5. Filter cell culture supernatant through 0.2 μm PES filters (we r-ecommend Millipore Steritop-GP fast flow filters) and store at 4 °C until ready to purify protein.

3.6 Purification

1. Equilibrate a 5 mL IgG column (GE Healthcare) with at least 20 mL of 20 mM sodium phosphate pH 7.0.
2. During the equilibration, transfer up to 2 L of harvested supernatant to a beaker. Add 1 mL of 1 M NaN_3 to help prevent contamination of the supernatant during the purification process and to extend the life of the column.
3. After the column is equilibrated, the cell supernatant may be applied to the column at a flow rate of 0.5–1 mL/min. Collect flow-through for later biochemical analysis.
4. After loading the cell supernatant, wash the column with at least 10 column volumes of 20 mM sodium phosphate pH 7.0.
5. While the column is being washed, prepare collection tubes, each containing 400 μL of 1 M Tris pH 9.0. If you are using a fraction collector, prepare the fraction collector to dispense 2 mL per tube.
6. Elute the protein with 0.1 M sodium citrate pH 3.0 and 20 mM KCl at 2 mL/min. If collecting manually, we recommend collecting 2 mL fractions. If collecting larger fractions, you will need to increase the volume of 1 M Tris pH 9.0 in each collection tube to ensure full neutralization of the eluted fraction.
7. Wash the column with at least 6 column volumes (30 mL) of 20 mM sodium phosphate pH 7.0 to immediately restore the pH of the resin.

8. Analyze fractions via SDS-PAGE, including samples of your preloaded cell supernatant, flow through, and washes (*see* **Note 3**). Pool all fractions containing your protein of interest to prepare for dialysis.
9. Store the IgG column in 50 mM Tris pH 7.6 and 150 mM NaCl supplemented with 0.05% Tween and 10–20% ethanol.
10. Prepare a beaker for dialysis containing chilled (4 °C) 20 mM HEPES pH 7.5, 250 mM NaCl, and 5% glycerol buffer. 100 mL of dialysis buffer is used per milliliter of protein elution (e.g., for 5 mL of pooled elution, prepare at least 500 mL of dialysis buffer).
11. Prepare the dialysis tubing according to the manufacturer's recommendations. For SnakeSkin (ThermoFisher) dialysis tubing, rinse the appropriate length of tubing for your elution volume in dialysis buffer.
12. Add the appropriate volume of enzyme to cleave the protein-A tag according to the manufacturer's instructions (*see* **Note 4**).
13. Dialyze protein overnight at 4 °C.
14. On the following day, confirm complete cleavage using SDS-PAGE and proceed to affinity or ion exchange chromatography to further purify the target protein (*see* **Note 5**).
15. *Alternate approach.* Cleavage of the tag can be performed while the protein is bound to the IgG column (after Subheading 3.6, **step 4**) (*see* **Note 6**). This approach avoids exposing the target protein to low pH and can produce purified target protein with fewer contaminants.

4 Notes

1. The protein A and other desired biochemical tags, your chosen protease cleavage site, and any signal sequence that may be necessary for proper folding and processing of your particular protein, are included in the cloning step.
2. Caution: Cell culture supernatant containing lentivirus should be handled with care in an approved cell culture facility, following safety guidelines. All materials (i.e., tubes, flasks, tips) which come in contact with lentivirus should be properly disinfected before disposal. Please check with your institution's safety office before handling lentivirus.
3. The FBS component of the DMEM may obstruct visualization by nonspecific staining techniques depending on the size of your eluted protein.

4. We use a rhinovirus 3C protease cleavage site and corresponding enzyme (produced by our lab) at a 1:50 (w/w) ratio.
5. We prefer to load the sample onto an IgG column a second time to remove any uncleaved protein and free protein-A tag before proceeding to any further downstream purification steps.
6. After extensive washing with 20 mM phosphate pH 7.0 (after Subheading 3.6, **step 4**), equilibrate the column with 20 mM HEPES pH 7.5, 250 mM NaCl, and 5% glycerol. In a separate tube, add the enzyme of your choice in ~4.5 mL of the previously noted buffer and mix gently. Using a syringe and an appropriate adaptor, gently inject the enzyme mix onto the column with a steady flow rate. For rhinovirus 3C protease, the column must be incubated at 4 °C overnight. On the following day, equilibrate the purification system with 20 mM HEPES pH 7.5, 250 mM NaCl, and 5% glycerol before connecting the column. Connect the column and begin collecting the flow-through immediately as cleaved protein will elute as soon as the buffer flows through the column.

References

1. Jordan M, Kohne C, Wurm FM (1998) Calcium-phosphate mediated DNA transfer into HEK-293 cells in suspension: control of physiological parameters allows transfection in stirred media. Cytotechnology 26:29–47
2. Schlaeger EJ, Christensen K (1999) Transient gene expression in mammalian cells grown in serum-free suspension culture. Cytotechnology 30:71–83
3. Wurm FM (2004) Production of recombinant protein therapeutics in cultivated mammalian cells. Nat Biotechnol 22:1393–1398
4. Chaudhary S, Pak JE, Gruswitz F, Sharma V, Stroud RM (2012) Overexpressing human membrane proteins in stably transfected and clonal human embryonic kidney 293S cells. Nat Protoc 7:453–466
5. Khan AG et al (2014) Structure of the core ectodomain of the hepatitis C virus envelope glycoprotein 2. Nature 509:381–384
6. Naldini L et al (1996) In vivo gene delivery and stable transduction of nondividing cells by a lentiviral vector. Science 5259:263–267
7. Wang IK et al (2006) A novel control scheme for inducing angiostatin-human IgG fusion protein production using recombinant CHO cells in a oscillating bioreactor. J Biotechnol 121:418–428
8. Zufferey R et al (1998) Self-inactivating lentivirus vector for safe and efficient in vivo gene delivery. J Virol 72:9873–9880
9. Dull T et al (1998) A third-generation lentivirus vector with a conditional packaging system. J Virol 72:8463–8471
10. Zufferey R, Donello JE, Trono D, Hope TJ (1999) Woodchuck hepatitis virus posttranscriptional regulatory element enhances expression of transgenes delivered by retroviral vectors. J Virol 73:2886–2892
11. Mancia F et al (2004) Optimization of protein production in mammalian cells with a coexpressed fluorescent marker. Structure 12:1355–1360
12. Elbein AD, Tropea JE, Mitchell M, Kaushal GP (1990) Kifunensine, a potent inhibitor of the glycoprotein processing mannosidase I. J Biol Chem 265:15599–15605
13. Chang VT et al (2007) Glycoprotein structural genomics: solving the glycosylation problem. Structure 15:267–273
14. North SJ et al (2010) Glycomics profiling of Chinese hamster ovary cell glycosylation mutants reveals N-glycans of a novel size and complexity. J Biol Chem 285:5759–5775

Part IV

Methods for Studying Immune Responses

Chapter 22

Monitoring of Interferon Response Triggered by Cells Infected by Hepatitis C Virus or Other Viruses Upon Cell–Cell Contact

Séverin Coléon, Sonia Assil, and Marlène Dreux

Abstract

Plasmacytoid dendritic cells (pDCs) constitute a unique DC subset specialized in rapid and massive secretion of cytokines, including type I interferon (i.e., IFNα and IFNβ), known to be pivotal for both innate immunity and the onset of adaptive response. The production of type I IFNs by pDCs is primarily induced by the recognition of viral nucleic acids through Toll-like receptor (TLR)-7 and -9 sensors located in the endolysosomal compartment. Importantly, in the context of hepatitis C virus (HCV) infection, pDC type I IFN response is triggered by the sensing of infected cells via physical cell–cell contact. Such a feature is also observed for many genetically distant viruses, including notably viruses of the *Retroviridae, Arenaviridae, Flaviviridae, Picornaviridaea, Togaviridae* families and observed for various infected cell types. Here, we described a set of experimental methods for the ex vivo studies of the regulation of pDC activation upon physical cell–cell contact with virally infected cells.

Key words Innate immunity, Hepatitis C virus (HCV), Plasmacytoid dendritic cells (pDCs), Interferon (IFN), Cell–cell contact, Inflammation, Toll-like receptor (TLR), Imaging flow cytometry, Confocal microscopy analysis, Coculture

1 Introduction

Dendritic cell (DC) subsets have varied functions including antigen presentation and cytokine production, and hereby they are essential players of innate and adaptive immunity. Among DCs, plasmacytoid DC (pDC) is thought to be a DC subset specialized for the production of type I interferons (IFN-I) (reviewed in [1, 2]). The IFN-I are pleiotropic cytokines with abilities to enhance both innate and adaptive immunity. Their antiviral activity involves notably the induction of interferon stimulated genes (ISGs), which shut-down viral replication. The pDC IFN-I response is mainly triggered by the endolysosomal sensors Toll-like receptors (TLR)-7 and -9 that recognize nucleic acids (i.e., RNA and DNA, respectively). These immunostimulatory nucleic acids could be derived

Mansun Law (ed.), *Hepatitis C Virus Protocols*, Methods in Molecular Biology, vol. 1911,
https://doi.org/10.1007/978-1-4939-8976-8_22, © Springer Science+Business Media, LLC, part of Springer Nature 2019

from bacterial or viral genomes in the context of infections, or from cellular DNA or RNA when presented in immune complexes, formed by apoptotic or necrotic cellular materials along with autoantibodies (reviewed in [1, 3, 4]). Such self DNA- or RNA-containing immune complexes can act as endogenous inducers of IFN-I response contributing to the development of autoimmune diseases, such as systemic lupus erythematosus (SLE) (reviewed in [3, 4]).

The pDCs are able to rapidly produce large amount of IFN-I, in part, likely because they constitutively express high level of the transcriptional activator interferon regulatory factor 7 (IRF7) [1, 5]. In the context of viral infections, such a rapid activation of antiviral response might explain why they are refractory to infection by most viruses and consistently, pDC activation does not require their productive infection by viruses [1, 6–8]. The pDCs also produce type III IFNs, yet the contribution of this part of the pDC response in the outcome of infection is still elusive. As exemplified for different viruses, pDCs are likely an important component of the host antiviral response at the acute phase of infection [9–13]. The pDCs, via an early IFN-I production, are thought to modulate various innate and adaptive responses, yet the regulatory mechanisms and outcomes appear to differ for distinct viruses. For instance, specific depletion of pDCs in a transgenic mouse model, in which the diphtheria toxin receptor (DTR) is expressed under the control of the promoter of a specific pDC gene, (i.e., *BDCA-2*) revealed that pDC response limits the expansion of virus-specific NK cells in the context of murine cytomegalovirus (MCMV) infection, likely in keeping with the control via IFN production of viral load [11]. In contrast, this study, using the same mouse model, also uncovered that pDCs primarily influence the accrual of CD8+ T cells in response to vesicular stomatitis virus (VSV) infection [11]. Notably, pDCs, via the NFκβ-mediated signaling, are a source of proinflammatory cytokines and chemokines, that promote immune cell recruitment to infection sites and the adaptive immunity [14]. Importantly, a study using a mouse model in which IRF7 signaling is restricted to pDCs demonstrated that this pDC-restricted response is sufficient to control both dengue and Chikungunya viruses and was sufficient to protect mice from lethal Chikungunya infection [15].

Interestingly, recent works uncovered that pDCs preferentially respond to viral RNA transferred by noncanonical and/or noninfectious vesicles rather than through infection by conventional viral particles [8]. In the context of Hepatitis C virus (HCV), infected cells produce exosome-like vesicles, which contain HCV genome (referred to as HCV EXOs) [16]. The transfer of viral genomes via HCV EXOs activates IFN-I response by the pDCs. This transmission does not required the viral structural proteins (i.e., the surface glycoproteins E1 and E2, as well as the core protein), thus demonstrating that HCV-EXOs are structurally distinct from canonical HCV particles [16]. HCV EXOs also likely contribute to the spread of viral infection to neighboring naïve hepatocytes [17–20]. Interestingly, recent studies highlighted additional regulatory functions exerted by HCV-EXOs on the host responses, including monocyte activation and differentiation into macrophage [21] and the activation of profibrotic markers by hepatic stellate cells [22]. These observations imply that the modulation of cellular functions by HCV EXOs depends on the target cell type. The transfer of immunostimulatory RNAs to pDCs via exosome-like vesicles has been demonstrated for other genetically distant viruses, including Hepatitis A virus (HAV) and Lymphocytic choriomeningitis virus (LCMV) [23, 24].

Importantly, several recent in vitro studies uncovered that pDCs sense viral infections when in physical contact with infected cells [8, 15, 16, 23–30]. Additionally, studies suggested that such activation of pDCs by cell–cell contact with infected cells involves clathrin-mediated endocytosis of secreted PAMPs, in accordance with the involvement of a vesicular-mediated cell-to-cell transfer of the viral RNA to the pDCs (via exosomes or immature viral particles, depending on the viruses) [16, 23, 24, 26, 30]. The first study on the structures and components present at the contacts between pDCs and virally infected cells highlighted a local accumulation of the actin network [26], a hallmark of previously described synapses. Systemic and massive productions of IFNs and inflammatory cytokines can be detrimental to the host as they correlate notably with tissue damages [31, 32]. Therefore, one might speculate that the requirement for direct cell–cell contact to trigger a robust pDC response could have evolved in favor of the host fitness to locally respond at the infected sites, and hereby to thwart the otherwise harmful systemic IFN and inflammatory responses.

The requirement for cell–cell contact is a newly assigned primarily feature of pDC functionality. Therefore, defining the mechanism underlying the establishment of cell–cell contact and its impact on the antiviral state in vitro and in vivo is a key question in the field of innate immunity. Here, we described the methods for ex vivo studies of pDC response to virally infected cells, with a focus on the experimental methods to determine how pDC activation is regulated by cell–cell contact with infected cells and the current approaches developed to define the underlining molecular events.

2 Materials

The indicated quantity of buffers and reagents are relative to the procedure of pDC isolation from a 450 mL blood unit (*see* **Note 1**). The same experimental process can apply to pDC isolation from a cytapheresis unit.

2.1 Isolation of pDCs from Peripheral Blood Mononuclear Cells (PBMCs)

1. Ficoll Hypaque (GE Healthcare Life Sciences) stored away from light at room temperature.
2. HANK's 2 mM EDTA buffer: prepared with 0.5 M pH 8 EDTA solution (Life Technologies) diluted in HANK's Balanced Salt Solution (HBSS, Life Technologies) stored at room temperature.
3. Separation buffer: Phosphate-buffered saline (PBS, Life Technologies), supplemented with 0.5% bovine serum albumin (BSA, Sigma-Aldrich), 2 mM EDTA. Filtration with 0.22 μm filter unit and stored at 4 °C.
4. Cell culture medium for isolated pDCs: RPMI 1640 medium (reference: 31870-074, from Life Technologies), supplemented with 10% fetal calf serum (FCS), 10 mM hepes, 100 μg/mL streptomycin, 100 units/mL penicillin, 1 mM sodium pyruvate, 2 mM L-glutamine, and nonessential amino acids (Life Technologies), stored at 4 °C.
5. pDCs are positively selected from PBMCs using BDCA-4/Neuropilin-1-magnetic microbead kit (reference: 130-090-532, from MACS Miltenyi Biotec) and LS Columns (reference: 130-042-401, from MACS Miltenyi Biotec) and magnetic stand (MACS Miltenyi Biotec).
6. The antibodies against pDC markers used to assess the purity of pDC isolation by immunostaining followed by FACS analysis were: PE-conjugated anti-CD123 and APC-conjugated anti-BDCA-2 (MACS Miltenyi Biotec).

2.2 Biological Reagents: Infected Cells

1. Huh-7, the derivative Huh-7.5.1 and a subclone of Huh-7.5.1 cells (Huh-7.5.1c2) [16, 33, 34] are maintained in Dulbecco's modified Eagle medium (DMEM, high glucose, GlutaMAX, reference 61965059, from Life Technologies), supplemented with 10% FCS, 10 mM hepes, 100 μg/mL streptomycin, 100 units/mL penicillin, 2 mM L-glutamine, and nonessential amino acids (Life Technologies) at 5% CO_2/37 °C. Huh-7.5.1c2 cells that harbor the JFH-1 HCV subgenomic replicon (SGR) [35] were maintained in the same medium with neomycin selection by adding G418 (at 500 μg/mL final concentration, Sigma-Aldrich).
2. Cells were infected by the JFH-1 infectious molecular clone of HCV [36–38]. Viral stocks of cell-culture adapted JFH-1 virus (D183) [39] were prepared by infection of Huh-7.5.1c2 cells at a multiplicity of infection (moi) of 0.01, as described previously [16, 40].

2.3 Imaging of the Cell–Cell Contacts by Immunofluorescent Staining and/or Fluorescent In Situ Hybridization (FISH) Analyzed by Confocal Microscopy

1. 96-well optical-bottom plates (Thermo Fisher Scientific).
2. Polylysine solution: dissolved polylysine (reference: P6282, from Sigma-Aldrich) at final concentration 8 μg/mL in H_2O.
3. Vibrant CM-DiI cell-labeling solution (reference V22888, from Thermo Fisher Scientific) and CellTrace Violet Cell Proliferation Kit (reference C34571, from Thermo Fisher Scientific) used for the staining of pDC prior to coculture with infected cells, with fluorescence absorption/emission at approximately 553/570 nm and 405/450, respectively. Store at −20 °C.
4. Fixation with 4% paraformaldehyde (PFA) solution (Sigma-Aldrich).
5. Reagents for immunofluorescent staining include Hoechst solution (Life Technologies); Alexa-conjugated secondary antibodies (Life Technologies); CF488A-conjugated phalloidin (Biotium); mouse APC conjugated anti-IFNα (MACS Miltenyi Biotec); mouse IgG2a monoclonal Anti-β-Actin (Sigma); mouse anti-NS5A (HCV protein)—clone 9E10 (kindly provided by Dr. C Rice, Rockefeller University).
6. Detection of HCV RNA by FISH performed using reagents from Panomics/Affymetrix [16].
7. Confocal microscopy analysis performed using Zeiss LSM 710 laser scanning confocal microscope and analyzed with Image J (http://rsb.info.nih.gov/ij) and IMARIS (Bitplane Inc.) software packages.

2.4 Imaging Combined with Flow Cytometry Analysis of pDC/HCV Cell Conjugates

1. Micro-plate designed for cell harvesting by temperature reduction (Nunc UpCell 96F Microwell Plate from Thermo Scientific).
2. Analysis by a technology combining imaging and flow cytometry analysis (Image Stream X technology, Amnis) using IDEAS software.

2.5 Enzyme-Linked Immunosorbent Assay (ELISA)

The cytokine productions by cocultured pDCs are quantified using commercially available ELISA kits specific for IFNα and IFNβ (PBL Interferon Source), TNFα and IL-6 (Affymetrix), following the manufacturer's instructions.

3 Methods

The workflow of the experimental procedure to study the activation of pDCs by infected cells includes the isolation of pDCs from PBMCs and the infection of cells by HCV, followed by the setup of coculture. The study of the response of pDCs to infected cells

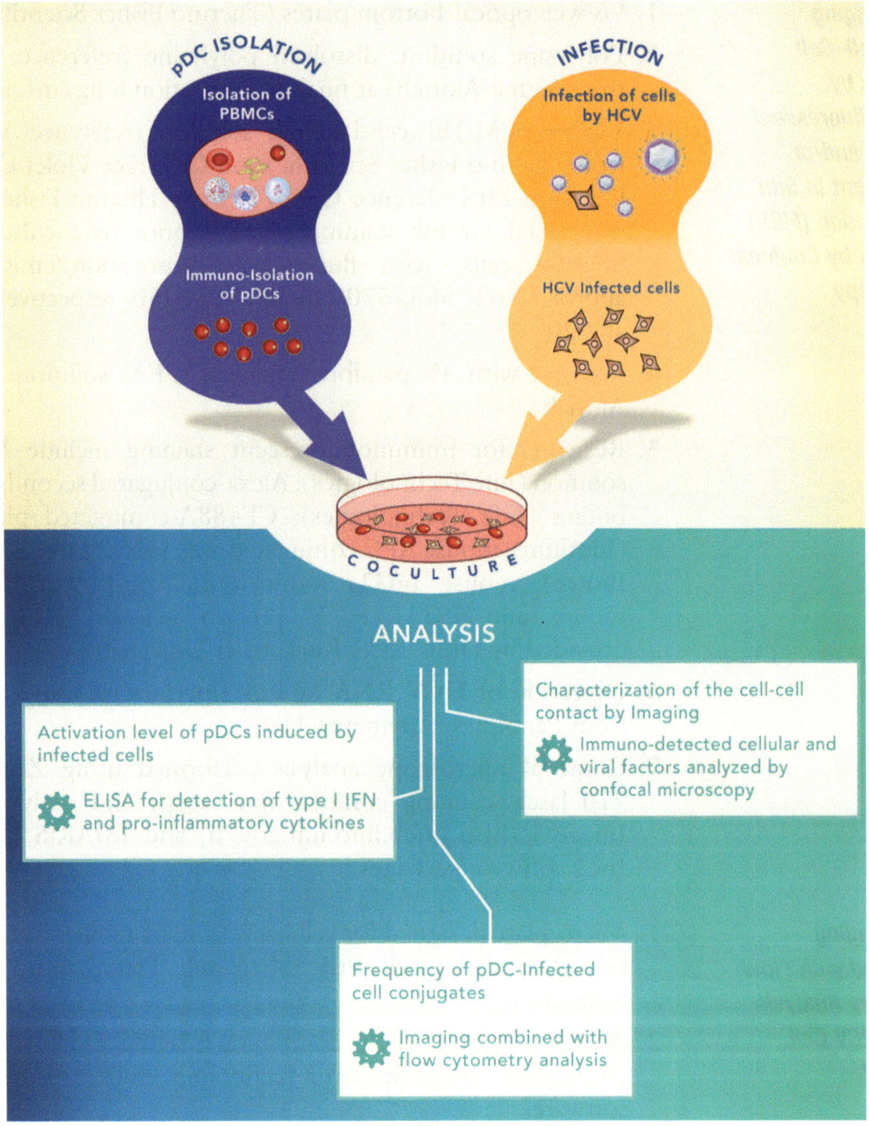

Fig. 1 Workflow of the analysis of pDC activation by cell–cell contact with infected cells. The pDCs are isolated from PBMCs issued from healthy blood donors (Subheadings 3.1 and 3.2) and cocultured with infected cells (Subheadings 3.3 and 3.4). The levels of pDC activation by infected cells can be assessed by the quantification of cytokine levels in the supernatant of cocultures (Subheading 3.3). The features of this cell–cell communication can be determined by confocal microscopy analysis of cellular and viral determinants, including the transfer of viral genomes (Subheading 3.4), along with the impact of the experimental modulations of cellular factors on the frequency of pDC/infected cell conjugates determined by imaging flow cytometry technology (Subheading 3.5)

can include the analysis of the production of type I IFNs and other cytokines, the confocal imaging of the cell–cell contact and transfer of the viral genome to the pDCs and the quantification of the formation of pDC/infected cell conjugates by imaging flow cytometry (Fig. 1).

3.1 Isolation of PBMCs from Blood from Healthy Adult Human Volunteers

PBMCs are separated using density gradient medium (ficoll).

1. Dispense 15 mL Ficoll-Hypaque in 50 mL tubes. Prepare 16 tubes for a blood unit.
2. Carefully overlay 20–30 mL of blood on 15 mL Ficoll-Hypaque by inclining the tube (*see* **Note 2**).
3. Centrifuge the tubes at 500 × *g* for 30 min at 20 °C (without brake).
4. During centrifugation, distribute 30 mL of HANK's 2 mM EDTA buffer into 50 mL tubes. Prepare 12 tubes for a blood unit.
5. After the centrifugation, discard about 11–12 mL of the upper layer (i.e., corresponding to plasma, yellow to translucid phase).
6. Collect and pool mononuclear cells at the cloudy interface by pipetting carefully on the side of the tube. Distribute the collected cells into the 50 mL tube containing the HANK's 2 mM EDTA buffer (prepared at **step 4**).
7. Centrifuge at 300 × *g* for 10 min at 20 °C (with brake).
8. Discard the supernatants and resuspend cell pellets in 8 mL HANK's 2 mM EDTA buffer par tube. Combine all cell suspensions in two 50 mL tubes.
9. Centrifuge at 300 × *g* for 6 min at room temperature.
10. Discard the supernatants and resuspend cell pellets in 25 mL HANK's 2 mM EDTA buffer. Centrifuge tubes at 300 × *g* for 6 min at RT.
11. Discard the supernatants and resuspend cell pellets into 10 mL of Separation buffer. Combine all cell suspensions in one 50 mL tube. Keep cells on ice from now on.
12. Determine viable cell count with Trypan Blue (*see* **Note 3**).

3.2 Isolation of pDCs from PBMCs

The pDC isolation from PBMCs for most steps is performed by positive selection using BDCA-4/Neuropilin-1-magnetic microbead kit and LS Columns accordingly to the manufacturer's instructions (MACS Miltenyi). The steps of the protocol that were optimized and thus different from the manufacturer' instructions are indicated by asterisks (*).

1. Pellet the PBMCs by centrifugation at 300 × *g* for 6 min at 4 °C.
2. Discard the supernatant and resuspend cell pellet in 1 mL of separation buffer (*). Add 500 μL of FcR blocking antibody.

3. Resuspend cells with 1 mL tips pipetting to disrupt cell aggregates (*), then add 500 μL of CD304 (BDCA-4) antibody microbeads.
4. Mix well and incubate for 15 min at 4 °C in the dark.
5. Resuspend cells in 50 mL of separation buffer. Centrifuge at 300 × g for 8 min at 4 °C.
6. During centrifugation, set the LS column into magnetic stand above a 15 mL tube and load 3 mL of separation buffer (do not let the column dry).
7. After centrifugation, discard the supernatant and resuspend cell pellet in 5 mL of separation buffer.
8. Load cell suspension onto the LS column. Let pass the whole cell suspension through the column.
9. Wash the column three times with 3 mL of Separation buffer. The through-out of the column can be collected, i.e., 12 mL of cell suspension in separation buffer, corresponding to PBMCs depleted of pDCs (*see* **Note 4**).
10. Remove the LS column from the magnet/stand and place it on a new 15 mL tube.
11. Load 5 mL separation buffer onto the column and immediately flush out the cells by firmly pushing the plunger into the column.
12. (*) Place a new LS column into the stand and add 3 mL separation buffer and repeat **steps 8-11** (*see* **Note 5**). pDCs are collected by a flush out of the column harvested in a 15 mL tube (as in **step 11**).
13. Centrifuge the collected pDCs at 300 × g for 8 min at 4 °C. Discard the supernatant and resuspend cell pellet in 1 mL of cell culture medium for pDCs are kept on ice.
14. Determine viable cell count with Trypan Blue (*see* **Note 3**).

3.3 Infection of the Cells and Coculture with Isolated pDCs for the Analysis of Cytokine Production

The infection of Huh-7.5.1 cells (or derivate) are performed using the JFH1 molecular clone of HCV [36–38] or adapted viral stock, as previously described [16].

1. Seed Huh-7.5.1 cells (or derivate) at 5 × 10^5 in T25 culture flask.
2. Six to eight hours later, inoculate cells with viral stock at MOI of 1.
3. Next day, change culture medium of infected cells.
4. The infected cells are harvested for coculture with pDCs 48 h post-infection. After one wash with PBS, infected cells are detached by incubation with trypsin/EDTA (Life technologies) at 37 °C for 2–3 min and pipetting up and down.

5. Determine viable cell count with Trypan Blue.
6. Centrifuge the collected infected cells at 300 × *g* for 8 min at 4 °C. Discard supernatant and resuspend cell pellet in cell culture medium for pDCs at the appropriated cell density (e.g., 10^6 cells per mL, in accordance with indication below).
7. Cocultures are performed in 96-well round-bottom plate with 2×10^4 pDCs and 10^5 infected cells (e.g., by adding 100 μL of cell suspension at 10^6 cells per mL for infected cells and 100 μL at 2×10^5 for pDCs) or uninfected parental cells used as negative controls, in a 200 μL final volume, incubated at 37 °C/5% CO_2.
8. Eighteen to twenty hours later, cell-culture supernatants are collected. Levels of IFNα, IFNβ, TNFα, and IL-6 are measured by ELISA, following the manufacturer's instructions. The cell-culture supernatants can be stored at −80 °C prior to analysis.

   Cocultured cells can be collected in guanidinium thiocyanate citrate (GTC) solution to determine of the upregulation of IFN-I and ISG expression by RT-qPCR analysis.

To demonstrate the requirement of cell–cell contact for pDC activation by infected cells, 10^5 infected cells or uninfected cells are cocultured with 2×10^4 pDCs in 96-well format transwell chambers separated by a 0.4 μm membrane (Corning) [26]. Similar results were obtained when pDCs were cultured in either the upper or the lower compartment [26].

The impact of HCV-EXOs is analyzed by the isolation of exosome-enriched fraction from the supernatants of SGR or infected cells as previously described [16, 41] (*see* **Note 6**).

3.4 Analysis of Cell–Cell Contacts by Immunofluorescence and Confocal Microscopy Analysis

Cell culture wells are coated with polylysine to improve cell adherence. Prior to coculture with infected cells (and uninfected cells, used as controls), isolated pDCs are stained with lipophilic dyes (e.g., CM-Dil staining solution) to facilitate the identification of pDCs in the coculture. The pDC/infected cell coculture should be set up rapidly after pDC isolation.

3.4.1 Setup of Cocultures with pDCs and Infected Cells

1. Culture plates are coated by distributing about 50–100 μL of polylysine solution per well (i.e., enough volume to cover the bottom of the well) in a 96-Well optical-bottom plates, overnight incubation at 37 °C.
2. Remove the polylysine solution from the wells and rinse twice with 100 μL of PBS.
3. Discard supernatant and rinse wells once with 200 μL of cell culture medium for pDCs. Add 100 μL cell culture medium for pDCs in the wells surrounding the wells of interest to prevent border effect and evaporation. The plate is kept at room temperature until the addition of cells.

4. The infected cells are prepared as in Subheading 3.3.
5. Prior to coculture with infected cells, isolated pDCs are stained by using a CM-Dil staining solution (at the final concentration of 0.5 μM in PBS), with gentle pipetting of the pDC suspension in 1 mL of PBS-CM-DiI solution and incubate for 10 min at 37 °C and then 15 min at 4 °C, in the dark.
6. Spin down pDCs at 300 × *g* for 6 min at 4 °C and remove supernatant.
7. Resuspend pDCs at the require concentration in cell culture medium for pDCs.

   Alternatively, for compatibility with other fluorochromes used for the immunostaining and/or FISH detection, pDCs can be stained with CellTrace Violet Cell Proliferation (*see* **Note** 7).
8. The infected cells and pDCs are cocultured in the plates prepared as described in **steps 1–3**. 6×10^4 pDCs and 2×10^4 infected cells are distributed in a final volume of 200 μL per well, and incubated for 4–8 h at 37 °C/5% CO_2.

3.4.2 Immunostaining and Fluorescent In Situ Hybridization (FISH) Detection for the Analysis of Cell–Cell Contact and Viral Product Transfer Analyzed by Confocal Microscopy

All the procedures of immunostaining should be performed by gentle addition of solutions by pipetting on the side of the wells and not directly on the bottom of the plate.

1. Remove culture medium and add 100 μL of 4% PFA solution. Incubate for 20 min at room temperature, in the dark. Wash cells three times with PBS. At this stage, the plate can be kept at 4 °C in the dark for a couple of days prior to immunostaining.
2. The cocultured cells can be permeabilized with 100 μL of 0.1% Triton X-100 in PBS for 7 min at room temperature followed by a washing step with PBS.
3. Blocking is performed by incubation with 50 μL of PBS supplemented with 3% BSA for 15 min at room temperature, followed by a PBS washing step.
4. Incubation with primary antibodies diluted in PBS supplemented with 3% BSA is performed for 1 h at room temperature with final volume of 50 μL per well, followed by three washes with PBS.
5. Incubation with secondary antibodies and Hoechst dye diluted in PBS supplemented with 3% BSA is performed for 1 h at room temperature (final volume of 50 μL per well), followed by three washes with PBS.
6. The detection of positive-strand HCV RNA is performed using a probe set that targets a region between nucleotide positions 3733-4870 in the JFH-1 genome [16] and a fluorochrome amplifying system designed by Panomics/Affymetrix (QuantiGene ViewRNA ISH Cell Assay kit) with incubation steps

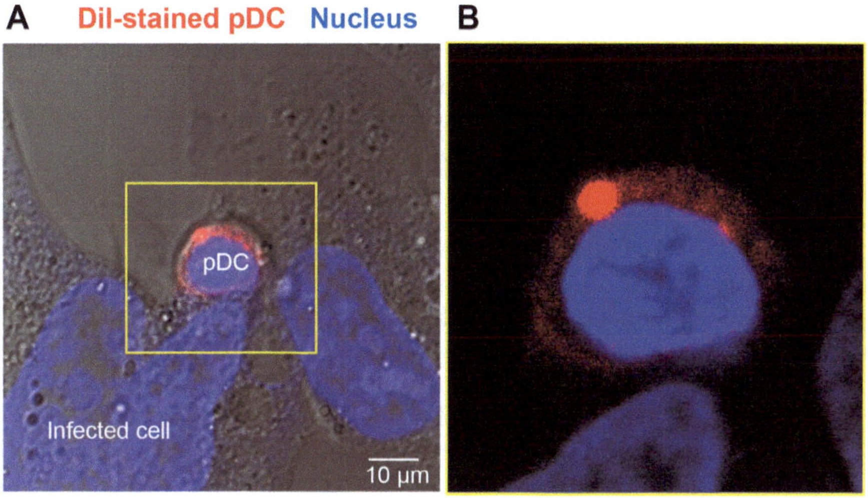

Fig. 2 Representative imaging of a pDC in contact with infected cells analyzed by confocal microscopy. The pDCs are stained with Dil lipophilic dye prior to coculture with HCV infected cells, nuclei stained with Hoechst. A flattening of pDCs, as reflected also by the shape of the nucleus, toward the infected cells is a hallmark of cell–cell contact between pDCs and infected cells. (**a**) Imaging of nucleus (blue) and Dil stained pDC (red) on the phase contrast. (**b**) Magnification of yellow-boxed pDC, shown in (**a**)

according to the manufacturer's instructions, except for the protease digestion step that was omitted. The omission of protease digestion step does not affect detection of HCV RNA. The steps for HCV RNA FISH labeling can be performed either prior to the immunostaining, or after. In the latter case, a PFA 4% fixation step is included at the end of the immunostaining procedure, as previously reported [16, 42]. Note the recognition of antigens by some antibodies can be compromised by the procedure for FISH detection. The negative-strand HCV RNA can be detected using a similar procedure (*see* **Note 8**). At this stage, the plate can be stored at 4 °C in the dark prior to confocal microscopy analysis.

7. Images are acquired with a Zeiss LSM 710 laser scanning confocal microscope and analyzed with Image J (http://rsb.info.nih.gov/ij) and IMARIS (Bitplane Inc.) software packages.

   The cell–cell contacts between pDC and infected cell are identified by the flattening of pDC toward the side of HCV infected cells (Fig. 2). Additionally, the flattening of pDC nucleus is also observed in front of the infected cells, a characteristic previously described for other types of physical contact, in relation with mechanical signaling [43]. These features are defined by the careful inspection of consecutive *Z*-axis sections (with 1–2 μm interspace) of the cell–cell contact.

8. Clustering and polarization of cellular and viral elements at the contact are assessed by two blind inspections of the acquired *Z*-stacks of the contact and with counting of the defined phenotypes, e.g., clusters or coclusters of elements at contact, outside of the contact or absence of clustering.

3.5 Quantification of pDC/Infected Cell Conjugates by Imaging Combined with Flow Cytometry Analysis

For analysis of the establishment of cell conjugates by imaging combined with flow cytometry technology, cocultures are performed in microplate designed for cell harvesting by temperature reduction (*see* Subheading 2). The pDCs are identified in the coculture by staining with CellTrace Proliferative Kit prior to coculture, as described in Subheading 3.4 and **Note** 7, or by antibody against specific pDC markers, e.g., CD123 [26]. Further, to accurately distinguish infected cells from pDCs, infected cells express GFP marker. This stable GFP expression is obtained thanks to their transduction using retroviral vector, as previously reported [26]. As mentioned in the previous sections, infected cell/pDC cocultures should be set up rapidly after pDC isolation. The frequency of cell conjugates is determined using Image stream X technology (Amnis).

1. GFP-expressing infected cells are harvested as in Subheading 3.3. Prior to coculture, the isolated pDCs are stained using CellTracer, as described in Subheading 3.4.
2. Cocultures are set-up with 4×10^4 pDCs and 1×10^5 infected cells in a final volume of 100 μL per well, and cells are incubated for 4–8 h at 37 °C/5% CO_2.
3. At the end of the coculture, culture plates are incubated at room temperature for 20 min to enable detachment of the cells from the support.
4. Cocultures are transferred into V-bottom 96 plates. The fixation of the cell culture is performed by incubation for 20 min at 4 °C with PFA 4% solution (addition of 100 μL per well).
5. Cells are washed three times in PBS (with spinning of cell pellet at 600 × *g* for 2 min). Culture plate is stored at 4 °C in the dark prior to analysis.
6. Cell suspensions in a final volume of 30 μL in PBS in 1.5 mL eppendorf tubes are analyzed using the Image stream X Technology.
7. The cell population defined as to pDC/HCV cell conjugates comprises conjugates of one stained pDC (by CellTrace Proliferative Kit or antibody against CD123) and one cell solely GFP+ cell (Fig. 3). Cell populations are thus sorted by using a series of masks (IDEAS software), notably to eliminate nonspecific signals, i.e., double positive single cells and to select for

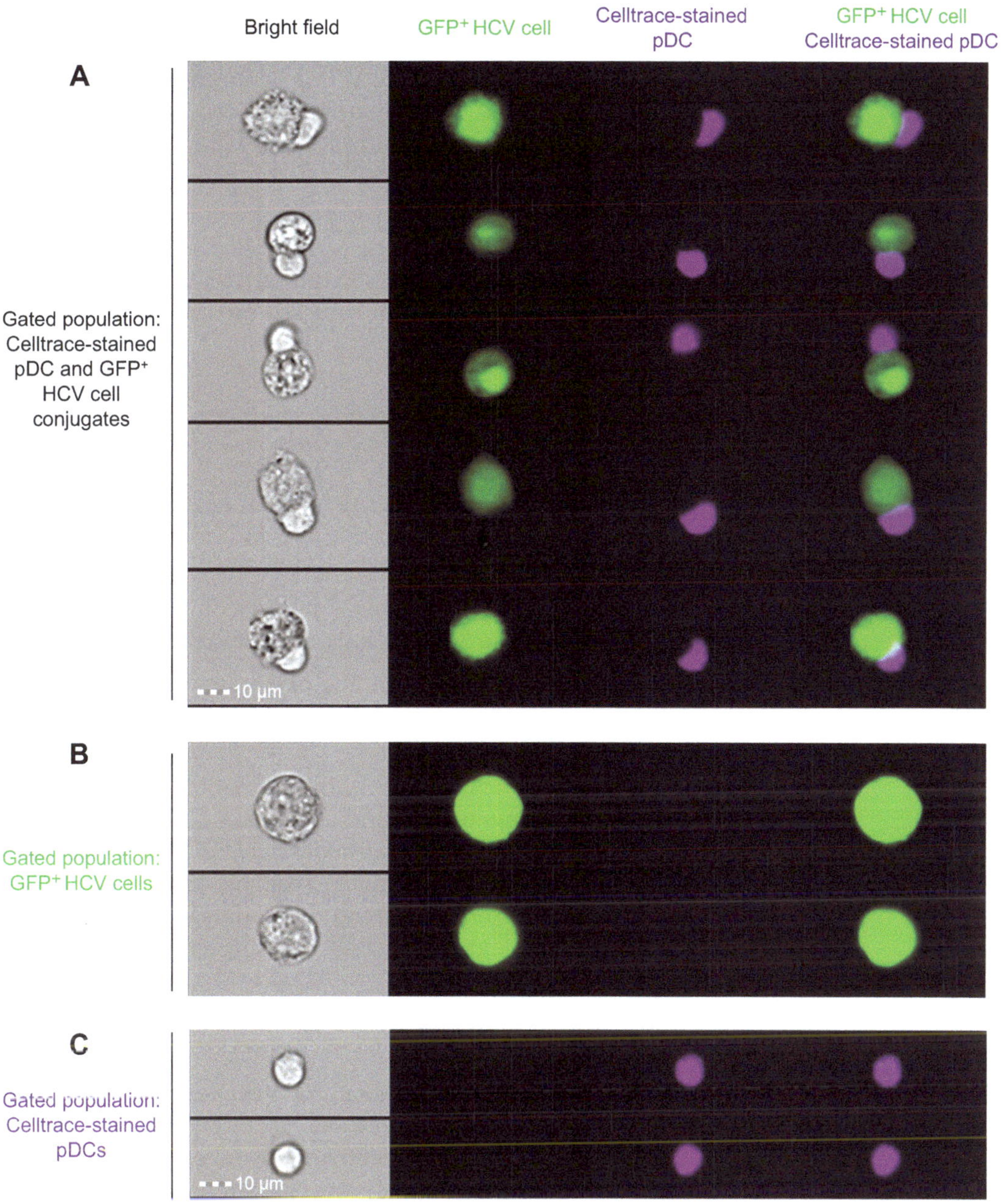

Fig. 3 Analysis of pDC/infected cell conjugates by imaging flow cytometry technology. Imaging flow cytometry analysis (Image stream X technology) of HCV infected Huh-7.5.1 cells, which stably express GFP marker (green), were cocultured with pDCs for 4-5 h. pDCs are detected by using CellTrace Violet Cell Proliferation solution (purple). Representative pictures of the cell populations gated as conjugates between pDCs and GFP expressing HCV infected cells as well as single cell populations. Panels, as displayed from the left to the right, Bright field; GFP field; Celltrace-stained pDC field; and Merge

cells in direct contact using automatic analysis of the overlay of GFP and pDC staining/marker at the cell periphery.

8. The frequency of pDC/HCV cell conjugates is determined relatively to total cell number of cells solely CD123 (or CellTrace staining) positive, solely GFP positive and the conjugates (Fig. 3).
9. Post-cell sorting, the accuracy of the gated cell population in regards to the defined criteria is controlled by a visual inspection of the individual pictures in the gated cell population (i.e., assessment with 100 randomly picked pictures of the population defined as conjugates).

4 Notes

1. Blood units and cytapheresis units from healthy adult human volunteers are obtained according to procedures approved by the "Etablissement Français du sang" (EFS) Committee.
2. The blood should not be mixed with Ficoll and carefully layered without disturbing the interface between the two layers.
3. The typical yields of PBMCs and pDCs are 5-to-8 $\times$ 10^8 and 1-to-3 $\times$ 10^6 cells, respectively, with an expected purity of >95% for pDCs.
4. The analysis of the efficiency of pDC isolation from PBMCs is performed by immunostaining and FACS analysis of PBMCs before and after pDC isolation and isolated pDCs [16, 26]. The pDCs are detected by combining staining of the CD123 and BDCA2 specific markers using PE-conjugated anti-CD123, mouse APC-conjugated anti-BDCA-2, respectively (MACS Miltenyi Biotec).
5. Two rounds of passage of the cell suspension through the column increase the purity with a minimal loss of pDCs.
6. The HCV-EXOs can be isolated from the supernatant of SGR and HCV infected cells, as previously described [16, 41]. Briefly, supernatants harvested from 80% confluent cells are precleared by sequential low- and high-speed centrifugations. Exosomes in the clarified supernatants are then pelleted at 110,000 $\times$ g for 2 h at 4 °C using a SW28 rotor. The pelleted exosome preparations are resuspended in exosome-depleted medium. Approximately 0.25–0.5 mg of exosomal proteins are obtained from 1-to-2 $\times$ 10^6 cells. The purity of the preparation can be determined by western blot analysis using exosome markers (e.g., CD81, CD63) and nonexosome markers (e.g., the calnexin ER protein and the COX IV mitochondrial protein) [16].

7. Isolated pDCs are stained with CellTrace Violet Cell Proliferation solution (prepared at the final concentration of 2.5 μM in PBS) by incubation for 20 min at 37 °C, in the dark. After centrifugation (600 × *g* for 8 min), stained pDCs are resuspended at the required cell density in cell culture medium for pDCs by pipetting, followed by an incubation for 5 min at 37 °C.
8. Detection of the negative strand HCV RNA is performed using a probe set that target regions between nucleotide positions 4904 and 5911 in the JFH1 complementary strand with experimental procedures similar to those described for the positive strand HCV RNA and as previously reported [42].

Acknowledgments

We are grateful to Helana Paidassi (CIRI, INSERM U1111, Lyon) for critical reading of the manuscript and to our colleagues for encouragement and help. We are grateful to Camille Demure for the graphical design of Fig. 1. This work was supported by grants from the French "Agence Nationale pour la Recherche" (ANR-13-JSV3-0004-01-EXAMIN) and the "Agence Nationale pour la Recherche contre le SIDA et les Hépatites Virales" (ANRS-AO 2017-01, ECTZ35316) and the LabEx Ecofect (ANR-11-LABX-0048).

References

1. Swiecki M, Colonna M (2015) The multifaceted biology of plasmacytoid dendritic cells. Nat Rev Immunol 15:471–485
2. Merad M, Sathe P, Helft J, Miller J, Mortha A (2013) The dendritic cell lineage: ontogeny and function of dendritic cells and their subsets in the steady state and the inflamed setting. Annu Rev Immunol 31:563–604
3. Gilliet M, Cao W, Liu YJ (2008) Plasmacytoid dendritic cells: sensing nucleic acids in viral infection and autoimmune diseases. Nat Rev Immunol 8:594–606
4. Panda SK, Kolbeck R, Sanjuan MA (2016) Plasmacytoid dendritic cells in autoimmunity. Curr Opin Immunol 44:20–25
5. Izaguirre A, Barnes BJ, Amrute S, Yeow WS, Megjugorac N et al (2003) Comparative analysis of IRF and IFN-alpha expression in human plasmacytoid and monocyte-derived dendritic cells. J Leukoc Biol 74:1125–1138
6. Kumagai Y, Kumar H, Koyama S, Kawai T, Takeuchi O et al (2009) Cutting Edge: TLR-Dependent viral recognition along with type I IFN positive feedback signaling masks the requirement of viral replication for IFN-{alpha} production in plasmacytoid dendritic cells. J Immunol 182:3960–3964
7. Thitithanyanont A, Engering A, Ekchariyawat P, Wiboon-ut S, Limsalakpetch A et al (2007) High susceptibility of human dendritic cells to avian influenza H5N1 virus infection and protection by IFN-alpha and TLR ligands. J Immunol 179:5220–5227
8. Assil S, Webster B, Dreux M (2015) Regulation of the host antiviral state by intercellular communications. Viruses 7:4707–4733
9. Bruel T, Dupuy S, Demoulins T, Rogez-Kreuz C, Dutrieux J et al (2014) Plasmacytoid dendritic cell dynamics tune interferon-alfa production in SIV-infected cynomolgus macaques. PLoS Pathog 10:e1003915
10. Pichyangkul S, Endy TP, Kalayanarooj S, Nisalak A, Yongvanitchit K et al (2003) A blunted blood plasmacytoid dendritic cell response to an acute systemic viral infection is

associated with increased disease severity. J Immunol 171:5571–5578

11. Swiecki M, Gilfillan S, Vermi W, Wang Y, Colonna M (2010) Plasmacytoid dendritic cell ablation impacts early interferon responses and antiviral NK and CD8(+) T cell accrual. Immunity 33:955–966
12. Swiecki M, Wang Y, Gilfillan S, Colonna M (2013) Plasmacytoid dendritic cells contribute to systemic but not local antiviral responses to HSV infections. PLoS Pathog 9:e1003728
13. Smit JJ, Rudd BD, Lukacs NW (2006) Plasmacytoid dendritic cells inhibit pulmonary immunopathology and promote clearance of respiratory syncytial virus. J Exp Med 203:1153–1159
14. Sozzani S, Vermi W, Del Prete A, Facchetti F (2010) Trafficking properties of plasmacytoid dendritic cells in health and disease. Trends Immunol 31:270–277
15. Webster B, Werneke SW, Zafirova B, This S, Coleon S, Decembre E, Paidassi H, Bouvier I, Joubert PE, Duffy D et al (2018) Plasmacytoid dendritic cells control dengue and Chikungunya virus infections via IRF7-regulated interferon responses. Elife 7:e34273. https://doi.org/10.7554/eLife.34273.001
16. Dreux M, Garaigorta U, Boyd B, Decembre E, Chung J et al (2012) Short-range exosomal transfer of viral RNA from infected cells to plasmacytoid dendritic cells triggers innate immunity. Cell Host Microbe 12:558–570
17. Longatti A, Boyd B, Chisari FV (2015) Virion-independent transfer of replication competent HCV RNA between permissive cells. J Virol 89:2956–2961
18. Ramakrishnaiah V, Thumann C, Fofana I, Habersetzer F, Pan Q et al (2013) Exosome-mediated transmission of hepatitis C virus between human hepatoma Huh7.5 cells. Proc Natl Acad Sci U S A 110:13109–13113
19. Bukong TN, Momen-Heravi F, Kodys K, Bala S, Szabo G (2014) Exosomes from hepatitis C infected patients transmit HCV infection and contain replication competent viral RNA in complex with Ago2-miR122-HSP90. PLoS Pathog 10:e1004424
20. Liu Z, Zhang X, Yu Q, He JJ (2014) Exosome-associated hepatitis C virus in cell cultures and patient plasma. Biochem Biophys Res Commun 455:218–222
21. Saha B, Kodys K, Adejumo A, Szabo G (2017) Circulating and exosome-packaged hepatitis C single-stranded RNA induce monocyte differentiation via TLR7/8 to polarized macrophages and fibrocytes. J Immunol 198:1974–1984
22. Devhare PB, Sasaki R, Shrivastava S, Di Bisceglie AM, Ray R et al (2017) Exosome-mediated intercellular communication between hepatitis C virus-infected hepatocytes and hepatic stellate cells. J Virol 91: e02225–e02216
23. Feng Z, Li Y, McKnight KL, Hensley L, Lanford RE et al (2014) Human pDCs preferentially sense enveloped hepatitis A virions. J Clin Invest 125:169–176
24. Wieland SF, Takahashi K, Boyd B, Whitten-Bauer C, Ngo N et al (2014) Human plasmacytoid dendritic cells sense lymphocytic choriomeningitis virus-infected cells in vitro. J Virol 88:752–757
25. Takahashi K, Asabe S, Wieland S, Garaigorta U, Gastaminza P et al (2010) Plasmacytoid dendritic cells sense hepatitis C virus-infected cells, produce interferon, and inhibit infection. Proc Natl Acad Sci U S A 107:7431–7436
26. Decembre E, Assil S, Hillaire ML, Dejnirattisai W, Mongkolsapaya J et al (2014) Sensing of immature particles produced by dengue virus infected cells induces an antiviral response by plasmacytoid dendritic cells. PLoS Pathog 10:e1004434
27. Bruni D, Chazal M, Sinigaglia L, Chauveau L, Schwartz O et al (2015) Viral entry route determines how human plasmacytoid dendritic cells produce type I interferons. Sci Signal 8: ra25
28. Lepelley A, Louis S, Sourisseau M, Law HK, Pothlichet J et al (2011) Innate sensing of HIV-infected cells. PLoS Pathog 7:e1001284
29. Python S, Gerber M, Suter R, Ruggli N, Summerfield A (2013) Efficient sensing of infected cells in absence of virus particles by plasmacytoid dendritic cells is blocked by the viral ribonuclease E(rns.). PLoS Pathog 9:e1003412
30. Garcia-Nicolas O, Auray G, Sautter CA, Rappe JC, McCullough KC et al (2016) Sensing of porcine reproductive and respiratory syndrome virus-infected macrophages by plasmacytoid dendritic cells. Front Microbiol 7:771
31. Costa VV, Fagundes CT, da Gloria de Souza D, Teixeira MM (2013) Inflammatory and innate immune responses in dengue infection: protection versus disease induction. Am J Pathol 182:1950–1961
32. Tomasello E, Pollet E, Vu Manh TP, Uze G, Dalod M (2014) Harnessing mechanistic knowledge on beneficial versus deleterious IFN-I effects to design innovative immunotherapies targeting cytokine activity to specific cell types. Front Immunol 5:526

33. Pedersen IM, Cheng G, Wieland S, Volinia S, Croce CM et al (2007) Interferon modulation of cellular microRNAs as an antiviral mechanism. Nature 449:919–922
34. Gastaminza P, Dryden KA, Boyd B, Wood MR, Law M et al (2010) Ultrastructural and biophysical characterization of hepatitis C virus particles produced in cell culture. J Virol 84:10999–11009
35. Kato T, Date T, Miyamoto M, Furusaka A, Tokushige K et al (2003) Efficient replication of the genotype 2a hepatitis C virus subgenomic replicon. Gastroenterology 125:1808–1817
36. Wakita T, Pietschmann T, Kato T, Date T, Miyamoto M et al (2005) Production of infectious hepatitis C virus in tissue culture from a cloned viral genome. Nat Med 11:791–796
37. Zhong J, Gastaminza P, Cheng G, Kapadia S, Kato T et al (2005) Robust hepatitis C virus infection in vitro. Proc Natl Acad Sci U S A 102:9294–9299
38. Lindenbach BD, Evans MJ, Syder AJ, Wolk B, Tellinghuisen TL et al (2005) Complete replication of hepatitis C virus in cell culture. Science 309:623–626
39. Zhong J, Gastaminza P, Chung J, Stamataki Z, Isogawa M et al (2006) Persistent hepatitis C virus infection in vitro: coevolution of virus and host. J Virol 80:11082–11093
40. Dreux M, Gastaminza P, Wieland SF, Chisari FV (2009) The autophagy machinery is required to initiate hepatitis C virus replication. Proc Natl Acad Sci U S A 106:14046–14051
41. Thery C, Amigorena S, Raposo G, Clayton A (2006) Isolation and characterization of exosomes from cell culture supernatants and biological fluids. Curr Protoc Cell Biol Chapter 3:Unit 3 22
42. Boson B, Denolly S, Turlure F, Chamot C, Dreux M et al (2017) Daclatasvir prevents hepatitis C virus Infectivity by blocking transfer of the viral genome to assembly sites. Gastroenterology 152:895–907
43. Cho S, Irianto J, Discher DE (2017) Mechanosensing by the nucleus: From pathways to scaling relationships. J Cell Biol 216:305–315

Chapter 23

Complement Regulation and Immune Evasion by Hepatitis C Virus

Young-Chan Kwon and Ranjit Ray

Abstract

A prominent role for complement has been identified in the linkage of innate and adaptive immunity. The liver is the main source of complement and hepatocytes are the primary sites for synthesis of complement components in vivo. We have discovered that hepatitis C virus (HCV) impairs C4 and C3 synthesis. Liver damage may diminish capacity of complement synthesis in patients. However, we observed that the changes in measured complement components in chronically HCV infected patients do not correlate with liver fibrosis or rheumatoid factor present in the blood, serum albumin, or alkaline phosphatase levels. Complement component C3 is of critical importance in B cell activation and T cell-dependent antibody responses. C3 activity is required for optimal expansion of $CD8^{+}$T cells during a systemic viral infection. Deficiencies in complement may predispose patients to infections via ineffective opsonization, and defects in lytic activity via membrane attack complex. Interestingly, C9 is significantly reduced at the mRNA level in chronically HCV infected liver biopsy specimens, while many hepatocyte derived complement components (C6, C8, Factor B, MASP1, and MBL) and unrelated genes remain mostly unaffected. This implies an HCV specific effect, not a global effect from liver disease.

Key words Hepatitis C virus, Complement, Cascade, Convertase, Opsonin, Lysis, siRNA, Liver disease

1 Introduction

HCV is an important cause of morbidity and mortality worldwide, promoting a spectrum of liver disease ranging from an asymptomatic carrier state to end stage liver disease. Approximately 3% (170 million) of the world's population is estimated to be infected with HCV [1]. HCV establishes chronic infection in more than 70% of infected individuals. Persistent HCV infection is associated with hepatic fibrosis, cirrhosis, and hepatocellular carcinoma.

The complement system performs a vital effector function in the innate immune system by providing an efficient means for targeting and eliminating infected cells and invading microorganisms, including free viral particles [2–4]. Activation of the complement cascade occurs primarily via the classical, alternative, or lectin

Mansun Law (ed.), *Hepatitis C Virus Protocols*, Methods in Molecular Biology, vol. 1911,
https://doi.org/10.1007/978-1-4939-8976-8_23, © Springer Science+Business Media, LLC, part of Springer Nature 2019

pathway [2, 5]. These three pathways activate C3 via cleavage to C3a and C3b by the C3 convertases. C5 convertases are generated by the association of C3b with the C3 convertases, which in turn cleaves C5 into C5a and C5b. The release of C5b initiates the nonenzymatic process of membrane-attack complex (MAC) formation that then sequentially recruits C6, C7, C8, and C9 proteins [3, 4, 6]. The MAC forms a pore-like structure within the lipid envelope of the pathogen or the membrane of the infected cells that ultimately leads to lysis. In order to avert damage from excessive complement activation and MAC formation, host cells express membrane-bound regulatory proteins to limit these processes [7]. Complement regulatory proteins (CRP) are expressed on the surfaces of host cells and include CD46, DAF/CD55, and CD59 [8–10].

HCV has evolved mechanisms to evade immune activation, including complement response. We have shown a regulatory role of HCV on complement related functions [11–14]. HCV suppresses C3, C4, and C9 synthesis in hepatocytes, and induces DAF/CD55, a negative regulator of complement activation. DAF/CD55 inhibits the formation and dissociation of C3/C5 convertases. HCV core protein upregulates DAF/CD55 expression on the cell surface and inhibits complement-dependent cytolysis [13]. Further, we have recently reported that a secreted isoform of CD55 (sCD55) is induced in HCV-infected cells and might play a role in cancer immune evasion [15]. Modulation of CRP by HCV may limit immune attack on cancer cells. Complement elements promote tumor growth [16]. Clearance of HCV may help in reconstituting inflammatory response and complement functions [17–19]. Higher C3 and C4 serum concentrations positively correlated to the end of IFN and ribavirin treatment response in patients with chronic HCV infection [17, 19]. In fact, serum level of complements C3 in cases of liver cirrhosis without HCC were lower than in HCC cirrhotic patients [18]. On the other hand, serum levels of C3 & C4 were significantly higher in HCC group than in controls. Differentiation of regulatory T cells is correlated with C5a concentration within the tumor. In a lung cancer mouse model, the blockade of C5a signaling downregulates the expression of key immunosuppressive molecules within the tumors [16]. Thus, recovery of an HCV-specific suppressed immunity may unintentionally lead to a tumor microenvironment more suitable to neoplastic growth as an underlying disease pathology.

2 Materials

1. HCV infected patient sera (*see* **Note 1**) and normal human serum (NHS) (Complement Technology).

2. Cell lines: Human liver hepatocellular carcinoma cell line (Huh7.5), Immortalized human hepatocytes (IHH).
3. NK3.3 cell line.
4. Fetal bovine serum.
5. Cell culture grown HCV 2a (Clone: JFH1).
6. C3 or C4 promoter-luciferase reporter plasmid.
7. Luciferase assay Kit (Promega).
8. Dulbecco's modified Eagle medium and RPMI1640 medium.
9. 1× Phosphate buffered saline (1× PBS).
10. Calcein AM (Molecular Probe).
11. Recombinant IL-2 (R&D Systems).
12. CD55 antibody (BRIC216, EMD Millipore), CD59 antibody (BRIC229, rILYd4 or YTH53.1, Santa Cruz Biotechnology).
13. Anti-HCV E2 pAbs or irrelevant pAbs.
14. QuickTiter HCV Core ELISA Kit.
15. Triton X-100 in PBS.
16. Solid-phase enzyme immunoassay (EIA) plate.
17. Mouse monoclonal antibody (MAb) against human C9 (Santa Cruz Biotechnology).
18. *E. coli* (DH5α).
19. Bovine serum albumin (BSA).
20. Purified human complement component C3 protein (Quidel).
21. Mouse monoclonal antibody to human C3 (Abcam).
22. Anti-mouse HRP conjugated secondary antibody (Sigma-Aldrich).
23. 10% H_2SO_4.
24. LB plate.
25. Amboceptor sensitized sheep erythrocytes (EA) (Complement Technology).
26. Rabbit erythrocytes (rE) (Complement Technology).
27. GVB++ buffer (0.1% gelatin, 5 mM Veronal, 145 mM NaCl, 0.025% NaN_3, 1 mM $MgCl_2$, 0.15 mM $CaCl_2$, pH 7.3).
28. EDTA-GVB buffer (40 mM EDTA, 5 mM veronal buffer, 0.1% gelatin, 145 mM NaCl, 0.025 NaN_3, pH 7.3).
29. 96-well V-shape microplate (Nunc).
30. C3-depleted normal human serum (Quidel).
31. C5-depleted serum (Complement Technology).
32. Guinea pig serum (Complement Technology).

3 Methods

3.1 HCV Infection Inhibits the Expression of C3 and C4 Complement Component

The third component of human complement (C3) plays a central role in innate immune function as its activation is required to trigger classical as well as alternative complement pathways. The complement component C4 contributes to the clearance of several viral pathogens through its functions as an opsonin and by virtue of its central role in the classical and lectin complement activation pathways [20]. The sera from patients chronically infected with HCV displayed significantly lower C3 levels than sera from healthy individuals. HCV core and NS5A proteins transcriptionally down-regulate C4 expression by modulating the expression of upstream stimulating factor 1 (USF-1) and interferon regulatory factor 1 (IRF-1), respectively. Thus, HCV infection appears to play a significant role in innate immune modulation, favoring establishment of chronic infection and liver disease progression.

3.1.1 C3 and C4 Promoter Analysis

To examine whether HCV transcriptionally inhibits complement components (C3 and C4):

1. Huh7.5 cells are grown to ~70–85% confluence in a 24-well plate overnight.
2. Cells are transfected with a C3 or C4 promoter-luciferase reporter plasmid (500 ng) with or without the plasmid encoding the HCV full-length gene, the specific genomic region (NS2, NS3/4A, or NS5A), or 50 nM specific small interfering RNA (siRNA).
3. The cells are incubated at 37 °C for 36 h.
4. The cells are lysed with reporter lysis buffer (Promega).
5. The samples are measured using a luminometer (OpticompII; MGM Instrument).

3.2 Impairment of Complement-Mediated Bacterial Killing by HCV-Infected Patient Sera

The membrane attack complex (MAC) is composed of C5b to C9 (C5b-9) and mediates cell lysis of invaded pathogens. Complement C9 is one of the components of MAC and forms a pore on the bacterial cell surface. MAC is typically formed on the surface of bacteria and enveloped virus as a result of the activation of the complement system. MAC-mediated formation of transmembrane channels disrupts the phospholipid bilayer of target cells, leading to cell lysis and death [21]. We examined whether HCV can interfere with MAC-mediated Gram-negative bacterial lysis [12].

3.2.1 C9 Deposition Assay

1. *E. coli* DH5α (1×10^7 cells) is incubated with HCV infected patient sera or NHS (20% sera in a final volume of 100 μl) at 37 °C for 60 min.

2. After a wash with phosphate-buffered saline (PBS), opsonized log-phase bacteria are plated onto 96-well enzyme immunoassay (EIA) plates at 5×10^6 per well and adhered to them by dry desiccation.
3. Nonspecific binding sites are blocked with 200 μl of 0.5% bovine serum albumin (BSA)–PBS and incubated for 30 min at 37 °C.
4. After a wash, the plates are sequentially incubated at room temperature with 100 μl of 1:100 (vol/vol) C9 antibody for 60 min, 100 μl of 1:1000 horseradish peroxidase (HRP)-conjugated secondary antibody for 30 min and finally, 100 μl of substrate solution for visualizing HRP activity.
5. The reaction is stopped with 50 μl of 10% H_2SO_4 and the color reaction is detected at an optical density at 490 nm (OD_{490}).

3.2.2 Bactericidal Assay

1. *E. coli* DH5α (1×10^4 cells) is incubated with 100 μl of a 1:500 dilution of HCV-infected patient sera or normal human sera (NHS) at 37 °C for 60 min.
2. After washing, serial tenfold dilutions of bacterial suspensions into saline solution are made, and 100 μl is spread on LB plates.
3. After 16 h of incubation at 37 °C, the CFU are counted.

3.3 HCV Inhibits Complement-Mediated Cytolysis (CDC), Antibody-Dependent Cell Mediated Cytolysis (ADCC), or Virolysis

CD55 limits excessive complement activation on the host cell surface by accelerating the decay of C3 convertases. HCV infection of hepatocytes or HCV core protein expression in transfected hepatocytes induces CD55 expression at the mRNA and protein levels. The enhanced CD55 expression in hepatoma cells infected with HCV can inhibit complement-mediated cytolysis and thereby promote cell survival. Other studies have suggested that incorporation of the complement-regulatory protein CD59 in HCV in plasma as well as proteins derived from cell lines [22] protected HCV from antibody-dependent complement-mediated lysis (*see* **Note 2**). The CDC and ADCC can be measured by Calcein AM release methods [23].

3.3.1 The Complement Dependent Cytolysis

1. Cells (2×10^4 cells/ml) are seeded in a 96-well plate and grown to confluency.
2. Cells are incubated with Calcein AM (5 μg/ml) for 30 min at 37 °C, washed twice prior to treatment with cell surface antigen specific antibody with anti-CD55 or anti-CD59 for 20 min at 37 °C (*see* **Note 3**).
3. Cells are incubated in the presence of normal human serum (NHS) as a source of complement at 37 °C for 4 h.
4. The supernatant is transferred to a new 96 well plate to represent the complement-mediated release of Calcein AM fraction.

5. The Calcein AM remaining in the cells are released by incubation with 0.1% Triton X-100 for 30 min as maximum release.
6. Samples are measured using a Multimode Plate Reader using an excitation filter 485 nm and emission filter 530 nm.
7. Percent lysis is calculated according to the formula [(test release – spontaneous release)/(maximum release – spontaneous release)] × 100.

3.3.2 NK Cell Mediated Antibody-Dependent Cell Mediated Cytolysis (ADCC)

1. Cells (2×10^4 cells/ml) are seeded in a 96-well plate and separately NK3.3 cells are grown to confluency (*see* **Note 4**).
2. Cells are incubated with Calcein AM (5 μg/ml) for 30 min at 37 °C, washed twice prior to treatment with cell surface antigen specific antibody with anti-CD55 or anti-CD59 for 20 min at 37 °C.
3. Cells are incubated with NK 3.3 effector cells at 37 °C for 4 h (effector–target cells = 2.5:1, 5:1, or 10:1).
4. The supernatant is transferred and the fluorescence is measured similar to CDC assay.
5. Percent lysis is calculated using the same formula with CDC.

3.3.3 Complement-Dependent Virolysis

1. HCV-containing supernatants (50 μl) are preincubated with each blocking antibody (for CD55–BRIC216, CD59–BRIC229, or rILYd4) at 37 °C for 30 min.
2. After preincubation, anti-HCV E2 pAbs or irrelevant pAbs are added, followed by exposure to either complement-component human sera or heat-inactivated complement diluted in GVB buffer.
3. Virolysis of HCV are quantified by measuring HCV core release using the QuickTiter HCV Core ELISA Kit as the manufacturer's description.
4. For 100% lysis, Triton X-100 can be added and for background, PBS can be used.
5. The percentage of virolysis can be calculated as followed: (core released by blocking antibody – core released by PBS)/(core released by Triton X-100 – core released by PBS) × 100%.

3.4 Conditioned Medium from HCV Infected Cells Inhibits Convertase Activity

Two isoforms of CD55 are known in humans, a glycosyl phosphatidylinositol (GPI)-anchored form (gDAF or CD55) and a soluble form (sDAF or sCD55) [24, 25]. CD55 or gDAF is the major form, which is composed of four N-terminal complement control protein (CCP) domains, a heavily glycosylated serine, threonine, and proline (STP)-rich domain, and a C-terminal GPI-anchored portion [26]. Unlike CD55, sCD55 is present in body fluids, including urine and plasma [27], and extracellular matrix; and its levels are much lower than CD55. sCD55 is generated from the CD55 gene by alternative usage of an optional exon and lacks the GPI-anchored portion in the C-terminal [28]. sCD55 is upregulated by HCV. In addition, HCV infection induces Factor H and C4BP expression which inhibits C3 convertase activity [15, 29]. Thus, we can examine whether conditioned medium from HCV infected cells inhibits C3 convertase activity with hemolytic assay [15, 29, 30].

3.4.1 Classical Pathway Using Sheep Erythrocytes (EA)

1. Amboceptor sensitized sheep erythrocytes (EA) (5×10^8) are washed with GVB++ buffer (*see* **Note 5**).
2. EA suspensions are adjusted and lysed with 90 μl of water to have an absorbance value of ~1.2 at 405 nm.
3. 10 μl of EA are placed in a 96-well V-shape microplate and 40 μl of GVB++ buffer containing a 2% C3-depleted NHS (Quidel) is added to assemble C3 convertases.
4. The plate is kept in an incubator shaker (300 rpm) at 30 °C for 5 min for convertase formation.
5. After removing the supernatant, complete medium (DMEM+ 10% FBS) containing purified sCD55 or conditioned medium from each cell line is added.
6. To initiate complement mediated lysis from existing convertase complexes, 50 μl of 40 mM EDTA-GVB buffer containing 1:40 guinea pig serum as a source of complement is added.
7. The plate is incubated for an additional 30 min at 37 °C on a 300 rpm shaker.
8. Cells are pelleted and 80 μl of the supernatant containing hemoglobin released from lysed EA are measured at 405 nm.

3.4.2 Alternative Pathway Using Rabbit Erythrocytes (rEA)

1. Rabbit erythrocytes (rE) (~4×10^8 cells/ml) are washed with GVB++ buffer including 10 mM EGTA until there is no visible hemoglobin in the supernatant (*see* **Note 5**).
2. To ensure an equal number of rE in each experiment, cell suspensions are standardized so that 10 μl of such suspension lysed with 90 μl of water has an absorbance of ~1.0 at 405 nm.
3. 10 μl of rE are placed in wells of a 96-well V-shape microplate (Nunc) and 40 μl of GVB++ buffer including 10 mM EGTA containing a given concentration of C5-depleted serum is added in order to assemble convertase.
4. Complement-mediated lysis resulting from assembled convertase complexes is initiated and measured as described for the classical pathway assay.

3.5 HCV Infected Patient Sera Exhibit Reduced C3 Convertase Formation and C3b or iC3b Level

HCV may inhibit C3 convertase activity at multiple steps in the infected host. HCV downregulates C2, C3, and C4 expression, which contributes to C3 convertase formation. HCV induces CD55/sCD55 expression, which contribute C3 convertase activation. To verify the impairment of C3 convertase activity in HCV-infected patients, we can analyze the C3 convertase assembly of alternative pathway by analyzing C3b deposition on bacterial surface and the C3 cleavage by C3 convertase activity. C3 convertase cleaves C3 into C3a and C3b. iC3b is generated by Factor I after formation of C3b. iC3b is a proteolytically inactive form of C3b that retains the ability to opsonize microbes, but cannot participate in convertase function. Thus, we can examine that serum from HCV infected patients inhibits the production of C3b / iC3b [15, 29]. These means that HCV inhibits C3 convertase activity in order to escape from complement response.

3.5.1 C3 Deposition Assay

1. *E. coli* DH5α (~1 × 10^7) is incubated with HCV infected patient serum or normal human serum (NHS) (10% sera in a final volume of 50 ml) at 37 °C for 30 min.
2. Purified C3 protein (20 mg/ml) is added in the bacteria and serum mixture.
3. After washing with PBS, opsonized log-phase bacteria are plated onto 96-well EIA plates at ~5 × 10^6 per well and adhered by dry desiccation.
4. Nonspecific binding sites are blocked with 0.5% BSA in PBS and incubated for 30 min at 37 °C. After washing, the plates are sequentially incubated a room temperature with C3 antibody for 60 min, HRP-conjugated secondary antibody for 30 min.
5. After washing with PBS, substrate solution is finally added for visualizing HRP activity.
6. The reaction is stopped with H_2SO_4 and color intensity measured at OD_{490}.

3.5.2 C3 Cleavage Assay

1. 50 mg/ml of C3 purified protein is incubated with different dose of normal human serum or serum from HCV patients at 37 °C for 30 min.
2. Samples are separated under reducing condition by SDS-PAGE.
3. Proteins are transferred onto nitrocellulose membrane.
4. Anti-human C3 antibody is treated at 4 °C for overnight.
5. Anti-mouse HRP conjugated secondary antibody is treated.
6. C3 will be detected in 115 kDa as C3α and 75 kDa as C3β. C3bα will be detected in 104 kDa and iC3bα will be in 45 kDa.

4 Conclusion

The consequences on complements and related protein modulations by HCV will impair HCV clearance and inhibit HCV-associated hepatocellular carcinoma cell killing.

5 Notes

1. All patient samples are collected from chronically HCV-infected patients and sera prepared after storage on ice within a short time (2–3 h) to retain complement activity. Sera are aliquoted and stored at −70 °C. Each aliquot is thawed and tested once for complement-related activity.
2. HCV does not express viral protein on the infected host cell surface. Thus, the Ab against HCV Ag does not itself activate CDC of infected cells.
3. Calcein-AM (calcein-acetoxymethyl) is a fluorescent and a cell-permeant dye which has been used in cytotoxicity assay. Inside the cells it is converted by intracellular esterases into a polar lipid-insoluble fluorescent product (Calcein) but that is released by damaged ones (similarly to ^{51}Cr), producing an intense green signal. Release of Calcein in the supernatants recovered from cytotoxicity assays can be measured rapidly and with a high level of sensitivity by a fluorimeter.
4. NK3.3 cells are grown in RPMI1640 medium containing 15% FBS and 200 IU/ml IL-2.
5. Amboceptor sensitized sheep erythrocytes (EA) and rabbit erythrocytes (rE) are used immediately after purchasing to avoid nonspecific hemolysis.

Acknowledgments

Our research was supported by research grants DK080812 from NIDDK, U54-AI057160 from the NIAID to the Midwest Regional center of Excellence (MRCE) for Biodefense and Emerging Infectious Diseases Research, and from the Presidential and Liver Center Research Funds of Saint Louis University.

References

1. Negro F, Alberti A (2011) The global health burden of hepatitis C virus infection. Liver Int 31(Suppl 2):1–3
2. Gasque P (2004) Complement: a unique innate immune sensor for danger signals. Mol Immunol 41:1089–1098
3. Kim DD, Song WC (2006) Membrane complement regulatory proteins. Clin Immunol 118:127–136
4. Mollnes TE, Song WC, Lambris JD (2002) Complement in inflammatory tissue damage and disease. Trends Immunol 23:61–64
5. Blue CE, Spiller OB, Blackbourn DJ (2004) The relevance of complement to virus biology. Virology 319:176–184
6. Walport MJ (2001) Complement. First of two parts. N Engl J Med 344:1058–1066
7. Pangburn MK, Ferreira VP, Cortes C (2008) Discrimination between host and pathogens by the complement system. Vaccine 26(Suppl 8): I15–I21
8. Hourcade D, Liszewski MK, Krych-Goldberg-M, Atkinson JP (2000) Functional domains, structural variations and pathogen interactions of MCP, DAF and CR1. Immunopharmacology 49:103–116
9. Seya T, Atkinson JP (1989) Functional properties of membrane cofactor protein of complement. Biochem J 264:581–588
10. Williams P, Chaudhry Y, Goodfellow IG, Billington J, Powell R, Spiller OB et al (2003) Mapping CD55 function. The structure of two pathogen-binding domains at 1.7 A. J Biol Chem 278:10691–10696
11. Banerjee A, Mazumdar B, Meyer K, Di Bisceglie AM, Ray RB, Ray R (2011) Transcriptional repression of C4 complement by hepatitis C virus proteins. J Virol 85:4157–4166
12. Kim H, Meyer K, Di Bisceglie AM, Ray R (2013) Hepatitis C virus suppresses C9 complement synthesis and impairs membrane attack complex function. J Virol 87:5858–5867
13. Mazumdar B, Kim H, Meyer K, Bose SK, Di Bisceglie AM, Ray RB et al (2013) Hepatitis C virus infection upregulates CD55 expression on the hepatocyte surface and promotes association with virus particles. J Virol 87:7902–7910
14. Mazumdar B, Kim H, Meyer K, Bose SK, Di Bisceglie AM, Ray RB et al (2012) Hepatitis C virus proteins inhibit C3 complement production. J Virol 86:2221–2228
15. Kwon YC, Kim H, Meyer K, Di Bisceglie AM, Ray R (2016) Distinct CD55 isoform synthesis and inhibition of complement-dependent cytolysis by hepatitis C virus. J Immunol 197:1127–1136
16. Pio R, Corrales L, Lambris JD (2014) The role of complement in tumor growth. Adv Exp Med Biol 772:229–262
17. Dumestre-Perard C, Ponard D, Drouet C, Leroy V, Zarski JP, Dutertre N et al (2002) Complement C4 monitoring in the follow-up of chronic hepatitis C treatment. Clin Exp Immunol 127:131–136
18. Ali OS, Abo-Shadi MA, LN H (2005) The biological significance of serum complements C3 and C4 in HCV-related chronic liver diseases and hepatocellular carcinoma. Egypt J Immunol 12:91–99
19. El-Fatah Fahmy Hanno A, Mohiedeen KM, Deghedy A, Sayed R (2014) Serum complements C3 and C4 in chronic HCV infection and their correlation with response to pegylated interferon and ribavirin treatment. Arab J Gastroenterol 15:58–62
20. Ricklin D, Hajishengallis G, Yang K, Lambris JD (2010) Complement: a key system for immune surveillance and homeostasis. Nat Immunol 11:785–797
21. Peitsch MC, Tschopp J (1991) Assembly of macromolecular pores by immune defense systems. Curr Opin Cell Biol 3:710–716
22. Amet T, Ghabril M, Chalasani N, Byrd D, Hu N, Grantham A et al (2012) CD59 incorporation protects hepatitis C virus against complement-mediated destruction. Hepatology 55:354–363
23. Roden MM, Lee KH, Panelli MC, Marincola FM (1999) A novel cytolysis assay using fluorescent labeling and quantitative fluorescent scanning technology. J Immunol Methods 226:29–41
24. Medof ME, Walter EI, Rutgers JL, Knowles DM, Nussenzweig V (1987) Identification of the complement decay-accelerating factor (DAF) on epithelium and glandular cells and in body fluids. J Exp Med 165:848–864
25. Osuka F, Endo Y, Higuchi M, Suzuki H, Shio Y, Fujiu K et al (2006) Molecular cloning and characterization of novel splicing variants of human decay-accelerating factor. Genomics 88:316–322
26. Coyne KE, Hall SE, Thompson S, Arce MA, Kinoshita T, Fujita T et al (1992) Mapping of epitopes, glycosylation sites, and complement

regulatory domains in human decay accelerating factor. J Immunol 149:2906–2913
27. Miot S, Crespo S, Schifferli JA (2002) Distinct forms of DAF in urine and blood. J Immunol Methods 260:43–53
28. Caras IW, Weddell GN (1989) Signal peptide for protein secretion directing glycophospholipid membrane anchor attachment. Science 243:1196–1198
29. Kim H, Meyer K, Di Bisceglie AM, Ray R (2014) Inhibition of c3 convertase activity by hepatitis C virus as an additional lesion in the regulation of complement components. PLoS One 9:e101422
30. Okroj M, Holmquist E, King BC, Blom AM (2012) Functional analyses of complement convertases using C3 and C5-depleted sera. PLoS One 7:e47245

Chapter 24

Direct Ex Vivo Functional Analysis of HCV-Specific T Cells

Su-Hyung Park and Eui-Cheol Shin

Abstract

$CD8^+$ and $CD4^+$ T cells play a critical role in viral clearance during self-limited HCV infection. Moreover, induction of robust and polyfunctional memory T cells is a primary goal in the development of T-cell-based HCV prophylactic vaccines. Direct ex vivo function of HCV-specific T cells without in vitro expansion can be assessed by measuring cytokine secretion or cell proliferation upon in vitro stimulation with HCV antigens using enzyme-linked immunospot (ELISpot) assays, intracellular cytokine staining (ICS) assays, and flow cytometry-based T-cell proliferation assays. In particular, polyfunctionality of T cells can be assessed by ICS assays using multicolor flow cytometry. Here, we provide protocols to measure direct ex vivo functions of HCV-specific T cells.

Key words Hepatitis C virus, T cells, Cytokines, Proliferation, Polyfunction, Flow cytometry

1 Introduction

$CD8^+$ and $CD4^+$ T cells play an important role in HCV infection. Robust and multiple epitope-specific T-cell responses are present in patients or chimpanzees with a self-limited HCV infection, whereas weak HCV-specific T-cell responses are observed in hosts who develop a chronic persistent infection [1]. If HCV infection progresses to a chronic infection, the function of HCV-specific T cells is impaired by two different mechanisms. Virus escape mutations abrogate antiviral functions of HCV-specific T cells. If T-cell epitopes are preserved without mutations, HCV-specific T cells become exhausted and functionally impaired because of sustained antigenic stimulation [1, 2].

$CD8^+$ and $CD4^+$ T cells are also important for protective immunity in the setting of HCV challenge following spontaneous recovery or prophylactic vaccination. An essential role of memory $CD8^+$ and $CD4^+$ T cells in HCV clearance was clearly demonstrated by in vivo depletion of $CD8^+$ or $CD4^+$ T cells in rechallenge studies of HCV-recovered chimpanzees [3, 4]. Moreover, early proliferation of polyfunctional $CD8^+$ T cells induced by a T-cell-based

Mansun Law (ed.), *Hepatitis C Virus Protocols*, Methods in Molecular Biology, vol. 1911,
https://doi.org/10.1007/978-1-4939-8976-8_24,

vaccine was observed in chimpanzees with protective immunity upon HCV challenge [5]. In a clinical trial, a T-cell-based HCV prophylactic vaccine elicited sustained HCV-specific $CD8^+$ and $CD4^+$ T-cell responses with enhanced functionality and proliferative capacity [6].

Precise monitoring of HCV-specific $CD8^+$ and $CD4^+$ T-cell responses is required for understanding the mechanism of HCV clearance and the development of prophylactic HCV vaccines. Direct ex vivo functional analysis of HCV-specific T cells without in vitro expansion is performed by measuring the cytokine secretion or cell proliferation upon in vitro stimulation with HCV antigens. Direct ex vivo cytokine production of HCV-specific T cells can be assessed by enzyme-linked immunospot (ELISpot) and flow cytometry-based intracellular cytokine staining (ICS) assays (*see* **Note 1**). The proliferative capacity of HCV-specific T cells can be analyzed by a flow cytometry-based assay using fluorescent dyes that become diluted during cell proliferation. Flow cytometry-based assays have the advantage of analyzing $CD8^+$ and $CD4^+$ T-cell responses separately in a single assay. Polyfunctional T cells exert several effector functions simultaneously [7], and the induction of polyfunctional T cells is considered a practical goal in the development of T-cell-based HCV prophylactic vaccines [5, 6]. Polyfunctionality of HCV-specific T cells can be assessed by measuring the production of multiple cytokines in a single cell in ICS assays using multicolor flow cytometry (*see* **Note 2**). Here, we describe standard protocols for direct ex vivo functional assays, including cytokine ELISpot assays, ICS assays, and flow cytometry-based T-cell proliferation assays.

2 Materials

All materials should be of cell culture grade.

2.1 Ex Vivo IFN-γ ELISpot Assay with Overlapping HCV Peptides (See Notes 1 and 2)

1. Complete RPMI medium: RPMI-1640 medium with 10% FBS, 2 mM L-glutamine, 100 IU/ml penicillin, and 100 μg/ml streptomycin.
2. Overlapping peptides (OLPs) spanning HCV proteins: 15-mer peptides overlapping by 10 amino acids covering the desired HCV polyprotein sequences (*see* **Note 3**). To prepare OLP pools, mix 50 μl of each OLP solution (1 mg/ml in 5% DMSO/PBS) and adjust the total volume with 5% DMSO/PBS to 2000 μl. Then, the concentration of each individual OLP in the pool will be 25 μg/ml.
3. Positive control: phytohemagglutinin (PHA; Sigma) as mitogen.
4. Coating antibody: anti-IFN-γ monoclonal antibody (Clone 1-D1K, MabTech).

5. Detection antibody: biotinylated anti-IFN-γ monoclonal antibody (Clone 7-B6-1, MabTech).
6. Alkaline phosphatase (AP)-conjugated streptavidin (Mabtech) (*see* **Note 4**).
7. AP-conjugate substrate kit (Bio-Rad).
8. Sterile phosphate-buffered saline (PBS).
9. PBS with 0.05% Tween 20 as washing buffer (PBS/Tween).
10. PBS with 1% BSA as coating buffer (PBS/BSA).
11. PBS with 1% BSA and 0.05% Tween 20 as substrate dilution buffer (PBS/Tween/BSA).
12. ELISpot plates (Cat. no. MSIP S4510, Millipore).
13. ELISpot reader (AID).
14. Class II biosafety cabinet (BSC).
15. Tabletop centrifuge.
16. Microplate washer.
17. CO_2 incubator.

2.2 Ex Vivo ICS Assay with Overlapping HCV Peptides

1. Complete RPMI medium: RPMI-1640 medium with 10% FBS, 2 mM L-glutamine, 100 IU/ml penicillin and 100 μg/ml streptomycin.
2. OLPs spanning HCV proteins: 15-mer peptides overlapping by 10 amino acids covering the desired HCV polyprotein sequences (*see* **Note 3**). To prepare OLP pools, mix 50 μl of each OLP solution (1 mg/ml in 5% DMSO/PBS) and adjust the total volume with 5% DMSO/PBS to 2000 μl. Then, the concentration of each individual OLP in the pool will be 25 μg/ml.
3. Anti-CD3 (clone OKT3) and anti-CD28 (clone CD28.2) antibodies.
4. Brefeldin A-containing transport inhibitor, BD GolgiPlug and monensin containing transport inhibitor, BD GolgiStop.
5. LIVE/DEAD Fixable Red Dead Cell Stain Kit for 488 nm excitation (Molecular Probes).
6. BD Horizon Brilliant Stain Buffer (BD) (*see* **Note 5**).
7. Surface staining antibodies: anti-CD3-BV605 (clone SK7), anti-CD4-PerCP-Cy5.5 (clone RPA-T4), anti-CD8-BV711 (clone SK1), anti-CD14-PE-Texas Red (clone 61D3), and anti-CD19-PE-Texas Red (clone HIB19) antibodies.
8. Intracellular staining antibodies: anti-IFN-γ-BV421 (clone 4S. B3), anti-TNF-α-APC (clone MAb11), and anti-MIP-1β-PE-Cy7 (clone D21-1351) antibodies (*see* **Notes 6** and **7**).
9. Anti-CD107a-PE (clone H4A3) antibody.

10. Fixation and permeabilization buffer: Foxp3/Transcription Factor Staining Buffer Set (Thermo Fisher).
11. FACS staining buffer: PBS with 2% FBS.
12. White-capped FACS tubes.
13. Flow cytometry (LSRII).
14. CO_2 incubator.
15. Class II biosafety cabinet (BSC).
16. Tabletop centrifuge.

2.3 T-Cell Proliferation Assay Using Flow Cytometry

1. CellTrace Violet Cell Proliferation Kit (Molecular Probes).
2. LIVE/DEAD Fixable Red Dead Cell Stain Kit (Molecular Probes).
3. Complete RPMI medium: RPMI-1640 medium with 10% FBS, 2 mM L-glutamine, 100 IU/ml penicillin, 100 μg/ml streptomycin.
4. CellTrace buffer: 5% FBS in sterile PBS.
5. FACS staining buffer: PBS with 2% FBS.
6. Culture plate: 96-well flat-bottom culture plate.
7. Surface staining antibodies: anti-CD3-BV605 (clone SK7), anti-CD4-BV786 (clone RPA-T4), anti-CD8-APC-Cy (clone SK1), anti-CD14-PE-Texas Red (clone 61D3), and anti-CD19-PE-Texas Red (clone HIB19) antibodies (*see* **Note 8**).
8. Fluorochrome-conjugated MHC class I multimers (Optional).
9. Class II biosafety cabinet (BSC).
10. Tabletop centrifuge.
11. CO_2incubator.

3 Methods

3.1 Ex Vivo IFN-γ ELISpot Assay with Overlapping HCV Peptides

3.1.1 Day 1: Coating of Plate with Anti-IFN-γ Coating Antibody

Perform experiments in a class II biosafety cabinet.

1. Prepare anti-IFN-γ coating antibody solution at 2 μg/ml in sterile PBS (*see* **Note 9**).
2. When using PVDF-membrane plates, the membranes should be prewetted with 50 μl 70% ethanol at room temperature for 1 min. After the wells are thoroughly washed three times with sterile PBS, add a proper volume of coating antibodies to the wells (*see* **Note 10**).
3. Add 100 μl of coating antibody solution to each well.
4. Incubate the plate at 4 °C overnight in a humidified box.

3.1.2 Day 0: Assay Set Up (Start in the Morning)

Perform experiments in a class II biosafety cabinet.

1. Wash the coated plate four times with sterile PBS.
2. Block the plate with 200 μl of 1% BSA/PBS for 2 h at room temperature in a class II biosafety cabinet.
3. Wash the plate twice with 200 μl of sterile PBS.
4. Block the plate with 200 μl of complete RPMI medium for 1 h at room temperature.
5. Discard medium, and blot the plate once on a paper towel.
6. Prepare a single-cell suspension of peripheral blood mononuclear cells (PBMCs) at 3×10^6 cells/ml in complete RPMI medium (*see* **Note 11**).
7. Add 100 μl of prediluted PBMC suspension to each well (0.3×10^5 cells/well) (*see* **Note 12**).
8. Predilute HCV OLP pools by adding 8 μl of each OLP pool to 92 μl of compete RPMI medium. Each OLP in the OLP pool is diluted to a concentration of 2 μg/ml.
9. Add 100 μl of prediluted HCV OLP pool to the plate. The final concentration of each OLP will be 1 μg/ml (*see* **Note 13**).
10. Set up negative control wells using 5% DMSO/PBS (final concentration of DMSO should be 0.2%) and positive control wells using 1 μg/ml of PHA.
11. Incubate the plate at 37 °C in a CO_2 incubator for 20–48 h (*see* **Note 14**). Avoid stacking or moving the plate during incubation (*see* **Note 15**).

3.1.3 Day 1 or 2: Development of the ELISpot Plate

Aseptic experimental conditions are no longer required.

1. Prepare biotinylated anti-IFN-γ detection antibody solution in PBS/Tween/BSA buffer at a final concentration of 0.25 μg/ml (*see* **Note 9**).
2. Wash the plate three times with 200 μl of PBS and four times with 200 μl of PBS/Tween. Blot the plate on a paper towel after each wash.
3. Add 100 μl of diluted detection antibody solution to each well.
4. Incubate the plate at room temperature for 4 h or at 4 °C overnight in a humidified box.
5. Wash the plate six times with 200 μl of PBS/Tween.
6. Prepare streptavidin-AP solution by adding 5 μl of streptavidin-AP to 10 ml of PBS/Tween/BSA (1:2000 dilution).
7. Add 100 μl of diluted streptavidin-AP to each well.
8. Incubate the plate at room temperature for 1 h (*see* **Note 16**).

9. Wash the plate eight times with PBS (no Tween 20, no BSA) (*see* **Note 17**). Take the back cover off, and rinse the back of the membrane with PBS using a squirt bottle. After drying, reassemble the plate.
10. Prepare the AP development solution following the manufacturer's instructions.
11. Add 100 μl of AP development solution to each well.
12. Wait for ~10 min until spots appear in wells.
13. Rinse the plate under running tap water. Take the back cover off, and rinse the back of the membrane under running tap water.
14. Dry the plate completely in the dark. The plate is now ready to read (*see* **Note 18**).

3.1.4 Data Analysis

1. Read the plate and count the spots using an ELISpot reader.
2. Data can be reported as the number of spot-forming cells (SFCs) per 3×10^5 or 1×10^6 PBMCs.
3. Numbers of SFC against specific antigens can be calculated as follows: mean SFC of experimental wells—mean SFC of background wells.

3.2 Ex Vivo ICS Assay with Overlapping HCV Peptides

3.2.1 Stimulation

1. Prepare a single-cell suspension of PBMCs at 4×10^6 cells/ml in complete RPMI medium (*see* **Note 11**).
2. Add 250 μl of prediluted PBMC suspension to each FACS tube (1×10^6 cells/tube).
3. Predilute HCV OLP pools by adding 20 μl of each OLP pool to 230 μl of complete RPMI medium. Each OLP in the OLP pool is diluted to a concentration of 2 μg/ml.
4. Add 250 μl of prediluted HCV OLP pool to the cell-containing FACS tube. Final concentration of each OLP will be 1 μg/ml (*see* **Note 13**).
5. Add anti-CD28 antibodies at a final concentration of 1 μg/ml.
6. Set up negative control tubes using 5% DMSO/PBS (final concentration of DMSO should be 0.2%) and positive control tubes using 0.1 μg/ml of anti-CD3 antibodies and 1 μg/ml of anti-CD28 antibodies (*see* **Note 19**).
7. Add PE-conjugated anti-CD107a antibodies (*see* **Note 20**).
8. Incubate the tubes at 37 °C in a CO_2 incubator for 1 h.
9. Add Brefeldin A containing transport inhibitor (BD GolgiPlug) and Monensin containing transport inhibitor (BD GolgiStop) to each cell culture tube and mix thoroughly (*see* **Note 21**).
10. Incubate the tubes at 37 °C in a CO_2 incubator for 12 h.

3.2.2 Cell Surface Staining

1. Wash cells three times with FACS buffer.
2. After the last wash, discard the supernatant and pulse-vortex the sample to completely dissociate the cell pellet. Resuspend the cells in 100 μl of FACS buffer.
3. Stain cells with LIVE/DEAD dye at room temperature for 20 min. Protect from light.
4. Add 2 ml of FACS buffer to each tube. Centrifuge cells at 300 × *g* for 8 min and carefully remove supernatant completely.
5. Resuspend the cells in 100 μl of FACS buffer.
6. Premix the cell surface antibodies to CD3, CD4, CD8, CD14, and CD19 with Brilliant Stain Buffer.
7. Add the mixture of fluorochrome-conjugated antibodies for surface staining.
8. Stain cells with cell surface antibody mixture at 4 °C for 20 min. Protect from light.
9. Add 2 ml of FACS buffer to each tube. Centrifuge cells at 300 × *g* for 8 min and carefully remove supernatant completely. Then, pulse-vortex the sample to completely dissociate the cell pellet.

3.2.3 Fixation/ Permeabilization and Intracellular Staining

1. Fix cells by adding 1 ml of fixation buffer and pulse-vortex to mix.
2. Incubate at room temperature for 30 min. Protect from light.
3. Add 2 ml of 1× permeabilization buffer and centrifuge at 400 × *g* at room temperature for 8 min. Discard the supernatant.
4. Repeat **step 3**.
5. Resuspend cell pellet in 100 μl of 1× permeabilization Buffer.
6. Prepare a mixture of intracellular antibodies to IFN-γ, TNF-α, and MIP-1β (*see* **Notes** 7 and **22**).
7. Stain the cells with the mixture of intracellular antibodies at 4 °C for 30 min. Protect from light.
8. Add 3 ml of 1× permeabilization buffer to each tube and centrifuge at 400 × *g* at room temperature for 8 min. Discard the supernatant.
9. Add 2 ml of FACS buffer to each tube and centrifuge at 400 × *g* at room temperature for 8 min. Discard the supernatant.
10. Resuspend cells in an appropriate volume of FACS buffer.
11. Analyze cells by flow cytometry.
12. $CD8^+$ and $CD4^+$ T-cell responses can be analyzed separately by gating for each cell population.

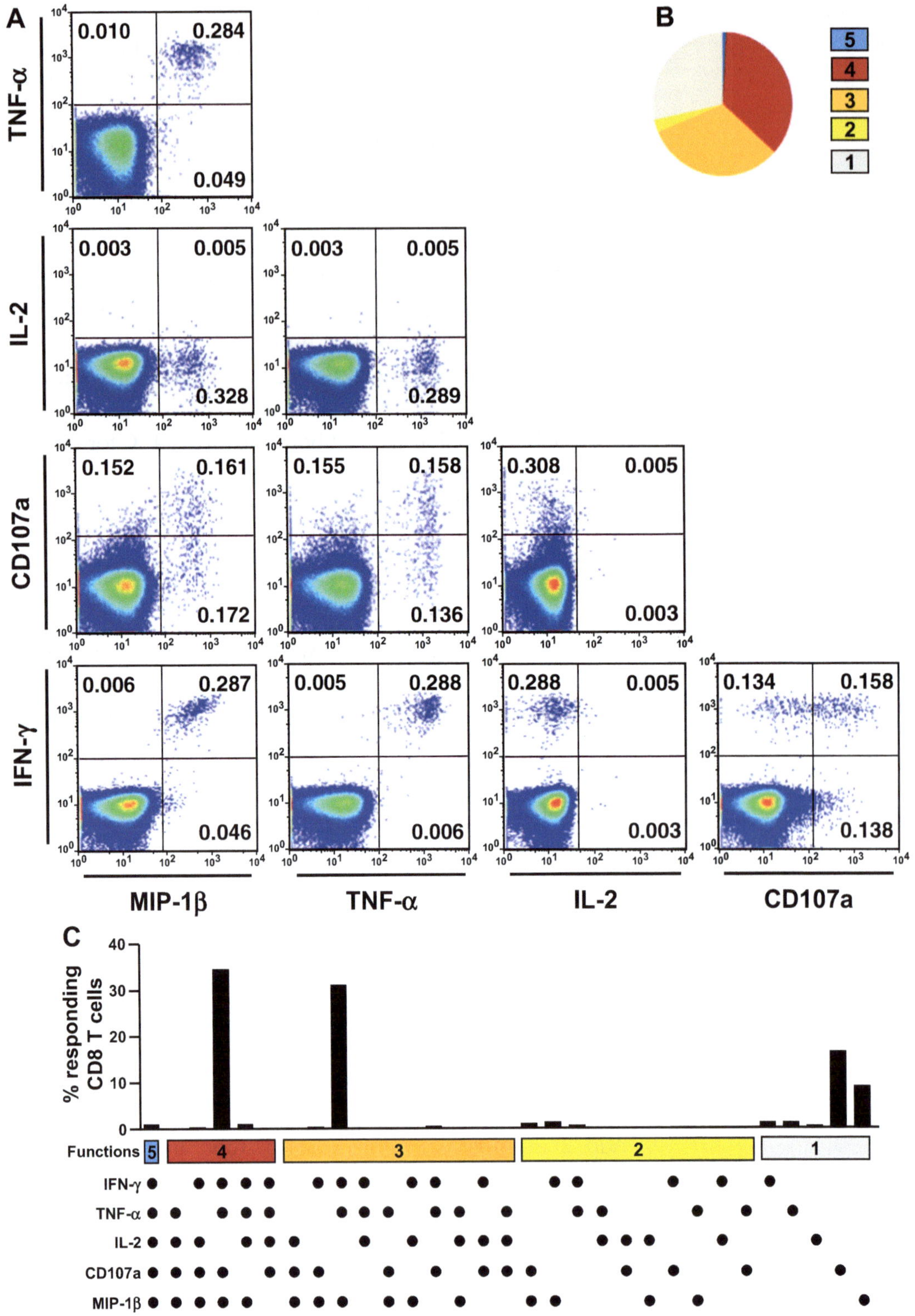

Fig. 1 Polyfunctionality assay of HCV-specific T cells. A representative example of polyfunctionality assays of $CD8^+$ T cells is presented. Patient PBMCs were stimulated with a single T-cell epitope peptide. For the gating of $CD8^+$ T cells, the production of IFN-γ, TNF-α, MIP-1β, and IL-2 was simultaneously assessed by ICS assay.

13. Polyfunctionality of T cells can be assessed by analyzing IFN-γ, TNF-α, MIP-1β, and/or IL-2 simultaneously. In the $CD8^+$ T cell population, cytotoxic function can be assessed by CD107a staining (*see* **Note 7**) (Fig. 1).

3.3 T-Cell Proliferation Assay Using Flow Cytometry

3.3.1 Day 1: Assay Set Up

1. Prepare a single-cell suspension of PBMCs, and wash PBMCs (centrifuge at 300 × *g* for 8 min and remove the supernatant) twice with CellTrace buffer and resuspend at concentration of 1×10^6 cells/ml in CellTrace buffer.
2. Prepare a working solution of CellTrace™ Violet dye immediately prior to the next step (*see* **Note 23**).
3. Stain cells with prediluted CellTrace™ Violet solution at 37 °C for 20 min (1 μl of prediluted CellTrace dye solution per one million cells in 1 ml of prewarmed CellTrace buffer). Protect from light.
4. Add four times the original staining volume of CellTrace buffer to cells and incubate for 5 min.
5. Centrifuge cells at 300 × *g* for 8 min and carefully remove the supernatant completely.
6. Suspend cells at 4×10^6 cells/ml in fresh prewarmed RPMI culture medium.
7. Incubate cells for at least 10 min.
8. Add 100 μl of cell suspension to 96-well culture plate.
9. Predilute HCV OLP pools by adding 8 μl of each OLP pool to 92 μl of compete RPMI medium. Each OLP in the OLP pool will be diluted to a concentration of 2 μg/ml.
10. Add 100 μl of prediluted HCV OLP pool to the plate. The final concentration of each OLP will be 1 μg/ml (*see* **Note 13**).
11. Culture cells at 37 °C in a CO_2 incubator for 4–7 days (*see* **Note 24**).

3.3.2 Day 5–8: FACS Staining and Analysis

1. Harvest cells into FACS tubes by pipetting.
2. Wash cells and suspend in 100 μl of FACS staining buffer.
3. Stain cells with LIVE/DEAD® dye at room temperature for 20 min. Protect from light.
4. Wash cells and suspend in 100 μl of FACS staining buffer.

Fig. 1 (continued) In addition, the cytotoxic degranulation activity was measured by CD107a staining. The data are presented as FACS dot plots (**a**). The pie graph shows the fraction of T cells positive for a given number of functions (**b**). Detailed analysis of polyfunctionality is presented with every possible combination of functions (**c**). The figure is modified from a ref. 8 according to the "Creative Commons Attribution (CC BY) license"

5. (Optional) Add pretitrated fluorochrome-conjugated MHC multimers (*see* **Note 25**). Incubate at room temperature for 20 min, and then wash the cells and resuspend them in 100 μl of FACS staining buffer.
6. Add fluorochrome-conjugated antibodies to CD3, CD4, CD8, CD14 and CD19.
7. Incubate for 20 min on ice in the dark.
8. Wash the cells and resuspend them in 500 μl of FACS staining buffer (*see* **Note 26**).
9. Analyze the cells by flow cytometry in the gated population.
10. Cells with lower CellTrace Violet intensity than the control represent proliferating cells. Assess the proliferative response by measuring the percentage of CellTrace Violet-low cells, calculating the stimulation index or calculating the mitotic index (*see* **Note 27**).

4 Notes

1. It is generally considered that the sensitivity of the IFN-γ ELISpot assay is 10- to 100-fold greater than that of the ICS assay [9].
2. In the ELISpot assay, only one analyte can be measured at a time. To overcome this limitation, FluoroSpot assays have been developed by several companies, including Mabtech, R&D Systems, Abcam, and U-Cytech. In FluoroSpot assays, multiple fluorescent-labeled anti-cytokine detection antibodies are used to measure multiple analytes in a single assay. For detailed information on the commercially available FluoroSpot assay kits, please refer to the manufacturer's instructions.
3. Alternatively, the use of 20-mer peptides overlapping by 15 amino acids can be considered. Both 15-mer and 20-mer peptides can be used to detect both $CD8^+$ and $CD4^+$ T-cell responses, but 20-mer peptides are considered to be more useful than 15-mer peptides for the detection of $CD4^+$ T-cell responses.
4. Two enzyme choices are available: alkaline phosphatase (AP) and horseradish peroxidase (HRP). Since AP and HRP catalyze different reactions, AP and HRP require BCIP/NBT substrate and TMB substrate, respectively. Note that enzyme-linked immunosorbent assay (ELISA) substrates cannot be used in ELISpot assays.
5. Because of nonspecific interactions between the polymer dyes themselves, which may affect data interpretation, Brilliant Stain Buffer should be used when two or more Brilliant dyes are used in the same experiment.

6. If ICS assays are performed to assess the production of a single cytokine, antibodies to the interesting cytokine can be chosen from the list of 'intracellular staining antibodies' and used following cell staining with 'surface staining antibodies' according to the protocol described in Subheading 3.2.
7. In polyfunctionality assays of T cells, simultaneous production of multiple cytokines in a single cell is assessed by ICS assays. In this case, multiple antibodies to the interesting cytokines can be chosen from the list of 'intracellular staining antibodies'.
8. Employing a dump channel to gate out dead cells and unwanted cell populations is necessary to reduce the background staining. In particular, when working with MHC class I multimers, the exclusion of $CD19^+$ cells is necessary to reduce background staining. In this protocol, the dump channel (PE-Texas Red channel) includes viability (LIVE/DEAD Fixable Red Dead Cell Stain Kit) and CD14 and CD19 staining.
9. Titrate the coating and detection antibodies to find their optimal concentrations.
10. PVDF membranes have a porous hydrophobic structure that requires an activating step to acquire high protein-binding capacity. Activation is achieved by wetting the membrane with a short treatment with ethanol, which makes the membrane hydrophilic.
11. Freezing and thawing of cells leads to a varying number of dead cells that may not be visible immediately after thawing. If frozen PBMCs are used for ELISpot or ICS assays, it is recommended to rest the PBMCs after thawing in fresh culture medium for at least 1 h at 37 °C.
12. The number of cells in each well must be adapted to the expected frequency of the cytokine-secreting cells by conducting pilot experiments. If the frequency is low, a higher number of cells are needed. When the frequency is high, a lower number of cells are required. More than 500,000 cells/well is not generally recommended since this may create multiple cell layers.
13. If a single minimal epitope peptide is used as a stimulant, a final concentration of the peptide of 10 μg/ml is recommended.
14. Unnecessarily long incubation times may cause high background staining or merged/indistinguishable spots. Therefore, an optimal incubation time should be determined in advance by pilot experiments.
15. Avoid movements of the plates during cell incubation. Moving the plate during incubation may cause formation of irregular or double spots. Moreover, stacking the plates may prevent even heat distribution across the plates or individual wells.

16. An incubation time of more than 1 h may result in increased dark background color.
17. Detergents, such as Tween 20, can inhibit enzyme reactions. Use PBS only for final wash steps.
18. Spots may fade when the plates are exposed to light. The AP-developed ELISpot plates can be stored for at least 6 months in the dark.
19. Streptococcal enterotoxin B or PMA/Ionomycin can be used as a positive control instead of anti-CD3/anti-CD28 antibodies.
20. PE-conjugated anti-CD107a antibodies are added to assess the cytotoxic degranulation activity of $CD8^+$ T cells. It is important NOT to restain for CD107a with antibodies following the incubation period, especially not after permeabilization of cells.
21. Monensin helps to maximize the CD107a readout, whereas brefeldin A helps to maximize other cytokine (INF-γ, TNF-α, and MIP-1β) readouts.
22. If necessary, anti-IL-2 antibody can be included in the mixture of intracellular antibodies.
23. Prepare a fresh working solution for each experiment to avoid a rapid decrease in signal intensity. The optimal labeling concentration may vary according to the sample volume and cell density. Recommended concentrations for CellTrace staining are 1–10 μM. If tracking five or more generations of cell cycles, dye concentrations of 5–10 μM are recommended; otherwise, 1–2 μM may be sufficient.
24. Optimal incubation times should be determined in advance. If the incubation time is more than 4 days, adding 20 U/ml of IL-2 at day 4 can be considered.
25. Like antibodies, optimal concentration of MHC multimers (tetramer, pentamer or dextramer) should be determined in advance to maximize the separation and to minimize the background staining due to nonspecific binding.
26. If clumped cells are visible, pass the cell suspension through a cell strainer with a 40 μm mesh size.
27. Stimulation index (SI) measures the effects of stimulation on a relative number of cells that have divided at least once, calculated as:

    SI = proliferative fraction in antigen-stimulated cell population/proliferative fraction in unstimulated cell population

    Mitotic index estimates the total number of divisional events during the culture period and can be calculated by the dividing mitotic events by the absolute number of precursor cells based on the number of cells in each mitotic division [10].

References

1. Shin EC, Sung PS, Park SH (2016) Immune responses and immunopathology in acute and chronic viral hepatitis. Nat Rev Immunol 16:509–523
2. Park SH, Rehermann B (2014) Immune responses to HCV and other hepatitis viruses. Immunity 40:13–24
3. Grakoui A, Shoukry NH, Woollard DJ, Han JH, Hanson HL, Ghrayeb J et al (2003) HCV persistence and immune evasion in the absence of memory T cell help. Science 302:659–662
4. Shoukry NH, Grakoui A, Houghton M, Chien DY, Ghrayeb J, Reimann KA et al (2003) Memory CD8+ T cells are required for protection from persistent hepatitis C virus infection. J Exp Med 197:1645–1655
5. Park SH, Shin EC, Capone S, Caggiari L, De Re V, Nicosia A et al (2012) Successful vaccination induces multifunctional memory T-cell precursors associated with early control of hepatitis C virus. Gastroenterology 143:1048–1060 e1044
6. Swadling L, Capone S, Antrobus RD, Brown A, Richardson R, Newell EW et al (2014) A human vaccine strategy based on chimpanzee adenoviral and MVA vectors that primes, boosts, and sustains functional HCV-specific T cell memory. Sci Transl Med 6:261ra153
7. Precopio ML, Betts MR, Parrino J, Price DA, Gostick E, Ambrozak DR et al (2007) Immunization with vaccinia virus induces polyfunctional and phenotypically distinctive CD8(+) T cell responses. J Exp Med 204:1405–1416
8. Choi YS, Lee JE, Nam SJ, Park JT, Kim HS, Choi KH et al (2013) Two distinct functional patterns of hepatitis C virus (HCV)-specific T cell responses in seronegative, aviremic patients. PLoS One 8:e62319
9. Rehermann B, Naoumov NV (2007) Immunological techniques in viral hepatitis. J Hepatol 46:508–520
10. Wallace PK, Tario JD Jr, Fisher JL, Wallace SS, Ernstoff MS, Muirhead KA (2008) Tracking antigen-driven responses by flow cytometry: monitoring proliferation by dye dilution. Cytometry A 73:1019–1034

Chapter 25

Simultaneous Analysis of HCV-Specific CD4⁺ and CD8⁺ T Cells by Multicolor Flow Cytometry

David Wolski and Georg M. Lauer

Abstract

CD4 T cell responses are key to effective control of HCV infection. However, their precise mechanisms of action and the molecular programs leading to effective T help versus CD4 T cell failure remain elusive. Direct visualization of HCV-specific CD4 T cells using HLA class II tetramers holds the promise to better define the function and phenotype of these cells and to isolate them for further molecular analysis. Here we describe how to utilize peptide-MHC (pMHC) class II tetramers in multicolor flow cytometry to define the expression of molecules on the surface and within HCV-specific CD4 T cells, how to analyze HCV-specific CD4 and CD8 T cells in the same tube, and how to sort live HCV-specific CD4 T cells as single cells or T cell populations for further analysis by RNAseq or other methods.

Key words HCV-specific T cells, CD4, CD8, Cell sorting, Flow cytometry, Tetramer, Pentamer, Peptide-MHC, Intracellular staining, Transcription factor

1 Introduction

Many studies have demonstrated the importance of CD4 T cells for spontaneous resolution of HCV infection, where HCV-specific CD4 T cells with high proliferative potential can be found in early acute infection as well as many years after viral clearance. In contrast, HCV-specific CD4 T cells in patients with chronic infection are usually not detectable by standard proliferation assays [1]. This led to the hypothesis that most patients with chronically evolving HCV infection lack HCV-specific CD4 T cells or, at the very least, have very weak responses against a very narrow repertoire of targeted CD4 epitopes. Using enhanced in vitro proliferation assays and HLA class II tetramer staining we could show that HCV-specific CD4 T cells are present in all subjects during early infection and typically target a range of HCV antigen-derived epitopes, but become undetectable in peripheral blood within the first year of chronic infection [2]. These findings shift the investigative focus from an assumed primary failure to mount an HCV-specific

Mansun Law (ed.), *Hepatitis C Virus Protocols*, Methods in Molecular Biology, vol. 1911,
https://doi.org/10.1007/978-1-4939-8976-8_25,

CD4 T cells response to the mechanisms of secondary CD4 T cell failure. Answering the most pertinent questions about virus-specific CD4 T cell function and how it is regulated in different disease outcomes will require direct ex-vivo functional and phenotypic analyses, as well as isolation of specific cells for further analysis by RNA sequencing and other high-throughput omics approaches. Sourcing input cells for these experimental techniques can only reliably be accomplished by direct ex vivo identification and isolation of virus-specific T cells using pMHC class II tetramers by multicolor flow cytometry [3, 4]. While other methods of detection/isolation of virus-specific T cells exist, they each have considerable caveats that put them at a major disadvantage when compared to multicolor flow cytometry. Enrichment-based isolation strategies, such as those that employ magnetic particles, can prohibit the evaluation of additional parameters and assessment of population purity and are generally not well suited for exceedingly rare populations. In vitro expansion assays can circumvent the problem posed by cell population rarity, but they have the potential to fundamentally change the phenotypic and functional profiles of cells when compared to their ex vivo counterparts [5]. Approaches that rely on detection of HCV-specific T cells by intracellular cytokine staining of peptide-stimulated cells are often used in combination with cell expansion and share their limitation of altering the physiological state of detected cells. Furthermore, they are based on the often flawed assumption that all HCV-specific T cells express the cytokines that are used as readout of specificity [4]. Finally, while mass cytometric analysis of cells allows for inclusion of a much larger number of parameters, all cells are destroyed during the acquisition step and overall transmission rates are low, making this technique unsuitable for cell sorting or the detection of rare events in precious clinical samples [6].

The recent addition of many fluorochromes with excitation wavelengths in the ultraviolet and violet range of the electromagnetic spectrum and the continuous improvements of optical and signal processing components used in modern flow cytometers have made the design of serviceable flow cytometry panels with 18 or more parameters straightforward [7]. While these numbers cannot rival the theoretical number of parameters observable with mass cytometers, they come close to most real-world applications that utilize these techniques without destroying the cell material in the process.

These developments in multicolor flow cytometry together with advances in sensitivity and specificity of peptide-MHC (pMHC) multimers have enabled us to combine pMHC class II and pMHC class I multimer staining with many other markers of interest, allowing for simultaneous and relatively comprehensive assessment of HCV-specific CD4 and CD8 T cell responses.

While this protocol, in its entirety, focuses on the analysis of highly pure populations of viable HCV-specific CD4 and CD8 T cells in the context of immune checkpoints (PD-1, CTLA-4), important transcriptional regulators of T cell function (T-bet, Eomes), regulatory cell lineage markers (FOXP3, CD25), and memory phenotype (CD45RA, CCR7), we have indicated where minor variations in the experimental procedure make it compatible with sorting of live cell populations for downstream applications such as RNA sequencing. In general, these variations are largely focused on the omission of the intracellular staining steps, as they require fixation and permeabilization of cells, both of which are incompatible with live cell sorting. Note that this protocol does not call for the inclusion of commonly used markers of apoptosis, such as Annexin V, as these markers can be unsuited for use with protocols that include fixation and permeabilization of cells prior to staining for apoptotic cells. We do, however, recommend the use of such markers in cell sorting experiments where the explicit in- or exclusion of these cells is desired [8], since Fixable Live/Dead viability stains identify necrotic rather than apoptotic cells, which would therefore be identified as regular live cells.

When designing panels for multicolor flow cytometry experiments, several considerations need to be made. Since fluorophores vary greatly in brightness and different antibodies and even different antibody clones differ significantly in affinity and avidity, brighter fluorophores should be reserved for important markers of interest with low signal or populations of interest, like the tetramer positive population, whereas markers with good separation, such as T cell lineage markers, or markers for exclusion (dump channel) can be assigned to relatively dim fluorophores. At the same time, markers that need to be visualized against each other should not be assigned to fluorophores with known spectral overlap. Similarly, assigning fluorophores with known spectral overlap to markers that could confound the staining of a rare population of interest, such as choosing PE-Cy5 for CD3^{+} T cells and APC for tetramer^{+} cells, since cross-activation of PE-Cy5 by the red laser and consequent spill of fluorescence into the APC channel could lead to overestimation of the truly tetramer^{+} population. Inversely, cross-activation of a BV421-labeled tetramer by a 355 nm UV laser could result in falsely excluded tetramer^{+} cells, when used in combination with a UV-excitable blue dead cell stain. We highly recommend that the reader familiarize him- or herself with the finer technical aspects of flow cytometry and the underlying physical principles, as it will help to avoid pitfalls and help troubleshoot issues that may arise when designing and performing multicolor flow cytometry experiments [9].

When detecting virus-specific T cells by pMHC multimers, it is important to be mindful of the fact that the choice of antibody clones for T-cell coreceptors CD3, CD4, and CD8 can have profound effects on multimer staining, as they can (1) block or stabilize binding of pMHC complexes to the T-cell receptors, and (2) induce cellular signaling events that can lead to changes in cell phenotype [10–12]. While performing multimer staining prior to surface antibody staining greatly reduces the potential of coreceptor antibodies blocking TCR binding sites, we advise the readers to keep these caveats in mind, and perform experiments swiftly and in adherence with the guidelines laid out in this protocol to avoid unwanted and staining-induced phenotypic changes in the cells under study. We also remind the reader that the additional steps required for intracellular staining make use of harsh detergents that will significantly alter cell morphology and can lead to a slight loss of pMHC class II tetramer staining while simultaneously reducing background staining (Fig. 1c), when compared to cells stained for surface antigens only (Fig. 1b).

2 Materials

All materials should be of cell culture grade and stored at 4 °C, unless specified otherwise.

2.1 Buffer Solutions and Media

1. Dulbecco's phosphate buffered saline (PBS).
2. Sterile-filtered, heat-inactivated fetal bovine serum (FBS).
3. FACS Buffer: PBS with 2% FBS (*see* **Note 1**).
4. R10 Medium: RPMI-1640 with 10% FBS, 2 mM L-glutamine, 5 mL 10 mM HEPES buffer, 10,000 U/L penicillin, and 100 mg/L streptomycin.
5. Fixation Buffer: 4% Paraformaldehyde solution in PBS.
6. DNAse I stock solution: 10 mg/mL DNAse I in 0.15 M NaCl (5 mg/mL).

2.2 Antibodies and Staining Reagents

1. Anti-human CD4 antibody—Brilliant Ultraviolet 395 (SK3).
2. Anti-human PD-1 antibody—Brilliant Ultraviolet 737 (EH12.1).
3. Anti-human CTLA-4 antibody—Brilliant Violet 421 (BN13).
4. Anti-human CD45RA antibody—Brilliant™ Violet 605 (HI100).
5. Anti-human T-bet antibody—Brilliant Violet 711 (4B10).
6. Anti-human CD8 antibody—Brilliant Violet 785 (RPA-T8).
7. Anti-human Eomes antibody—FITC (W01928).
8. Anti-human CD3 antibody—PE/Dazzle 594 (UCHT1).
9. Anti-human CCR7 antibody—PerCP-Cy5.5 (G043H7).
10. Anti-human CD25 antibody—PE-Cy7 (BC96).

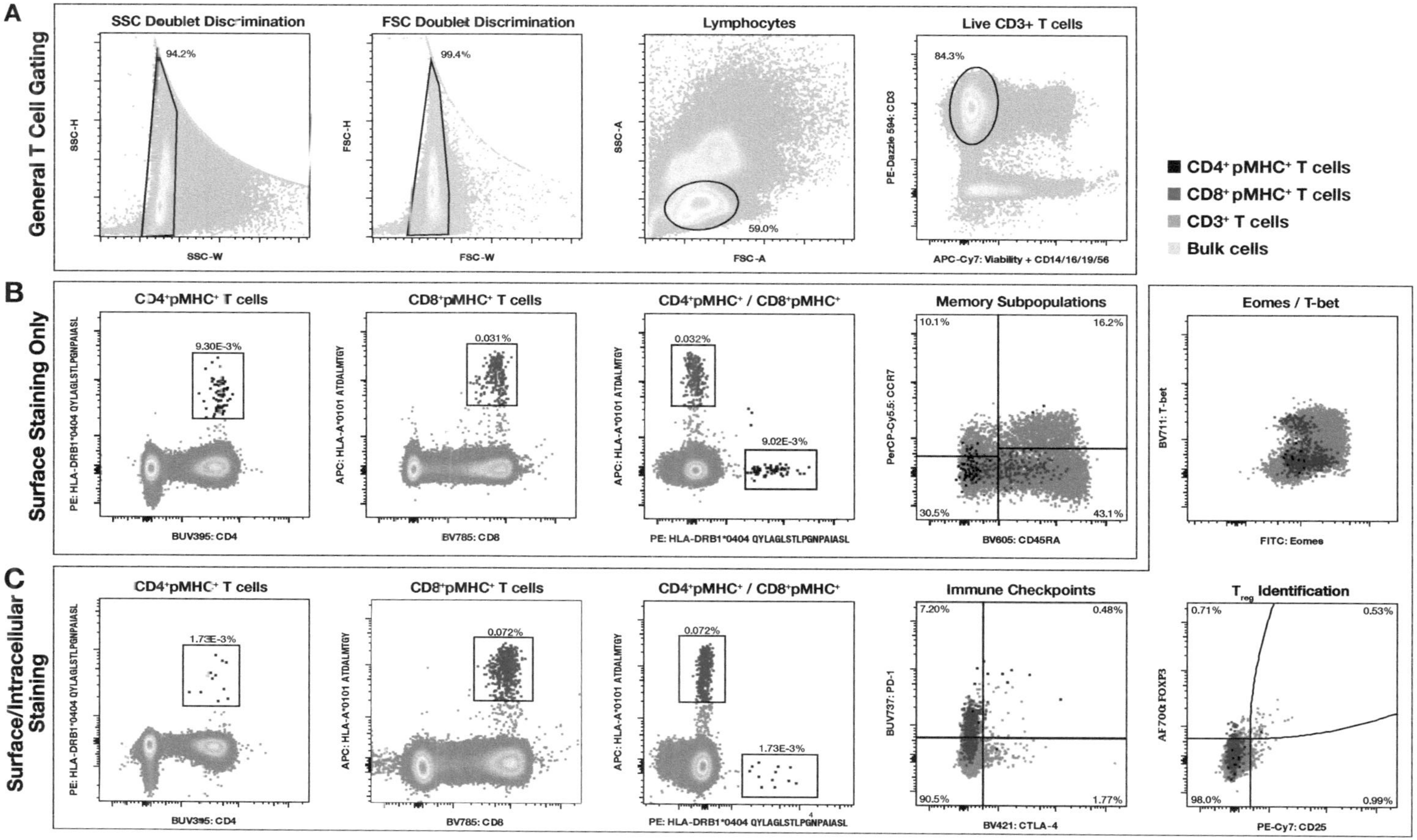

Fig. 1 Gating strategy for simultaneous detection of HCV-specific $CD4^+$ and $CD8^+$ T cells. (**a**) Cells are first gated to exclude double events based on side and forward scatter width and height (SSC-W, SSC-H, FSC-W, and FSC-H), after which lymphocytes are identified based on forward and side scatter area (FSC-A and SSC-A) and live $CD3^+$ T cells are identified by excluding dead cells (Live/Dead$^+$), monocytes ($CD14^+$), NK and NKT cells ($CD16^+$ and $CD56^+$), and B cells ($CD19^+$). (**b** and **c**) After identifying live $CD3^+$ cells, HCV-specific $CD4^+$ and $CD8^+$ T cells are identified by gating on $CD4^+pMHC^+$ and $CD8^+pMHC^+$ events. Other cell populations of interest can then be identified based on additional markers. While the panel was designed to accommodate all surface and intracellular markers in combination with pMHC multimers for $CD4^+$ and $CD8^+$ T cells, repeated plots show differences in frequencies and background staining for $pMHC^+$ populations when staining for (**b**) surface antigens only or for (**c**) both surface and intracellular antigens

11. Anti-human FOXP3 antibody—Alexa Fluor 700 (259D).
12. Anti-human CD14 antibody—APC-Cy7 (M5E2).
13. Anti-human CD16 antibody—APC-Cy7 (3G8).
14. Anti-human CD19 antibody—APC-Cy7 (H1B19).
15. Anti-human CD56 antibody—APC-Cy7 (HCD56).
16. LIVE/DEAD Fixable Near-IR Dead Cell Stain Kit, for 633 or 635 nm excitation.
17. PE-labeled peptide-MHC class II tetramer.
18. APC-labeled peptide-MHC class I pentamer (*see* **Note 2**).
19. eBioscience Foxp3/Transcription Factor Staining Buffer Set (*see* **Note 3**).

2.3 Fluorescent Particles

1. SPHERO EasyComp PE Fluorescent Particles (*see* **Note 4**).
2. SPHERO EasyComp APC Fluorescent Particles (*see* **Note 4**).
3. SPHERO EasyComp Blank Particles.
4. SPHERO Supra Rainbow Midrange Fluorescent Particles.
5. ArC Amine Reactive Compensation Bead Kit (*see* **Note 5**).

2.4 Consumables

1. Falcon 15 mL Polystyrene Centrifuge Tube, Conical Bottom, with Dome Seal Screw Cap, Sterile.
2. Falcon 5 mL Round Bottom Polystyrene Test Tube, with Snap Cap, Sterile.
3. Falcon 5 mL Round Bottom Polystyrene Test Tube, with Cell Strainer Snap Cap.

2.5 Equipment and Software

1. Class II biosafety cabinet (BSC).
2. Refrigerated centrifuge.
3. Microcentrifuge.
4. Flow cytometer with UV, violet, blue, yellow-green (optional), and red lasers, with FACSDiva software (e.g. LSRII or LSRFortessa from BD).
5. FlowJo flow cytometry analysis software.
6. CO_2 incubator.

3 Methods

HCV is a blood-borne pathogen that can cause inflammation of the liver and should be handled in accordance with Biosafety Level 2 (BSL2) containment criteria. All work with blood samples from HCV-infected donors should be carried out using additional primary containment and personnel precautions such as those described for BSL3.

Unless stated otherwise, all steps should be performed swiftly and at room temperature. All steps should be carried out with the BSC lights turned off to prevent light-induced degradation of photosensitive fluorophores used in the experiment.

This protocol uses peripheral blood mononuclear cells that were previously isolated from whole blood by Ficoll density gradient centrifugation and frozen in 2 mL cryogenic vials in 1 mL FBS with 10% DMSO at 10^7 mL.

3.1 Staining

1. Prepare Live/Dead Fixable Near-IR Dead Cell Stain by adding 50 μL of DMSO to lyophilized Live/Dead stain. Let ArC reactive beads warm up to room temperature. Prepare a 15 mL conical centrifuge tube with 9 mL of cold R10 medium for each cryopreserved sample. Add 30 μL of DNAse I stock solution to each vial to prevent clumping of cells (*see* **Note 6**).
2. Thaw cryo vial with cells in water bath at 37 °C until only a small pellet of ice remains. Quickly wipe outside of cryo vial with 70% ethanol and transfer to a biosafety cabinet (*see* **Note 7**).
3. Transfer 1 mL of cold R10 media from the 15 mL conical tube to the cryo vial and gently resuspend cells until remaining ice has disappeared. Immediately transfer entire volume from the cryo vial into the 15 mL conical tube. Gently mix cells in conical tube by inverting several times and incubate at room temperature for 5 min (*see* **Note 8**). After incubation, transfer tubes to a prechilled centrifuge and spin cells at 4 °C and 350 × *g* for 10 min.
4. Transfer tubes back to the BSC and aspirate as much volume as possible without disturbing the cell pellet. Resuspend cells in 10 mL of PBS, transfer tubes to a prechilled centrifuge and spin at 4 °C and 350 × *g* for 10 min. During centrifugation, prepare Live/Dead working solution, adding 1 μL of prepared Live/Dead Fixable Near-IR Dead Cell Stain to 1 mL of PBS, scaling up according to the number of samples and split vials that will be stained, aiming for a final concentration of 1–2 × 10^6 cells per tube (*see* **Note 9**).
5. After centrifugation, transfer tubes back to the BSC and aspirate as much volume as possible without disturbing the cell pellet. Resuspend cells in 1 mL of Live/Dead working solution and transfer cells into flow cytometer tubes at 1 mL per tube. Vortex ArC reactive beads and add 1 drop of beads to a separate flow cytometer tube.
6. Add 1 μL of prepared Live/Dead fixable blue dead cell stain to each cell-containing flow cytometer tube. Add 1 μL Live/Dead Fixable Near-IR Dead Cell Stain to reactive Arc Beads. Vortex

to mix. Cover all tubes with aluminum foil to shield them from light and incubate at room temperature for 30 min. Toward the end of the incubation time, centrifuge class II tetramers at 14,000 × *g* for 3 min (*see* **Note 10**). Let ArC negative beads come to room temperature.

7. Following Live/Dead staining, wash cells and ArC beads by adding 1 mL of PBS to each tube (2 mL to tube containing ArC reactive beads) and pipetting up and down several times. Transfer tubes to centrifuge and spin at 350 × *g* for 5 min.
8. After centrifugation, transfer tubes to BSC and decant liquid from flow cytometer tubes by inverting over a waste bucket in one swift motion. Pulse-vortex cells to resuspend (*see* **Note 11**). Vortex ArC negative beads and add 1 drop of beads to washed ArC reactive beads and place tubes on ice until acquisition (*see* **Note 12**). Add 2 μL of class II tetramer to each cell sample tube and add 100 μL of R10 medium to each tube, mixing cells by gently pipetting up and down multiple times. Cover tubes with aluminum foil and transfer to a CO_2 incubator.
9. Incubate cells at 37 °C and 5% CO_2 for 60 min, intermittently agitating tubes at 20 min intervals to keep cells in suspension. During the incubation step, prepare antibody master mixes for surface and intracellular stainings, as well as single color controls. For master mixes, combine appropriate antibody amounts in 1.5 mL Eppendorf tubes (*see* **Note 13**). For single color controls, vortex UltraComp compensation beads and add a single drop of beads to a flow cytometer tube for each antibody in the panel (*see* **Note 14**). For tetramer/pentamer single color controls, vortex PE/APC-labeled as well as blank EasyComp fluorescent particles and add a drop of labeled and blank beads to a flow cytometer tube (*see* **Note 15**). Transfer single color controls and master mixes to a refrigerator to keep cold and away from light. Toward the end of the incubation time, centrifuge class I pentamers at 14,000 × *g* for 3 min (*see* **Note 10**).
10. Following incubation, wash cells by resuspending in 2 mL of FACS Buffer and centrifuge at 350 × *g* for 5 min, decant and resuspend with 50 μL FACS Buffer, pipetting up and down several times to ensure that no cell clumps have formed. Add 2 μL of class I pentamer to each cell sample tube and mix by pipetting up and down several times (*see* **Note 16**). Incubate for 10 min at room temperature.
11. Wash cells by resuspending in 2 mL of FACS Buffer and centrifuge at 350 × *g* for 5 min, decant and pulse vortex to resuspend cells.

12. Add 50 μL of FACS buffer to each tube, mixing cells by gently pipetting up and down multiple times. Add appropriate amounts of cold antibody master mix to cell sample tubes and mix by pipetting. Incubate in the dark at 4 °C for 20 min. If full minus one (FMO) controls are being used, antibodies will have to be added to samples individually to achieve the desired combination of antibodies (*see* **Note 17**). If samples are meant to be used for fluorescent assisted cell sorting (FACS) of live cells and downstream processing of HCV-specific T cells for other procedures, skip intracellular staining and proceed with **step 17** (*see* **Note 18**).
13. Wash cells in 2 mL of FACS Buffer and centrifuge at 350 × *g* for 5 min, decant and vortex to resuspend cells.
14. Add 1 mL of prepared Foxp3 Fixation/Permeabilization working solution to each sample tube and mix by pipetting. Incubate cells at 4 °C in the dark for 45 min.
15. Wash cells by adding 2 mL of 1× permeabilization buffer to each sample tube, pipetting up and down to mix well. Centrifuge samples at 350 × *g* (room temperature) for 5 min, then decant the supernatant and resuspend pellet in 100 μL of 1× permeabilization buffer (*see* **Note 19**).
16. Add the appropriate amount of intracellular antibody master mix to cell samples and incubate in the dark at room temperature for 60 min.
17. Wash cells by adding with 2 mL of FACS buffer, pipetting up and down to mix. Centrifuge at 350 × *g* for 5 min, decant, vortex to resuspend, and repeat the wash step once.
18. After the second wash step, resuspend cells in 200 μL of FACS Buffer and keep in the dark at 4 °C until acquisition or sorting (*see* **Note 20**). If multiple experiments with the same panel are planned, prepare a tube of tracking beads by vortexing SPHERO Supra Rainbow Midrange Fluorescent Particles and adding 1 drop of beads to a flow cytometer tube. Keep tracking beads with samples and single-color controls at 4 °C and away from light (*see* **Note 21**).

3.2 Acquisition

1. Start up and prepare flow cytometer for acquisition according to manufacturer instructions (*see* **Note 22**). Keep all sample and control tubes cold and away from light until needed.
2. If this is the first experiment in a series, set up a new experiment in FACSDiva, and create global worksheets for acquisition, compensation control, and tracking (*see* **Note 23**). Set up compensation controls using the built-in option. If this is a repeat experiment proceed with **step 6**.

3. Acquire an unstained sample of cells to ensure that fluorescence intensity peaks and perform an initial adjustment of voltages to ensure that background fluorescence in all channels is at reasonable and equally low levels (*see* **Note 24**).
4. Acquire single color controls in ascending order of excitation and emission wavelength, adjusting voltages to minimize spectral overlap between channels (*see* **Note 25**).
5. When optimum voltages have been determined, acquire and record tracking beads. Using the tracking worksheet, set tight single-color gates for each channel using histogram plots and export the experiment as a template for future use.
6. If this is the first experiment in a series, proceed with **step** 7. If this is a repeat experiment, acquire tracking beads, adjust voltages, aiming for fluorescence intensity peaks of tracking beads to fall within the gates set on the tracking worksheet.
7. Acquire and record single color controls for antibodies, viability stains and tetramers. Adjust forward scatter (FSC) and side scatter (SSC) to accommodate for different particles (*see* **Note 26**).
8. Acquire and record cell samples and FMO controls (*see* **Note 27**).
9. Make sure to appropriately label channels for all samples and controls before exporting FCS files and experiment for analysis in FlowJo.

3.3 Analysis

1. Import FCS files into FlowJo.
2. Select compensation controls and create a new compensation (*see* **Note 28**). Apply compensation to all samples.
3. Select a fully stained sample to create a basic gating strategy (Fig. 1a).
4. First, gate on singlets by forward scatter height and width (FSC-H and FSC-W) and side scatter height and width (SSC-H and SSC-W) to exclude double events (*see* **Note 29**). Next, gate on lymphocytes by forward and side-scatter area (FSC-A and SSC-A).
5. Gate on live CD3 T cells by gating on cells that are positive for CD3 and negative for Live/Dead, CD14, CD16, CD19, and CD56 (*see* **Note 30**).
6. Identify populations of interest by gating on $CD4^+tet^+$ and $CD8^+pent^+$ cells (*see* **Note 31**).
7. Create gates for all markers of interest (*see* **Note 32**) (Fig. 1b, c).
8. For comprehensive analysis of marker coexpression on cell subpopulations using Boolean gates, first create single color gates on histogram plots using for the largest population of

interest. Next, copy gates to other subpopulations, select gates for markers of interest and create combination gates (*see* **Note 33**).

9. Frequencies or other summary statistics of combination gates can then be export using the FlowJo table editor and analyzed further.

4 Notes

1. We recommend that fresh FACS buffer be made with FBS and stored at 4 °C. If desired, fetal bovine serum (FBS) can be substituted with appropriate amounts of bovine serum albumin (BSA). While some commercial versions of this buffer, such as Brilliant Stain Buffer (BD Biosciences) and the Cell Staining Buffer (Biolegend), contain sodium azide as a metabolic inhibitor and preservative, we advise against its use due to its high toxicity and limited benefit when working with freshly prepared reagents in the context of this protocol.
2. Labeled pMHC class I pentamers often exhibit slightly superior staining intensities than their unlabeled counterparts. The use of unlabeled pentamers can be advantageous if the pentamer in question is only rarely used or is purchased in bulk, since unlabeled pentamers can be kept at −80 °C for long term storage without significant loss of quality. While we use APC-labeled pMHC class I pentamer in this protocol, costaining of HCV-specific CD4 and CD8 T cells can be performed using unlabeled pentamer, in which case APC Pro5 Fluorotag is added during the surface antibody staining step of the protocol.
3. While other manufacturers offer similar kits for transcription factor staining, this kit has yielded superior results in our hands. If a different kit is chosen, we ask the reader to refer to the manufacturer's instructions for any changes to incubation steps and required buffers that might need to be made.
4. Since multimers are not antibodies, they will not bind to regular compensation beads. Instead, antibodies labeled with corresponding fluorophores are often used for compensation. This, however, can result in improper compensation if the fluorescence intensity of the used antibody either does not match that of the multimer in magnitude or changes between lots. The latter is a relatively common occurrence, making this method of compensation particularly unsuitable for experiments that need to be tracked over longer periods of time. Therefore, we recommend the use of single color fluorescent particles for compensation, which feature consistently bright

fluorescent peaks that are compatible with long-term tracking of experiments.

5. Amine reactive compensation beads were developed for use with Live/Dead Fixable dead cell stain kits and present a much more reliable and repeatable single-color control than cells. Using cells for compensation of Live/Dead stains should be generally avoided and considered a last resort.
6. Addition of DNAse helps to prevent the forming of clumps caused by large amounts of DNA from cells that ruptured during the freezing/thawing process.
7. Cells are cryopreserved in the presence of DMSO. Since DMSO is toxic to cells, the amount of time that cells are exposed to high concentrations must be kept at a minimum. Keeping the temperature of the cell mixture as cold as possible during the thawing process helps to minimize toxic effects of DMSO by reducing cellular metabolic rate.
8. Rapid dilution after thawing helps to decrease the concentration of cytotoxic DMSO. However, the enzymatic activity of DNAse I is low at the low temperatures required in the thawing step. Added time after dilution of cells allows for more DNA cleavage and helps to reduce cell clumping.
9. Whether a cell sample is split into multiple tubes or multiple samples are being stained, it makes sense to be mindful of limiting the total number of tubes in one experiment to a manageable number so as not to compromise the timing of individual steps. To this end, we suggest that the readers monitor the time it takes to perform multimer staining steps to ensure that incubation times are followed as precisely as possible and are similar for all samples. When working with pMHC multimers, we recommend that cells be counted after thawing during the first or second wash step to allow for splitting of cell samples into the appropriate number of tubes, as too few or too many cells can lead to improper staining. Cell counts after thawing are usually good approximations of the final number of events that can be acquired. It is important to consider that some cell loss will occur with each washing step when adjusting samples for optimal cell concentration.
10. Prior to both pMHC class I and II staining steps, multimers should be centrifuged to pellet aggregates that can form over time and negatively affect staining quality because of nonspecific binding. After centrifugation, the appropriate volume of multimer should be taken from the very top, avoiding the aggregate pellet.
11. Flow cytometer tubes are designed to allow for decanting of excess liquid after centrifugation. Decant liquid by inverting tubes over a waste bucket in a swift motion and keep tube

inverted. In its inverted state, carefully dab the edge of the flow cytometer tube to remove droplets of remaining liquid. Depending on the number of cells used, a small cell pellet may be visible after returning the tube to its upright state, serving as a confirmation of proper decanting. It important to not invert the tube again after returning it to its upright state, since the pellet can separate from the bottom of the tube and can be lost upon a second inversion. After centrifuging, cells in flow cytometer tubes can be resuspended in remaining liquid by vortexing or by adding 50–100 μL of the appropriate buffer and vigorously pipetting up and down multiple times. Any cell clumps that occur and cannot be broken up by pipetting should be carefully discarded as they will continue to form larger aggregates and result in greater cell loss down the line.

12. ArC beads serve as single color controls for the Live/Dead fixable blue dead cell stain. Since the ArC beads are based on a different particle than UltraComp and EasyComp beads, the appropriate negative beads need to be added to this population directly.
13. While some antibodies require the use of exact manufacturer-recommended test volumes, antibodies should be titrated to their optimum concentration prior to finalizing a multicolor flow cytometry panel. In our experience, recommended test volumes for commercial antibodies are often conservative estimates, and proper titration can lead to greater separation of cell populations and lesser background staining. If multiple Brilliant Dyes are used in the staining, add 50 μL of Brilliant Staining Buffer or FACS buffer per sample to the master mix.
14. UltraComp beads contain both negative control and positive capture populations of particles that can be used with antibodies raised in humans, rats, and hamsters. They do not require the use of additional blank beads or specific beads for non-mouse antibodies and therefore simplify the experimental procedure, but compensation control kits from other companies can be used in their place if so desired.
15. Like ArC beads, the particles used in EasyComp beads are different from UltraComp and ArC beads and require the manual addition of both positive and negative beads. While antibodies with the same fluorophore as the tetramer can be used for compensation, we advise against it. The fluorescence intensity of the single-color control needs to be greater or equal to that of the tetramer for proper compensation and considerable differences in brightness of the same fluorophore conjugated to different antibodies makes it hard to guarantee that this requirement is met.

16. Like antibodies, tetramer and pentamer test volumes should be titrated to their optimum concentrations and adjusted to sample cell numbers to maximize separation and minimize background staining due to non-specific binding.
17. Full minus one (FMO) controls are recommended to establish proper gating, particularly for antibodies with poor separation between negative and positive cell populations. For FMO controls, cells undergo the complete staining protocol, omitting only the antibody or viability/tetramer stain in question, to establish an appropriate gate for a negative cell population. There will be cases where negative and positive populations will separate well and the resulting gate differs from the gate determined by an FMO control. In these cases, where gating based on an FMO control would result in inclusion of part of the negative cell population, we recommend to base gating on the fully stained sample, as increase in fluorescence of the negative population can occur due to very low levels of antigen expression and/or nonspecific binding of antibodies to antigen-negative cells.
18. Downstream applications such as genomics and transcriptomics require live cells. Intracellular staining requires fixing and permeabilization of cells and therefore cannot be combined with live sorting.
19. Fixation and permeabilization will make the cell pellet hard to detect with the human eye. This lack of visual confirmation makes it important to be very diligent in decanting, as cell loss during this step will not be apparent.
20. In general, cells can be kept in their fixed state overnight for data acquisition on the next day or even later. However, delaying acquisition will result in increasing dissociation of tetramer from cells and considerably worsen results and is therefore not recommended. When staining cells for live cell sorting, it is important not to delay acquisition, since antibodies and tetramers have the potential to stimulate cells and/or to be internalized by them, causing likely undesired changes in the biological state of the cells in question. If samples are not being used for live cell sorting, but the intracellular staining step was skipped, it is important to fix cells in the final staining step, by resuspending them in Fixation buffer instead of FACS buffer.
21. Ideally, single-color controls and tracking beads should be prepared fresh prior to each experiment. If experiment is repeated within 48 h, remaining single color controls and tracking beads can be used again, but we caution the reader to be mindful of possible degradation of fluorescent reagents due to light exposure with repeated use.

22. When sharing an instrument with other people, make sure to check that the instrument is clean and in proper working condition, with all lasers running at optimal temperature and detector arrays outfitted with the appropriate filters for your experiment. Remember to run a tube of deionized water in between samples and/or controls to avoid crossover between them. Should unexpected issues occur, the technician or technologist in charge of the machine should be consulted.
23. Optimum voltages determined for an experiment, and resulting fluorescence intensities, are usually not very stable to changes in instrument settings. To make data and results comparable when performing a set of experiments over longer periods of time, tracking of cytometer settings is highly recommended. SPHERO Supra Rainbow Midrange Fluorescent Particles provide a single fluorescent peak for all channels used in common flow cytometric setups, including the UV spectrum, that is of similar intensity to that of a positive cell population, and can therefore be used to track cytometer settings. Once optimum voltages for an experimental setup are determined, tracking beads can be recorded and gates can be drawn for each channel used in the experiment and the experiment can be exported as a template. The next time the experiment is performed, the template can be imported and voltages can be adjusted for fluorescence intensity peaks to match up with the template gates to recover optimum settings. When setting up a new experiment template unused channels can be deleted to reduce resulting file size and keep the experiment file more organized. Make sure that recording of forward and side scatter height and width (FSC-H, FSC-W, SSC-H, SSC-W) is enabled, as they are needed for successful differentiation of single and double events.
24. Always vortex flow cytometer tubes before acquiring on a cytometer, as cells and control beads will sink to the bottom and lead. Mixing before acquisition will result in more even flow rates and avoid clogs. We also strongly recommend filtering of cell samples through the filters of the filtered flow cytometer tube snap caps immediately prior to acquisition to remove unwanted cell aggregates.
25. When working with many colors we recommend a systematic way of organizing channels and corresponding voltage adjustments. A good method is to order channels by detector array (excitation wavelength) and emission wavelength, starting out with relatively low voltages and adjusting upward. Finding optimum voltages for easy and good compensation, especially when working with larger numbers of colors, requires patience and diligent iterative (re-)adjustment of voltages. To simplify the process, we recommend that the experimenter set up a

compensation control worksheet plotting colors that are known to exhibit spectral overlap against each other and try to aim for separation of an order of magnitude between them. It is important to note that filter choice can heavily influence spectral overlap of colors and that channels on different detector arrays can overlap despite their physical separation, if the fluorochromes in question exhibit cross-excitation by other lasers outside of their optimum excitation wavelength.

26. When using controls with positive and negative populations in a single tube, we recommend acquiring slightly more than the default number of events for controls suggested by FACSDiva, to ensure that a sufficiently large number of events is available for compensation.

27. Always run cell samples at a low flow rate. Higher stream rates will cause the stream to be slightly wider resulting in sub-optimal positioning of cells and a higher rate of aborted events. If the event rate is low, is it advisable to aim for a higher concentration of cells by diluting samples with less FACS buffer at the last step of staining. After acquiring and recording cell samples, discard flow cytometer tubes in the appropriate waste receptacles. Reuse of single color controls and tracking beads for follow-up experiments within 2 days is possible, but not recommended. Some fluorophores are susceptible to light-induced degradation, which can result in dimmer signals and off-target signals for tandem dyes.

28. Newer versions of FlowJo (>v10) will recognize tubes labeled as compensation controls and automatically create a group for them. When using FACSDiva defaults for compensation controls, FlowJo will also detect the right fluorochrome from the filename and try to create proper gates for positive and negative controls, provided the populations are large and distinct enough. Overall, autocompensation in FlowJo can be very straightforward, but should always be checked manually. We advise the experimenters to familiarize themselves with the details of manual compensation in both FACSDiva and FlowJo, so that they can better detect improper compensation.

29. Despite filtering and mixing of cells by vortexing prior to acquisition, events of two cells will always occur to varying degrees. Gating out doublets is an effective method to increase purity of the detected tetramer positive population and decrease background noise.

30. Employing a dump channel to gate out unwanted cell populations is an easy method to significantly improve tetramer staining and is always recommended when feasible, especially when sorting live cells or working with very rare populations of interest. Particularly when working with pMHC class I

pentamers, the exclusion of CD19^{+} events can significantly reduce background staining related to nonspecific pentamer binding. When designing flow cytometry panels with larger numbers of parameters, a dump channel comes at the cost of occupying a valuable channel that could be used for a marker of interest. In this case, dump channels can be combined with viability staining by matching Live/Dead Fixable Dead Cell Stain kits to fluorochromes for exclusion markers, such as Live/Dead Fixable Near-IR and APC-Cy7, as exemplified in this protocol. However, if excluded cell populations are to be used to establish proper gating strategies or for bulk cell analysis, they should be assigned a separate channel.

31. Since HCV-specific CD4 T cells are very rare, assessing the quality of tetramer staining is often difficult. Generally, CD4^{+}tetramer^{+} cells should exhibit a higher fluorescence intensity than unspecific background events (CD4^{-}tetramer^{+}), separate clearly from tetramer^{-} cells, and form a dense cloud rather than being highly dispersed. When in doubt over the quality of a staining, we recommend consulting with someone experienced in tetramer staining.
32. When creating gates for markers of interest with poor separation of positive and negative cell populations, it is important to be as unbiased as possible and very consistent with gating. We therefore strongly recommended the use FMO controls to establish proper negative gates for these markers.
33. Once proper gates have been established for all channels, marker coexpression can be analyzed with combination gates based on Boolean logic in a highly efficient manner. However, we recommend that the experimenter familiarize himself or herself with the analysis software to avoid mistakes when copying single color gates and creating combination gates for multiple markers, as the number of resulting combination gates will increase exponentially for each marker.

References

1. Schulze zur Wiesch J, Lauer GM, Day CL et al (2005) Broad repertoire of the CD4+ Th cell response in spontaneously controlled hepatitis C virus infection includes dominant and highly promiscuous epitopes. J Immunol 175:3603–3613. https://doi.org/10.4049/jimmunol.175.6.3603
2. Schulze zur Wiesch J, Ciuffreda D, Lewis-Ximenez L et al (2012) Broadly directed virus-specific CD4+ T cell responses are primed during acute hepatitis C infection, but rapidly disappear from human blood with viral persistence. J Exp Med 209:61–75. https://doi.org/10.1084/jem.20100388
3. Altman JD, Davis MM (2003) MHC-peptide tetramers to visualize antigen-specific T cells. Curr Protoc Immunol Chapter 17:Unit 17:3. https://doi.org/10.1002/0471142735.im1703s53
4. Klenerman P, Cerundolo V, Dunbar PR (2002) Tracking T cells with tetramers: new tales from new tools. Nat Rev Immunol 2:263–272. https://doi.org/10.1038/nri777
5. Vargas AL, Lechner F, Kantzanou M et al (2001) Ex vivo analysis of phenotype and

TCR usage in relation to CD45 isoform expression on cytomegalovirus-specific CD8+ T lymphocytes. Clin Exp Immunol 125:432–439. https://doi.org/10.1046/j.1365-2249.2001.01614.x

6. Bendall SC, Nolan GP, Roederer M, Chattopadhyay PK (2012) A deep profiler's guide to cytometry. Trends Immunol 33:323–332. https://doi.org/10.1016/j.it.2012.02.010

7. Perfetto SP, Chattopadhyay PK, Roederer M (2004) Seventeen-colour flow cytometry: unravelling the immune system. Nat Rev Immunol 4:648–655. https://doi.org/10.1038/nri1416

8. Schmid I, Ferbas J, Uittenbogaart CH, Giorgi JV (1999) Flow cytometric analysis of live cell proliferation and phenotype in populations with low viability. Cytometry A 35:64–74. https://doi.org/10.1002/(sici)1097-0320(19990101)35:1<64::aid-cyto9>3.3.co;2-p

9. Shapiro HM (2005) Practical Flow Cytometry. https://doi.org/10.1002/0471722731

10. Hoffmann TK, Donnenberg VS, Friebe-Hoffmann U et al (2000) Competition of peptide-MHC class I tetrameric complexes with anti-CD3 provides evidence for specificity of peptide binding to the TCR complex. Cytometry A 41:321–328. https://doi.org/10.1002/1097-0320(20001201)41:4<321::aid-cyto11>3.3.co;2-0

11. Wooldridge L, Scriba TJ, Milicic A et al (2006) Anti-coreceptor antibodies profoundly affect staining with peptide-MHC class I and class II tetramers. Eur J Immunol 36:1847–1855. https://doi.org/10.1002/eji.200635886

12. Clement M, Ladell K, Ekeruche-Makinde J et al (2011) Anti-CD8 Antibodies Can Trigger CD8+ T Cell Effector Function in the Absence of TCR Engagement and Improve Peptide–MHCI Tetramer Staining. J Immunol 187:654–663. https://doi.org/10.4049/jimmunol.1003941

Chapter 26

Probing the Antigenicity of HCV Envelope Glycoproteins by Phage Display Antibody Technology

Erick Giang, Fernando Aleman, and Mansun Law

Abstract

The envelope glycoproteins E1 and E2 of hepatitis C virus form a heterodimeric complex on the viral surface. They are the targets of neutralizing antibodies and are being investigated as potential vaccine antigens. Because of the high level of cysteine residues and N-glycosylation sites in the polypeptide sequences, it is technically challenging to produce pure, folded recombinant E1, E2, and E1E2 complex for downstream analysis. In this chapter, the methods we used to isolate a panel of human antibodies specific to diverse antigenic regions on the glycoproteins are discussed. The antibodies have been found to be valuable reagents for the study of HCV envelope glycoproteins, including the determination of the first E2 core domain structure.

Key words Phage display, Antibody, Epitope, Antigenic site, Antigenic region, Glycoproteins

1 Introduction

Phage display is a robust technique to screen interactions between proteins and other ligands. DNA encoding for protein populations of interest (e.g., antibody repertoire) is inserted into the phage genome in frame with one of the viral coat protein genes for surface display. With the phenotype of the displayed protein linked to the genotype of phage, and the extremely efficient replication of phage, one may sample billions and trillions of interactions within a standard academic laboratory [1–4].

Phage display has been applied successfully to antibody discovery and affinity improvement because of its robustness and lower cost compared to other methods [5]. Phage-display antibody libraries have been generated from B cells of naïve donors and from immunized donors, or generated synthetically and semi-synthetically [6]. In all cases, the antibody variable domains, either in the format of single-chain variable fragment (scFv) or antigen-binding fragment (Fab), are expressed on the surface of the phage. Through a procedure called "panning," phage virions bearing

Mansun Law (ed.), *Hepatitis C Virus Protocols*, Methods in Molecular Biology, vol. 1911,
https://doi.org/10.1007/978-1-4939-8976-8_26,

1. **Cloning of combinatorial antibody repertoire using donor sample (PBMC or bone marrow) into a phage display vector**
2. **Preparation of phage display antibody library for panning**
3. **Antibody isolation by panning:**
   - Binding
   - Washing
   - Eluting
   - Amplification of eluted phage
4. **Antibody isolation by exhaustive panning:**
   - Epitope masking
   - Competitive Binding
   - Washing
   - Elution
   - Amplification of eluted phages
5. **DNA isolation and sequencing**
6. **Functional validation**

Fig. 1 Flowchart of exhaustive panning of phage-display antibody libraries

antibodies are allowed to bind to the panning antigen, usually presented on a solid surface. With increasingly stringent washing conditions in multiple rounds of panning, antibodies with high affinity can be isolated (Fig. 1).

Despite a number of caveats associated with phage-display (*see* **Note 1**), it has been proven particularly useful for antibody discovery when the target antigens cannot be easily produced in pure form. In the case of HCV neutralizing antibodies, the source of panning antigens, HCV E1E2 envelope glycoproteins, are produced as intracellular membrane proteins by transient transfection and the highly glycosylated recombinant antigens in cell lysate can be enriched onto a solid support with lectin. Combining with the exhaustive panning strategy, antibodies specific to diverse antigenic sites can be isolated from the immune antibody library [7] (Fig. 2; Table 1).

2 Materials

2.1 General Preparations

1. Prepare 1% bleach in a spray bottle.
2. Designate work area only for phage panning. Clear and clean area with 1% bleach before and after panning.
3. Clean plasticware/glassware and dry autoclave (e.g., centrifuge bottles, Oakridge tubes, etc.) for panning experiments.
4. Use filtered pipette tips.
5. 10% bleach waste bucket in work area to decontaminate waste (e.g., tips, conical tubes, etc.) in contact with bacteria and phage.

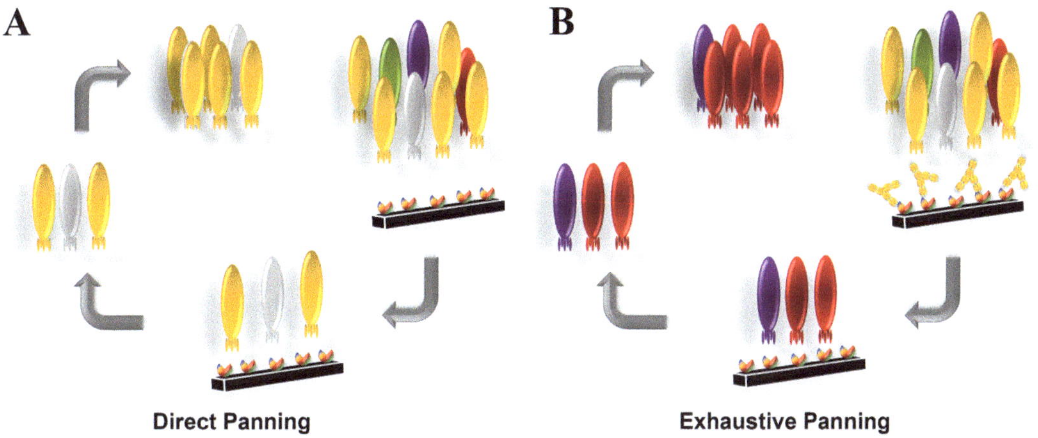

Fig. 2 Direct panning versus exhaustive panning. (**a**) Direct panning starts with the incubation of a phage library with a coated antigen (rainbow colored representing multiple epitopes in the antigen). Grey phage particles represent non-specific binding phage that are isolated and amplified occasionally during the process. In repeated round of washes, specific binders, particularly those of high affinity and frequency in the antibody library (yellow phage), will be selected by the standard panning strategy. (**b**) Exhaustive panning strategy uses epitope-masked antigens to repeatedly pan a given antibody library until all binding antibodies are isolated. Antibody/antibodies isolated from the previous panning experiments (yellow antibody) can be used to mask the dominant epitopes on the antigens before panning. Consequently, antibodies of sufficient binding affinity and recognizing non-overlapping epitopes can be isolated even if they are minor populations in the antibody library

Table 1
List of HCV antibodies obtained by exhaustive panning of an HCV immune antibody library by phage display

Strategy	Glycoprotein	Antigenic region	mAb	Epitope	Binding region
Direct panning	E2	AR1	AR1A AR1B	Discontinuous	E2 non-neutralizing face involving residues 495, 519, 544, 545, 547, 548, 549, and 632
		AR2	AR2A	Discontinuous	E2 back layer region involving residues 625 and 628
		AR3	AR3A AR3B AR3C AR3D	Discontinuous	E2 neutralizing face involving residues 427–443, 529–530. Residues 459, 499, 503, 558, and 616 influence folding of front layer, and residues 424, 425, 517, 518, 520, 523, 535, 536 influence folding of CD81-binding loop
Exhaustive panning	E1E2	AR4	AR4A AR4B	Discontinuous	E1E2 interface with specific residue 698
		AR5	AR5A	Discontinuous	E1E2 interface with specific residues 639 and 665

6. 10% bleach bucket to decontaminate reusable plasticware and glassware.
7. Bunsen burner always on in work area during panning.

2.2 Helper Phage Preparation

1. VCSM13 Helper Phage (Agilent, 200,251): >1.0 × 10^{11} PFU/mL.
2. XL1-Blue Electrocompetent *E. coli* cells (Agilent, 200249).
3. Super Broth (SB): 30 g tryptone, 20 g yeast extract, 10 g MOPS into 1 L water pH 7.0, autoclaved.
4. Top Agar: 10 g Bacto tryptone, 5 g NaCl, 6.5 g Bacto agar into 1 L water, autoclaved.
5. Luria–Bertani (LB) agar plates.
6. Kanamycin (50 mg/mL in water).

2.3 Library Panning on Immobilized Antigens

1. Antibiotics: carbenicillin (100 mg/mL), tetracycline (5 mg/mL), kanamycin (50 mg/mL) in water.
2. Wash Buffer (DPBS-T): 1× DPBS (14190144, Gibco) + 0.5% Tween 20, pH 7.0.
3. Super Broth (SB).
4. XL1-Blue Electrocompetent cells (Agilent, 200249).
5. Preparation Antibiotic Agar Plates: Tetracycline (10 μg/mL); carbenicillin (100 μg/mL)/tetracycline (10 μg/mL).
6. Fresh nonfat milk (*see* **Note 2**).
7. SOC medium [Super Optimal Broth (SOB) + 20 mM Glucose].
8. 20% PEG/2.5 M NaCl.
9. 0.1 M HCl, pH adjusted to 2.2 with solid glycine (Elution Buffer).
10. 2 M Tris base pH 8.0 (Neutralization Solution).
11. 96-well ELISA plate (Costar 3590) and plate sealers.
12. Electroporator: BioRad Gene Pulser Xcell.
13. Gene Pulser/MicroPulser Electroporation Cuvettes, 0.1 cm gap (BioRad, 1652089).
14. Oakridge Tubes (3119-0050, Nalgene).
15. 500 mL Bottles (3122-0500, Nalgene).

2.4 Expression of Soluble Antibody Fragment

1. Antibiotics: Carbenicillin (100 mg/mL).
2. Super Broth (SB).
3. XL10-Gold Ultracompetent cells (Agilent, 200315).
4. SpeI and NheI restriction enzymes (NEB).
5. T4 DNA ligase (NEB).

6. Nunc 96-Well Polypropylene DeepWell Storage Plates (Fisher).
7. Isopropyl β-D-1-thiogalactopyranoside (IPTG).
8. BugBuster 10× Protein Extraction Reagent (Millipore).

3 Methods

This chapter focuses on the strategy and detailed methods for antibody discovery from an HCV-immune antibody DNA library. Methods for the construction of antibody library can be found elsewhere [8]. Panning consists of multiple rounds (usually 3–5) of binding phage to an antigen immobilized onto a solid support (e.g., microwells of an ELISA plate), a number of washing steps, phage elution by low pH and phage amplification (Fig. 1). The first two rounds of panning enrich phage populations that bind the targeted antigen, followed by subsequent rounds with increased stringency of washes to select for high-affinity binders. The usage of "epitope-masked" antigens in panning can allow full exploration of the antibody repertoire in the phage library thus improving the diversity of antibodies to be isolated (Fig. 2).

3.1 Helper Phage Preparation

The pComb3 phagemid [8] vector used in this study contains only the origin of replication of the filamentous bacteriophage f1 and lacks the genes required for replication and assembly of phage particles. Therefore, a helper phage is required to provide the structural components for phage assembly for library screening. The VCSM13 Helper Phage is especially helpful because it contains a kanamycin resistance gene to select for *E. coli* positively infected with VCSM13. This part can be divided in three consecutive days.

1. *On day 1*, inoculate 2 mL SB with 2 μL XL1-Blue cells and incubate with shaking at 250 RPM (rotations per minute), 37 °C for 1 h.
2. Melt Top Agar by microwave and keep liquefied by leaving it in a 56 °C water bath until use.
3. Prepare 10^{-6}, 10^{-7}, and 10^{-8} dilutions of VCSM13 Helper Phage (usually >1.0×10^{11} PFU/mL) in SB media.
4. Add 1 μL of each dilution to 50 μL of the XL1-Blue culture and incubate for 15 min at room temperature (R.T.).
5. Mix the infected XL1-Blue culture with 4 mL liquefied Top Agar (cooled to below 50 °C) and add to an LB plate. Incubate overnight (O/N) at 37 °C to develop individual phage plaques.
6. *On day 2*, in a 50 mL conical tube, inoculate 10 mL prewarmed SB with 10 μL XL1-Blue cells and incubate with shaking at 250 RPM, 37 °C for 1 h.

7. After 1 h incubation, use a pipet tip to transfer a single VCSM13 plaque to the XL1-Blue culture and incubate with shaking at 250 RPM, 37 °C for 2 h.
8. Inoculate the infected 10 mL culture into a baffled flask containing 500 mL SB. Add 700 μL of 50 mg/mL kanamycin to a final concentration of 70 μg/mL and incubate with shaking at 250 RPM, 37 °C overnight.
9. On day 3, transfer the O/N culture into ten 50 mL conical tubes and centrifuge at 2500 × *g* for 15 min.
10. Transfer clarified supernatant into new 50 mL conical tubes, incubate in a water bath at 70 °C for 20 min, centrifuge, and transfer the clarified supernatants into new 50 mL conical tubes and store at 4 °C.

3.2 Titration of VCSM13 Helper Phage Preparation

It is important to titrate the helper phage to verify each batch is similar to the commercial helper phage (>1.0 × 10^{11} PFU/mL) for full coverage of the antibody library.

1. Repeat Subheading 3.1, **steps 1–5** to titrate the new helper phage preparation. Good titer is in the range of 10^{12}–10^{13} PFU/mL. Preparations are stable for months stored at 4 °C.

3.3 Phage Antibody Library Amplification and Library Panning

3.3.1 Panning Antigen Preparation

1. A day before library panning, the panning antigen is coated onto a solid support. Use a 96-well ELISA plate, coat four wells with 150 μL of *Galanthus nivalis* lectin (GNL) (10 μg/mL) or an anti-E1/E2 antibody at 10 μg/mL (two for antigen of interest and two for preabsorption) for capturing E1E2 in transfected cell lysate (*see* **Note 3**). Cover the plate with a plate sealer to prevent evaporation and incubate overnight at 4 °C (*see* **Note 4**).
2. Wash the microwells four times with DPBS-T and block them with 250 μL 50% nonfat milk (diluted in DPBS/0.5% Tween 20) for 1 h with rocking at R.T.
3. After blocking, blot out the 50% nonfat milk and add 150 μL to 2 wells with undiluted E1E2 cell lysate and the other two wells (preabsorption wells) with control cell lysate only. Incubate the plate with rocking at 37 °C for 1 h.
4. Wash wells once with DPBS, repeat antigen capture as in **step 3** as needed to fully saturate all binding sites.
5. To isolate antibodies to diverse epitopes, immunodominant/undesired epitopes can be masked by well-characterized antibodies or antibodies isolated in previous rounds of panning (Fig. 2). For example, E1E2 can be captured using mAb AR3A coated on ELISA plate at 10 μg/mL, and blocked with 150 μL mAb AR1B (20 μg/mL rocking at 37 °C for 1 h) for isolation of antibodies outside these two E2 antigenic regions.

3.3.2 Amplification of Phage Antibody Library

1. A day before library panning, electroporate a total of 9 μg library DNA into 300 μL XL1-Blue cells in three different electroporation reactions (add 3 μg DNA to 100 μL cells in each electroporation cuvette and pulse at 1.8 V/0.1 cm). Immediately flush the cuvettes twice with 1 mL SOC and combine entire transformation into a 50 mL conical tube. Add SOC to a final volume of 10 mL and incubate with shaking at 250 RPM, 37 °C for 1 h.
2. Add SB to a final volume of 20 mL with antibiotics (20 μg/mL carbenicillin; 12.5 μg/mL tetracycline). Take out 50 μL of culture for spot titration to determine the transformation efficiency (*see* **Note 5**). Incubate culture with shaking at 250 RPM, 37 °C for 1 h.
3. Add 1 mL VCSM13 helper phage from Subheading 3.1 to the bacteria culture (titer at 10^{12}–10^{13} PFU/mL range). Incubate the tubes at 250 RPM, 37 °C for 2 h.
4. Add SB to a final volume of 100 mL with antibiotics (50 μg/mL carbenicillin; 12.5 μg/mL tetracycline; 70 μg/mL kanamycin). Incubate culture with shaking at 250 RPM, 30 °C overnight.
5. On the same day, streak a tetracycline-LB agar plate with XL1-Blue cells and incubate at 37 °C in a bacterial incubator to form individual colonies. The next day, pick a colony and start growing XL1-Blue cells in SB (three cultures of 3 mL for pre-Absorb, post-Absorb, and elution) with antibiotics (12.5 μg/mL tetracycline) (*see* **Note 6**).

3.3.3 Panning of Phage Antibody Library

1. Centrifuge the 100 mL overnight phage culture from Subheading 3.3.2, **step 4** in a 500 mL centrifuge bottle at 4800 × *g*, 4 °C for 15 min.
2. Decant clarified media containing the phage library into a new 500 mL bottle and add 25 mL 20% PEG/2.5 M NaCl. Mix and incubate on ice for 30 min to precipitate the phage.
3. Centrifuge at 13,500 × *g*, 4 °C for 20 min to pellet the phage. Carefully, decant media and leave bottle upside down on paper towel to drain residual liquid. A visible pellet should be found at the bottom of the bottle.
4. Add 2 mL 10% nonfat milk/DPBS to the phage pellet and resuspend. Transfer the resuspension into a 2 mL microcentrifuge tube and centrifuge for 5 min at 15,500 × *g* and transfer to a new tube (*see* **Note 7**). Take 1 μL for spot titration (*see* **Note 8**).
5. Wash the two control wells (cell lysate without E1E2 from Subheading 3.3.1) four times with DPBS-T and add 150 μL library phage suspension to preadsorb nonspecific phage populations.

6. Incubate the plate with rocking at 37 °C for 1 h.
7. Remove unbound antigens and blocking antibodies from the E1E2 antigen-coated wells (from Subheading 3.3.1) by washing with DPBS-T four times and transfer the phage suspension from the pre-Absorb wells into the E1E2 antigen-coated wells. Take 1 μL for spot titration (*see* **Note 8**). Incubate the plate with rocking at 37 °C for 2 h.
8. Remove the phage suspension and wash the two wells twice with DPBS-T by pipetting up and down ten times, incubating for 5 min during each wash (*see* **Note 9**).
9. Elute bound phage by adding 175 μL Elution Buffer and incubate for 10 min at R.T. Transfer the eluted phage to a microcentrifuge tube and neutralize the acidic pH with 6 μL of 2 M Tris Base. Repeat elution with another 175 μL Elution Buffer, incubate for 1 min at R.T. and neutralize as above (*see* **Note 10**).
10. Infect 3 mL XL1-Blue cells from Subheading 3.3.2 with the eluted phage for 15 min at R.T. After infection, take 50 μL of infected culture for spot titration (*see* **Note 11**).
11. Add 7 mL prewarmed SB with antibiotics (20 μg/mL carbenicillin and 12.5 μg/mL tetracycline) to the infected culture, transfer to a 50 mL conical tube and incubate with shaking at 250 RPM, 37 °C for 1 h.
12. Add 40 mL prewarmed SB to a final volume of 50 mL with antibiotics (50 μg/mL carbenicillin and 12.5 μg/mL tetracycline). Transfer the content into a larger bottle or flask and add 1 mL VCS M13 Helper phage. Incubate culture at 37 °C with shaking at 250 RPM for 2 h.
13. Add kanamycin to a final concentration of 70 μg/mL and incubate culture at 30 °C with shaking at 250 RPM overnight.
14. Streak new Tetracycline LB agar plate with XL1-Blue cells and incubate at 37 °C to form new individual colonies for growing new XL1-Blue cells for next round of panning as in Subheading 3.3.2, **step 5**.
15. Centrifuge the overnight 50 mL bacterial culture in a conical tube at 3,500 × *g*, 4 °C for 15 min. Collect the clarified supernatant and split into two Oakridge tubes (25 mL each). Save the bacterial pellets by freezing it at −20 °C for DNA phagemid analysis if needed for that panning round. Add 6.25 mL 20% PEG/2.5 M NaCl per 25 mL supernatant in each Oakridge tube to precipitate phage as in **steps 2** and **3**. Add 1 mL 10% nonfat milk/DPBS to resuspend the phage pellet and repeat panning as in **steps 4–14** except the number of washing steps in each round of panning is different (*see* **Note 12**).

3.4 Expression of Soluble Antibody Fragments

Screen clones of bacteria pellets from rounds 3 and 4 of the above panning experiment.

1. Thaw frozen pellet in water bath and prepare phagemid DNA using any commercial DNA midiprep kit (from overnight 50 mL culture).
2. Digest 2 μg of purified DNA with SpeI and NheI restriction enzymes according to the manufacturer's protocol to excise the phage capsid protein Gene III fragment (~650 bp) fused with the antibody fragment (*see* **Note 13**).
3. Analyze by DNA gel electrophoresis. Excise the cut vector and purify using a DNA gel extraction kit.
4. Religate 0.2 μg of cut DNA using T4 DNA ligase according to the manufacturer's protocol.
5. Transform 2 μL ligation mixture with XL10-gold competent cells according to the manufacturer's protocol.
6. Inoculate up to 94 individual colonies per 2-mL deep-well plate with 0.5 mL SB with 50 μg/mL carbenicillin and incubate at 37 °C with shaking at 250 RPM overnight (*see* **Note 14**).
7. Next day, inoculate 50 μL of the overnight culture into a new 2-mL deep well plate with 1 mL SB with 50 μg/mL carbenicillin and incubate at 37 °C with shaking at 250 RPM until OD600 reaches ~0.5. Induce with IPTG to final concentration of 0.5 mM and incubate at 20 °C with shaking at 250 RPM overnight (*see* **Note 15**).
8. Generate bacterial lysates containing soluble Fab fragments (crude Fab) using a commercial protein extraction reagent (e.g., BugBuster Protein Extraction Reagent by Novagen) according to manufacturer's protocol.

3.5 Functional Validation of Isolated Antibodies

3.5.1 Preliminary Screening of Fab Clones by ELISA

1. Refer to Chapter 28 for the ELISA method with a few changes (*see* **Note 16**).

3.5.2 Testing Functional Fab Specificity to E1E2 Antigen

1. Refer to Chapter 28 for the ELISA method to determine Fab specificity to the E1E2 antigen.
2. After determining E1E2 specificity, ELISA competition is done to further identify antibodies binding to overlapping antigenic sites (*see* **Note 17**).

4 Notes

1. Antibody libraries are usually generated by PCR and combinatorial in nature. The antibodies in an immune library may not fully recapitulate heavy and light chain pairing and the actual frequency in the original repertoire. Furthermore, antibodies that express or fold poorly in *E. coli* will unlikely be isolated by the phage display method.
2. Nonfat milk must be fresh and prepared by centrifugation at $6000 \times g$ for 30 min in Oakridge tubes. After centrifugation, carefully decant into 50 mL conical tubes, this fresh prepared nonfat milk is ready to use. Can be stored at 4 °C for a week by adding sodium azide to 0.02%.
3. E1E2 antigens can be produced by transient transfection in 293T cells. Transfected cell monolayers are lysed with lysis buffer (25 mM Tris–HCl, 1% Triton X-100, 0.5% NP-40, 140 mM NaCl, 0.02% sodium azide, pH 7.6 and 0.2 μm filtered; 0.5–1 mL per well of a 6-well plate). Lysed cells are incubated at 4 °C for 2 h or overnight with rocking. Cell debris is pelleted by centrifugation and the clarified cell lysate is analyzed by lectin-capture ELISA with anti-HCV mAbs to determine antigen levels before use for panning.
4. Coat as many plates needed for the entire panning experiment. Always use plate sealers during incubation for any steps (including wash steps) throughout the panning experiment to prevent evaporation and cross contamination of the other wells.
5. Calculation of transformants: Titrate the 50 μL culture tenfold (six dilutions); 5 μL spots onto carbenicillin/tetracycline agar plates. Incubate overnight at 37 °C. The next day, count the number of colonies for the spot dilution that is countable. Calculate the number of transformants:

$$(\#\text{ of colonies}) \times (\text{Dilution factor}) \times \left(\frac{\text{Culture vol.}(\mu L)}{\text{Plating vol.}(\mu L)}\right) = (\#\text{ of transformants})$$

6. Number of XL1-Blue culture tubes required is dependent on how many pre-Absorb, post-Absorb, and output for each panning condition. Bacterial cultures can be infected once OD_{600} reaches about 1.
7. Be careful not to disturb residual cell debris pellet when pipetting after centrifugation. Phage suspension can be stored at 4 °C or long term in 0.02% sodium azide.

8. Determining the phage titer is important in that it must be large enough to adequately represent all the clones of interest from the library for antibody selection (panning). Calculate phage titer for pre-Absorb and post-Absorb steps: Dilute the 1 μL phage suspension into 999 μL SB (10^{-3} dilution), mix, dilute 1 μL of the 1/1000 dilution into 99 μL (10^{-5} dilution). Take 50 μL of the 10^{-5} dilution to infect the 3 mL XL1-Blue cultures from Subheading 3.3.2 (1/60 dilution) for 15 min at R.T. (dilution at this point: 6×10^{-6} dilution). After 15 min, take out 50 μL of infected culture for spot titration starting from undiluted to a series of 6 tenfold dilution in SB. Spot 5 μL each of the dilution onto Carbenicillin/Tetracycline agar plates. Incubate overnight at 37 °C. Calculate the phage titer:

$$(\#\text{ of colonies}) \times (\text{Phage dilution}) \times \left(\frac{\text{Culture vol.}(\mu L)}{\text{Plating vol.}(\mu L)}\right) \times \left(6 \times 10^{6}\right) = (\text{Phage titer}).$$

9. During the wash steps, be extra careful when pipetting to minimize creating bubbles. When they burst, it can generate airborne/droplets of phage causing cross-contamination. Working next to a Bunsen burner and keeping the other wells sealed when not in use helps minimize contamination.
10. Eluted phage can be stored in 1 mL SB at 4 °C for a few days.
11. Calculate phage titer for Output/Eluted step: Use the 50 μL of infected culture for spot titer starting undiluted ten-fold (6 dilutions in SB); 5 μL spots onto Carbenicillin/Tetracycline agar plates. Incubate overnight at 37 °C. Calculate the phage titer:

$$(\#\text{ of colonies}) \times (\text{Phage dilution}) \times \left(\frac{\text{Culture vol.}(\mu L)}{\text{Plating vol.}(\mu L)}\right) = (\text{Phage titer}).$$

12. Increase wash stringency in subsequent rounds of panning to remove nonspecific phage bound to the antigen and amplify highly specific binders (e.g., two washes in Round 1 of panning, five washes in Round 2, and ten washes in Rounds 3 and 4).
13. If DNA yield is very low, or if after restriction digestion there is no cut vector and Gene III fragment, it can be caused by the over-packaging of the library DNA by the helper phage. One way to troubleshoot this is to use the resuspended phage from a particular round to infect XL1-Blue cells and expand the culture to 50 mL without adding Helper phage. Phagemid DNA can be prepared from the culture using a commercial plasmid purification kit.

14. Dedicate two wells for controls: one well will be for negative control bacteria only (no Fab) and the other well positive control of a known HCV Fab to be used later in ELISA.
15. Very important: For the overnight 2-mL deep-well plate with ~0.45 mL culture leftover, add 100 μL 80% Glycerol (pipet up and down to mix) and freeze entire plate with adhesive seal at −80 °C. Any Fab clones of interest can be reinoculated later for further analysis (e.g., DNA preparation/sequencing, Fab expression for purification, etc.)
16. This ELISA will determine presence of soluble Fab in bacteria lysate. Instead of Lectin coating the wells, coat microwells with unconjugated goat anti-human IgG, $F(ab')_2$ fragment-specific antibody to capture soluble Fab from crude bacterial lysate, and detect captured Fab with peroxidase-conjugated goat anti-human IgG, $F(ab')_2$ fragment-specific antibody as the secondary antibody. Make sure to have a negative control of crude bacteria lysate and/or irrelevant crude non-HCV Fab for determining background and cutoff values in ELISA.
17. ELISA competition is done similarly in Subheading 3.5.2, **step 1** ELISA to determine crude Fab specificity but with the following modifications. After antigen capture, undiluted crude Fab lysate is added to block all its epitopes for 30 min rocking before addition of purified IgGs as the primary antibody binding (no wash step in between). Also include a no blocking well (negative control crude bacteria lysate) for each purified IgG mAb being tested. Secondary antibody is peroxidase conjugated Goat Anti-Human IgG, Fc specific antibody. The percentage change in mAb binding is calculated as the OD450 signals in the presence of blocking soluble Fab divided by that of the negative control. % reduction in mAb binding indicates the level of epitope overlap between the blocking Fab and the binding mAb. The panel of purified mAbs to screen should represent the different antigenic regions (AR) or other epitope cluster groups (e.g., mAbs that represent the AR1, 2, 3, 4, 5 groups). This will aid in identifying Fab clones overlapping with known antigenic regions.

Acknowledgments

This work was supported by NIH grants AI079031, AI106005, AI123365, and AI123861.

References

1. Aghebati-Maleki L et al (2016) Phage display as a promising approach for vaccine development. J Biomed Sci 23:66
2. Schirrmann T et al (2011) Phage display for the generation of antibodies for proteome research, diagnostics and therapy. Molecules 16:412–426
3. Georgiou G et al (2014) The promise and challenge of high-throughput sequencing of the antibody repertoire. Nat Biotechnol 32:158–168
4. Omidfar K, Daneshpour M (2015) Advances in phage display technology for drug discovery. Expert Opin Drug Discov 10:651–669
5. Strohl WR (2014) Antibody discovery: sourcing of monoclonal antibody variable domains. Curr Drug Discov Technol 11:3–19
6. Zhao AZ et al (2016) Phage antibody display libraries: a powerful antibody discovery platform for immunotherapy. Crit Rev Biotechnol 36:276–289
7. Giang E et al (2012) Human broadly neutralizing antibodies to the envelope glycoprotein complex of hepatitis C virus. Proc Natl Acad Sci U S A 109:6205–6210
8. Barbas Iii CF, Wagner J (1995) Synthetic human antibodies: selecting and evolving functional proteins. Methods 8:94–103

Chapter 27

Isolation of HCV Neutralizing Antibodies by Yeast Display

Zhen-yong Keck, Yong Wang, Patrick Lau, and Steven K. H. Foung

Abstract

Yeast surface display (YSD) enables efficient screening and selection of single chain variable fragments (scFvs) of heavy (V_H) and light (V_L) chains that bind to target antigen with different affinities. Assembly of a scFv library from cDNA usually involves adding different primers and linkers $(Gly_4/Ser)_3$ through multiple rounds of PCR amplification and purification. We describe here a simplified scFv assembly method by creating a modified YSD vector with a built-in linker that reduces the time of assembly and decreases accumulated base exchanges due to PCR errors. In addition, we describe a bias screening strategy toward maximizing novel antibodies of interest by a combination of memory B cell selection and depletion by binding to mutant antigens that do not bind to previously identified monoclonal antibodies.

Key words Yeast surface display, Human monoclonal antibody isolation, Hepatitis C virus, Virus neutralization, Mutant antigen selection

1 Introduction

Comprehensive reviews and protocols describe yeast surface display (YSD) [1] as a powerful tool for isolating monoclonal antibodies and engineering these antibodies to fine tune their binding properties [2–5]. This protocol is focused on YSD application on the isolation of human monoclonal antibodies (HMAbs) from hepatitis C virus (HCV) infected individuals with modifications from previously described methods, which have been incorporated in three sections of the YSD system.

To facilitate scFv assembly in YSD library construction, we first modified a yeast display vector wherein a flexible linker region $(Gly_4/Ser)_3$ is presented in the form of a preassembled scFv. By digesting this vector at various restriction digestion sites located inside and outside of the scFv insertion, the variable regions of immunoglobulin gene pool amplified from cDNA are cloned directly into the vector. This allows all of the gene fragments to be cloned in-frame with its own linker in a single step, without the need for multiple steps of adding separate linkers to V_H and V_L, and

Mansun Law (ed.), *Hepatitis C Virus Protocols*, Methods in Molecular Biology, vol. 1911,
https://doi.org/10.1007/978-1-4939-8976-8_27,

then connecting the two as scFv, which are described in other scFv assembly protocols [3, 6–8]. In this way, the scFv assembly process timeline can be shortened and accumulated base exchanges due to PCR errors and/or PCR biased amplification can be minimized. Second, to increase the probability of isolating neutralizing HMAbs against HCV, we enriched a subset of B cells that is more likely to secrete these antibodies by screening the supernatants of small pools of activated B cells in an infectious cell culture-derived HCV virion (HCVcc) neutralization assay [9, 10]. This step allows for the collection of cells of interest for the initial RNA extraction. As a result, the YSD library size can be smaller and screening by fluorescence-activated cell sorting (FACS) will be more efficient. Third, to have a greater bias for novel neutralizing HMAbs, we developed a series of HCV envelope mutant constructs that are not able to bind nonneutralizing HMAbs or previously identified neutralizing HMAbs [11, 12]. The sequential FACS separation of scFvs with these antigens increases the likelihood of discovering new antibodies. Using this method, we obtained higher affinity neutralizing HCV HMAbs to a new cluster of overlapping epitopes, designated as antigenic domain D, that previously were masked by more immunodominant clusters, designated as antigenic domains A and B [10].

2 Materials

2.1 Reagents

2.1.1 Cells and Plasmids

1. EBY100 cells (*see* **Note 1**).
2. pYD2.A2 plasmid (*see* **Note 2**).
3. DH5α electroporation competent cells.

2.1.2 Chemicals, Reagents and Buffers

1. YPD media (Yeast Extract Peptone Dextrose media): Add 10 g yeast extract and 20 g peptone to 600 mL of deionized water (*see* **Note 3**). Adjust the final volume to 900 mL with deionized water. Autoclave and cool to 55 °C. Add 100 mL filtered 20% dextrose to bottle. Store at 4 °C.
2. YPD plates: Add 10 g yeast extract, 20 g peptone, and 17 g agar to 600 mL of deionized water (*see* **Note 3**). Adjust the final volume to 900 mL with deionized water. Autoclave and cool to 55 °C. Add 100 mL filtered 20% dextrose to bottle. Pour into 100 mm plates. Store at 4 °C.
3. SD-CAA media (Synthetic Dextrose minimal media with Casein Amino Acids): Add 7 g yeast nitrogen base (*see* **Note 4**), 5.4 g sodium phosphate dibasic, 7.4 g sodium phosphate monobasic anhydrous and 5 g casein amino acids to 600 mL of deionized water. Adjust the final volume to 888 mL with deionized water (*see* **Note 3**). Autoclave and cool to 55 °C

(*see* **Note 5**). Add 100 mL filtered 20% dextrose, 10 mL filtered 0.6% leucine, 1 mL filtered 50 mg/mL kanamycin, and 1 mL filtered 15 mg/mL tetracycline to the contents in the bottle. Store at 4 °C.

4. SD-CAA plates: Add 7 g yeast nitrogen base (*see* **Note 4**), 5.4 g sodium phosphate dibasic, 7.4 g sodium phosphate monobasic anhydrous, 5 g casein amino acids, and 17 g agar to 600 mL of deionized water. Adjust the final volume to 888 mL with deionized water (*see* **Note 3**). Autoclave and cool to 55 °C (*see* **Note 5**). Add 100 mL filtered 20% dextrose, 10 mL filtered 0.6% leucine, 1 mL filtered 50 mg/mL kanamycin, and 1 mL filtered 15 mg/mL tetracycline to the contents in the bottle. Pour into 150 mm plates. Store at 4 °C.
5. SG-CAA induction liquid media (Synthetic Galactose Casein Amino Acids): Add 7 g yeast nitrogen base (*see* **Note 4**), 5.4 g sodium phosphate dibasic, 7.4 g sodium phosphate monobasic anhydrous and 5 g casein amino acids to 600 mL of deionized water. Adjust the final volume to 888 mL with deionized water (*see* **Note 3**). Autoclave and cool to 55 °C (*see* **Note 5**). Add 100 mL filtered 20% galactose, 10 mL filtered 0.6% leucine, 1 mL filtered 50 mg/mL kanamycin, and 1 mL filtered 15 mg/mL tetracycline to the contents in the bottle. Store at 4 °C.
6. Yeast Freezing Media: Add 30 mL glycerin to 70 mL autoclaved water. Store at room temperature.
7. LB carbenicillin Media: Add 10 g tryptone, 5 g yeast extract, and 10 g NaCl to 600 mL of deionized water. Adjust the final volume to 1 L with deionized water. Autoclave and cool to 55 °C (*see* **Note 6**). Add 1 mL of filtered 100 mg/mL carbenicillin to contents in the bottle. Store at 4 °C.
8. LB carbenicillin plates: Add 10 g tryptone, 5 g yeast extract, 10 g NaCl, and 15 g agar to 600 mL of deionized water. Adjust the final volume to 1 L with deionized water. Autoclave and cool to 55 °C (*see* **Note 3**). Add 1 mL of filtered 100 mg/mL carbenicillin to contents in the bottle. Pour into 100 mm plates. Store at 4 °C.
9. Yeast electroporation buffer: 1 M sorbitol and 1 mM calcium chloride. Add 91.1 g sorbitol and 50 mL 100 mM calcium chloride to 400 mL of deionized water. Adjust the final volume to 500 mL with deionized water. Filter contents with 0.22 μM filter. Store at room temperature.
10. Yeast resuspension media: 0.1 M lithium acetate (LiAc) and 10 mM dithiothreitol (DTT). Add 1 mL 1 M LiAc and 100 μL 1 M DTT to 8.9 mL of deionized water. Filter contents with 0.22 μM filter. Store at room temperature.

11. PBS (Phosphate-buffered saline): Add 8 g sodium chloride, 0.2 g potassium chloride, 1.44 g disodium phosphate, and 0.24 g monopotassium phosphate to 600 mL of deionized water. Adjust the final volume to 1 L with deionized water and adjust pH to 7.4. Filter contents with 0.22 μM filter. Store at room temperature.
12. PBSB (Magnetic-activated cell sorting buffer): Add 8 g sodium chloride, 0.2 g potassium chloride, 1.44 g disodium phosphate, 0.24 g monopotassium phosphate, 4 mL 0.5 M EDTA, and 5 g bovine serum albumin to 600 mL of deionized water. Adjust the final volume to 1 L with deionized water and adjust pH to 7.4. Filter contents with 0.22 μM filter. Store at 4 °C.
13. TAE (Tris–acetate–ethylenediaminetetraacetic acid) buffer: 40 mM Tris base (pH 7.6), 20 mM glacial acetic acid, 1 mM Ethylenediaminetetraacetic acid (EDTA). Add 242 g Tris base, 57.1 mL glacial acetic acid, and 100 mL 0.5 M EDTA (pH 8.0) to 600 mL of deionized water. Adjust the final volume to 1 L with deionized water to prepare a liter of 50× TAE. Store at room temperature. Prepare 1× TAE by adding 20 mL 50× TAE to 980 mL deionized water. Store at room temperature.
14. DNA Gel: Add 1 g of agarose to 100 mL TAE buffer. To prepare 1% DNA gel. Microwave to dissolve agarose into TAE buffer (*see* **Note 7**). Before pouring, add SYBR Safe DNA gel stain to the agarose at a 1:10,000 dilution. Swirl to incorporate the gel stain and then pour.
15. Restriction enzyme NcoI.
16. Restriction enzyme SalI.
17. Restriction enzyme BspEI.
18. Restriction enzyme NotI.
19. T4 DNA Ligase.
20. Phusion polymerase.
21. 10 mM dNTP mix.
22. Magnetic-activated Cell sorting beads (*see* **Note 8**).
23. Anti-V5.
24. Fluorescence labeled antibodies (*see* **Note 9**).
25. Biotinylated CBH-4G.
26. CBH-4G (available upon request).
27. Biotinylated HCV-E2.
28. S.O.B. (Super Optimal Broth) media.
29. S.O.C. (Super Optimal broth with Catabolite repression) media.

30. Molecular grade water.
31. Autoclaved water.
32. Double distilled water.

2.2 Kits

1. QIAGEN RNeasy Mini kit.
2. Invitrogen Superscript III First-Strand Synthesis System.
3. QIAGEN QiAquick Gel Extraction kit.
4. QIAGEN QiAquick PCR purification kit.
5. QIAGEN QiAprep Spin Miniprep kit.
6. QIAGEN QiAfilter Midi kit.

2.3 Material and Equipment

1. 0.22 μM filter.
2. PCR machine.
3. PCR tubes.
4. Microcentrifuge tubes.
5. Cryotubes.
6. 17 × 100 mm polypropylene round-bottom tubes.
7. 50 mL polypropylene conical tubes.
8. Gel electrophoresis unit.
9. SYBR safe gel stain.
10. Microwave.
11. Autoclave.
12. Spectrophotometer.
13. MACS equipment: MidiMACS separation unit, LS Columns and magnetic stand.
14. 5 mL round bottom polystyrene test tube with cell strainer snap cap.
15. 96-deep well plate.
16. Pre-sterilized breathable sealing film.
17. 96 V-bottom well plate.
18. 3 mm glass beads.
19. 425–600 μm acid-washed glass beads.
20. Electroporation equipment: Electroporator and cuvettes.
21. 37 °C incubator shaker.
22. 37 °C water bath.
23. 30 °C incubator shaker.
24. Centrifuge.
25. 18 °C incubator shaker.
26. Tube shaker and rotator.

27. Vortex.
28. Flow cytometry machine.
29. Vector NTI software.

3 Methods

3.1 Construction of Yeast Display Library

A schematic overview describing all stages in this section is shown in Fig. 2.

3.1.1 Gene Cloning: RNA Extraction and cDNA Synthesis

1. Extract total RNA from selected B cells isolated from HCV infected individuals (*see* **Note 10**) using Qiagen RNeasy Mini kit according to the manufacturer's instruction. Store RNA sample at −80 °C or use for cDNA synthesis immediately.
2. Set up five cDNA synthesis reactions for each library using Invitrogen kit according to the manufacturer's instruction. The following mixes are for a single reaction.
3. Denature 1–1.5 μg RNA in a 10 μL reaction containing 1 mM dNTP and 50 μM oligo$(dT)_{20}$ for 5 min at 65 °C and immediately place the reaction on ice for 1 min to relax the secondary structures of RNA, which often inhibit reverse transcriptase and primer annealing.
4. Add 10 μL of reverse transcriptase mix to each denatured RNA/oligo$(dT)_{20}$ mixture as described in the table below. Incubate for 50 min at 50 °C.

Reagent	1 Rxn (μL)
10× RT buffer	2
25 mM $MgCl_2$	4
0.1 M DTT	2
RNaseOUT (40 U/μL)	1
SuperScript III RT (200 U/μL)	1
Final vol	10

5. Heat-inactivate reverse transcriptase for 5 min at 85 °C to terminate the reaction.
6. Add 1 μL (2 U) of RNase H to each reaction and incubate the reactions for 20 min at 37 °C to remove the RNA template from the cDNA/RNA products to achieve more efficient gene amplification by PCR.
7. Combine the five reactions to a total of 100 μL of cDNA and store at −20 °C or use for PCR immediately.

3.1.2 Gene Cloning: Amplification of Heavy and Light Chain Variable Genes (V_H, V_κ, and V_λ) (Fig. 2b: Gene Amplification)

1. Prepare separate V_H, V_κ, and V_λ PCR master mixes for 8× 50 μL reactions, as described in the table below, to generate a minimum of 10 μg PCR products for each gene. A total of seven to nine primers is included to amplify all of the heavy and light chain family genes.

Reagent	1 Rxn (μL)
5× Phusion buffer	10
10 mM dNTP mix	1
Phusion polymerase	1
Primer forward/primer reverse[a]	2
cDNA	2
Molecule grade water	34
Final vol	50

[a]*See* Table 1

2. Aliquot 50 μL for each PCR tube and run the following conditions for V_H and V_κ: 95 °C for 30 s, 25 cycles of 95 °C for 10 s, 55 °C for 20 s, 72 °C for 20 s, and 1 cycle of 72 °C for 5 min for final extension. Run the following condition for V_λ amplification: 95 °C for 30 s, 25 cycles of 95 °C for 10 s, 48 °C for 20 s, 72 °C for 20 s, and 1 cycle of 72 °C for 5 min for final extension (*see* **Note 11**).
3. Combine the eight PCR reactions and run 2 μL of PCR product on a 1% agarose gel (wt/vol) to ensure the desire bands are present, approximately 350 bp band for V_L and 390 bp band for V_H various with the length of CDR3 junction.
4. To purify PCR products, run all the samples on a 1% agarose gel (wt/vol). Excise the correct size bands and elute from agarose using QiAquick Gel Extraction Kit according to the manufacturer's instruction.
5. Measure purified DNA concentrations from the A_{260} reading of a spectrophotometer. For pure double stranded DNA: A_{260} of 1.0 = 50 mg/mL.

3.1.3 scFv Assembly: Digestion and Purification of Candidate Insert Genes (V_H, V_κ, and V_λ) and Vector pYD2.A2 (Fig. 2b: Gene Assembly)

1. Prepare separate digestion master mixes for three purified PCR products from Subheading 3.1.2, **step 4** and vector pYD2.A2, as described in the table below.

V_H digestion		V_κ/V_λ digestion		
Vector[a]	V_H insert[b]	Vector	V_κ insert	V_λ insert
10 μg	3 μg	10 μg	3 μg	3 μg
NcoI/SalI	NcoI/SalI	BspEI/NotI	BspEI/NotI	BspEI/NotI

[a]Vector pYD2.A2, *see* Subheading 2.1.1

[b]Inserts of V_H, V_κ, and V_λ are the purified PCR products from Subheading 3.1.2, **steps 4** and **5**

A

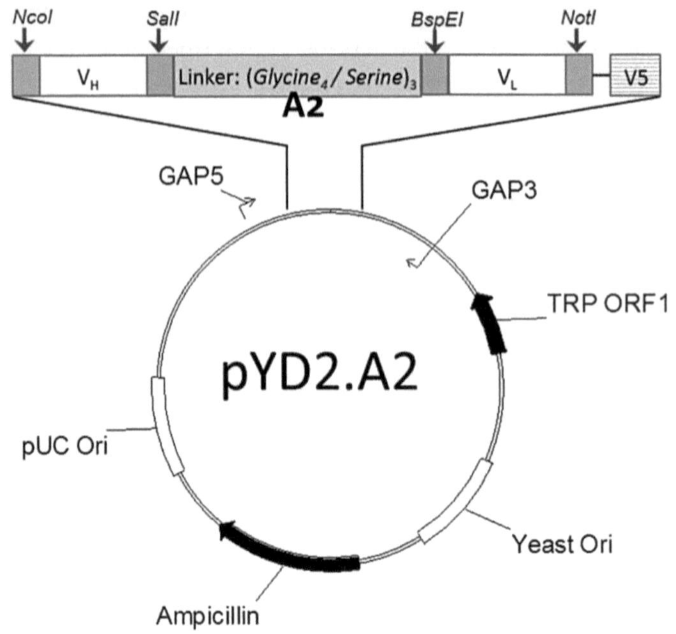

B

```
---------Aga2-------------GAP5---------><------------------
ATAAACACACAGTATGTTTTTAAGCTTCTGCAGGCTAGTGGTGGTGGTGGTTCTGGTGGT
                                         G  G  G  G  S  G  G

-(G4S)3 Linker--------->-----------NcoI---<-----VH----/ /---
GGTGGTTCTGGTGGTGGTGGTTCTGCTAGCGGGGCCATGGCCGAGGTGCAGCTG/ /ACC
 G  G  S  G  G  G  G  S

--VH---->--SalI--HuJHF-------(G4S)3 Linker----BspEI---------
GTCTCCTCAGGGTCGACAGGTGGAGGCGGTTCAGGCGGAGGTGGCTCCGGAGGTGGCGGA
                       G  G  G  G  S  G  G  G  G  S  G  G  G  G

--><----VL-----/ /--VL-->--NotI-----------------<-----------
TCGGACATCGTGATG/ /TCAACGTGCGGCCGCAGCGGGCCCTTCGAAGGTAAGCCTATC
 S                                                G  K  P  I

-----GAP3---V5 tag----------->------
CCTAACCCTCTCCTCGGTCTCGATTCTACGCGTACC
 P  N  P  L  L  G  L  D  S  T
```

Fig. 1 Vector map and sequence of pYD2.A2 plasmid. (a) Vector map of pYD2.A2 plasmid showing the relative position of genes, linkers, primer sites and restriction digestion sites. **(b)** Sequence showing the insert of A2 region in the vector. The bold nucleotides in the middle row indicate 1) GAP5, GAP3 and HuJHF primer binding sites and 2) the restriction digestion sites for NcoI, SalI, BspEI and NotI. The grey amino acids on the bottom row indicate the (G4S)3 linkers and V5 tag. The dashed line with arrows on the top row indicate the span of the (G4S)3 linkers, V_H insert, V_L insert and V5 tag

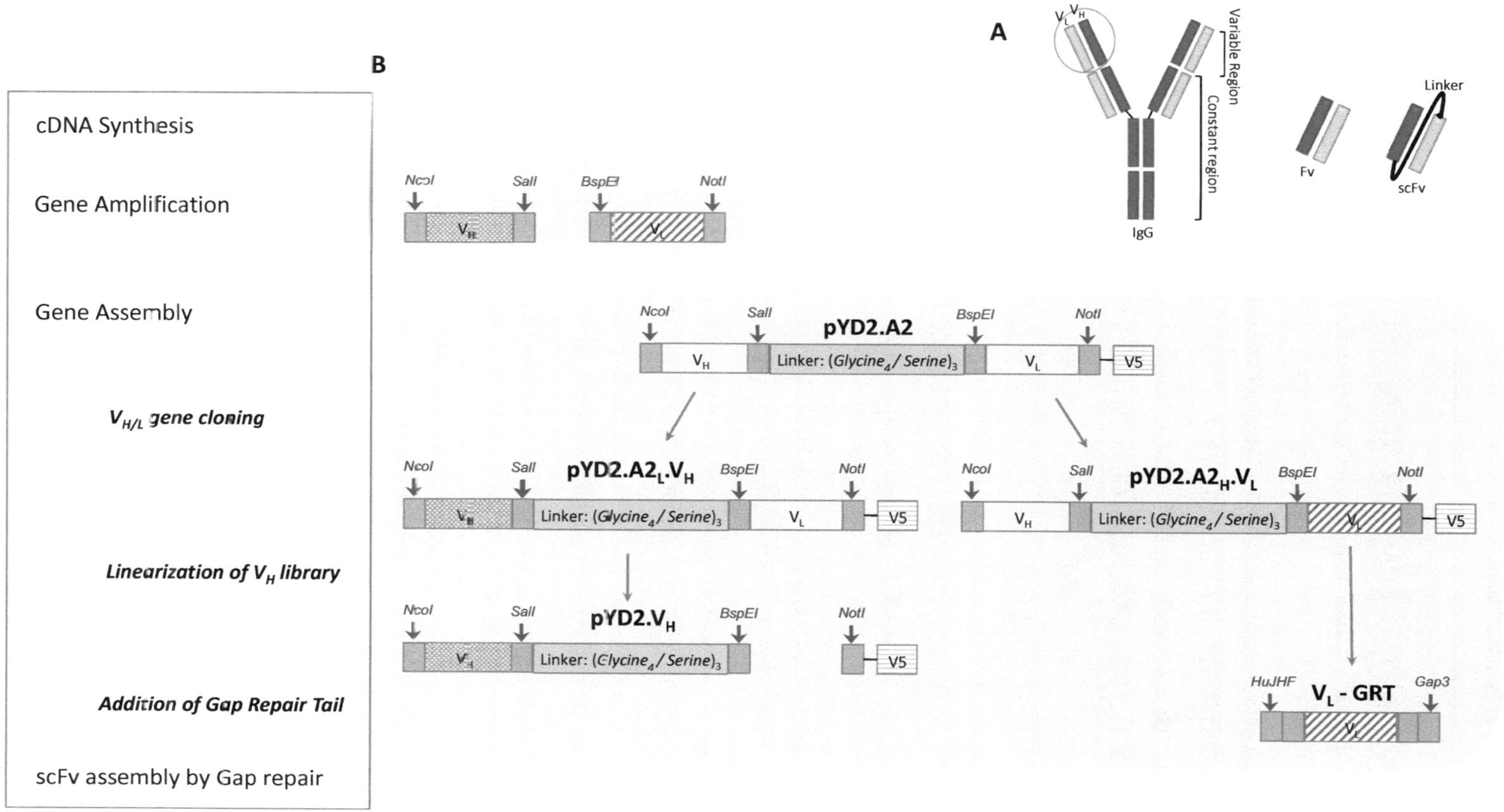

Fig. 2 Illustration of method 3.1 section: *Construction of yeast display library*. **(a)** Schematic representation of a scFv fragment. During library construction, VH and VL are randomly paired via a flexible linker to form scFvs. The quality of a library construction is dependent on the degree of scFv pairing diversity. **(b)** Schematic overview of scFv construction is boxed on the left, with a list of major steps involved in scFv construction. The corresponding diagram on the right illustrates the designated products derived from PCR, restriction digestion and ligation reactions. Vector pYD2.A2 displaying A2-scFv was created by modifying the pYD2 vector, which comprises a (Gly4/Ser)3 linker region carrying SalI and BspEI restriction sites, and NcoI and NotI restriction sites flanking the inserted A2-scFv. An in-frame V5 epitope allows detection of scFv fusion product and normalization of scFv surface expression through immunofluorescence labeling of V5-tag. Patterned filled bars represent amplified V_H and V_L from all of the heavy and light chain family genes using partially degenerated primers

Table 1
Primers for gene cloning and scFv assembly

V_H-forward	
HuVH1a	CGGGGCCATGGCCCAGGTGCAGCTGGTGCAGTCTGG
HuVH2b	CGGGGCCATGGCCCAGGTCACCTTGAAGGAGTCTGG
HuVH3a	CGGGGCCATGGCCGAGGTGCAGCTGGTGGAGTCTGG
HuVH4a	CGGGGCCATGGCCCAGGTGCAGCTGCAGGAGTCGGG
HuVH5b	CGGGGCCATGGCCAGGTGCAGCTGGTGCAGTCTGG
HuVH6a	CGGGGCCATGGCCCAGGTACAGCTGCAGCAGTCAGG
HuVH7a	CGGGGCCATGGCCCAGGTGCAGCTGGTGCAATCTGG
V_H-reverse	
HuJH1,2	CACCTGTCGACCCTGAGGAGACGGTGACCAGGGTGCC
HuJH3	CACCTGTCGACCCTGAAGAGACGGTGACCATTGTCCC
HuJH4,5	CACCTGTCGACCCTGAGGAGACGGTGACCAGGGTTCC
HuJH6	CACCTGTCGACCCTGAGGAGACGGTGACCGTGGTCCC
V_κ-forward	
HuVk1a	GTGGCTCCGGAGGTGGCGGATCGGACATCCAGATGACCCAGTCTCC
HuVk2a	GTGGCTCCGGAGGTGGCGGATCGGATGTTGTGATGACTCAGTCTCC
HuVk2b	GTGGCTCCGGAGGTGGCGGATCGGATATTGTGATGACCCAGATCCC
HuVk3a	GTGGCTCCGGAGGTGGCGGATCGGAAATTGTGTTGACGCAGTCTCC
HuVk4a	GTGGCTCCGGAGGTGGCGGATCGGACATCGTGATGACCCAGTCTCC
HuVk5a	GTGGCTCCGGAGGTGGCGGATCGGAAACGACACTCACGCAGTCTCC
HuVk6a	GTGGCTCCGGAGGTGGCGGATCGGAAATTGTGCTGACTCAGTCTCC
V_κ-reverse	
HuJk 1	CGCCTGCGGCCGCACGTTTGATTTCCACCTTGGTCCC
HuJk 2	CGCCTGCGGCCGCACGTTTGATCTCCAGCTTGGTCCC
HuJk 3	CGCCTGCGGCCGCACGTTTGATATCCACTTTGGTCCC
HuJk 4	CGCCTGCGGCCGCACGTTTGATCTCCACCTTGGTCCC
HuJk 5	CGCCTGCGGCCGCACGTTTAATCTCCAGTCGTGTCCC
V_λ-forward	
HuVl1	GTGGCTCCGGAGGTGGCGGATCGCAGTCTGTSBTGACGCAGCCGCC
HuVl3	GTGGCTCCGGAGGTGGCGGATCGTCCTATGWGCTGACWCAGCCAC
HuVl38	GTGGCTCCGGAGGTGGCGGATCGTCCTATGAGCTGAYRCAGCYACC
HuVl4	GTGGCTCCGGAGGTGGCGGATCGCAGCCTGTGCTGACTCARYC

(continued)

Table 1
(continued)

HuVl78	GTGGCTCCGGAGGTGGCGGATCGCAGDCTGTGGTGACYCAGGAGCC
HuVl9	GTGGCTCCGGAGGTGGCGGATCGCAGCCWGKGCTGACTCAGCCMCC
HuVl11	GTGGCTCCGGAGGTGGCGGATCGTCCTCTGAGCTGASTCAGGASCC
HuVl13	GTGGCTCCGGAGGTGGCGGATCGCAGTCTGYYCTGAYTCAGCCT
HuVl15	GTGGCTCCGGAGGTGGCGGATCGAATTTTATGCTGACTCAGCCCC
V_λ-Reverse	
HuJl12	CGCCTGCGGCCGCTAGGACGGTSASCTTGGTCC
HuJl7	CGCCTGCGGCCGCGAGGACGGTCAGCTGGGTGC
Light chain amplification	
HuJH-F	TCAGGGTCGACAGGTGGAG
Sequence	
PYDF	AGTAACGTTTGTCAGTAATTGC
PYDR	GTCGATTTTGTTACATCTACAC
GAP5	TTAAGCTTCTGCAGGCTAGTG
GAP3	GAGACCGAGGAGAGGGTTAGG

2. Set up 10 μg of vector pYD2.A2 and 3 μg of each PCR products of V_H, V_κ, and V_λ in microcentrifuge tubes.
3. For V_H and its corresponding vector pYD2.A2 digestion, add 10 μL of 10× NEB buffer #3, 10 μL of 10× BSA, 4 μL of each NcoI and SalI to preset microcentrifuge tubes (**step 2**), and then add molecular grade water to a final volume of 100 μL.
4. For V_L and its corresponding vector pYD2.A2 digestion, add 10 μL of 10× NEB buffer #3, 10 μL of 10× BSA, 4 μL of each BspEI and NotI to preset microcentrifuge tubes (**step 2**), and then add molecular grade water to a final volume of 100 μL.
5. Incubate the vector digestion reaction overnight in a 37 °C water bath and incubate the insert gene digestion reactions of V_H, V_κ, and V_λ for 5 h in a 37 °C water bath.
6. V_H digestion reactions generate vector designated as pYD2A2$_L$ that is ready for ligation with V_H insert at NcoI/SalI restriction sites. V_L digestion reactions generate vector designated as pYD2A2$_H$ that is ready for ligation with V_κ or V_λ inserts at BspEI/NotI restriction sites.
7. Purify the digested DNA fragments as the same in Subheading 3.1.2, **step 4**.
8. Measure purified linearized DNA concentration of vector by the same calculation as in Subheading 3.1.2, **step 5**.

3.1.4 scFv Assembly: Ligation

1. A threefold molar excess of insert to vector ratio are used for ligation. As vector pYD2.A2 is approximately 5670 bp and the average insert of V_H, V_κ, or V_λ is 400 bp, for 1 μg of vector ligation reaction add 0.2 μg insert DNA (*see* **Note 12**).
2. Prepare separate ligation master mixes (2 × 50 μL) for each digested V_H, V_κ, or V_λ from Subheading 3.1.3, **steps 3** and **4** as described in the table below.

Ligation Rxn	Vector*	Insert*	Final product
V_H	pYD2.A2$_L$	Steps 6-8	pYD2.A2$_L$.V_H
V_κ	pYD2.A2$_H$	Steps 6-8	pYD2.A2$_H$.V_κ
V_λ	pYD2.A2$_H$	Steps 6-8	pYD2.A2$_H$.V_λ

*Both digested vectors and inserts are purified products from Subheading 3.1.3 **Steps 6–8**.

3. Add 5 μL of T4 DNA ligase buffer, 1 μL of vector (1 μg/μL), 0.2 μg insert DNA (applicable volume), 2 μL of T4 ligase, and molecular grade water to a final volume of 50 μL.
4. Incubate the ligation reactions overnight at 16 °C (*see* **Note 13**).
5. Purify ligation reactions using QiAquick PCR purification kit according to the manufacturer's instruction (*see* **Note 14**).
6. Measure purified ligated DNA concentration by the same calculation as in Subheading 3.1.2, **step 5**. Adjust DNA concentration to 0.1 μg/μL for transformation.

3.1.5 scFv Assembly: Transformation and Diversity Evaluation of Prepared V_H, V_κ, and V_λ

1. Ligation reactions are transformed into DH5α cells via electroporation using ElectroMAX DH5α-E cells according to the manufacturer's instruction with the following modifications.
2. Set up four prechilled microcentrifuge tubes for each V_H, V_κ, and V_λ transformation reaction.
3. Mix 1 μL of ligation reaction (0.1 μg/μL) and 20 μL electrocompetent DH5α cells in each tube.
4. Transfer the cell–DNA mixture into 0.2 cm cuvette and electroporate using BioRad GenePulser II electroporator with the following conditions 2.5 kV, 200 Ω, 25 μF for 0.2 cm cuvette.
5. To the cells in the cuvette, add 1.0 mL of S.O.B. medium and transfer the solution to a 15 mL snap-cap tube. Flush the cuvette with an additional 1.0 mL of S.O.C.
6. Shake at 225 rpm (37 °C) for 1 h.
7. Add 2 mL of transformed cells into prewarmed 50 mL LB containing 100 μg/mL of carbenicillin.

8. Incubate overnight with shaking at 225 rpm and at 37 °C. This results in plasmid pools designated as pYD2.A2$_L$.V$_H$, pYD2. A2$_H$ V$_κ$, and pYD2.A2$_H$.V$_λ$.
9. To titer the transformation, plate 1.0, 10, and 50 μL of transformed culture from **step** 7 on LB plate containing 100 μg/mL of carbenicillin.
10. Incubate the plates overnight at 37 °C.
11. For diversity evaluation, pick ten colonies from V$_H$, V$_κ$, and V$_λ$ transformed plates and grow them individually in a 15-mL round bottom tube containing 5 mL of LB and carbenicillin (100 μg/mL) overnight with shaking at 225 rpm and at 37 °C.
12. Isolate plasmid DNA from the overnight 5 mL culture using QiAprep Spin Miniprep kit according to the manufacturer's instruction.
13. Sequence eluted plasmid DNA with the pYD2.A2 forward primer (GAP5, *see* Table 1). Compare the CDRs sequences derived from randomly picked ten plasmid insert sequences using IMGT/V-QUEST: http://www.imgt.org/IMGT_vquest/share/textes/imgtvquest.html and Vector NTI software.
14. For library size (diversity) determination, count total colonies from the plates (**steps 9–11**) and calculate library size using the general formula: cfu (colonies on the plate, **step 9**) × 10^1, 10^2 or 10^3 (dilution factor) × total vol (total volume in the LB flask, **step** 7) = library size (diversity) (*see* **Note 15**).

3.1.6 scFv Assembly: Generation of Gap Repair Ready V$_κ$ and V$_λ$

1. Purify plasmid DNA from Subheading 3.1.5, **steps** 7 and **8** (4× 50 mL overnight bacterial cells cultures) using QiAfilter Midi kit according to the manufacturer's instruction.
2. Measure purified plasmid DNA concentration by the same calculation as in Subheading 3.1.2, **step 5**.
3. Amplify V$_κ$ and V$_λ$ genes from prepared plasmid pYD2$_H$.A2.V$_l$ to add nucleotides (GRT primers, Table 1) that are homologs to pYD2.A2 for gap-repair mediated integration of the light chain genes into the vector plasmid carrying V$_H$ genes.
4. Set up PCR master mixes (4× 50 μL) for amplification of V$_κ$ and V$_λ$ from **steps 1** and **2**, as described in the table below.

Reagent	1 Rxn (μL)
5× Phusion buffer	10
10 mM dNTP mix	1
Phusion polymerase	1
HuJH-F	2

(continued)

Reagent	1 Rxn (μL)
Gap3	
Plasmids[a] (350 ng)	2
Molecule grade water	34
Final vol	50

[a]Plasmids pYD2.A2H.Vκ and pYD2.A2H.Vλ from **steps 1** and **2**

5. Aliquot 50 μL for each PCR tube and run the following conditions for V_κ: 98 °C for 30 s, 25 cycles of 98 °C for 10 s, 55 °C for 20 s, 72 °C for 30 s, and 1 cycle of 72 °C for 5 min for final extension. Run the following condition for V_λ amplification: 98 °C for 30 s, 25 cycles of 98 °C for 10 s, 48 °C for 20 s, 72 °C for 20 s, and 1 cycle of 72 °C for 5 min for final extension.
6. Isolate the 400 bp PCR product as described in Subheading 3.1.2, **steps 3–5**.

3.1.7 scFv Assembly: Digestion of Vector pYD2.A2$_L$.V$_H$ for Gap Repair

1. Set up 20 μg of pYD2.A2$_L$.V$_H$ plasmid prepared from Subheading 3.1.6, **steps 1** and **2** in a microcentrifuge tube.
2. Add 10 μL of 10× NEB buffer #3, 10 μL of 10× BSA, 7 μL of each BspEI and NotI to the tube and add molecular grade water to a final volume of 100 μL.
3. Incubate the digestion reaction overnight in a 37 °C water bath.
4. Purify the digested DNA fragments by the same protocol as in Subheading 3.1.2, **step 4**.
5. Measure purified linearized DNA concentration of vector by the same calculation as in Subheading 3.1.2, **step 5**.

3.1.8 Yeast Transformation, Tittering and Freezing

1. EBY100 *Saccharomyces cerevisiae* is the host strain for transformation. Streak from frozen EBY100 stock onto an YPD plate and grow at 30 °C for 48 h.
2. Inoculate single or a few colonies from the newly streaked YPD plate into 10 mL of YPD medium. The starting concentration should range between 0.05 and 0.1 OD_{600}, which can be performed by using a spectrophotometer (*see* **Note 16**). Grow overnight the yeast culture in the shaker incubator at 250 rpm and 30 °C.
3. After 12–16 h of growth, determine the concentration (OD_{600}) of the yeast culture by spectrophotometer (*see* **Note 17**).
4. Inoculate an aliquot of the overnight culture in 100 mL YPD medium to achieve an OD_{600} of 0.3.

5. Incubate the flask in the shaking incubator at 250 rpm and 30 °C until OD_{600} is at 1.6 (*see* **Note 18**).
6. Pellet the cells at 3000 × *g* for 5 min at 20 °C and aspirate the supernatant. Wash the cell pellet twice by resuspending cells first in 25 mL autoclaved water, centrifuge and aspirate the supernatant, and once in 50 mL of ice-cold electroporation buffer (1 M sorbitol–1 mM $CaCl_2$).
7. The cell pellet is resuspended in 20 mL 0.1 M LiAc/10 mM DTT, transferred to a 100 mL culture flask and then incubated in a shaker incubator at 250 rpm and 30 °C for 30 min.
8. Collect the cells by centrifugation again, wash once in 50 mL ice-cold electroporation buffer and then resuspend the cell pellet in 100–200 μL electroporation buffer, then adjust to a final volume of 1 mL (*see* **Note 19**).
9. Combine 4 μg of digested vector and 6 μg of DNA insert for each 400 μL electroporation reaction. The DNA mixture in water should be less than 50 μL. Reduce the volume by precipitation and resuspend in a smaller volume if necessary. See table below.

Reagent	$pYD2.V_H.V_\kappa$[a]	$pYD2.V_H.V_\lambda$	Control
$pYD2.A2_L.V_H$/BspEI/NotI[b]	4 μg	4 μg	4 μg
V_κ with gap tail[c]	6 μg		
V_λ with gap tail[c]		6 μg	
EBY100	400 μL	400 μL	400 μL
Final vol	<450 μL	<450 μL	<450 μL

[a]The volume of DNA mixture should be less than 50 μL
[b]$pYD2.A2_L.V_H$/BspEI/NotI is the product from Subheading 3.1.7, **step 5**
[c]V_κ and V_λ with gap tails are the purified PCR products from Subheading 3.1.6, **steps 3–6**

10. Gently mix 400 μL electrocompetent cells and DNA and transfer to a prechilled BioRad GenePulser cuvette (0.2 cm electrode gap). Tap the suspension to the bottom and keep on ice for 5 min until electroporation.
11. Set the Gene Pulser Xcell conditions as the following: $C = 25$ μF, $R = 200$ Ω, and $V = 2.5$ kV. Before pulsing, prepare 8 mL of 1:1 mixture of 1 M sorbitol and YPD medium in a sterile 17 × 100 mm tube per sample. Place the cuvette in the ShockPod. Push the chamber lid down to close and pulse once.
12. Remove the cuvette from the chamber and immediately add 1.0 mL of the 1 M sorbitol–YPD medium to the cuvette. Gently transfer the diluted cells into the tube with the remaining 7 mL of sorbitol–YPD medium.

13. Check and record the pulse parameters. The voltage should be approximately 2.5 kV. Typical time constant ranges 3.5–4.5 ms.
14. Incubate the transformant tube on a platform shaker at 225 rpm and 30 °C for 1 h.
15. Collect cells by centrifugation and discard the supernatant. For the negative control sample, the transformants can be resuspended in 220 μL autoclaved water and then plate 20 and 200 μL onto SD-CAA plates. For the library construction samples, resuspend the pellet in 300 mL SD-CAA medium in a 1.5 L flask, which is the total library of transformants. To titer the transformation, dilute the transformants at $1{:}10^4$, $1{:}10^5$, and $1{:}10^6$ by plating 30 μL, 3.0 μL, or 0.3 μL from the 300 mL transformation to SD-CAA agar plates. The liquid cultures are incubated with shaking at 250 rpm and 30 °C for 16–24 h or longer until the OD_{600} reaches 2.5–3. The titration plates are incubated at 30 °C for 2–3 days.
16. For library titer determination, count the colonies on the titer plates and multiply by the dilution factor (10^4, 10^5, or 10^6) to obtain the yeast library size (*see* **Note 20**). The negative control plate may show some colonies but should be less than 0.1% of the library size.
17. To preserve the library, pool 100 OD_{600} yeast cells into 50 mL conical tubes and collect the pellet by centrifugation at 3000 × *g* for 5 min at 20 °C and aspirate the supernatant. Prepare the freezing buffer by mixing sterile 50% glycerol with SD-CAA medium at a ratio of 2:3 to get 20% glycerol/SD-CAA media. Resuspend the pellet to a concentration of 100 OD_{600} yeast cell/mL, aliquot into cryotubes, and store at −80 °C.

3.2 Library Sorting

3.2.1 Library Growth and Induction

1. Quickly thaw aliquots of frozen scFv yeast library (*see* **Note 21**) at 30 °C and resuspend in SD-CAA media to obtain an OD_{600} of 0.2–0.5. Grow the culture overnight with shaking at 250 rpm and 30 °C.
2. Determine the growth of yeast by measuring the OD_{600}. Pellet at least a 10× representation of library diversity from freshly passaged culture at 3000 × *g* for 5 min in 50 mL conical tubes. For a library size of 1×10^8, a total of 1×10^9 yeast cells are needed for induction, which equals to 100 OD_{600} yeast cells.
3. For induction, resuspend the cells in SG-CAA media to an absorbance of 0.5 OD_{600}/mL (*see* **Note 22**) and grow the cells for 48 h at 18 °C and 250 rpm (*see* **Note 23**).

3.2.2 Magnetic-Activated Cell Sorting (MACS)

1. Pellet freshly induced yeast cells of at least 10× representation of library diversity by centrifugation at 3000 × *g* for 5 min at 20 °C and aspirate the supernatant. Wash cells once by resuspending in 25 mL PBSB buffer, centrifuge as described and aspirate the supernatant.
2. Wild-type (wt) HCV E2 glycoprotein or E2 mutant proteins are used for selection. The preparation of secreted E2 glycoprotein or mutants is described in our publication [10] (*see* **Note 24**). Prepare dilution of E2 in 10 mL of PBSB buffer. Resuspend the yeast cells in the E2 solution by gentle mixing.
3. Incubate cell suspension at room temperature with gentle rotation for 60 min and then followed by 10 min incubation on ice. All subsequent steps should be done with ice-cold buffer or at 4 °C (*see* **Note 25**).
4. Pellet and wash the cells with 25 mL PBSB buffer. Repeat wash one more time. Resuspend the cell pellet in 5 mL PBSB buffer with biotinylated E2-specific detection antibodies (*see* **Note 26**). Incubate on ice for 30 min. If the biotinylated wt E2 or mutant E2 is used to label the library in **step 2**, then skip **step 4** and directly go to **step 5**.
5. Pellet and wash cells with 5 mL PBSB buffer. Repeat wash one more time.
6. Resuspend the cell pellet in 5 mL PBSB buffer with 25 μL of MACS streptavidin magnetic microbeads from Miltenyi (*see* **Note 27**).
7. Incubate on ice for 10 min with gentle mixing by inverting the tube several times.
8. Pellet cells at 3000 × *g* for 5 min at 4 °C, aspirate supernatant and resuspend the pellet in 50 mL PBSB buffer.
9. Place a LS column onto the magnetic stand. Equilibrate the column by flowing through the column by gravity 3 mL of ice-cold PBSB buffer.
10. Vortex and pass the cells through a cell strainer cap tube to achieve a single-cell suspension and immediately load onto the Miltenyi LS column. Add 7 mL of cell suspension to the column. After the cells have flowed through the column, briefly remove the column from the magnet and immediately place back on the magnet (*see* **Note 28**). With the column back in the magnet, add 1 mL of PBSB wash buffer and let it flow through. Repeat until all of the cells have been loaded.
11. Once the cells are loaded on the column, wash the column with 3 mL of PBSB wash buffer. Remove column from magnet and immediately place it back, as noted in **step 10**. Repeat this wash two additional times.

12. To elute the cells, remove the column and place it over a collection tube. Add 7 mL SD-CAA medium to the column and use the plunger supplied with the column to push the remaining cells through.
13. Propagate eluted yeast for subsequent rounds of sorting. Add SD-CAA medium to the eluted cell suspension to a final volume of 50 mL in a 250 mL flask and grow overnight at 30 °C. An optional step is another round of MACS by repeating **steps 1–13**.

3.2.3 FACS Library Sorting

1. After expansion, the MACS output cells are further enriched by cell sorting.
2. MACS usually enriches at least tenfold of the original library. Figure 3 illustrates an example [10]. For FACS sorting, 10^8 cells can be used as the input, which should cover at least tenfold of the diversity of the MACS output. Pellet and wash the cells once with PBSB buffer. Resuspend the pellet in 1 mL PBSB including both HCV wt E2 glycoprotein (or E2 mutant) at 5 μg/mL and anti-V5 antibody (1:5000). Besides the E2 staining, other controls can be employed, as described in the table below. This will help to monitor specific binding against the E2 antigen (*see* **Note 29**).

Yeast	Sample	Ctrl 1	Ctrl 2	Ctrl 3	Ctrl 4	Unstained
Biotin-E2	+	–	–	–	–	–
Anti-V5	+	+	+	–	–	–
SA-APC	+	+	–	+	–	–
Anti-mouse-FITC	+	+	+	–	+	–

3. Incubate at room temperature for 1 h and then followed by 10 min on ice.
4. Pellet and wash the cells with 1 mL PBSB. Incubate the cells with anti-mouse-FITC and streptavidin-APC, both at 1:200 dilution, for 30 min at 4 °C.
5. Pellet the labeled cells, wash and sort. In a typical quadrant gated setting, the RU region corresponds to specific antigen binding and fully scFv expressed cells. Draw an appropriate region to gate in the double-positive quadrant for cell sorting. Apply "Purity mode" for the sorting. An example of enrichment of positive binders after going through MACS and FACS is shown in Fig. 3.
6. Grow the collected cells in 5 mL SD-CAA medium that is used for analysis or the next round of sorting after induction in SG-CAA, as described above.

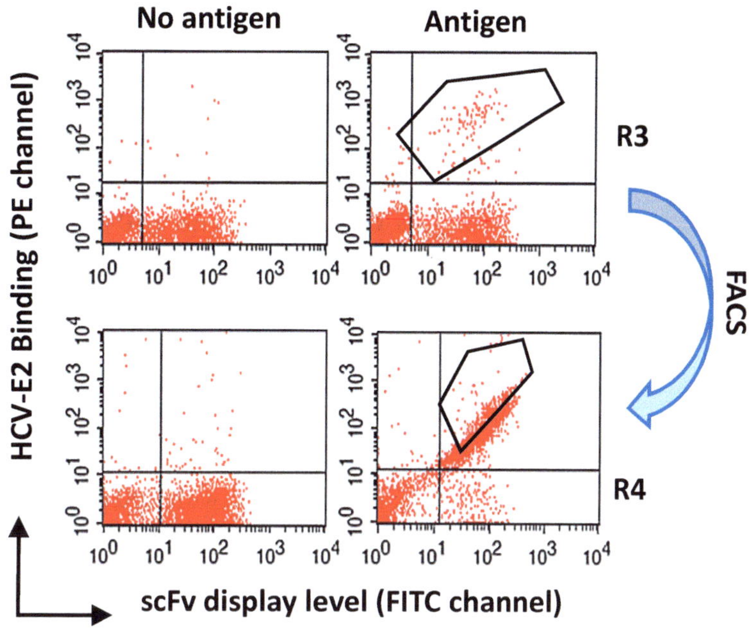

Fig. 3 An example of cell sorting on the sample after two rounds of MACS enrichment (R1 and R2). To perform 3rd round sorting (R3): 107 MACS isolated cells (from R2) were incubated with E2 mutant (E2Y632A), and then incubated with anti-V5 and HC-33.1 (an anti-E2 HMAb to a defined epitope) at 10 μg/ml. The cells were then labeled with FITC-anti-mouse and PE-anti-human (Fc-specific). The labeled cells, 107 cells/ml, were used for sorting by flow cytometry. The sorting gates were set to collect the desired double positive cells (the upper right graph). The "no antigen" control (the upper left graph) provided guidance on setting the sorting gates. After R3 selection, the double positive cells are highly enriched as showing in the lower right graph

7. The enrichment of double stained cells can be analyzed on the collected samples (presorted and postsorted samples) by staining as outlined in **step 2**. Repeat **steps 2–7** until the population of double positive binding cells is above 20% of total cells. Typically, this will require 2–3 rounds of FACS sorting. At this point, the sorted cells are ready for monoclonal yeast screening in 96-well plates.

3.2.4 Clonal Analysis

1. The cell sorter will calculate the sorted yeast cell number during the process of sorting. After the final round of sorting, directly plate a few hundreds to one thousand cells on a SD-CAA plate. At least several plates can be prepared.
2. Incubate the plates at 30 °C for 48 h or until the yeast colonies can be seen and easily picked.
3. Prepare 96-deep well (2 mL vol/well) plates with 0.6 mL/well of SD-CAA medium. Inoculate freshly grown single yeast cells

into each well. For each plate, inoculate a negative control well of yeast transformed with pYD2.A2 (see reagent). Seal the plate with breathable sealing film and shake at 250 rpm at 30 °C for 16 h until the OD_{600} is >5.

4. Similar to **step 3**, prepare another 96-deep well plate with 0.54 mL/well of SG-CAA medium. Subculture 60 μL SD-CAA culture into 540 μL SG-CAA medium, which will make a 1:10 dilution of each individual yeast culture. The yeast density will be 0.5–1 OD_{600}, and under good condition for induction. Grow at 30 °C for 20 h to induce expression of scFv.
5. We routinely use 96 V-bottom well plates for monoclonal screening by flow cytometry. Prepare 1×10^6 yeast (~50 μL) in each well. Pellet the cell by 3000 × *g* for 5 min. Resuspend the induced yeast with 200 μL of PBSB buffer, wash once and pellet again as described.
6. Resuspend the yeast cells in 100 μL PBSB buffer containing anti-V5 mAb (1:5000) and biotinylated HCV-E2 (5 μg/mL).
7. Incubate for 60 min at room temperature followed by 10 min on ice. Pellet by 3000 × *g* for 5 min and wash once with 200 μL of ice-cold PBSB buffer.
8. Resuspend pellet in 100 μL of secondary reagents: anti-mouse-FITC and streptavidin-APC, each at 1:200 dilution. Incubate for 30 min on ice, pellet the cells and wash two times in cold buffer.
9. Resuspend the cells in 100 μL of ice-cold PBSB buffer and keep on ice in the dark until analysis by flow cytometry. E2-specific binders will show both APC and FITC reactivity.

3.2.5 Plasmid Recovery and Conversion to Full Length IgG (See Note 30)

1. The method here is based on the Qiagen spin miniprep kit and is described in the literature [2]. The buffers used (P1, P2, N3, EB) are provided in the kit. Pellet approximately 250 μL of yeast cell culture centrifuge at 3000 × *g* for 5 min, aspirate the supernatant and wash once with double distilled water.
2. Resuspend the pellet in 250 μL Qiagen buffer P1 and add approximately 100 μL of glass beads. Vortex at high speed for 5 min.
3. Add 250 μL buffer P2, gently mix by inverting, and incubate at room temperature for 5 min.
4. Add 350 μL buffer N3 (*see* **Note 31**) and spin at 16,000 × *g* in a microcentrifuge for 10 min.
5. Apply supernatant to a Qiagen spin miniprep column and spin at 16,000 × *g* for 1 min. Discard the flow-through.

6. Add 750 μL buffer PE, and spin at 16,000 × *g* for 1 min. Discard flow-through and spin at 16,000 × *g* for 2 min to remove residual buffer PE. Replace collection tube with a clean eppendorf tube and add 50 μL of elution buffer EB and spin at 16,000 × *g* for 1 min to elute.
7. The miniprep can be used as template for PCR amplification as in Subheading 3.1.2, **step 1**. The primer set is PYDF/PYDR, and 5 μL of miniprep DNA is used as template. The PCR product is suitable for DNA sequencing. Use the Gap5 or Gap3 primer to sequence.
8. After sequencing, the individual antigen-specific yeast clones can be cloned into full IgG mammalian expression vector for IgG production and further characterization [13].

4 Notes

1. EBY100 can be purchased from ATCC with ATCC number: MYA-4941.
2. Vector pYD2 was kindly provided by J.D. Marks (UCSF) [14] and we made further modifications to vector pYD2.A2 (Fig. 1).
3. Use a stir bar to help incorporate the different components. If pouring plates, it is recommended to use a conical flask because the narrow top makes it easier to hold and pour.
4. Use yeast nitrogen base without amino acids.
5. Salts will precipitate at higher temperatures and dissolve back into solution as the temperature drops.
6. A 0.22 μM filter can be used in lieu of autoclaving.
7. Use proper PPE gear (i.e., eye protection and hot gloves). Microwave for short durations (i.e., 15–30 s) and swirl the bottle in between heating to see if the agarose is dissolved. Allow the bottle to cool down to 55 °C or cool enough to hold without hot gloves before pouring in gel tray. If the agarose is too hot, it will warp the walls of the gel tray.
8. Various beads have been used depending on the experimental setup (such as anti-biotin magnetic microbeads and streptavidin magnetic microbeads).
9. Some detection fluorescence labelled antibodies that we used are FITC-anti-mouse IgG, PE-anti-human IgG (Fcγ specific), PE-streptavidin, and APC-streptavidin.
10. Peripheral blood memory B cells are plated at different densities (100–1000 cells/well) and then activated by EBV as described [9]. Supernatants are screened by HCVcc

neutralization assay [10] and cells from neutralization positive wells are pooled and used for total RNA isolation.

11. Setting up multiple PCR reactions with minimal cycles of amplification help to obtain 10 μg of final PCR products and reduce biased selection due to over amplification.
12. 0.2 μg of insert in this reaction was calculated using the formula [$3/1 \times 1$ μg $\times$ 400 bp/5670 bp = 0.2 μg] used: required mass insert (μg) = desired insert–vector molar ratio × mass of vector (μg) × ratio of insert to vector lengths. Please note, in comparison to the gap-repair method in Subheading 3.1.8, **step 9**, in which 6 μg/reaction of PCR product is required for the library construction, here 0.2 μg/reaction of PCR product is much easier to obtain. It is one of the reasons that we first clone the $V_H/V_\kappa/V_\lambda$ into an *E. coli* plasmid rather than directly gap-repair into yeast.
13. Ligation reaction can be also carried out for 4 h at room temperature.
14. Gel purification of ligation reaction is very important because salts and buffers in ligation reaction can severely inhibit electroporation.
15. Example for calculation of the library size: (a) 1 mL transformed cells are added into 100 mL of medium for bacterial growth, (b) plate 100 μL from 100 mL of diluted transformed cells on to LB plate, (c) 74 colonies appear on the LB plate after overnight incubation, (d) calculation of library size is as following: [$74/100$ μL $\times 10^3 \times 100{,}000$ μL $= 7.4 \times 10^7$]. At this stage, three libraries are constructed: pYD2.A2$_L$.V_H, pYD2.A2$_H$.V_κ, and pYD2.A2$_H$.V_λ. pYD2.A2$_L$.V_H is used as a backbone to be opened for the V_κ/V_λ fragment to be cloned into to form a full scFv. pYD2.A2$_H$.Vκ and pYD2.A2$_H$.V_λ serve as template for the V_κ/V_λ amplification in Subheading 3.1.6, **step 3**. Another application for V_H/V_L libraries is that it can be directly used for light chain shuffling for affinity maturation of lower affinity scFvs.
16. Cell numbers are generally determined by an absorbance reading at 600 nm on a spectrophotometer (OD_{600}). Blank the spectrophotometer with the same buffer without cells. An absorbance of 1 OD_{600} contains about 1×10^7 cells per mL, hereafter, the yeast OD means the absorbance at 600 nm.
17. Due to the high density of overnight culture, a 1:10 or 1:20 dilution needs to be prepared for the OD measurement; thus, the OD readout will in a range of 0.5–1. Pipet 50 μL of cells into 0.95 mL of water in a spectrophotometer cuvette, mix thoroughly and measure the OD. Remember to multiply by the dilution factor to determine the titer in the cell culture.

18. It usually takes around 5 h, which corresponds to approximately 1.6×10^9 yeast cells.
19. It is not easy to estimate the volume of yeast pellet and residual buffer after centrifugation and aspiration. To prepare competent cells, first add 100–200 μL electroporation buffer to resuspend the pellet and then measure the volume with the pipettor. More buffer can be added to reach 1 mL of yeast suspension. For library construction, 50 OD is required for each transformation. Take into account cell loss during each wash step. The yeast competent cells are sufficient for two electroporation reactions of 400 μL. The cell culture and preparation can be proportionally scaled up if more electrocompetent cells are needed to make larger or more libraries.
20. Example for calculation of the library size: (a) transformed cells is resuspended and added into 300 mL of medium for yeast growth, (b) plate 3 μL from 300 mL of diluted transformed cells on to SD-CAA plate, (c) 200 colonies appear on the SD-CAA plate after 2–3 day incubation, (d) calculation of library size is as following: [200/3 μL × 300,000 μL $=2.0 \times 10^7$].
21. To maximize diversity, the starting SD-CAA culture should contain a minimum of a 10× representation of the diversity of the immunized yeast library. For a 10^8 diverse library, 10^9 yeast cells are needed to start the culture.
22. An easier way for yeast induction is to directly dilute the SD-CAA cultured yeast in SG-CAA medium. Please note that the dilution factor should be above 1:10 since 2 g/L dextrose will improve scFv display on yeast surface (higher dextrose concentrations may inhibit scFv display).
23. The induction temperature can be raised to 30 °C after the first MACS selection.
24. Selection antigen is based on the experimental design. Besides wt E2 protein, we have employed E2 mutants to avoid the selection of previously identified antibodies [10]. Purified E2 antigens are preferred reagents in this application. Before MACS enrichment, the antigen concentration suitable for MACS needs to be determined by flow cytometry analysis with a positive control of E2 proteins binding scFv displayed yeast cells. The concentration showing a high signal by flow cytometry can be used for MACS selection. In our lab, 5 μg/mL of purified E2 proteins works well.
25. Completion of secondary labeling at 4 °C to minimize antigen dissociation from yeast-displayed scFvs.

26. To isolate neutralizing antibodies, we have employed E2 mutants to avoid the selection of nonneutralizing scFvs. For details, please refer to related publications [10].
27. As an option, anti-biotin magnetic microbeads can be used. Alternating between different microbeads during subsequent rounds of MACS selection will decrease the probability of obtaining unwanted secondary reagent-specific clones.
28. This rearranges the iron beads in the column and allows the unbound cells that are physically trapped between the beads to pass through.
29. It is important to use different secondary antigen-labeling reagents (streptavidin-APC or streptavidin-PE) between FACS sorts to eliminate or reduce unwanted secondary reagent-specific scFv. If PE and FITC are used for double staining the cells, the flow cytometer should be properly compensated to eliminate the overlap between these two fluorophores.
30. While the E2 binding scFvs are identified by flow cytometry analysis, the unique sequences among the binders will need to be examined. There are different PCR methods to directly amplify the scFv fragment from the yeast colony. We found the most reliable way is to extract yeast plasmid first and then perform PCR as described in the following.
31. A cloudy precipitate will form.

Acknowledgments

This work was support in part by National Institute of Allergy and Infectious Diseases/NIH grants R41-AI108024, U19-AI123862, R01-AI132213 and R21-AI126582 to SKHF.

References

1. Boder ET, Wittrup KD (1997) Yeast surface display for screening combinatorial polypeptide libraries. Nat Biotechnol 15:553–557
2. Bidlingmaier S, Su Y, Liu B (2015) Combining phage and yeast cell surface antibody display to identify novel cell type-selective internalizing human monoclonal antibodies. Methods Mol Biol 1319:51–63
3. Miller KD, Pefaur NB, Baird CL (2008) Construction and screening of antigen targeted immune yeast surface display antibody libraries. Curr Protoc Cytom Chapter 4:Unit 4 7
4. Sheehan J, Marasco WA (2015) Phage and yeast display. Microbiol Spectr 3:AID-0028-2014
5. Van Deventer JA, Wittrup KD (2014) Yeast surface display for antibody isolation: library construction, library screening, and affinity maturation. Methods Mol Biol 1131:151–181
6. Chao G, Lau WL, Hackel BJ, Sazinsky SL, Lippow SM, Wittrup KD (2006) Isolating and engineering human antibodies using yeast surface display. Nat Protoc 1:755–768
7. Feldhaus MJ, Siegel RW, Opresko LK, Coleman JR, Feldhaus JM, Yeung YA et al (2003) Flow-cytometric isolation of human antibodies from a nonimmune Saccharomyces cerevisiae surface display library. Nat Biotechnol 21:163–170

8. Feldhaus MJ, Siegel RW (2004) Yeast display of antibody fragments: a discovery and characterization platform. J Immunol Methods 290:69–80
9. Perkins S, Foung SK (1995) Stabilizing antibody secretion of human Epstein Barr virus-activated B-lymphocytes with hybridoma formation by electrofusion. Methods Mol Biol 48:295–307
10. Keck ZY, Xia J, Wang Y, Wang W, Krey T, Prentoe J et al (2012) Human monoclonal antibodies to a novel cluster of conformational epitopes on HCV E2 with resistance to neutralization escape in a genotype 2a isolate. PLoS Pathog 8:e1002653
11. Keck ZY, Saha A, Xia J, Wang Y, Lau P, Krey T et al (2011) Mapping a region of hepatitis C virus E2 that is responsible for escape from neutralizing antibodies and a core CD81-binding region that does not tolerate neutralization escape mutations. J Virol 85:10451–10463
12. Hadlock KG, Lanford RE, Perkins S, Rowe J, Yang Q, Levy S et al (2000) Human monoclonal antibodies that inhibit binding of hepatitis C virus E2 protein to CD81 and recognize conserved conformational epitopes. J Virol 74:10407–10416
13. Smith K, Garman L, Wrammert J, Zheng NY, Capra JD, Ahmed R et al (2009) Rapid generation of fully human monoclonal antibodies specific to a vaccinating antigen. Nat Protoc 4:372–384
14. Razai A, Garcia-Rodriguez C, Lou J, Geren IN, Forsyth CM, Robles Y et al (2005) Molecular evolution of antibody affinity for sensitive detection of botulinum neurotoxin type A. J Mol Biol 351:158–169

Chapter 28

Detection of Antibodies to HCV E1E2 by Lectin-Capture ELISA

Marian Major and Mansun Law

Abstract

Enzyme-linked immunosorbent assays (ELISAs) enable rapid detection and quantitation of antibodies in samples. Such assays can be highly sensitive and can be performed in most laboratories with basic equipment. Although detecting binding antibodies to the surface proteins of most pathogens by ELISA is not always indicative of antibody function, i.e., neutralizing activity of antibodies, the results can be used as a first step toward more in-depth analysis of antibody responses. Here we describe a method that can be used to standardize ELISAs for the detection of HCV envelope antibodies across laboratories and provide adaptations of the method to further characterize antibody responses in serum samples.

Key words ELISA, Lectin, Glycoproteins, Antibody titer, Affinity, Avidity

1 Introduction

The HCV envelope glycoproteins E1E2 represent a primary target for vaccines designed to induce neutralizing antibody responses. The role of anti-E1E2 antibodies in controlling HCV replication during primary or secondary infections is not fully understood. High antibody titers are not produced until the chronic phase of the infection [1], although the induction of neutralizing antibodies has been associated with clearance of the virus in a number of studies [2–4]. In addition, antibodies to surface proteins of HCV have been demonstrated to neutralize the virus in vitro and in vivo [5, 6]. Using in vitro cell culture systems, patient serum/plasma samples and monoclonal antibodies have been identified that are capable of cross-neutralizing a number of different genotypes of HCV [1, 7–9]. Thus, the induction of antibodies to HCV E1E2 remains a goal for many studies and screening for these antibodies in a quick, consistent manner is essential.

ELISAs for HCV E1E2 antibodies do not give a direct indication of neutralizing activity. However, the differences in signal strength obtained for different samples generally translate to similar

Mansun Law (ed.), *Hepatitis C Virus Protocols*, Methods in Molecular Biology, vol. 1911,
https://doi.org/10.1007/978-1-4939-8976-8_28,

differences in neutralizing activity [10]. An envelope ELISA could be used as a first step toward developing a functional-based neutralizing antibody test using a subset of samples with the strongest antibody signals. Such an ELISA could use a purified form of HCV E1E2 protein although a lectin capture approach can be used to enrich E1E2 from crude cell lysates onto the ELISA well surface to bypass the need for antigen purification [10]. Additionally, purification of the envelope proteins can be time consuming and cost prohibitive, particularly when multiple E1E2 antigens are required (e.g., different strains, genotypes, or mutated variants). Thus, a lectin capture ELISA can be considered as the most economical, rapid, and reproducible system for detecting antibodies to E1E2.

1.1 Lectin Type

There are several lectins available commercially that have been shown to bind to the carbohydrates on the HCV envelope proteins and inhibit infection [11–14]. Care should be taken to select a lectin most specific to the HCV envelope glycans in order to reduce the binding of cellular proteins. *Galanthus nivalis* lectin (GNL) antigen has been used previously to purify HCV E1E2 proteins for immunization studies in chimpanzees [15, 16]. GNL was shown to bind HCV E1E2 with a high affinity and in preference to cellular proteins compared to other well-characterized lectins [12]. Consequently, GNL is recommended as the capture lectin for use in this ELISA system.

1.2 Antigen Source

Multiple expression systems have been used to produce HCV envelope proteins. Some have used viral vectors for expression [12, 15–17], others have used transient DNA plasmid expression [18] or constitutively expressing cell lines [10, 19]. There are many E1E2 sequences in the PubMed database and the application of the assay may dictate the type of expression system to be used. For example, a transient gene expression system is rapid and could be used to generate a number of different E1E2 isolates if the goal of the study is to examine mutations or sequence diversity in antibody recognition. The disadvantage is that large amounts of DNA are usually needed for transfection together with large culture volumes. Alternatively, stable cell lines constitutively expressing the E1E2 proteins provide a consistent and reproducible source of antigen for large studies using specific envelope isolates but the time to development of such cell lines may not be feasible for some projects (Reviewed in [20]). Regardless of the system chosen it can be designed to express the E1E2 proteins together. Previous studies indicate that E2 folding is not dependent upon the presence of E1 [17, 18, 21], although E2 may be essential for correct folding of E1 [21]. For either E1 or E2 proteins a signal sequence should be included at the N terminus to facilitate secretion from the ER, this may be the signal sequence present in the HCV polyprotein or a heterologous signal sequence. Early work showed that E2 truncated at residue 661 appeared to be the only secreted form that was correctly folded [21]. Although truncated forms of E1E2 can result in a soluble form of the antigen which can be secreted from cells the amounts isolated from the supernatants may be low while the cell lysates can provide a rich source of antigen that can function equally well in lection capture ELISAs. Whether to isolate intracellular proteins or use the secreted form of the protein should be determined by the user. Finally, the cell line used to express the proteins does not seem to be essential for correct folding of the antigens. Envelope proteins have been expressed in several cell types including CHO cells, HEK 293T cells, HeLa cells, Huh7.5 cells, and HepG2 cells.

2 Materials

2.1 Buffers and Solutions

1. Phosphate buffered saline (PBS).
2. PBS plus 0.05% (v/v) Tween-20 (#P1379 Sigma-Aldrich) (PBST).
3. Lysis buffer: 0.5% Triton X-100 (#X100 Sigma) in 50 mM Tris–HCl, pH 7.5/150 mM NaCl/2 mM EDTA (TNE) plus protease inhibitor cocktail (#P8340 Sigma). Use 40 μL of cocktail/mL of lysis buffer (*see* **Note 1**).
4. Capture lectin: *Galanthus nivalis* lectin (GNL) (#L8275 Sigma Aldrich) at 5 μg/mL in PBS (*see* **Note 2**).
5. E1E2 antigen as a crude lysate from mammalian cells.
6. Blocking reagent: 5% nonfat dried milk powder in PBST (*see* **Note 3**).
7. Dilution buffer: 1% nonfat dried milk powder in PBST (*see* **Note 3**).
8. Primary antibodies: Monoclonal antibodies, serum or plasma samples representing test articles.
9. Secondary antibody conjugated to horseradish peroxidase (HRP) (Anti-species of the primary antibodies).
10. Substrate: HRP substrate.
11. Appropriate stopping solution to stop enzyme reaction prior to reading the signal.

2.2 Supplies and Instruments

1. 96-well Maxisorp Plates (#436110 Nunc) (*see* **Note 4**).
2. Standard round-bottom or deep-well 96-well plates for sample dilution.
3. 96-well plate cover or adhesive seal.
4. Micropipettes and pipette tips (to dispense 50–200 μL).
5. 96-well plate reader.
6. ELISA plate washer (optional).

3 Methods

3.1 Antigen Preparation

The protocol below describes the isolation of antigen from a cell lysate. If using a secreted form of the antigen the supernatant should be centrifuged to remove cell debris, filtered through a 0.22 μM filter, aliquoted and stored at −80 °C.

1. Isolate cells expressing E1E2 antigen either through trypsinization or through harvesting directly from the flask if cells are grown in suspension.
2. Pellet the cells by centrifugation at 400 × *g* for 5 min. Discard supernatant, wash with cold PBS and pellet again by centrifugation.
3. Resuspend the washed pellet in cold lysis buffer plus protease inhibitor cocktail at 10^6 cells per 100 μL.
4. Incubate on ice for 30 min.
5. Clarify by centrifugation at 20,000 × *g*, 4 °C for 15 min.
6. Clarified lysate can be aliquoted and stored at −80 °C.

3.2 Lectin Coating

1. Dilute GNL to 5 μg/mL in PBS and add 100 μL per well to a 96-well plate.
2. Seal the plate with an adhesive cover and incubate at room temperature for 1 h or overnight at 4 °C (*see* **Note 5**).

3.3 Blocking Step

1. Aspirate off the lectin and wash plate once with PBS. Add 200 μL per well of blocking reagent and incubate at room temperature for 60 min.
2. Aspirate off the blocking reagent and wash three times with 200 μL PBST per well either by hand using a multichannel pipette, wash bottle, or using an automatic ELISA plate washer (*see* **Note 6**).

3.4 Antigen Capture

1. Dilute antigen in dilution buffer to the optimal concentration (*see* Subheading 3.8).
2. Add 100 μL of diluted antigen into each well of a 96-well plate and incubate at room temperature for 2 h.
3. Aspirate off unbound antigen and wash three times with 200 μL PBST per well.

3.5 Primary Antibody Binding

1. Dilute the primary antibodies or serum samples in dilution buffer and add 100 μL per well. Serum samples should be diluted to a minimum of 1:50 to reduce nonspecific signal. In many cases a titration of samples by checkboard titration may be necessary (*see* Subheading 3.8).

2. Testing should always include a negative sample, preferably a preimmune sample, diluted to the same level as the test sample (*see* **Note 7**).
3. Seal the plate with an adhesive cover and incubate at room temperature for 60 min.
4. Aspirate off the primary antibody and wash five times with 200 μL PBS per well.

3.6 Secondary Antibody Binding

1. Dilute the HRP-conjugated secondary antibody in dilution buffer to the concentration recommended by the manufacturer.
2. Add 100 μL of the HRP-conjugated diluted secondary antibody to each well.
3. Seal the plate with an adhesive cover and incubate at room temperature for 60 min.
4. Aspirate off the secondary antibody and wash three times with 200 μL PBS per well.

3.7 Apply Enzyme Substrate

1. Prepare HRP substrate according to the manufacturer's instructions.
2. Add 100 μL of substrate to each well and incubate according to the manufacturer's instructions.
3. Add 100 μL of required stop solution (determined by the substrate used) after a predetermined time (e.g., 2–10 min) to ensure consistency between assay plates and experiments (*see* **Note 8**). Read the OD signals on an ELISA plate reader at the absorbance determined by the substrate used.

3.8 Optimization of Antigen and Antibody Reagents

For initial characterization of an antigen preparation, normal cell lysates (i.e., those without E1E2) should be included to assess background reactivity and establish a cutoff value for specific signal. The E1E2 antigen used for lectin capture ELISAs should be titrated for each batch to determine the dilution factor needed to achieve optimal signal. To optimize a batch of E1E2 lysate the user will need an anti-E1E2 antibody. This can be a monoclonal antibody or a defined polyclonal serum but should be available in a good enough supply such that it can be used to qualify new preparations at future dates. If the optimal dilution of antibody is unknown a matrix setup, or checkerboard titration, can be used with twofold dilutions of antigen in the vertical direction on the plate and twofold dilutions of antibody in the horizontal direction.

1. Coat, block, and wash the plate as described in Subheadings 3.2 and 3.3.
2. To serially dilute the antigen: Place 100 μL of dilution buffer in columns 2–12 of a clean 96 well plate.

3. Add 200 μL of the antigen prep diluted 1:10 in dilution buffer into all wells of column 1.
4. Using a multichannel pipette remove 100 μL of diluted antigen from column 1 and transfer to wells in column 2. Gently pipet up and down to mix.
5. Repeat **step 4** transferring 100 μL of diluted antigen from one column to the next. Pipet gently up and down to mix contents and change pipette tips between each transfer. Leave column 12 with dilution buffer only as blank wells.
6. Transfer 100 μL from each well of the dilution plate to the corresponding well of the coated, blocked and washed ELISA plate.
7. Seal the plate with an adhesive cover and incubate at room temperature for 2 h.
8. Aspirate off the antigen and wash three times with 200 μL PBS per well.
9. To serially dilute the antibody: In a separate 96-well plate dispense 125 μL of dilution buffer into rows B to H.
10. Dilute the primary antibody in dilution buffer to the preferred starting dilution (minimum 1:50).
11. Add 250 μL diluted antibody to all 12 wells of row A.
12. Using a multichannel pipette transfer 125 μL of diluted antibody from all wells of row A to row B. Gently pipet up and down three times to mix.
13. Change pipette tips and repeat **step 12** transferring antibody from row B to row C. Repeat this transfer from one row to the next leaving row H with only dilution buffer.
14. Using a multichannel pipette transfer 100 μL from each well of the antibody dilution plate to the coated ELISA plate beginning with row H and working back to the highest antibody concentration in row A.
15. Seal the plate and incubate at room temperature for 60 min.
16. Aspirate off the antibody and wash five times with 200 μL of PBS per well.
17. Continue with the secondary antibody binding and signal development as described in Subheadings 3.6 and 3.7.

The final antigen dilution selected should be that which achieves maximum signal, to obtain the greatest sensitivity, or the dilution that returns a signal most similar to that obtained for the previous batch of antigen.

3.9 Assays to Characterize HCV E1E2 Antibodies

3.9.1 Identification of Positive Samples Using Cutoff Values

In order to identify positive samples a cutoff should be established based on the values of the negative controls. Each plate should contain its own negative and positive controls and blank wells (dilution buffer in place of primary antibody with all other components included for the remainder of the assay). Include at least three negative control wells on each plate.

1. Calculate the mean OD signal for blank wells. Subtract this value from all other sample wells. Use these subtracted values for all future calculations.
2. Calculate the mean OD signal for negative controls.
3. Calculate the mean OD values for positive controls and test samples.
4. Calculate a P/N value (positive to negative ratio) by dividing the positive OD value by the OD value for the negative control.
5. A P/N value >2 is considered positive. All samples with a P/N value less than 2 should be considered negative.

An alternative method to calculate the cutoff for the plate is to use the value of the negative control plus 4 standard deviations (SD) of the negative control values.

1. Calculate the mean OD signal for blank wells. Subtract this value from all other sample wells. Use these subtracted values for all future calculations.
2. Calculate the mean OD signal and SD for negative control wells (Neg-SD).
3. Multiply the Neg-SD by 4 to obtain 4σ-NEG-SD.
4. Add the value of 4σ-NEG-SD to the mean of the negative control samples to obtain NEG + 4σ. This value represents the cutoff for the assay.

In our experience the cutoff value is very similar whether the user employs the P/N ratio approach or the 4 SD approach and in most cases the P/N approach has been found to be more stringent.

3.9.2 Endpoint Titration

Lectin capture ELISA is a useful tool for assessing antibody titers quickly. The titers obtained can provide an indicator for starting dilutions to use in the more labor-intensive neutralization assay. As for a single dilution ELISA always include a negative control, positive control and blank wells. The negative control should be included in the dilution series. The positive control can be used at a single dilution/concentration as an indicator for the development and termination of the HRP reaction at equivalent OD values to generate consistent results between ELISA plates/experiments.

1. Coat plates with lectin, block, and coat with antigen as described above in Subheadings 3.2–3.4.

2. Perform twofold serial dilutions of the test antibody starting with a 1:50 dilution. Higher serial dilutions, e.g., threefold, fivefold, or tenfold, may be used for sample batches that are strongly reactive to the antigens.
3. Aspirate antigen from the antigen-coated ELISA plate (Subheading 3.4) and wash three times with 200 μL PBS per well.
4. Using a multichannel pipette transfer 100 μL of diluted serum samples from the dilution plate to each respective well on the ELISA plate.
5. Seal the plate with an adhesive cover and incubate at room temperature for 60 min.
6. Aspirate off the antibody and wash five times with 200 μL PBS per well.
7. Apply the secondary antibody and enzyme substrate as described above in Subheadings 3.6 and 3.7.
8. Calculate the mean OD for each pair of replicate wells after subtracting values for the blank wells.
9. To determine the endpoint titer of a given antibody sample, it can be defined as the last dilution of a serum sample that returns a P/N ratio of >2 (*see* **Note 9**).

3.9.3 Antibody Avidity

Antibody avidity can be assessed using increasing amounts of sodium isothiocynate (NaSCN) on the ELISA plate following antibody binding. This reagent disrupts the antibody–antigen bond and the concentration of NaSCN required to reduce the signal can provide information on antibody avidity. These studies are performed using a single antibody dilution that gives an OD signal of 0.5–1.

1. Coat plates with lectin, block, and coat with antigen as described above in Subheadings 3.2–3.4.
2. Aspirate off the antigen and wash three times with 200 μL PBS per well.
3. Dilute the primary antibodies or serum samples in dilution buffer to a dilution that gives an OD signal of 0.5–1. Prepare enough antibody to test six different concentrations of NaSCN (0–5 M) in duplicate. A volume of 1500 μL will easily give 12 × 100 μL aliquots.
4. Add 100 μL of antibody per well into 12 wells.
5. Seal the plate with an adhesive cover and incubate at room temperature for 60 min.
6. Prepare solutions of NaSCN in PBS at 0, 1, 2, 3, 4, and 5 M.
7. Aspirate off the antibody and wash five times with 200 μL PBS per well.

8. Add 100 μL of increasing concentrations of NaSCN in duplicate to the plate for each antibody tested.
9. Incubate the plate for 15 min at room temperature.
10. Aspirate off the NaSCN solution and wash five times with 200 μL PBS per well.
11. Continue with the secondary antibody binding and signal development as described in Subheadings 3.6 and 3.7.
12. Calculate the mean OD for each set of duplicate wells.
13. Calculate the relative avidity as follows: [OD of NaSCN-treated wells/OD of untreated wells] × 100.
14. Relative avidity results for each NaSCN concentration can be plotted and the amount of NaSCN needed to reduce binding by 50% can be calculated.
15. Alternatively, data for different antibodies can be compared by plotting the % binding for each antibody at a specific NaSCN concentration, e.g., 1 or 2 M NaSCN.

4 Notes

1. To make 100 mL of Triton X-100 solution add 90 mL of TNE to a conical flask. Add 500 μL of Triton X-100 to the TNE. Use a cutoff pipette tip to facilitate pipetting the viscous solution. Rinse the pipette tip by gently pipetting up and down two to three times. Add a stir bar and stir the solution at room temperature for 10 min to dissolve the Triton X-100. Make the volume up to 100 mL with TNE and store at room temperature.
2. Concentrations of GNL from 1 to 5 μg/mL can be used depending on the source of the lectin. We recommend determining the saturating concentration of each lectin from a new supplier for maximal sensitivity and conservation of reagents.
3. To make 50 mL of blocking reagent add 45 mL of PBS to a 50 mL graduated cylinder or a conical screw cap tube. Transfer 2.5 g of skim milk powder to the cylinder, cover the top with Parafilm and invert gently until the milk is dissolved. Add 25 μL of Tween-20 (cut off the end of the pipette tip to facilitate pipetting the viscous solution), invert gently to mix and dissolve the Tween-20. Make up to 50 mL with PBS. Use this solution as the blocking reagent or dilute fivefold with PBST to make dilution buffer.
4. The immunoassays can also be performed using half-area assay plates (Corning 3690 Costar 96 Well Assay Plate). In such a case the volume of all reagents used (e.g., GNL, antigens and

antibodies) can be reduced to half of the volumes stated in the chapter.

5. Incubations for all stages can be performed overnight at 4 °C or at room temperature for 60 min. However, the incubation time and temperature chosen should remain consistent between assays.
6. Following all washes and just prior to adding the next reagent invert plate onto a paper towel and gently tap to remove residual buffer in the wells.
7. The choice of negative controls for this assay are critical to correct interpretation of results. Ideally, a preimmune serum sample should be used as a negative control. However, preimmune serum samples may not be available, e.g., if a set of sera from chronically infected patients is being tested. In this instance a normal human (or appropriate animal) serum sample should be used. This could be obtained from a commercial source or from a known uninfected subject. An alternative negative control would be the use of normal cell lysate (i.e., cell lysate without E1E2) tested against the positive control sample on the plate. If this type of negative control is used the positive control should be of the same type as the test article, serum or monoclonal antibody, and at the same dilution. A cutoff can then be calculated using an OD value three times above the result obtained in the assay with the negative cell lysate incubated with the positive control. Whichever type of negative control is used it should be maintained consistently from one assay to the next.
8. The positive control antibody is important for monitoring data quality. Their OD readings should not vary greatly (<20%) between assay plates and ELISA experiments to ensure consistent titration results.
9. There are several ways to determine antibody titer of an immune serum. The most commonly used are mid-point titration (50% maximal binding level) and end-point titration. We prefer end-point titration because it can be applied to samples with relatively low antibody titers whereas for mid-point titration saturating (plateau) binding is required to determine mid-points. Overall, it is necessary to clearly define how antibody titers are determined in scientific reports and avoid direct comparison of data when antibody titers are measured by different methods.

Acknowledgments

Financial support was provided by Food and Drug Administration intramural funds [Program Number Z01 BK04010-11 LHV to M. M.] and by the National Institutes of Health [grant numbers AI079037, AI106005 and AI123861 to M.L.]. We thank Yusra Gimie and Kenna Nagy for comments and proofreading of the manuscript.

References

1. Logvinoff C, Major ME, Oldach D, Heyward S, Talal A, Balfe P et al (2004) Neutralizing antibody response during acute and chronic hepatitis C virus infection. Proc Natl Acad Sci U S A 101:10149–10154
2. Pestka JM, Zeisel MB, Blaser E, Schurmann P, Bartosch B, Cosset FL et al (2007) Rapid induction of virus-neutralizing antibodies and viral clearance in a single-source outbreak of hepatitis C. Proc Natl Acad Sci U S A 104:6025–6030
3. Dowd KA, Netski DM, Wang XH, Cox AL, Ray SC (2009) Selection pressure from neutralizing antibodies drives sequence evolution during acute infection with hepatitis C virus. Gastroenterology 136:2377–2386
4. Osburn WO, Snider AE, Wells BL, Latanich R, Bailey JR, Thomas DL et al (2014) Clearance of hepatitis C infection is associated with the early appearance of broad neutralizing antibody responses. Hepatology 59:2140–2151
5. Houghton M, Abrignani S (2005) Prospects for a vaccine against the hepatitis C virus. Nature 436:961–966
6. Law M, Maruyama T, Lewis J, Giang E, Tarr AW, Stamataki Z et al (2008) Broadly neutralizing antibodies protect against hepatitis C virus quasispecies challenge. Nat Med 14:25–27
7. Hsu M, Zhang J, Flint M, Logvinoff C, Cheng-Mayer C, Rice CM et al (2003) Hepatitis C virus glycoproteins mediate pH-dependent cell entry of pseudotyped retroviral particles. Proc Natl Acad Sci U S A 100:7271–7276
8. Owsianka A, Tarr AW, Juttla VS, Lavillette D, Bartosch B, Cosset FL et al (2005) Monoclonal antibody AP33 defines a broadly neutralizing epitope on the hepatitis C virus E2 envelope glycoprotein. J Virol 79:11095–11104
9. Keck Z, Wang W, Wang Y, Lau P, Carlsen TH, Prentoe J et al (2013) Cooperativity in virus neutralization by human monoclonal antibodies to two adjacent regions located at the amino terminus of hepatitis C virus E2 glycoprotein. J Virol 87:37–51
10. Wong JA, Bhat R, Hockman D, Logan M, Chen C, Levin A et al (2014) Recombinant hepatitis C virus envelope glycoprotein vaccine elicits antibodies targeting multiple epitopes on the envelope glycoproteins associated with broad cross-neutralization. J Virol 88:14278–14288
11. Helle F, Wychowski C, Vu-Dac N, Gustafson KR, Voisset C, Dubuisson J (2006) Cyanovirin-N inhibits hepatitis C virus entry by binding to envelope protein glycans. J Biol Chem 281:25177–25183
12. Kachko A, Loesgen S, Shahzad-Ul-Hussan S, Tan W, Zubkova I, Takeda K et al (2013) Inhibition of hepatitis C virus by the cyanobacterial protein MVL: mechanistic differences between the high-mannose specific lectins MVL, CV-N, and GNA. Mol Pharm 10:4590–4602
13. Pohlmann S, Zhang J, Baribaud F, Chen Z, Leslie GJ, Lin G et al (2003) Hepatitis C virus glycoproteins interact with DC-SIGN and DC-SIGNR. J Virol 77:4070–4080
14. Takebe Y, Saucedo CJ, Lund G, Uenishi R, Hase S, Tsuchiura T et al (2013) Antiviral lectins from red and blue-green algae show potent in vitro and in vivo activity against hepatitis C virus. PLoS One 8:e64449
15. Choo QL, Kuo G, Ralston R, Weiner A, Chien D, Van Nest G et al (1994) Vaccination of chimpanzees against infection by the hepatitis C virus. Proc Natl Acad Sci U S A 91:1294–1298
16. Puig M, Major ME, Mihalik K, Yu MY, Feinstone SM (2004) Immunization of chimpanzees with an envelope protein-based vaccine enhances specific humoral and cellular immune responses that delay hepatitis C virus infection. Vaccine 22:991–1000
17. Khan AG, Whidby J, Miller MT, Scarborough H, Zatorski AV, Cygan A et al (2014) Structure of the core ectodomain of

the hepatitis C virus envelope glycoprotein 2. Nature 509:381–384

18. Kong L, Giang E, Nieusma T, Kadam RU, Cogburn KE, Hua Y et al (2013) Hepatitis C virus E2 envelope glycoprotein core structure. Science 342:1090–1094

19. Whidby J, Mateu G, Scarborough H, Demeler B, Grakoui A, Marcotrigiano J (2009) Blocking hepatitis C virus infection with recombinant form of envelope protein 2 ectodomain. J Virol 83:11078–11089

20. Hacker DL, Balasubramanian S (2016) Recombinant protein production from stable mammalian cell lines and pools. Curr Opin Struct Biol 38:129–136

21. Michalak JP, Dubuisson J, Rice CM, Ung S, Meunier JC, Choukhi A et al (1997) Characterization of truncated forms of hepatitis C virus glycoproteins. J Gen Virol 78:2299–2306

Chapter 29

In Vitro Neutralization Assay Using Cultured Hepatitis C Virus

Jannick Prentoe and Jens Bukh

Abstract

The method outlined here enables evaluation of the neutralization potency of monoclonal and polyclonal antibodies against in vitro cultured hepatitis C virus (HCV). The high variation in envelope protein sequence among HCV isolates necessitates the inclusion of several isolates, spanning the major genotypes of HCV, in order to make strong conclusions concerning the cross-reactive neutralization potential of a given antibody. This would be particularly relevant for any neutralization experiments aimed at uncovering novel therapeutic- or vaccine-relevant antibodies. In addition, these assays can also be used to compare neutralization sensitivity of novel cultured HCV to that of previously characterized isolates.

Key words Hepatitis C virus, HCV, Neutralization, Antibodies, Immunostaining, Vaccine

1 Introduction

Infectious JFH1-based Core-NS2 recombinants of HCV have greatly impacted our ability to evaluate the HCV-specific neutralization efficacy of purified antibodies, as well as sera containing anti-HCV antibodies. In these systems the Core, Envelope 1 and 2 (E1 and E2), p7 and NS2 regions of HCV are combined with the 5′ and 3′ untranslated regions and NS3-NS5B regions of the unique cell-culture-replication-capable isolate, JFH1 [1, 2]. The neutralizing epitopes of HCV are present on E1 and E2 and thus these JFH1-based Core-NS2 recombinants allow for evaluation of the neutralization sensitivity of divergent HCV isolates. We developed numerous cell-culture-adapted JFH1-based recombinants with genotypes 1–7 sequence in the Core-NS2 region [3–10]; only the genotype 6 recombinant virus has adaptive substitutions in the envelope proteins that could influence neutralization. We have used these infectious culture systems to evaluate the cross-genotype reactivity of several human and murine monoclonal antibodies [7, 11–14]. It is noteworthy that even though some patient sera might have limited or no detectable neutralizing antibodies against culture

Mansun Law (ed.), *Hepatitis C Virus Protocols*, Methods in Molecular Biology, vol. 1911,
https://doi.org/10.1007/978-1-4939-8976-8_29,

viruses with wild-type envelope sequences this might merely indicate that their targeted epitopes are shielded; such antibodies can potentially be detected by using the corresponding recombinant viruses deleted for the hypervariable region 1 [9], which have increased neutralization sensitivity [15, 16].

Three different readouts have been devised for this neutralization method. Two of these are based on the insertion of a reporter in the NS5A region (either EGFP or luciferase) in which case the degree of infection is determined by fluorescence or luminescence measurements, respectively [17]. The final approach is based on immunostaining of foci of infected cells (FFUs), which are counted, preferably in an automated process [18]. This method is aided by immunostaining an HCV protein encoded by the JFH1 part of the viral genome to ensure uniform staining efficiency across JFH1-based HCV recombinants with highly divergent sequences in the Core-NS2 region, and this is the approach that will be explained in detail in this methods chapter [4, 5]. The antibody used here, 9E10 [1], displays superior specificity in immunostaining of the HCV NS5A protein of the JFH1 isolate. Alternatively, tagged viral genomes (such as flag-tagged E2 [19]) would also allow for similar staining patterns for all the used HCV recombinants.

2 Materials

All cell work described involve the human hepatoma 7.5 cells (Huh7.5 cells [20]), which are highly susceptible to infection with the aforementioned cell culture adapted JFH1-based Core-NS2 HCV recombinants. The cells can be cultured in DMEM (GIBCO) supplemented with 10% fetal calf serum. Cells are always incubated at 37 °C in a CO_2 incubator set at 5%.

2.1 Neutralization

1. Virus stocks.
2. Poly-D-lysine (PDL)-coated 96-well plates (NUNC, 152038).
3. Multichannel pipette (preferably automated enabling smooth consistent pipetting to minimize cell loss).

2.2 Immunostaining of FFUs

1. PBS–Tween (0.1% Tween 20).
2. Methanol.
3. Hydrogen peroxide.
4. Blocking buffer (PBS/Tween with 1% bovine serum albumin and 0.2% Skim milk powder).
5. Primary antibody against HCV protein. In the method described here the mouse monoclonal antibody 9E10 against NS5A is used.

6. Secondary antibody coupled to horse radish peroxidase (ECL anti-mouse from GE Healthcare, NA931V).
7. 3,3′-Diaminobenzidine (DAB) substrate (Liquid DAB+ from Dako, K3468).

3 Methods

3.1 Neutralization

The preparation of authentic virus stocks is essential for the procedure. Standard methods are used to culture HCV on Huh7.5 cells [3]. The final HCV Core-NS2 recombinant viruses should have their envelope protein sequences confirmed to ensure identity of the virus and to identify potential substitutions, which can arise by cell culture passaging of HCV. Whether such substitutions are acceptable depends on the specific application of the method, but since they can influence results by changing the inherent HCV particle neutralization sensitivity [21–23] they should always be identified and reported. The extraction of HCV RNA from virus stocks, prepared from cell culture supernatants collected at peak of infection, can be done by several methods including various commercial columns or solvent based RNA extraction kits. Traditional Sanger sequencing is sufficient and the reactions can be procured commercially.

The virus stocks should be infectivity titrated such that 20–200 FFUs per well in the positive control wells can be ensured. This is done by performing a variant of the neutralization assay described below in which no antibody is added to the wells (6 wells for each virus is sufficient). The desired number of FFUs per well will depend on whether manual or automated counting is used. Additional handling tips are provided in **Notes 1–6**.

1. A twofold, threefold, or fivefold dilution series of the antibody or serum to be tested is prepared (a good starting dilution for monoclonal antibodies is 100 μg/ml and for sera a 1:50 dilution; these will be further diluted 1:2 in the steps below where the antibody is mixed with virus). The choice of fold dilution series should be based on previous knowledge of the efficacy of the antibody. If nothing is known a fivefold dilution series is a good starting point as it almost certainly ensures that the shift from high to no neutralization will be within the tested concentrations. Smaller steps (such as twofold or threefold) increase the accuracy of the assay results.
2. 50 μl of the dilution series is added to wells in a sterile 96-well plate (four replicates is common for statistical reasons) along with at least 4 wells in which only medium is added (to evaluate background and for background subtraction if automated counting is used). Subsequently, 50 μl of a virus suspension is

added to each well (containing the aforementioned 20–200 FFUs/well - final readout).

3. The virus/antibody mixtures are incubated at 37 °C for 1 h and then added onto a 96-well PDL plate of Huh7.5 cells plated at 7000 cells/well the previous day and incubated overnight in a 37 °C, 5% CO_2 incubator.
4. Following a 4-h infection with virus the cells are gently washed (preferably with an automated multichannel pipette to increase between-row reproducibility). In our experience the highest cell loss is observed at this washing step, which is minimized by pipetting evenly and slowly.
5. Next the cells are incubated for 48 h in a 37 °C, 5% CO_2 incubator, at which time the number of FFUs are visualized by immunostaining.

3.2 Immunostaining of FFUs

1. The 96-well plates are emptied by rapid inversion of the plate thus dumping the culture medium into a container filled with a medium for enveloped virus inactivation. Subsequently, the cells in the plates are fixed for 5 min using ice-cold methanol (lowering the entire plate and also the lid into a small tub with methanol is the most efficient technique to ensure complete HCV inactivation).
2. Following fixation the plates are rehydrated using PBS/Tween and endogenous cellular peroxidases are inactivated by treating the plates with a 3% peroxide solution for 5 min.
3. The plates are washed once in PBS/Tween and blocked for 20 min at room temperature in blocking buffer. The primary antibody (should target one of the HCV proteins encoded by the JFH1 isolate to ensure uniform staining across recombinants; here the NS5A-specific antibody, 9E10) is diluted to 0.1 μg/ml and 40 μl per well is added to each well of the 96-well plates and incubated overnight in blocking buffer at 4 °C in a sealed box containing several wet napkins to limit evaporation.
4. The following day the plates are washed three times in PBS/Tween and incubated for 40 min at room temperature with 1:300 secondary antibody coupled with HRP diluted in blocking buffer (this incubation can be performed at a 1:500 dilution of antibody overnight to increase staining brightness). Then the plates are washed three times in PBS/Tween and the FFUs are visualized using liquid DAB+ substrate as per the manufacturer's instructions.
5. The plates are washed in demineralized water and left to dry out completely. The signal will not decrease noticeably for a long time. The FFUs are observed as foci of cells with brown-colored cytoplasm and can either be counted manually with a

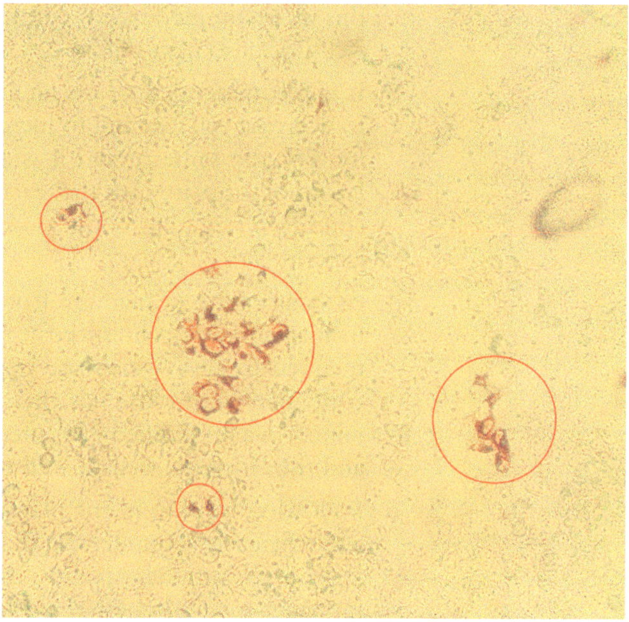

Fig. 1 DAB staining pattern of HCV-infected cells. Foci outlined in red circles adding up to four foci in this field of view. Cellular stain is typically only cytoplasmic with a clear nucleus

light microscope or through automated counting equipment such as the CTL analyzer with modified biospot software [18]. Figure 1 shows an area of a microwell, illustrating a typical staining pattern.

6. The number of FFUs per well is normalized to the wells with only virus. A four-parameter sigmoidal dose–response curve is fitted to the data (e.g., using the commercial software Graphpad PRISM). IC_{50} (50% inhibitory concentration) values are a common parameter used for comparing neutralization sensitivities, but attention should be paid to steepness of the transition of the dose–response curves as determined by the Hill slope coefficient.

4 Notes

1. If the Huh7.5 cells are not evenly distributed in the wells of the PDL-coated 96-well plate, this could mean that the plate was not completely level in the 5–10 min post plating. To address this we suggest placing a level on top of the plate.
2. Once Huh7.5 cells are methanol fixed it is our experience that they stick very well to the support. Therefore, cell loss occurs primarily prefixation during addition of virus/antibody and the subsequent washes and addition of fresh medium. To minimize

this we recommend slow, continuous pipetting into the microwells. *See* also next bullet.

3. It is recommended to use an automated multichannel pipette to decrease cell loss and to increase between-row reproducibility. For the same reason it is also recommended to plan the layout of the plate such that each row contains one well for each condition (e.g., antibody concentration) of the neutralization experiments.
4. As evaporation of medium from the microwells in the 96-well format is more pronounced at the rim of the plate, it is recommended to always incubate open cell culture containers (i.e., containers with loose lids that are not screwed onto the container) in a humidified CO_2 incubator to minimize evaporation and thus increase well-to-well reproducibility.
5. Following fixation of the plates, washing can be done quickly and efficiently by immersing the entire plate in PBS/Tween the appropriate number of times. It is not necessary to dry the plate on paper towels in between washes.
6. Using automated counting of FFUs any number of FFUs below 200 and preferably above 20 per well should be acceptable. However, if manual counting is performed it is highly recommended to aim for <50 FFUs/well to enable accurate and efficient counting. In manual counting using a light microscope you will also need a stage capable of moving the plate in increments to facilitate counting.

References

1. Lindenbach BD, Evans MJ, Syder AJ, Wolk B, Tellinghuisen TL, Liu CC et al (2005) Complete replication of hepatitis C virus in cell culture. Science 309:623–626
2. Wakita T, Pietschmann T, Kato T, Date T, Miyamoto M, Zhao Z et al (2005) Production of infectious hepatitis C virus in tissue culture from a cloned viral genome. Nat Med 11:791–796
3. Gottwein JM, Scheel TK, Hoegh AM, Lademann JB, Eugen-Olsen J, Lisby G et al (2007) Robust hepatitis C genotype 3a cell culture releasing adapted intergenotypic 3a/2a (S52/JFH1) viruses. Gastroenterology 133:1614–1626
4. Jensen TB, Gottwein JM, Scheel TK, Hoegh AM, Eugen-Olsen J, Bukh J (2008) Highly efficient JFH1-based cell-culture system for hepatitis C virus genotype 5a: failure of homologous neutralizing-antibody treatment to control infection. J Infect Dis 198:1756–1765
5. Scheel TK, Gottwein JM, Jensen TB, Prentoe JC, Hoegh AM, Alter HJ et al (2008) Development of JFH1-based cell culture systems for hepatitis C virus genotype 4a and evidence for cross-genotype neutralization. Proc Natl Acad Sci U S A 105:997–1002
6. Gottwein JM, Scheel TK, Jensen TB, Lademann JB, Prentoe JC, Knudsen ML et al (2009) Development and characterization of hepatitis C virus genotype 1-7 cell culture systems: role of CD81 and scavenger receptor class B type I and effect of antiviral drugs. Hepatology 49:364–377
7. Carlsen TH, Pedersen J, Prentoe JC, Giang E, Keck ZY, Mikkelsen LS et al (2014) Breadth of neutralization and synergy of clinically relevant human monoclonal antibodies against HCV genotypes 1a, 1b, 2a, 2b, 2c, and 3a. Hepatology 60:1551–1562
8. Scheel TK, Gottwein JM, Carlsen TH, Li YP, Jensen TB, Spengler U et al (2011) Efficient culture adaptation of hepatitis C virus

recombinants with genotype-specific core-NS2 by using previously identified mutations. J Virol 85:2891–2906
9. Pedersen J, Carlsen TH, Prentoe J, Ramirez S, Jensen TB, Forns X et al (2013) Neutralization resistance of hepatitis C virus can be overcome by recombinant human monoclonal antibodies. Hepatology 58:1587–1597
10. Prentoe J, Verhoye L, Velazquez Moctezuma R, Buysschaert C, Farhoudi A, Wang R et al (2016) HVR1-mediated antibody evasion of highly infectious in vivo adapted HCV in humanised mice. Gut 65:1988–1997
11. Sabo MC, Luca VC, Prentoe J, Hopcraft SE, Blight KJ, Yi M et al (2011) Neutralizing monoclonal antibodies against hepatitis C virus E2 protein bind discontinuous epitopes and inhibit infection at a postattachment step. J Virol 85:7005–7019
12. Giang E, Dorner M, Prentoe JC, Dreux M, Evans MJ, Bukh J et al (2012) Human broadly neutralizing antibodies to the envelope glycoprotein complex of hepatitis C virus. Proc Natl Acad Sci U S A 109:6205–6210
13. Keck ZY, Xia J, Wang Y, Wang W, Krey T, Prentoe J et al (2012) Human monoclonal antibodies to a novel cluster of conformational epitopes on HCV E2 with resistance to neutralization escape in a genotype 2a isolate. PLoS Pathog 8:e1002653
14. Keck Z, Wang W, Wang Y, Lau P, Carlsen TH, Prentoe J et al (2013) Cooperativity in virus neutralization by human monoclonal antibodies to two adjacent regions located at the amino terminus of hepatitis C virus E2 glycoprotein. J Virol 87:37–51
15. Prentoe J, Jensen TB, Meuleman P, Serre SB, Scheel TK, Leroux-Roels G et al (2011) Hypervariable region 1 differentially impacts viability of hepatitis C virus strains of genotypes 1 to 6 and impairs virus neutralization. J Virol 85:2224–2234
16. Prentoe J, Velazquez-Moctezuma R, Foung SK, Law M, Bukh J (2016) Hypervariable region 1 shielding of hepatitis C virus is a main contributor to genotypic differences in neutralization sensitivity. Hepatology 64:1881–1892
17. Gottwein JM, Jensen TB, Mathiesen CK, Meuleman P, Serre SB, Lademann JB et al (2011) Development and application of hepatitis C reporter viruses with genotype 1 to 7 core-nonstructural protein 2 (NS2) expressing fluorescent proteins or luciferase in modified JFH1 NS5A. J Virol 85:8913–8928
18. Gottwein JM, Scheel TK, Callendret B, Li YP, Eccleston HB, Engle RE et al (2010) Novel infectious cDNA clones of hepatitis C virus genotype 3a (strain S52) and 4a (strain ED43): genetic analyses and in vivo pathogenesis studies. J Virol 84:5277–5293
19. Prentoe J, Bukh J (2011) Hepatitis C virus expressing flag-tagged envelope protein 2 has unaltered infectivity and density, is specifically neutralized by flag antibodies and can be purified by affinity chromatography. Virology 409:148–155
20. Blight KJ, Kolykhalov AA, Rice CM (2000) Efficient initiation of HCV RNA replication in cell culture. Science 290:1972–1974
21. Grove J, Nielsen S, Zhong J, Bassendine MF, Drummer HE, Balfe P et al (2008) Identification of a residue in hepatitis C virus E2 glycoprotein that determines scavenger receptor BI and CD81 receptor dependency and sensitivity to neutralizing antibodies. J Virol 82:12020–12029
22. Bailey JR, Wasilewski LN, Snider AE, El-Diwany R, Osburn WO, Keck Z et al (2015) Naturally selected hepatitis C virus polymorphisms confer broad neutralizing antibody resistance. J Clin Invest 125:437–447
23. Pedersen J, Jensen TB, Carlsen TH, Schonning K, Christensen PB, Laursen AL et al (2013) Neutralizing antibodies in patients with chronic hepatitis C, genotype 1, against a panel of genotype 1 culture viruses: lack of correlation to treatment outcome. PLoS One 8:e62674

Chapter 30

Standardized Method for the Study of Antibody Neutralization of HCV Pseudoparticles (HCVpp)

Justin R. Bailey, Richard A. Urbanowicz, Jonathan K. Ball, Mansun Law, and Steven K. H. Foung

Abstract

Hepatitis C virus (HCV) pseudoparticles (HCVpp) are generated by cotransfection of HCV envelope (E1 and E2) genes along with a retroviral packaging/reporter construct into HEK293T cells. Enveloped particles bearing HCV E1E2 proteins on their surface are released through a retroviral budding process into the supernatant. Viral E1E2 glycoproteins facilitate a single round of receptor-mediated entry of HCVpp into hepatoma cells, which can be quantified by reporter gene expression. These HCVpp have been employed to study mechanisms of HCV entry into hepatoma cells, as well as HCV neutralization by immune sera or HCV-specific monoclonal antibodies.

Key words Hepatitis C virus, Flaviviridae, HCV pseudoparticles, Neutralizing antibodies, Neutralizing breadth

1 Introduction

The development of HCVpp [1, 2] has advanced studies of HCV entry and neutralization by antibodies for more than a decade. HCVpp were first used to identify HCV cell surface receptors required for viral entry [3–6]. More recently, studies using HCVpp have also advanced understanding of HCV neutralization by antibodies [7–13], establishing that neutralizing antibodies can drive evolution in vivo of HCV E1E2, leading to neutralization escape [11, 14]. These particles have been produced with a variety of E1E2 variants from multiple isolates and genotypes [15]. Diverse HCVpp panels have been used to define neutralizing breadth of antibodies and to show that development of broadly neutralizing antibodies is associated with spontaneous clearance of HCV infection [10, 11], and reduced liver fibrosis in chronic infection [16].

Mansun Law (ed.), *Hepatitis C Virus Protocols*, Methods in Molecular Biology, vol. 1911,
https://doi.org/10.1007/978-1-4939-8976-8_30, © Springer Science+Business Media, LLC, part of Springer Nature 2019

Key issues to allow comparison of HCVpp neutralization among different laboratories are quality of pseudoparticles that are generated from clinical isolates and use of appropriate controls in neutralization assays. The following protocol addresses these issues.

2 Materials

Recommended vendors are listed in parentheses. Unless otherwise specified, products of equal or better quality than the recommended ones can be used whenever necessary.

1. HEK293T cells (ATCC).
2. Huh7 (APATH, L.L.C.) or Hep3B2.1-7 ("Hep3B", ATCC) hepatoma cells.
3. HEK293T/Huh7 growth medium: Dulbecco's modified Eagle medium (DMEM, Gibco), supplemented with 2 mM L-glutamine (Gibco), 1× nonessential amino acids (Sigma-Aldrich), 1 mM Sodium Pyruvate (Gibco), and 10% heat-inactivated fetal bovine serum (Gibco).
4. HEP3B growth medium: Modified Eagle medium (MEM, Gibco), supplemented with 2 mM L-glutamine (Gibco), 1× nonessential amino acids (Sigma-Aldrich), 1 mM sodium pyruvate (Gibco), and 10% heat-inactivated fetal bovine serum (Gibco).
5. Opti-MEM medium (Invitrogen).
6. 6-well or 10 cm Primaria-coated tissue culture dishes (Corning).
7. 96-well white flat bottom tissue culture plates (Corning).
8. 0.45 micron filters (Millipore).
9. 37 °C, 5% CO_2 incubator.
10. 5× Cell Culture Lysis Reagent (Promega).
11. Luciferase Assay System reagent (Promega).
12. Luminometer (Berthold or BMG Labtech).
13. HCV E1E2 in pcDNA3.1 or pcDNA3.2 expression plasmids (Invitrogen).
14. Murine Leukemia Virus (MLV) Envelope expression plasmid or Vesicular Stomatitis Virus-G (VSV-G) Envelope expression plasmid (NIH AIDS Reagent Program).

2.1 For HIV Gag-Packaged HCVpp Production

1. Lipofectamine 2000 (Invitrogen).
2. pNL4.3.Luc.R-E-plasmid (NIH AIDS Reagent Program).
3. pAdvantage plasmid (Promega).

2.2 For MLV Gag-Packaged HCVpp Production

1. Polyethylenimine (PEI), MW 25,000 (Polyscience).
2. phCMV MLV Gag/Pol packaging construct (phCMV-5349).
3. pTG126 Luciferase encoding reporter.

3 Methods

3.1 Generation of HCVpp by Transfection

Two different transfection protocols in Subheading 3.1 have been optimized for generation of either HIV Gag-packaged or MLV Gag-packaged pseudoparticles.

3.1.1 Generation of HIV Gag-Packaged Pseudoparticles

1. *Day 1*: For each HCVpp to be produced, seed 1×10^6 HEK293T cells into each well of a 6-well tissue culture plate in 3 mL of HEK293T growth medium and incubate overnight at 37 °C and 5% CO_2. Culture size can be scaled up as needed.
2. *Day 2*: Check the density of cells seeded the day before. They should be well spaced over the entire dish and at a confluency of around 70%.
3. For each HCVpp to be produced, dilute 2 μg pNL4.3.Luc.R-E- plasmid, 1 μg E1E2 expression plasmid, and 0.5 μg pAdvantage plasmid in 250 μL Opti-MEM medium.
4. For each HCVpp to be produced, dilute 12 μL Lipofectamine 2000 in 250 μL Opti-MEM.
5. Generation of control pseudoparticles:
   (a) To generate mock pseudoparticles without E1E2 (a control for nonspecific entry) repeat **steps 3** and **4**, but do not add any E1E2 expression plasmid.
   (b) To generate MLV or VSV-G-enveloped pseudoparticles (a control for nonspecific neutralization), repeat **steps 3** and **4**, but replace HCV E1E2 with MLV or VSV-G-envelope expression plasmid.
6. Mix diluted plasmids with diluted Lipofectamine and incubate for 20–30 min at room temperature, then add drop-wise to HEK293T cells without removing growth medium.
7. Incubate overnight at 37 °C in a humidified incubator, then remove medium and replace with 2 mL of fresh HEK293T growth medium.
8. Incubate at 37 °C in a humidified incubator for 48 h. Harvest media and store at 4 °C. Replace media in wells with fresh growth medium and incubate at 37 °C in a humidified incubator for 24 more hours.
9. *Day 5*: Pool 48 h and 72 h supernatants from each well and pass it through a 0.45 μM filter. This supernatant contains the pseudoparticles and can be used either immediately, stored at 4 °C for a maximum of 2 weeks, or frozen at −80 °C indefinitely (*see* **Note 1**).

3.1.2 Generation of MLV Gag-Packaged Pseudoparticles

1. *Day 1*: For each HCVpp to be produced, seed 1.2×10^6 HEK293T cells into a 10 cm diameter Primaria coated dish (Corning) in 10 mL of HEK293T growth medium and incubate overnight at 37 °C and 5% CO_2.
2. *Day 2*: Check the density of cells seeded the day before. They should be well spaced over the entire dish and at a confluency of around 50%.
3. For each HCVpp to be produced, mix 2 μg of phCMV-5349 (MLV Gag/Pol packaging construct), 2 μg of pTG126 Luciferase reporter construct and 2 μg of E1E2 expression plasmid together in a 1.7 mL Eppendorf tube, topped up to 300 μL with Opti-MEM.
4. In a separate tube mix 276 μL of Opti-MEM with 24 μL cationic polymer transfection reagent (Polyethylenimine, PEI; Polyscience) (*see* **Note 2**).
5. Add the 300 μL PEI–Opti-MEM mix to the Eppendorf containing the plasmid mix to give a total of 600 μL.
6. Generation of control pseudoparticles:
   (a) To generate mock pseudoparticles (control for nonspecific entry), repeat **steps 3–5**, but do not add any E1E2 construct.
   (b) To generate MLV or VSV-G-enveloped pseudoparticles (control for nonspecific neutralization), repeat **steps 3–5**, but replace HCV E1E2 expression plasmid with MLV or VSV-G-envelope expression plasmid.
7. Incubate at RT for 1 h.
8. During this hour, remove the complete media with a serological pipette, taking care not to disrupt the cell layer. Add 7 mL of Opti-MEM by tilting the dish and pipetting into the bottom so as not to disturb the cells.
9. Add all 600 μL of the plasmid–PEI mix to the cells, dropwise across the top to ensure even spread.
10. Incubate for 6 h at 37 °C and 5% CO_2.
11. Remove the Opti-MEM and replace with 10 mL of complete medium.
12. Incubate at 37 °C in a humidified incubator for 72 h.
13. *Day 5*: Remove the supernatant from the dish and pass it through a 0.45 μM filter. This supernatant contains the pseudoparticles. This can be used either immediately or stored at 4 °C for a maximum of 2 weeks.

3.2 Single Round Infection of Hepatoma Cells

1. *Day 1*: Seed 1.5 × 10^4 Huh7 cells or 8x10^3 Hep3B cells per well into a 96-well white flat bottom tissue culture plate (Corning) in 100 μL of growth medium and incubate overnight at 37 °C and 5% CO_2 (*see* **Note 3**).
2. *Day 2*: Serially dilute HCVpp and mock pseudoparticles twofold into Huh7 or Hep3B growth medium, depending on the target cells to be infected.
3. Remove growth medium from cells and replace it with 100 μL of diluted HCVpp or mock pseudoparticles in duplicate or triplicate.
4. Incubate for 4–6 h at 37 °C and 5% CO_2.
5. Replace medium with a phenol-free growth medium and incubate cells at 37 °C and 5% CO_2 for 72 h.
6. *Day 5*: Discard the medium from each well and add 50 μL of 1× cell Lysis buffer (Promega) to each well.
7. Incubate at RT for 15 min.
8. Detect luminescence by adding 50 μL of luciferase reagent (Promega) per well and quantitate relative light units (RLU) for 1–5 s after a 0.2 s delay using a luminometer.
9. *Quality control of HCVpp*: HCVpp with functional E1E2 generally produce RLU values at least 10× greater than mock pseudoparticles, but some HCVpp produced using clinical isolate E1E2 may have lower infectivity. Neutralization results obtained using these low-infectivity HCVpp may be unreliable and should be interpreted with caution (*see* **Notes 4 and 5**).

3.3 Neutralization Assay

1. *Day 1*: Seed 1.5 × 10^4 Huh7 cells or 8 × 10^3 Hep3B cells per well per into a 96-well white flat bottom tissue culture plate (Corning) in 100 μL of growth medium and incubate overnight at 37 °C and 5% CO_2.
2. *Day 2*: If plasma or serum samples are to be tested for neutralizing activity, heat inactivate complement by incubating plasma/serum at 56 °C for 30 min. If white precipitate forms after heat inactivation, samples can be clarified by centrifugation at 1200 × *g* for 5 min.
3. Based on result of prior HCVpp infection, dilute HCVpp into HEK293T growth medium to a concentration that will result in infection in the linear range of the infectivity assay. Generally, all but the most infectious HCVpp supernatants can be used undiluted.
4. Dilute MLV-enveloped or VSV-G-enveloped pseudoparticles to similar level of infectivity to HCVpp to use as a control for nonspecific neutralization.

5. Neutralization assays are performed in duplicate or triplicate. Each well has a total volume of 100 μL, made up of 90 μL of pseudoparticle and 10 μL of mAb or plasma/serum. Once the highest concentration and dilution series has been calculated, dilute the mAb or plasma/serum accordingly into PBS or complete growth media. MAbs are generally tested at a maximum concentration of 50 μg/mL, and serum is generally tested at a minimum dilution of 1:50.
6. Dilute an isotype control mAb or nonspecific polyclonal human IgG as a noninhibiting control for neutralizing mAbs. Dilute HCV-negative plasma/serum as a noninhibiting control for neutralization by plasma/serum.
7. Mix diluted antibody and pseudoparticles and incubate for 1 h at 37 °C and 5% CO_2.
8. After 1 h, remove the media from the 96-well white plate seeded with Huh7 or Hep3B cells on Day 1, and add 100 μL of the pseudoparticle/antibody mixture per well, in duplicate or triplicate.
9. Incubate for 4–6 h at 37 °C and 5% CO_2.
10. Replace media with phenol-free growth media and incubate cells at 37 °C and 5% CO_2 for 72 h.
11. *Day 5*: Discard the media from each well and add 50 μL of 1× Cell Lysis Buffer (Promega) to each well.
12. Incubate at RT for 15 min.
13. Detect luminescence by adding 50 μL of luciferase reagent (Promega) per well and quantitating relative light units (RLU) for 1–5 s after a 0.2 s delay using a luminometer.
14. Neutralization at each mAb concentration, serum dilution, or plasma dilution can be calculated using HCVpp infectivity in the presence of test mAb, test serum, or test plasma (RLUtest) relative to infectivity of HCVpp alone (RLUcontrol). Neutralization values can be expressed as percent neutralization [Percent neutralization = (1 − RLUtest/RLUcontrol) × 100]. Fifty-percent inhibitory concentration (IC_{50}) of an inhibitor can be calculated from measurements of percent neutralization by serial dilutions of antibody (*see* **Note 6**).
15. Quality control of neutralization results:
    (a) Neutralization of MLV-pp or VSV-G-pp should be measured as a control for nonspecific neutralization or enhancement of infection. Neutralization by a test sample should only be considered HCV-specific if neutralization of HCVpp exceeds neutralization of MLV-pp or VSV-G-pp.

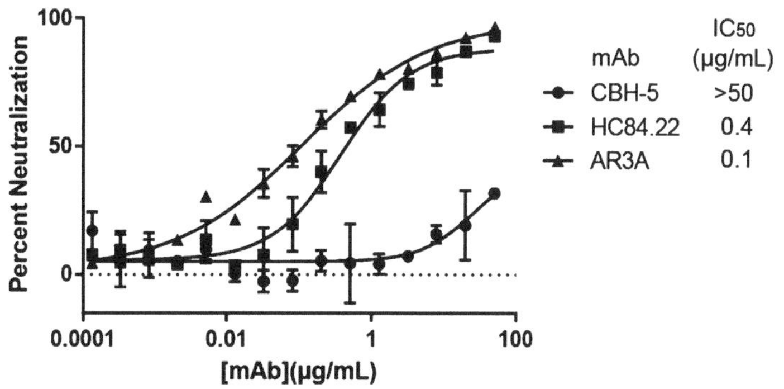

Fig. 1 Neutralization of H77 HCVpp by serial dilutions of three reference mAbs. Error bars indicate standard deviations between replicates. From Wasilewski et al. [17]

(b) Percent neutralization of HCVpp by a nonspecific control antibody (isotype control mAb or pooled human IgG), HCV negative serum pooled from multiple donors, or HCV negative plasma pooled from multiple donors, as appropriate, should also be measured as a control for nonspecific neutralization or enhancement of HCVpp infection. Optionally, RLU values in the presence of nonspecific control antibody, pooled HCV negative serum, or pooled HCV negative plasma can be used as RLUcontrol in the equation [Percent neutralization = (1 − RLUtest/RLUcontrol) × 100] to adjust percent neutralization of test samples for nonspecific neutralization or enhancement (*see* **Notes 7 and 8**).

(c) Optionally, neutralization of HCVpp with H77 strain E1E2 by well-characterized mAbs can be used to validate neutralization assay methods, since many previously described broadly neutralizing mAbs have activity against this viral strain (Fig. 1 and Table 1).

3.4 Measurement of Neutralizing Breadth

Neutralization of an HCVpp is generally considered positive if it is neutralized >50% by 50 μg/mL or less of a mAb or a 1:50 dilution of plasma or serum, with appropriate controls for nonspecific neutralization. Cross-neutralization is generally defined as neutralization of at least one heterologous HCV isolate. Broad neutralization is generally defined as neutralization of the majority of isolates in a diverse panel of HCVpp (*see* **Note 9**). Neutralization sensitivity of distinct HCVpp isolates varies widely both within and across HCV genotypes [13, 15], and further study is needed to better define the relationship between neutralizing breadth measured using diverse isolates from a single HCV genotype and neutralizing breadth measured using isolates from multiple genotypes.

Table 1
Percent neutralization of strain H77 HCVpp by reference mAbs, measured at a single concentration of mAb (10 μg/mL). Values are the average of replicate wells. From Bailey et al. [13]

mAb (10 μg/mL)	H77 HCVpp % Neutralization
HC84.26	99.7
HC33.4	99.6
HC33.8	95.7
AR4A	95.4
AR5A	91.4
AR3C	86.9
AR2A	82.8
HC84.22	82.6
AR3A	76.9
HC-1	67.3
AR3D	67.1
AR4B	65.7
AR3B	62.8
CBH-7	55.2
AR1A	36.2
CBH-2	35.4
CBH-5	33
CBH-4B	11.3

4 Notes

1. HCVpp infectivity decreases by two- to fivefold with each freeze–thaw cycle.
2. In MLV-Gag pseudoparticle transfection protocol, add PEI to Opti-MEM, not the other way around.
3. A confluent T175 flask has ~1×10^7 Hep3B cells, enough to seed ~10–12 96-well plates.
4. Infectivity can vary significantly between different preparations (transfections) of the same HCVpp, and some E1E2 variants are more likely to produce functional HCVpp when packaged with MLV-Gag rather than HIV-Gag, or vice versa. Some

E1E2 variants also require nonstandard ratios of E1E2 plasmid and packaging construct to produce functional pseudoparticles [19].

5. Inadequate infectivity of HCVpp can sometimes be improved through codon optimization of E1E2 genes. Infectivity may also be improved by switching from HIV Gag-packaged pseudoparticles to MLV Gag-packaged pseudoparticles, or vice versa.
6. Percent neutralization of the same HCVpp variant by the same antibody at the same concentration can vary by about 1.5-fold between independent experiments (an independent experiment is defined by a different HCVpp transfection and different neutralization assay). Therefore, small differences in neutralization resistance between HCVpp should be confirmed with repeated independent experiments.
7. Nonspecific enhancement or inhibition of HCVpp by sera may be corrected through the use of protein A purified serum IgG in neutralization experiments.
8. Heparin can inhibit HCV infection, so plasma isolated with heparin used as an anticoagulant should be avoided or used with appropriate controls.
9. Multiple different panels of HCVpp are commonly used for neutralization breadth testing [7, 15, 18]. A standardized panel of genotype 1 HCVpp that would allow for accurate comparision of neutralization breadth results across laboratories is currently under development.

Acknowledgments

The authors would like to acknowledge Dr. Stuart Ray for useful discussions. This work was funded by NIH grants R01AI127469, R01AI079031, R01AI106005, U19AI123861, and U19AI088791, Medical Research Council (MRC) grant G0801169 and EU Seventh Framework Programme (FP7) grant "HepaMab" 305600.

References

1. Bartosch B, Dubuisson J, Cosset FL (2003) Infectious hepatitis C virus pseudo-particles containing functional E1- E2 envelope protein complexes. J Exp Med 197:633–642
2. Hsu M, Zhang J, Flint M, Logvinoff C, Cheng-Mayer C, Rice CM et al (2003) Hepatitis C virus glycoproteins mediate pH-dependent cell entry of pseudotyped retroviral particles. Proc Natl Acad Sci U S A 100:7271–7276
3. Bartosch B, Vitelli A, Granier C, Goujon C, Dubuisson J, Pascale S et al (2003) Cell entry of hepatitis C virus requires a set of co-receptors that include the CD81 tetraspanin and the SR-B1 scavenger receptor. J Biol Chem 278:41624–41630
4. Cormier EG, Tsamis F, Kajumo F, Durso RJ, Gardner JP, Dragic T (2004) CD81 is an entry coreceptor for hepatitis C virus. Proc Natl Acad Sci U S A 101:7270–7274

5. McKeating JA, Zhang LQ, Logvinoff C, Flint M, Zhang J, Yu J et al (2004) Diverse hepatitis C virus glycoproteins mediate viral infection in a CD81-dependent manner. J Virol 78:8496–8505
6. Zhang J, Randall G, Higginbottom A, Monk P, Rice CM, McKeating JA (2004) CD81 is required for hepatitis C virus glycoprotein-mediated viral infection. J Virol 78:1448–1455
7. Giang E, Dorner M, Prentoe JC, Dreux M, Evans MJ, Bukh J et al (2012) Human broadly neutralizing antibodies to the envelope glycoprotein complex of hepatitis C virus. Proc Natl Acad Sci U S A 109:6205–6210
8. Law M, Maruyama T, Lewis J, Giang E, Tarr AW, Stamataki Z et al (2008) Broadly neutralizing antibodies protect against hepatitis C virus quasispecies challenge. Nat Med 14:25–27
9. Morin TJ, Broering TJ, Leav BA, Blair BM, Rowley KJ, Boucher EN et al (2012) Human monoclonal antibody HCV1 effectively prevents and treats HCV infection in chimpanzees. PLoS Pathog 8:e1002895
10. Osburn WO, Snider AE, Wells BL, Latanich R, Bailey JR, Thomas DL et al (2014) Clearance of hepatitis C infection is associated with the early appearance of broad neutralizing antibody responses. Hepatology 59:2140–2151
11. Pestka JM, Zeisel MB, Blaser E, Schurmann P, Bartosch B, Cosset FL et al (2007) Rapid induction of virus-neutralizing antibodies and viral clearance in a single-source outbreak of hepatitis C. Proc Natl Acad Sci U S A 104:6025–6030
12. Keck ZY, Xia J, Wang Y, Wang W, Krey T, Prentoe J et al (2012) Human monoclonal antibodies to a novel cluster of conformational epitopes on HCV e2 with resistance to neutralization escape in a genotype 2a isolate. PLoS Pathog 8:e1002653
13. Bailey JR, Wasilewski LN, Snider AE, El-Diwany R, Osburn WO, Keck Z et al (2015) Naturally selected hepatitis C virus polymorphisms confer broad neutralizing antibody resistance. J Clin Invest 125:437–447
14. Dowd KA, Netski DM, Wang XH, Cox AL, Ray SC (2009) Selection pressure from neutralizing antibodies drives sequence evolution during acute infection with hepatitis C virus. Gastroenterology 136:2377–2386
15. Urbanowicz RA, McClure CP, Brown RJ, Tsoleridis T, Persson MA, Krey T et al (2016) A diverse panel of hepatitis C virus glycoproteins for use in vaccine research reveals extremes of monoclonal antibody neutralization resistance. J Virol 90:3288–3301
16. Swann RE, Cowton VM, Robinson MW, Cole SJ, Barclay ST, Mills PR et al (2016) Broad anti-hepatitis C virus (HCV) antibody responses are associated with improved clinical disease parameters in chronic HCV infection. J Virol 90:4530–4543
17. Wasilewski L, Ray S, Bailey JR (2016) Hepatitis C virus resistance to broadly neutralizing antibodies measured using replication competent virus and pseudoparticles. J Gen Virol 97 (11):2883–2893
18. Bailey JR, Flyak AI, Cohen VJ, Li H, Wasilewski LN, Snider AE et al (2017) Broadly neutralizing antibodies with few somatic mutations and hepatitis C virus clearance. JCI Insight 2(9):92872
19. Urbanowicz RA, McClure CP, King B, Mason CP, Ball JK, Tarr AW (2016) Novel functional hepatitis C virus glycoprotein isolates identified using an optimized viral pseudotype entry assay. J Gen Virol 97:2265–2279

Part V

Small Animal Models

Chapter 31

Detection of Antigen-Specific CD8$^+$ T Cells in the Livers of HCV Core Transgenic Mice

Dustin A. Cobb, Aditya P. Dandekar, and Young S. Hahn

Abstract

Hepatitis C virus-mediated immune suppression is an underlying feature leading to the establishment of viral persistence and chronic infection. In particular, HCV core protein has been shown to exhibit significant immunosuppressive activity of T cells and antigen presenting cells. Using an HCV core transgenic mouse system, in which liver hepatocytes express core protein, it is possible to study the effects of core-mediated immune suppression in vivo during viral infection. In this protocol, we describe the procedures for evaluating antigen-specific CD8$^+$ T cell responses in response to recombinant adenovirus infection in HCV core transgenic mice.

Key words Hepatitis C virus, CD8 T cells, Liver, Core protein, Antiviral T cells, Liver mononuclear cell isolation, Flow cytometry, Tetramer

1 Introduction

Hepatitis C virus (HCV) establishes persistent infection in the majority of individuals that contract the virus, leading to the development of severe liver complications such as fibrosis, cirrhosis, and hepatocellular carcinoma [1, 2]. Protection against viral infection, including HCV, is highly dependent upon robust and sustained T cell responses. However, in chronically infected HCV patients, T cell responses are significantly impaired and are characterized as being weak, dysfunctional, or completely absent [3]. In particular, cytotoxic T lymphocyte (CTL) effector functions (e.g., IFN-γ production and cytolytic activity) are significantly impaired in chronic HCV patients. Impairment of T cell responses during HCV infection stems from multiple pathways that the virus employs to evade and subvert protective immunity, leading to the establishment of persistence.

HCV core protein exhibits immunomodulatory activity by suppressing host immunity including T cell responses and antigen presenting cell function [4–8]. The liver is the major site of HCV

Mansun Law (ed.), *Hepatitis C Virus Protocols*, Methods in Molecular Biology, vol. 1911,
https://doi.org/10.1007/978-1-4939-8976-8_31,

infection; therefore, it is critical to understand the modulation of intrahepatic host immunity by HCV core-mediated immune regulation. By utilizing an HCV core protein transgenic mouse line, one can study the underlying mechanism(s) of HCV core-mediated immune suppression during viral infection in the liver. To recapitulate the presence of HCV core expression in mouse livers, a transgene encoding core protein was introduced behind the albumin promoter to achieve expression in hepatocytes [9]. Thus, this model can be utilized to study HCV core mediated dysregulation of intrahepatic T cell responses during acute viral infection.

This protocol details the procedure for evaluating suppression of intrahepatic antiviral $CD8^+$ T cells in mice expressing HCV core protein. Specifically, this model is used to study the suppression of antigen-specific $CD8^+$ T cells in HCV core mice infected with recombinant adenovirus expressing LacZ which encodes β-galactosidase (β-gal). Upon intravenous injection with recombinant adenovirus the majority of virus is delivered to the liver [10]. Importantly, mice can then be monitored for development of β-gal-specific $CD8^+$ T cell responses by MHC class I tetramer staining and subsequent flow cytometry analysis. In this protocol, we detail our method for the isolation of liver-infiltrating lymphocytes using density gradient centrifugation and tetramer staining for detection of antigen-specific $CD8^+$ T cells.

2 Materials

2.1 Adenovirus Infection and Liver Mononuclear Cell Isolation

1. HCV core+ transgenic and core(−) mice.
2. Recombinant adenovirus (Ad-LacZ) (adenovirus 5, replication-deficient).
3. Sterile instruments for dissection and removal of liver.
4. 5 mL syringes.
5. 23 gauge needles.
6. Collagenase type IV.
7. 15 mL polypropylene conical centrifuge tubes.

2.2 Antibody Staining

1. Anti-mouse CD16/CD32 antibody for blocking of Fc receptors.
2. Anti-mouse CD8 fluorescent antibody.
3. PE-labeled H2-K^b βgal tetramer (ICPMYARV) provided by the Tetramer Core Facility of the National Institutes of Health.
4. 5 mL polystyrene round bottom tubes or 96-microwell plates (V-bottom or U-bottom) for cell staining.

2.3 Buffers and Media

1. Phosphate buffered saline (Mg^{2+}-/Ca^{2+}-free).
2. Iscove's modified Dulbecco's medium (IMDM) supplemented with 10% fetal bovine serum (FBS) (*see* **Note 1**).
3. Perfusion buffer: PBS with 0.05% Collagenase type IV (*see* **Note 1**).
4. Histodenz stock solution: 20 g of Histodenz dissolved in 40 mL of PBS.
5. ACK lysing buffer for red blood cell lysis.
6. Staining buffer: PBS supplemented with 2% fetal bovine serum and 0.1% sodium azide.

3 Methods

3.1 Infection of HCV Core Transgenic Mice with Recombinant Adenovirus (Ad-LacZ)

1. Infect HCV core(+) transgenic and control core(−) mice with 5×10^8 PFU of Ad-LacZ by intravenous injection in the lateral tail vein.
2. Euthanize mice on day 7 or day 14 post-infection.

3.2 Isolation of Liver Mononuclear Cells

1. Using dissection instruments cut the peritoneum to expose the liver and carefully isolate the liver.
2. Remove the gall bladder (*see* **Note 2**).
3. Arrange the lobes of the liver to see the superior vena cava and portal vein. Cut open the diaphragm to expose the inferior vena cava.
4. Using a 23 gauge needle, slowly flush the liver with 3–5 mL PBS via the portal vein until fully blanched (*see* **Note 3**).
5. Immediately cut the inferior vena cava and portal vein/superior vena cava to drain perfusion buffer (*see* **Note 4**).
6. The liver was then perfused with perfusion buffer (PBS containing 0.05% Collagenase type IV).
7. Carefully remove the liver and exclude as much connective tissue/blood clots from the surface. Wash the liver with Iscove's modified Dulbecco's medium (IMDM) supplemented with 10% fetal bovine serum.
8. Using dissection instruments finely mince the liver tissue and pass it through a stainless steel mesh screen.
9. Wash the mesh screen with wash media and bring up the volume to 10 mL.
10. Transfer liver cell suspensions to 15 mL polypropylene conical centrifuge tubes.

11. Add 100 μL of Collagenase type IV for a final concentration 0.05%. Place the tubes in a 37 °C water bath and incubate for 20–30 min, mixing every 5 min (*see* **Note 5**).
12. Remove hepatocytes and debris after collagenase digestion by spinning tubes at 300 × *g* for 3 min at 4 °C.
13. Remove the supernatant which contains cells of interest and pellet cells by centrifugation at 300 × *g* for 5 min at 4 °C.
14. Resuspend the cell pellet of each liver in 5 mL Histodenz density gradient medium (prepared beforehand by mixing 2.7 mL of Histodenz stock solution with 2.3 mL of PBS) by gently vortexing or pipetting.
15. Next, slowly overlay the Histodenz–cell mixture with 4 mL of PBS or IMDM (no serum) (*see* **Note 6**).
16. Perform density gradient separation by spinning the tubes at 1500 × *g* for 20 min at 4 °C with no brake (*see* **Note 7**).
17. Collect the interface of the liver gradients and place in IMDM media.
18. Any remaining red blood cells can be removed with ACK lysing buffer.
19. Count viable cells on a hemocytometer using trypan blue staining.

3.3 Fluorescent Antibody and Tetramer Labeling for Flow Cytometry Analysis

1. Transfer 2 × 10^6 liver mononuclear cells to a 5 mL polystyrene round bottom tube or into wells of a 96-microwell plate (V-bottom or U-bottom) for antibody and tetramer labeling.
2. Pellet liver mononuclear cells and resuspend in cell staining buffer containing anti-mouse Fc block (anti-CD16/anti-CD32) and incubate for 10–15 min on ice or at 4 °C.
3. Incubate cells with anti-CD8 fluorescent antibody and PE-labeled H2-K^b βgal tetramer (ICPMYARV) for 30 min on ice or at 4 °C in staining buffer.
4. Wash cells with cell staining buffer (1 mL for tubes or 250 μL for plate), and pellet by centrifugation at 300–400 × *g* for 5 min at 4 °C.
5. Aspirate the supernatant. Resuspend cells in cell staining buffer to wash (1 mL for tubes or 250 μL for plate) and centrifuge at 300–400 × *g* for 5 min at 4 °C.
6. Aspirate the supernatant and resuspend liver mononuclear cells in 300–500 μL of cell staining buffer and proceed to flow cytometric analysis.

4 Notes

1. Having the perfusion buffer and IMDM media prewarmed at 37 °C right before the perfusion and digestion improves digestion efficiency of Collagenase type IV.
2. Prior to removing the liver, be sure that you cut off the gall bladder from the liver. Disruption of gall bladder during mincing can seriously affect the cell viability due to acidic content release.
3. While flushing the liver with PBS, make sure the flush intensity is steady. Vigorous flushing may damage the liver tissue internally and subsequently affect the downstream analysis. Alternatively, use of a peristaltic pump (Fischer Scientific) with variable flow helps and improves the perfusion efficiency by maintaining laminar flow of buffer.
4. During perfusion, do not allow the hydrostatic/buffer pressure to build within the liver. It is important to cut the inferior vena cava to release the incoming flushing buffer.
5. The temperature during the collagenase digestion is very important. Any deviation in digestion temperature will result in inefficient digestion of liver and subsequent loss in cell isolation efficiency.
6. While preparing density gradient for cell isolation, it is important to set the pipette ejection speed to the minimum setting. Also, tilting tubes at a 45°–60° angle can help to reduce mixing of the gradients.
7. It is important to centrifuge the tubes (cells + gradient) with no brake setting. This helps to keep the gradient intact.

Acknowledgments

This work was supported by the National Institutes of Health. (Grants AI098126 and U19 AI066328 to Y.S.H.; T32AI007496 to DAC)

References

1. Afdhal NH (2004) The natural history of hepatitis C. Semin Liver Dis 24(Suppl 2):3–8
2. Dustin LB, Rice CM (2007) Flying under the radar: the immunobiology of hepatitis C. Annu Rev Immunol 25:71–99
3. Abdel-Hakeem MS, Shoukry NH (2014) Protective immunity against hepatitis C: many shades of gray. Front Immunol 5:274
4. Large MK, Kittlesen DJ, Hahn YS (1999) Suppression of host immune response by the core protein of hepatitis C virus: possible implications for hepatitis C virus persistence. J Immunol 162:931–938
5. Sarobe P, Lasarte JJ, Casares N, Lopez-Diaz de Cerio A, Baixeras E, Labarga P et al (2002) Abnormal priming of CD4(+) T cells by

dendritic cells expressing hepatitis C virus core and E1 proteins. J Virol 76:5062–5070

6. Zhang B, Yin Y, Lai RC, Lim SK (2014) Immunotherapeutic potential of extracellular vesicles. Front Immunol 5:518
7. Yao ZQ, Nguyen DT, Hiotellis AI, Hahn YS (2001) Hepatitis C virus core protein inhibits human T lymphocyte responses by a complement-dependent regulatory pathway. J Immunol 167:5264–5272
8. Eisen-Vandervelde AL, Waggoner SN, Yao ZQ, Cale EM, Hahn CS, Hahn YS (2004) Hepatitis C virus core selectively suppresses interleukin-12 synthesis in human macrophages by interfering with AP-1 activation. J Biol Chem 279:43479–43486
9. Lukens JR, Cruise MW, Lassen MG, Hahn YS (2008) Blockade of PD-1/B7-H1 interaction restores effector CD8+ T cell responses in a hepatitis C virus core murine model. J Immunol 180:4875–4884
10. Krebs P, Scandella E, Odermatt B, Ludewig B (2005) Rapid functional exhaustion and deletion of CTL following immunization with recombinant adenovirus. J Immunol 174:4559–4566

Chapter 32

Mice with Chimeric Human Livers and Their Applications

Donna N. Douglas and Norman M. Kneteman

Abstract

The complete life cycle of the hepatitis C virus (HCV) can be recapitulated in vivo using immunodeficient mice that have had their livers extensively repopulated with human hepatocytes. These human liver chimeric mouse models have enabled the study of many aspects of the HCV life cycle, including antiviral interventions that have helped to shape the curative landscape that is available today. The first human liver chimeric mouse model capable of supporting the HCV life cycle was generated in SCID-uPA mice. Although other human liver chimeric mouse models have since been developed, the SCID-uPA mouse model remains one of the most robust in vivo systems available for HCV studies. This chapter reviews development, validation and application of the SCID-uPA mouse model, and discusses their potential application for studying other liver-centric diseases and pathogens and for the design and testing of vaccine candidates for the eradication of HCV.

Key words Viral hepatitis, Antiviral therapy, Animal models, Pathology

1 Introduction

With >150 million people infected with the hepatitis C virus (HCV), HCV is a global public health threat. HCV is a major cause of progressive hepatic cirrhosis highly correlated with the development of hepatocellular carcinoma, the number 3 cause of cancer-related death. End-stage liver disease due to HCV is the leading indication for liver transplantation [1], representing ~30% of liver transplantations performed in Europe and North America. It is estimated that ~3–4 million HCV infections occur each year. This is likely an underestimation since many are unaware they are infected until symptoms of advanced liver disease appear, which can take decades, and prevalence data is not available for some geographical areas due to limited availabilities of HCV testing and blood donation surveillance.

Within the last five years, there has been considerable progress with the introduction of breakthrough medicines that will cure chronic HCV infection within 12 weeks in the vast majority (>90%) of cases. Unfortunately, translating these breakthroughs into

Mansun Law (ed.), *Hepatitis C Virus Protocols*, Methods in Molecular Biology, vol. 1911,
https://doi.org/10.1007/978-1-4939-8976-8_32,

sustainable global health gains is an unrealistic expectation because these new drugs are cost-prohibitive, especially for those in developing countries where the disease burden is greatest. These high costs also threaten to cripple health care budgets in developed countries. In the United States, the high price tag for full treatment (~$80,000) is unaffordable for most and has caused many government bodies and private payers to limit financial coverage of treatment for people who already have advanced liver disease secondary to HCV infection. In the meantime, the infected individual becomes a rich source for HCV transmission via high-risk behavior, IV drug abuse, and blood transfusion or organ donations. Complicating matters further, there is still no vaccine that can prevent the spread of HCV. Eradication of infection and decreasing the impact of chronic HCV-related diseases at affordable costs are areas of intense interest.

In response to the scientific and social call for the development of a small animal model for HCV infection, we developed the first mouse model that could support the entire HCV life cycle [2]. Because a cell culture system capable of supporting the HCV life cycle had not yet been developed, this indeed was a very important milestone in the HCV field [3]. Our work with these mice has greatly contributed to our understanding of HCV biology [4–6], and perhaps most critically, to the development of novel anti-HCV therapies that helped shape today's curative landscape [7–9]. This chapter reviews development, validation and early application of this mouse model. The potential for using these mice for studying other liver-centric diseases and pathogens, and for the design and testing of vaccine candidates for the eradication of HCV are also discussed.

2 Establishment of the SCID-uPA Mouse Model of HCV Infection

Biomedical research has relied heavily on model organisms to understand the complex molecular mechanisms of the human body. Mice in particular are the most widely used experimental mammals, contributing immeasurably to our understanding of human biology. The importance of using mouse models to study certain human diseases is validated by the many aspects important to human biology that are similar between the two species. While this is predicated on genetic and physiological similarities [10], mice and man have evolved in and adapted to different environments, and so not all aspects of mouse biology reflect human biology. Consequently, there are numerous pathogens that cause human infectious diseases that simply cannot cross the species barrier to infect mice. Such is the case for HCV: humans are the natural hosts for HCV and HCV infection can be experimentally transmitted to chimpanzees. Sophisticated approaches have been devised to overcome the species barriers to infection. One approach that has shown promise involves mutational adaptation of the pathogen so that it can efficiently engage with or utilize murine orthologs of human factors that otherwise restrict murine-tropism [11, 12]. Another strategy that has proven successful is genetic humanization of the murine host [13, 14], involving the transient or stable expression of human factors that render the murine host permissive to infection and/or the inactivation of endogenous murine factors that otherwise restrict the murine host to infection. However, both of these approaches required an in-depth knowledge of the barriers that restrict species tropism, which were not understood until after studies on viral entry and the ability to propagate HCV in vitro were enabled through the development of HCVpps (HCV pseudoparticles) [15, 16] and HCVcc (cell-culture-derived HCV) [17–22], respectively. Prior to this there had been substantial progress in the development of a humanized mouse model whereby an immunodeficient mouse strain, that also provides a selective regenerative stimulus to donor hepatocyte cells, could be engrafted with functional human hepatocytes to support productive HCV infection in vivo [2]. These "human liver" chimeric mice have proven to be one of the most robust HCV mouse models to date [3].

The very first mouse model to support the entire HCV life cycle in vivo is the SCID-uPA mouse developed by Kneteman et al. in 2001 [2]. These mice are derived from the Alb-uPA transgenic mouse that had been originally developed in 1990 as a model to study bleeding disorders [23]. The Alb-uPA mouse carries a tandem array of four murine urokinase-type plasminogen activator (uPA) genes controlled by a murine albumin (Alb) promoter/ enhancer sequence (referred to as the uPA transgene hereafter) that targets uPA over-production to the mouse liver, resulting in a sustained liver injury and failure with a high risk of neonatal bleeding [23, 24]. Mice hemizygous for the uPA transgene were better able to survive both liver failure and perinatal bleeding than their homozygous counterparts [24]. This was attributed to spontaneous deletion of the uPA transgene array, which rendered hepatocytes transgene-deficient more readily in mice that were hemizygous for the Alb-uPA transgene. Once liberated from the toxicity associated with the uPA transgene, transgene-deficient hepatocytes were thought to have a significant replicative advantage over surrounding cells, thereby repopulating the mouse liver parenchyma with transgene-deficient hepatocytes, and resulting in a gradual return of urokinase-type plasminogen to normal levels within 2 months of age [24]. With reports that this survival advantage had also been extended to hepatocytes transplanted from syngeneic [25] or xenogeneic [26] donors, we shifted our efforts in transplanting human hepatocytes into immunodeficient mice toward this exciting approach to develop a mouse model with a humanized liver. After all, without an ongoing stimulus that would enable selective expansion of donor cells, while at the same creating "space" for donor expansion, we were only able to achieve low levels of human hepatocyte engraftment in immunodeficient mice that was lost within weeks, and did not sustain HCV infection. Furthermore, our efforts to stimulate in vivo expansion of the initial engraftment (partial hepatectomy, administration of human hepatocyte growth factor and irradiation) failed to achieve the levels of human hepatocyte repopulation needed to sustain HCV infection. Therefore, we crossed Alb-uPA and the highly immunodeficient C.b-17/SCID/beige mouse lines, and through selective backcrosses bred the SCID trait, characterized by the absence of functional T and B cells, to homozygosity.

Due to the high mortality initially reported for homozygous Alb-uPA mice, our early experiments utilized SCID mice that were hemizygous for the uPA transgene as breeders. Freshly isolated human hepatocytes having viabilities >90% were transplanted into the offspring of these breeders at an early age (<10 days old), before transgene-deficient murine hepatocytes would have a chance to compete with the human hepatocytes for replacement of the murine liver parenchyma. Upon evaluation of the initial series of human hepatocyte transplants, we could identify three different subpopulations: some mice never demonstrated appreciable levels of human albumin in their serum, others had appreciable levels of human albumin in their serum that was lost within a few months, while a third group exhibited high levels of human albumin in their serum that was sustained for several months. Furthermore, only the group that sustained high levels of human albumin in their sera was capable of supporting HCV infection. Retrospective genotype analysis revealed that the highest levels of human hepatocyte engraftment were with SCID mice that were homozygous for the uPA transgene: whereas mice that were hemizygous for the uPA transgene demonstrated only modest hepatocyte engraftment, homozygous animals supported improved liver repopulation to levels well beyond 50% and sustained chimerism, with a median graft duration of >30.5 weeks [2, 27]. We subsequently converted our colony to solely homozygous breeders.

While careful attention to husbandry helped to keep mortality to acceptable levels, our breeding colony consisted of homozygous uPA SCID mice that had already been transplanted with human hepatocytes. The rationale behind using transplanted mice for our breeding colony was twofold: First, we believed that homozygous breeders had to be "rescued" from the toxic effects of the uPA transgene, and second, we hypothesized that success might breed success and so breeders that had extensive human hepatocyte engraftment might impart a "winner effect" to the offspring. With this strategy, we were able to achieve successful high level human hepatocyte engraftment using both freshly isolated human

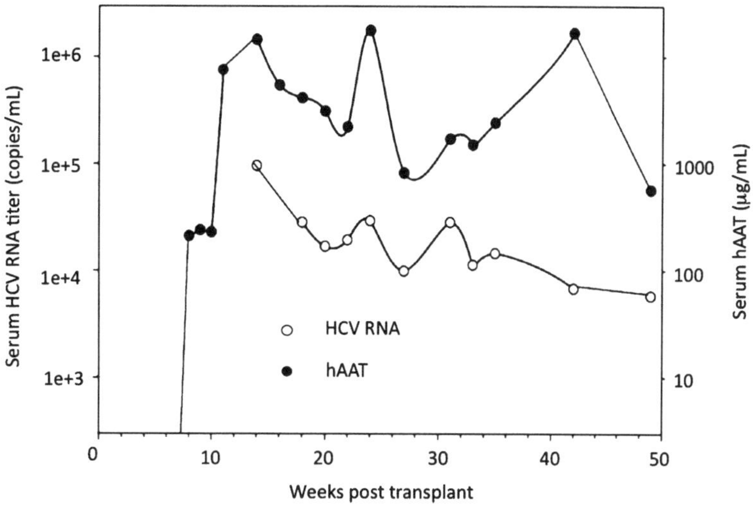

Fig. 1 The SCID-uPA mouse model is capable of supporting human hepatocyte chimerism and HCV infection beyond 1 year, as demonstrated by serum human alpha-1 antitrypsin (hAAT) and HCV RNA levels

hepatocytes and cryopreserved human hepatocytes [2, 28]. Successful engraftment was found to be critically dependent on high quality, high viability human hepatocytes [29]. Generally speaking, HCV infection can be experimentally transmitted to human liver chimeric mice through inoculation with serum obtained from HCV-infected individuals [2, 30, 31] or from cell cultures [32] or via intrahepatic injection with HCV RNA, enabling reverse genetic studies [33, 34]. The ability of human liver chimeric SCID-uPA mice to support productive HCV infection was initially supported by three important benchmarks. First, we observed an initial rise in the total viral load by up to 1950-fold in mice that had been inoculated with HCV [2]. Second, we were able to detect the negative HCV RNA strand (replicative intermediate of HCV) in the livers of infected mice using a primary hybridization method [2]. Finally, we were able to serially passage HCV infection through three generations of human liver chimeric mice [2]. The composition of the HCV particles produced in the human chimeric livers of SCID-uPA mice resemble those that are produced in human and chimpanzee livers [18]. We find that viral titers in SCID-uPA mice vary according to human hepatocyte engraftment levels, and therefore may vary considerably between infected mice. However, individual viral titers can be as high as those seen in the clinical setting (or higher) and usually remain quite stable for several weeks, and often for many months (Fig. 1). Others have explored the possibility that residual immune activity in homozygous uPA SCID mice might limit engraftment of human hepatocytes and have shown that natural killer cell depletion and/or macrophage depletion can improve engraftment levels [35, 36]. Indeed our own studies had

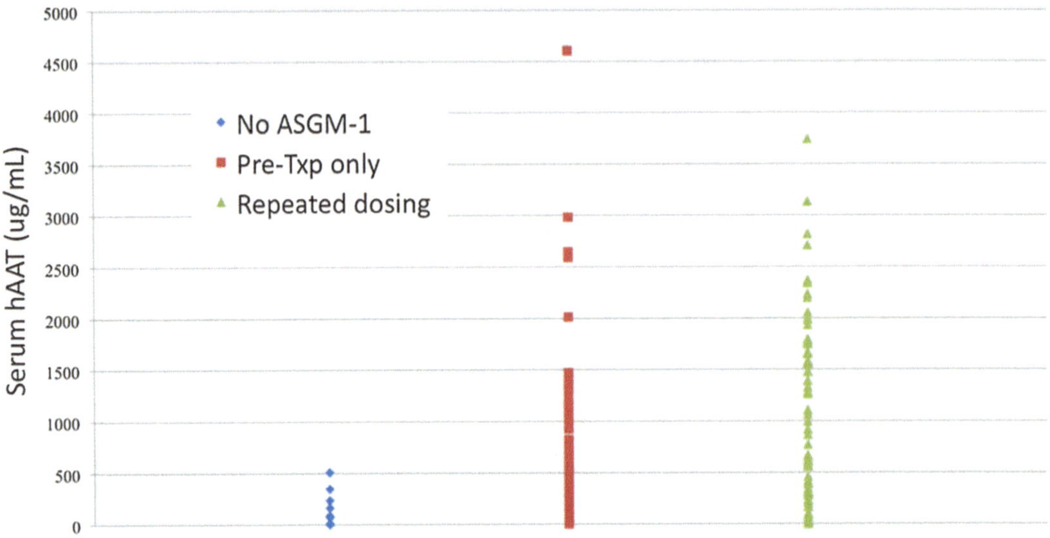

Fig. 2 Human hepatocyte engraftment in the SCID-uPA mouse model is enhanced by a single 20 mL (i.p.) administration of ASGM-1 (rabbit anti mouse asialo GM-1, polyclonal unpurified, 1 mL prep from Cedar Lane, CL8955) 24 h prior to transplant (Pre-Txp only). In some cases this initial dose was followed by repeated dosing (10 μL, i.p.) once every 4 days following transplant until weaning (6 weeks posttransplant), though this did not appear to improve outcomes relative to just a single dose prior to transplant

demonstrated a critical role for natural killer cells in the rejection of human hepatocytes in SCID-uPA mice [37]. We find that we can improve human hepatocyte engraftment in homozygous uPA mice by administering a single dose of an antibody that depletes NK cell activity (ASGM-1) 24 h prior to transplantation (Fig. 2).

After a decade of productivity, we found ourselves facing diminishing returns due to a decline in our breeding colony's overall performance. We therefore instituted several changes to how we manage our breeding colony. As a first step, we aggressively refreshed our breeding colony by bringing in SCID-beige female mice from Taconic for successive backcrossing to our inbred SCID-uPA males, with intermittent intercrossing of the F1 hybrid parents. These efforts yielded larger and more robust mice than previously seen. Another intriguing result of our efforts to restore colony vigor is that the coat colors of the SCID-uPA mice are exclusively white. This is different from the variety of coat colors that was previously displayed. We also recognized that there was no benefit to transplanting our breeding stock, which now consists of both hemizygous and homozygous uPA dams for crossing with homozygous uPA sires. We now find that hemizygous and homozygous uPA dams are just as likely to become pregnant, producing similar litter sizes. The advantage of using hemizygous uPA dams is that they tend to be larger and grow faster than homozygous uPA dams. Thus they can begin breeding at a younger age, are less likely to suffer birth complications such as dystocia, can undergo more

Table 1
Restoration of SCID-uPA colony vigor

	Number of crosses	Number of pregnancies	Pregnancy efficiency	Number of pups born	Average number of pups/litter	Number of pups weaned (post transplant)	Survival efficiency (post transplant)
Previous colony							
Hemi Dams	66	25	38%	131	5.7	95	73%
Homo Dams	37	8	22%	32	5.3	7	22%
Overall	103	33	32%	163	5.6	102	63%
Current colony							
Hemi Dams	52	43	83%	290	6.7	227	78%
Homo Dams	47	30	67%	167	5.8	123	74%
Overall	99	73	74%	457	6.3	350	77%

breeding cycles, and tend to generate offspring that are somewhat larger and more robust than those produced by homozygous dams, who are smaller and less robust themselves. For these reasons, hemizygous uPA dams are largely dedicated to amplifying our breeding stock, representing a giant leap forward in the overall productivity of our breeding colony. Credit must also be given to Meuleman et al. for developing a rapid multiplex PCR-based genotyping procedure [27] that has greatly facilitated management of our SCID-uPA mouse colony.

Table 1 (unpublished data) provides a comparison between the breeding efficiencies of SCID-uPA dams from our previous and current SCID-uPA breeding stocks. It shows a clear advantage in terms of breeding efficiency for both hemizygous and homozygous dams as well as a substantial increase in the numbers of pups that survive transplant with human hepatocytes and make it to the weaning period. To avoid the unforeseen selection pressures that may have been introduced to our previous breeding colony, mice that leave the breeding colony for transplantation are excluded from our breeding stock. With this dedicated breeding colony, we currently produce substantially more mice with high levels of human hepatocytes repopulation, and at a lower cost, than we have ever done in the past. A direct comparison of human hepatocyte engraftment in SCID-uPA mice derived from our previous and

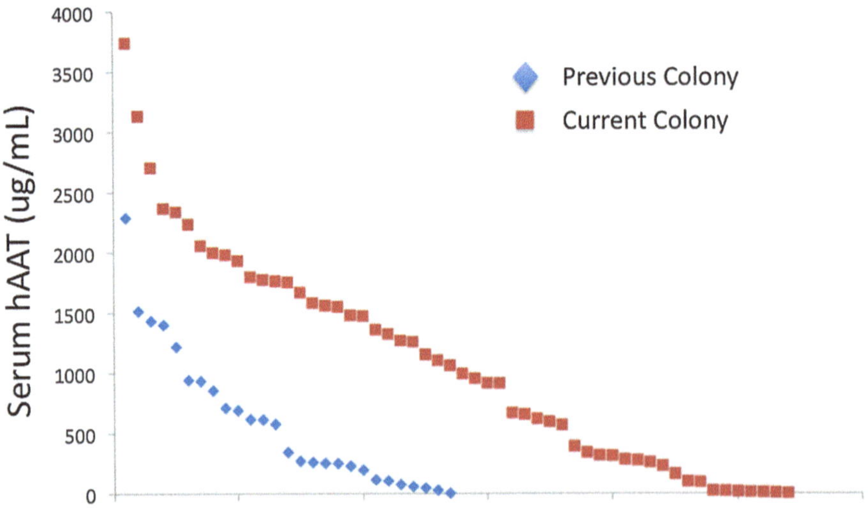

Fig. 3 Re-establishing colony vigor has a substantial impact on the numbers of SCID-uPA mice that achieve high levels of human hepatocyte engraftment, as determined by serum human anti-1 antitrypsin (hAAT) levels at weaning (6 weeks posttransplant)

current breeding stocks is shown in Fig. 3 (unpublished data). A greater proportion of transplanted mice exhibited near complete human hepatocyte repopulation of their livers in recipients derived from our current colony (27% achieving serum hAAT >1000 μg/mL vs. 5% seen with mice derived from our previous colony). As illustrated in Fig. 4 (unpublished data), this translates into a substantially higher proportion of mice sustaining high serum titers of HCV RNA following inoculation with patient derived serum. It is difficult to understand why we do not see the excessive mortality initially reported for homozygous Alb-uPA mice [24] or the infertility that was suggested to be the consequence of uPA-induced liver injury associated with homozygosity of the uPA transgene that was later reported for SCID-uPA mice [38]. Differences in the frequency of spontaneous uPA transgene deletion and/or in the toxicity associated with uPA transgene, perhaps due to excessive inbreeding, are possible explanations that require testing.

Another group that used the SCID-uPA platform to generate human liver chimeric mice has provided detailed histological assessment of the engrafted human hepatocytes in homozygous uPA SCID mice [31]. They demonstrated that regenerative nodules are comprised of both human and murine hepatocytes. The engrafted human hepatocytes themselves were reported to appear swollen and have a peculiar empty-appearing cytoplasm. These features are also supported by our own observations [39] (Fig. 5). However, we also find that large areas of the liver sections exhibit cytoplasmic vacuolations localized to areas dominated by engrafted human hepatocytes (brown nuclei). By contrast, these cytoplasmic vacuolations are absent from regions dominated by murine

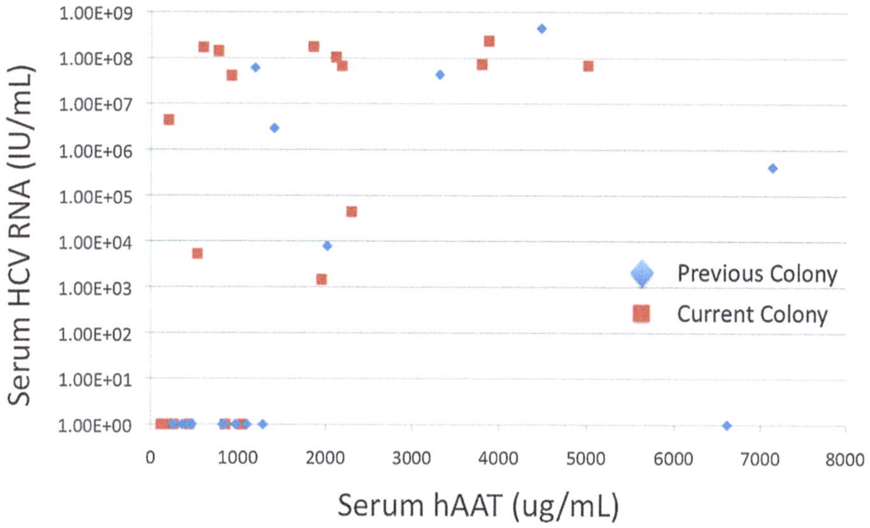

Fig. 4 Re-establishing colony vigor has a substantial impact on the numbers of SCID-uPA mice that achieve high HCV infection levels, as determined by serum levels of HCV RNA (qRT-PCR) at 8 weeks posttransplant

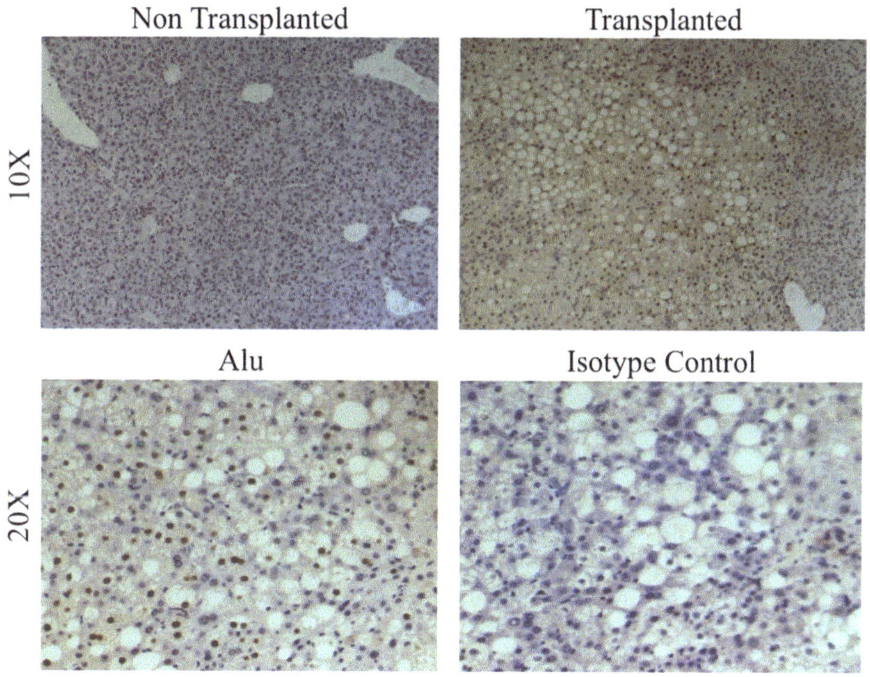

Fig. 5 Vast areas of liver tissue obtained from SCID-uPA mice having extensive human hepatocyte engraftment (transplanted) display large vacuoles localized to engrafted human hepatocytes (brown nuclei). These are absent in mice that did not receive transplant (nontransplanted). An in situ hybridization method for the detection of Alu repeats, which are present in histologically integrated primate hepatocytes, but are absent from murine hepatocytes (blue nuclei), is typically used to identify engrafted human hepatocytes in chimeric livers (Biogenex). Nontransplanted mice serve as a control. Isotype (mouse IgG) control is also used to confirm specificity of the secondary antibody detection system used to hybridize to the FITC-conjugated Alu probe (Biogenex)

constituents (blue nuclei) and are absent altogether from livers obtained from nontransplanted mice. These vacuolations have been characterized by Tateno et al. [40] and are due to hepatic steatosis arising from a deficiency in human growth hormone.

We have also shown that homozygous uPA SCID mice acquire more of a human-like plasma lipoprotein profile that correlates with the extent of human hepatocyte engraftment [41]. This is characterized by a shift in plasma lipoprotein profiles toward lipoprotein particles that resemble the very low density lipoproteins (VLDLs) found in human plasma, with fewer of the high density lipoproteins which are more characteristic of the majority of lipoproteins found in mouse plasma. In humans, VLDLs are synthesized and secreted exclusively by the liver. This is regulated by the synthesis of apolipoprotein B100, which is an obligatory constituent for VLDL production. The appearance of human apolipoprotein B100 in the sera from homozygous uPA SCID mice that have human hepatocyte engraftment parallels the production of VLDL-like particles [41]. In addition to apolipoprotein B100, cholesterol ester transfer protein (CETP) was also shown to accumulate in sera from SCID-uPA mice. CETP mediates the transfer of cholesteryl esters from HDL to VLDL (and low density lipoproteins, LDL) and the transfer of triglyceride from VLDL (and LDL) to HDL. Not only does this process decrease levels of antiatherogenic HDL cholesterol, but it also increases the levels of proatherogenic VLDL (and LDL) cholesterol. Importantly, mice are naturally deficient in CETP. In this context, the engrafted human hepatocytes are the only source for both CETP and human apolipoprotein B100. Overall, these reflect proatherogenic changes. As SCID mice retain elements of the innate immune system, including macrophages, the possibility that human liver chimeric SCID-uPA mice are susceptible to diet-induced atherosclerosis is intriguing and requires investigation. Finally, Yoshizato and coworkers conducted a series of functional assessments of human metabolic activity in homozygous uPA transgenic SCID mice with high levels of human hepatocyte engraftment [35, 42–47]. This body of work provided substantial evidence that these mice express constitutive and inducible human drug metabolizing enzymes and activities, highlighting the utility of these mice in areas of toxicology and drug design.

Although challenging, others have developed human liver chimeric mouse models on immunodeficient mouse backgrounds, either involving the uPA transgene [31, 48–50] or different liver injury models altogether [51–53], providing independent support for the utility of human liver chimeric mice for studying human liver infections caused by a range of hepatotropic organisms [54, 55].

3 HCV Studies in Human Liver Chimeric SCID-uPA Mice

There are many advantages in using human liver chimeric mouse models in the study of HCV. One clear advantage that this model has over the chimpanzee is that HCV infection and replication occur in human hepatocytes, which represent the natural environment for these processes to occur. In addition, HCV infection can be initiated by inoculation with serum obtained from an HCV-infected individual, demonstrating the susceptibility of the system to the usual route of transmission in humans. These models also enable the study of natural HCV infection with easily measurable readouts, including measurement of serum levels of viral proteins and RNA. With all stages of the HCV life cycle represented, these models are particularly well suited for studies of HCV biology and to generate new therapeutic approaches. The SCID-uPA human liver chimeric mouse model for example has contributed to our understanding of many aspects of HCV biology, including insights into viral entry [5, 56] and infectious virion composition [18]. With greater than 150 antiviral compounds having been tested in SCID-uPA human liver chimeric mice, the preclinical testing of novel anti-HCV therapies [8, 9, 57, 58] is also an important area where the SCID-uPA mouse model has a proven track record, helping to shape today's curative landscape for HCV therapeutics. In addition, studies performed in SCID-uPA human liver chimeric mice demonstrated for the first time the ability of monoclonal antibodies targeting highly conserved epitopes on the HCV envelope glycoprotein E2 to prevent HCV infection in vivo, renewing interest in vaccine development toward immunogens that generate such antibodies [5].

The selection of chimeric mice for HCV studies typically begins with the measurement of human protein products in the sera. At present we utilize an ELISA for human alpha-1 antitrypsin (hAAT) or for human albumin (hAlb) for these measurements. Serum levels of these proteins are analyzed at weaning (6 weeks posttransplant) to provide an initial screen for mice that may then undergo further serum analysis at 8 weeks posttransplant. Mice in our colony with serum hAAT levels below 80 μg/mL (corresponding to ~1 mg/mL hAlb) are rarely able to sustain infection. Mice with levels ranging from 100 to 200 μg/mL hAAT have a 20–30% chance of sustaining infection, whereas mice with levels above 500 μg/mL tend to exhibit near uniform infection. We typically enroll mice into therapeutic studies if they have at least 10^5 copies HCV RNA/mL serum HCV serum. This allows for evaluation of a 1–4 log reduction in HCV titer with intervention [57]. The serum titers of HCV within control mice are typically stable enough that groups of 5–6 test animals can provide statistical significance ($P < 0.05$) if an intervention results in a >0.5 log decline in titer.

4 Validation of the SCID-uPA Human Liver Chimeric Mouse Model of HCV Infection

Anti-HCV therapies are often advanced directly into clinical trials with limited or no in vivo efficacy data, limiting predictions of clinical efficacy and potentially exposing human subjects to unnecessary treatment or off-target side effects. The preclinical evaluation of putative anti-HCV therapies provided much of the impetus behind the development of the SCID-uPA mouse model of HCV infection. Several established and developing therapeutics that had undergone clinical evaluation, were studied in the chimeric mouse. These studies were instrumental in validating the utility of the model for the development of therapeutic products for use in those infected with HCV, since the virologic response to each of the antiviral regimens tested in the SCID-uPA mouse model was already known for the clinical setting. These studies include interferon alpha 2b (IFN-a2b) and molecules that target enzymes encoded by the viral genome (polymerase and protease inhibitors).

For examination of the virologic response of infected mice to IFN-a2b, several cohorts of HCV infected mice were studied during and after treatment with IFN-a2b [57]. Vehicle treated HCV infected mice served as control. Infected mice that had been treated with 1350 IU/g/day IFN-a2b intramuscularly for 10–28 days had significantly reduced viral titers (~1 log) compared to controls, with dose responsiveness demonstrated by reduced impact with a smaller dose (400 IU/g/day). The rates of viral titer decline suggested that a minimum of 20 weeks of therapy would be required for viral clearance from the genotype 1a infection. Similar responses were reliably observed in over ten cohorts of HCV infected mice. Consequently, we began to use IFN-a2b treatment routinely as a control treatment group for the study of other anti-HCV therapies. In contrast to the response of genotype 1a infected mice, but in keeping with the clinical setting, genotype 3a infected mice demonstrated a greater response upon treatment with IFN-a2b, with viral titers dropping 2–3 logs within 4–7 days of therapy. Furthermore, some titers dropped to below detection with no evident rebound after an additional 2 weeks of follow-up [57]. As such, the chimeric mice with genotype 3a infection were more amenable to viral titer reduction or clearance than were those with genotype 1 infection, mirroring the outcomes with clinical interventions with interferon based treatment.

The first small molecule to demonstrate anti-HCV efficacy in a clinical trial was the NS3/4A protease inhibitor BILN2061. Administration of BILN2061 in a q 12 h oral dosing regimen led to dramatic reductions in viral load (2–3 logs) in patients infected with genotype 1a/1b within 48 h or treatment. In a pilot experiment, 10 mg/kg of BILN2061 was administered to three HCV infected mice (genotype 1a) by oral gavage twice daily, resulting in a decline in viral titers by over 2 logs in all mice after just 4 days of treatment. Subsequent therapy in six HCV infected mice for 14 days revealed a drop of 1.4 logs at 7 days with a mean 1.1 log reduction from baseline after the 14 days due to rebound of HCV titers in three mice during the second week of therapy. This outcome may be explained by the development of drug resistance and/or a deterioration in pharmokinetics by inducible drug metabolizing enzymes. To explore these possibilities, sera and liver were recovered from two of the rebounding mice for recovery and sequencing of HCV RNA. While the sequence alterations that had been previously identified in replicon studies were not observed [59, 60], we did identify a viral variant characterized by a lysine at amino acid 80 instead of a glutamine (Q80K). Subgenomic replicons expressing the Q80K mutant revealed ~four-fold reduction in sensitivity to BILN2061 as compared to wild-type. Interestingly, HCV genotype 2 also has a lysine at the same position as the Q80K mutant and has demonstrated to have a reduced response to BILN2061 in clinical trials [61, 62]. The emergence of this mutation in infected mice is testimony to how rapidly resistance mutants can be selected in vivo. Although no deaths were noted during or after 14 days treatment regiment, another study using the SCID-uPA mouse model subsequently demonstrated signs of cardiotoxicity [63], an adverse effect originally observed in primates that led to discontinuation of clinical development of this drug by Boehringer-Ingelheim. Treatment-related deaths in the infected mice were associated with histological evidence of cardiotoxicity with plasma levels of 840 and 4400 nM. These plasma levels are approximately 20-fold higher than those seen in the clinical trials, and several folds higher than those demonstrated in our own mouse studies.

In collaboration with Wyeth Research and Viropharma, Inc., we conducted a series of experiments that further validated the application of the chimeric SCID-uPA mouse model for predicting the clinical impact of anti-HCV therapies [9]. These studies also demonstrate the range of studies possible in this mouse model that cannot be achieved with chimpanzees. In initial proof of concept studies with the HCV NS5B RNA dependent RNA polymerase inhibitor HCV796 (telaprevir) administered every 8 h by oral gavage, HCV infected mice demonstrated a 2.02 log reduction in viral titers after 5 days of therapy, with a rebound of 1 log within 1 week of stopping treatment. These outcomes were confirmed in a repeated study with a mean 1.78 ± 0.59 log drop after 10 days of treatment. However, viral titers rebounded in one of the five mice in the 2-week study during the second week of therapy. An additive effect of was observed when HCV796 was used in a combination approach with IFN-a2b, with a drop of 2.44 log after 10 days, with all mice declining throughout treatment. The outcomes achieved with both the HCV796 monotherapy and the HCV796-IFN-a2b combination therapy paralleled the outcomes that were achieved subsequently in clinical trials [64]. Another clinical correlate for eradicating HCV from human liver chimeric SCID-uPA mice was reported by Ohara et al. and involved combinations of direct-acting antivirals [65].

While these results provide strong validation for the positive predictive value of the SCID-uPA mouse model for therapies targeting HCV infection, support for their negative predictive value has also been provided. The polymerase inhibitor HCV 371 was the first lead molecule developed in the Wyeth–Virpharma collaboration, demonstrating an in vitro activity in the replicon model with an EC50 at 5 μM [66]. In vivo studies in chimeric SCID-uPA mice failed to demonstrate significant impact on infection over the 3-week course of oral administration [57]. The compound also failed to achieve efficacy targets in Phase Ib clinical trials and was dropped from further clinical evaluation [67].

We have also used the SCID-uPA mouse model of HCV infection to examine a gene therapy approach directed against the NS3/NS4A serine protease [7]. This approach involved a modified BH3 interacting death domain agonist (BID) molecule involved in the induction of apoptosis via activation of the mitochondrial caspase cascade. The BID molecule was engineered so that it could be activated to induce apoptosis following its cleavage via HCV NS3/NS4A protease then administered to HCV infected mice using a replication-defective adenovirus [7]. Serum ALT levels increased within 1–2 days after adenoviral administration of the modified BID and HCV titers decreased by over 2-logs. In subsequent studies, complete viral clearance was achieved in five of seven mice with repeated doses of modified BID at 7-day intervals.

Finally, the SCID-uPA mouse model of HCV infection has been used to test antibody-based strategies to prevent HCV infection or limit already established HCV infections. A substantial proportion of infected persons are able to clear HCV spontaneously during the acute phase of infection, and this ability has been linked to an early neutralizing antibody response [68]. Although their significance in controlling acute HCV infection is unclear, several recombinant vaccine candidates that have been tested in chimpanzees indicate that the ability of a vaccine to induce high titers of neutralizing antibodies will be important [69, 70]. However, a major problem HCV vaccine design faces is the extreme variability of the virus. In a collaboration with Mansun Law, we were able to demonstrate for the first time the ability of monoclonal antibodies directed against a highly conserved epitope within the HCV E2 glycoprotein to protect human liver chimeric SCID-uPA from heterologous HCV challenge [5]. These studies demonstrate that a prophylactic vaccine against HCV may be achievable through the design of immunogens that induce antibodies with these broadly neutralizing capabilities.

Others have performed a series of proof-of-concept studies in SCID-uPA mice whereby immunoglobulin that had been purified from a patient with chronic HCV genotype 1a was able to completely protect SCID-uPA mice from homologous HCV challenge [71, 72]. Yet, despite having higher in vitro neutralizing titers against heterologous HCV strains, only partial protection for heterologous HCV strains was observed with SCID-uPA mice. Thus neutralization titers observed in vitro do not always correlate with responses to in vivo challenge, highlighting the importance of using in vivo models like the SCID-uPA mouse model for assessing the neutralizing antibody responses induced by immunization protocols.

Finally, receptor-blocking antibodies targeting SR-B1 and CD81 that inhibit HCV infection in vitro have been identified and these have been tested for their abilities to prevent HCV infection or progression of an established infection in vivo using the SCID-uPA mouse model [56, 73]. Whereas monoclonal antibodies directed against either SR-B1 or CD81 were able to prevent HCV infection by heterologous HCV strains, only anti-SRB1 was able to control an already established infection [56, 73]. These experiments highlight the critical role of CD81 in HCV infection in vivo, offering an interesting perspective for novel antiviral strategies against HCV that have both a high genetic barrier to resistance and pan-genotypic antiviral activity. Furthermore, these strategies may improve clinical outcomes when combined with the current standard of care.

Aside from testing anti-HCV therapeutics, the SCID-uPA mouse model has also been used to examine the host genomic response to HCV infection, advancing our knowledge about the host innate immune response to HCV infection [4], potentially opening the door to additional therapeutic strategies.

5 Other Applications of the Chimeric Human Liver SCID-uPA Mouse Models

Although the main interest in the SCID-uPA mouse model has been for HCV-related studies, a mouse model that has an extensive repopulation of its liver by human hepatocytes has the potential for a wide range of applications spanning liver physiology, metabolism and toxicology, many of which have already been discussed. In addition, the human liver chimeric SCID-uPA mouse model is amenable to many pathogens for which the liver is an obligate part of their life cycle, including hepatitis B virus (HBV) and malaria infections [48, 55]. Like HCV, these also represent major global health problems.

HBV infects the liver causing the most serious type of viral hepatitis, killing nearly 750,000 people each year [74]. A highly effective HBV vaccine has prevented more than one million deaths worldwide, but this does not benefit the >250 million people who are chronically infected with the virus today [74] and who are at high risk of dying from serious liver disease or cancer [75]. Unfortunately, there is no HBV cure and most patients require life-long treatment to adequately manage their disease. Given that the HBV genome encodes just one viral protein that has enzymatic activity (HBV polymerase), the identification of host factors that can serve as therapeutic targets is critical for developing new classes of anti-HBV agents that may contribute to a cure when used in combination with the current standard of care. With the introduction of breakthrough medicines that in more than 90% of cases will cure chronic HCV infection within 12 weeks, the development of curative treatments for hepatitis B virus (HBV) is now a rapidly emerging area of interest for many researchers and companies. Given its proven track record for the preclinical testing of antivirals, the SCID-uPA mouse model of HBV infection will likely speed up these efforts.

Liver phase infection is an obligate, early step in *P. falciparum* development, making it an excellent target for intervention. However, among the different developmental stages of *P. falciparum*, the hepatic phase has proven most challenging to study owing to limited access to infected liver. Infection of humanized SCID-uPA mice with *P. falciparum* sporozoites produce mature liver-stage shizonts containing merozoites capable of invading human red blood cells [55]. As such, this model is useful for assessing the biology of the parasite and for testing antimalaria therapeutic strategies. Importantly, an effective vaccine against the liver stage is desirable since this would prevent maturation and subsequent development of the sexual stage of disease that is required for transmission of the parasite to others.

Finally, human liver-chimeric mice are generated on an immunodeficient background, precluding studies directed at understanding the adaptive immune response to infection and the testing of vaccine candidates. This is important because there is currently no vaccine for HCV. Without a vaccine to prevent HCV spread, the burden of HCV infection and costs to health care systems will continue to increase as thousands of new and current cases of chronically infected individuals develop severe illness, jeopardizing their capacity for employment and requiring extensive support from health and social assistance systems. Developing a vaccine against HCV is an area of intense interest. Novel strains of immunodeficient mice engrafted with functional human immune systems are rapidly gaining traction as important tools for the study of many human-specific infectious agents, with the NOD/SCID/interleukin receptor common g-chain (IL2rg) knockout (NSG) mouse having superior human immune system engraftment [76]. The ability to have a human immune system and a human liver in a single mouse offers an exciting opportunity to have a model that permits studying the immune response to HCV and HBV infection as well as for the pre-clinical testing of vaccine candidates both prophylactic and therapeutic. Dual humanization of both liver and hematopoiesis in otherwise immunodeficient mice has been reported [77–82], but productive HCV infection necessary for longitudinal analysis has not yet been achieved.

6 Concluding Remarks

The flexibility and utility of mice with chimeric human livers for the understanding of HCV biology and their demonstrated validity for the prediction of clinical treatment effects have played an important role in the development of the current generation of curative direct acting antivirals for HCV. This performance suggests promise for support in HCV vaccine development and for the provision of an important tool to support ongoing work to improve HBV therapeutics from their current capacity for achieving long term control of replication to a true cure, paralleled by a fall in the long term risk of hepatocellular carcinoma. Interest in the use of chimeric mice is also expanding in malaria research, as well as in the fields of human toxicology and drug metabolism. The emergence of three newer models for development of chimeric mice [49–52] further supports the usefulness of this approach to biological and therapeutic studies and should further broaden their application.

References

1. Alter MJ (1997) Epidemiology of hepatitis C. Hepatology 26:62S–65S
2. Mercer DF, Schiller DE, Elliott JF, Douglas DN, Hao C, Rinfret A et al (2001) Hepatitis C virus replication in mice with chimeric human livers. Nat Med 7:927–933
3. Bukh J (2012) Animal models for the study of hepatitis C virus infection and related liver disease. Gastroenterology 142(1279–1287): e1273
4. Walters KA, Joyce MA, Thompson JC, Smith MW, Yeh MM, Proll S et al (2006) Host-specific response to HCV infection in the chimeric SCID-beige/Alb-uPA mouse model: role of the innate antiviral immune response. PLoS Pathog 2:e59
5. Law M, Maruyama T, Lewis J, Giang E, Tarr AW, Stamataki Z et al (2008) Broadly neutralizing antibodies protect against hepatitis C virus quasispecies challenge. Nat Med 14:25–27
6. Brown RJ, Hudson N, Wilson G, Rehman SU, Jabbari S, Hu K et al (2012) Hepatitis C virus envelope glycoprotein fitness defines virus population composition following transmission to a new host. J Virol 86:11956–11966
7. Hsu EC, Hsi B, Hirota-Tsuchihara M, Ruland J, Iorio C, Sarangi F et al (2003) Modified apoptotic molecule (BID) reduces hepatitis C virus infection in mice with chimeric human livers. Nat Biotechnol 21:519–525
8. Kneteman NM, Asthana S, Lewis J, Dibben C, Douglas D, Nourbakhsh M et al (2012) Impact of calcineurin inhibitors with or without interferon on hepatitis C virus titers in a chimeric mouse model of hepatitis C virus infection. Liver Transpl 18:38–44
9. Kneteman NM, Howe AY, Gao T, Lewis J, Pevear D, Lund G et al (2009) HCV796: A

selective nonstructural protein 5B polymerase inhibitor with potent anti-hepatitis C virus activity in vitro, in mice with chimeric human livers, and in humans infected with hepatitis C virus. Hepatology 49:745–752

10. Waterston RH, Lindblad-Toh K, Birney E, Rogers J, Abril JF, Agarwal P et al (2002) Initial sequencing and comparative analysis of the mouse genome. Nature 420:520–562
11. Bitzegeio J, Bankwitz D, Hueging K, Haid S, Brohm C, Zeisel MB et al (2010) Adaptation of hepatitis C virus to mouse CD81 permits infection of mouse cells in the absence of human entry factors. PLoS Pathog 6: e1000978
12. Ploss A, Evans MJ, Gaysinskaya VA, Panis M, You H, de Jong YP et al (2009) Human occludin is a hepatitis C virus entry factor required for infection of mouse cells. Nature 457:882–886
13. Dorner M, Horwitz JA, Robbins JB, Barry WT, Feng Q, Mu K et al (2011) A genetically humanized mouse model for hepatitis C virus infection. Nature 474:208–211
14. Chen J, Zhao Y, Zhang C, Chen H, Feng J, Chi X et al (2014) Persistent hepatitis C virus infections and hepatopathological manifestations in immune-competent humanized mice. Cell Res 24:1050–1066
15. Bartosch B, Dubuisson J, Cosset FL (2003) Infectious hepatitis C virus pseudo-particles containing functional E1-E2 envelope protein complexes. J Exp Med 197:633–642
16. Hsu M, Zhang J, Flint M, Logvinoff C, Cheng-Mayer C, Rice CM et al (2003) Hepatitis C virus glycoproteins mediate pH-dependent cell entry of pseudotyped retroviral particles. Proc Natl Acad Sci U S A 100:7271–7276
17. Lindenbach BD, Evans MJ, Syder AJ, Wolk B, Tellinghuisen TL, Liu CC et al (2005) Complete replication of hepatitis C virus in cell culture. Science 309:623–626
18. Lindenbach BD, Meuleman P, Ploss A, Vanwolleghem T, Syder AJ, McKeating JA et al (2006) Cell culture-grown hepatitis C virus is infectious in vivo and can be recultured in vitro. Proc Natl Acad Sci U S A 103:3805–3809
19. Wakita T, Kato T (2006) Development of an infectious HCV cell culture system. Horizon Bioscience, Norfolk
20. Wakita T, Pietschmann T, Kato T, Date T, Miyamoto M, Zhao Z et al (2005) Production of infectious hepatitis C virus in tissue culture from a cloned viral genome. Nat Med 11:791–796
21. Zhong J, Gastaminza P, Cheng G, Kapadia S, Kato T, Burton DR et al (2005) Robust hepatitis C virus infection in vitro. Proc Natl Acad Sci U S A 102:9294–9299
22. Zhong J, Gastaminza P, Chung J, Stamataki Z, Isogawa M, Cheng G et al (2006) Persistent hepatitis C virus infection in vitro: coevolution of virus and host. J Virol 80:11082–11093
23. Heckel JL, Sandgren EP, Degen JL, Palmiter RD, Brinster RL (1990) Neonatal bleeding in transgenic mice expressing urokinase-type plasminogen activator. Cell 62:447–456
24. Sandgren EP, Palmiter RD, Heckel JL, Daugherty CC, Brinster RL, Degen JL (1991) Complete hepatic regeneration after somatic deletion of an albumin-plasminogen activator transgene. Cell 66:245–256
25. Rhim JA, Sandgren EP, Degen JL, Palmiter RD, Brinster RL (1994) Replacement of diseased mouse liver by hepatic cell transplantation. Science 263:1149–1152
26. Rhim JA, Sandgren EP, Palmiter RD, Brinster RL (1995) Complete reconstitution of mouse liver with xenogeneic hepatocytes. Proc Natl Acad Sci U S A 92:4942–4946
27. Meuleman P, Vanlandschoot P, Leroux-Roels G (2003) A simple and rapid method to determine the zygosity of uPA-transgenic SCID mice. Biochem Biophys Res Commun 308:375–378
28. Kneteman NM, Mercer DF (2005) Mice with chimeric human livers: who says supermodels have to be tall? Hepatology 41:703–706
29. Kawahara T, Toso C, Douglas DN, Nourbakhsh M, Lewis JT, Tyrrell DL et al (2010) Factors affecting hepatocyte isolation, engraftment, and replication in an in vivo model. Liver Transpl 16:974–982
30. Bissig KD, Wieland SF, Tran P, Isogawa M, Le TT, Chisari FV et al (2010) Human liver chimeric mice provide a model for hepatitis B and C virus infection and treatment. J Clin Invest 120:924–930
31. Meuleman P, Libbrecht L, De Vos R, de Hemptinne B, Gevaert K, Vandekerckhove J et al (2005) Morphological and biochemical characterization of a human liver in a uPA-SCID mouse chimera. Hepatology 41:847–856
32. Hiraga N, Imamura M, Tsuge M, Noguchi C, Takahashi S, Iwao E et al (2007) Infection of human hepatocyte chimeric mouse with genetically engineered hepatitis C virus and its susceptibility to interferon. FEBS Lett 581:1983–1987
33. Hiraga N, Imamura M, Abe H, Hayes CN, Kono T, Onishi M et al (2011) Rapid

emergence of telaprevir resistant hepatitis C virus strain from wildtype clone in vivo. Hepatology 54:781–788

34. Hiraga N, Abe H, Imamura M, Tsuge M, Takahashi S, Hayes CN et al (2011) Impact of viral amino acid substitutions and host interleukin-28b polymorphism on replication and susceptibility to interferon of hepatitis C virus. Hepatology 54:764–771
35. Tateno C, Yoshizane Y, Saito N, Kataoka M, Utoh R, Yamasaki C et al (2004) Near completely humanized liver in mice shows human-type metabolic responses to drugs. Am J Pathol 165:901–912
36. Yoshitsugu H, Nishimura M, Tateno C, Kataoka M, Takahashi E, Soeno Y et al (2006) Evaluation of human CYP1A2 and CYP3A4 mRNA expression in hepatocytes from chimeric mice with humanized liver. Drug Metab Pharmacokinet 21:465–474
37. Kawahara T, Douglas DN, Lewis J, Lund G, Addison W, Tyrrell DL et al (2010) Critical role of natural killer cells in the rejection of human hepatocytes after xenotransplantation into immunodeficient mice. Transpl Int 23:934–943
38. Brezillon NM, DaSilva L, L'Hote D, Bernex F, Piquet J, Binart N et al (2008) Rescue of fertility in homozygous mice for the urokinase plasminogen activator transgene by the transplantation of mouse hepatocytes. Cell Transplant 17:803–812
39. Douglas DN, Pu CH, Lewis JT, Bhat R, Anwar-Mohamed A, Logan M et al (2016) Oxidative stress attenuates lipid synthesis and increases mitochondrial fatty acid oxidation in hepatoma cells infected with hepatitis C virus. J Biol Chem 291:1974–1990
40. Tateno C, Kataoka M, Utoh R, Tachibana A, Itamoto T, Asahara T, Miya F, Tsunoda T, Yoshizato K (2011) Growth hormone-dependent pathogenesis of human hepatic steatosis in a novel mouse model bearing a human hepatocyte-repopulated liver. Endocrinology 152:1479–1491
41. Tachibana S, Watanabe T (2007) Sexual differences in the crucial environmental factors for the timing of postdiapause development in the rice bug Leptocorisa chinensis. J Insect Physiol 53:1000–1007
42. Katoh M, Matsui T, Nakajima M, Tateno C, Kataoka M, Soeno Y et al (2004) Expression of human cytochromes P450 in chimeric mice with humanized liver. Drug Metab Dispos 32:1402–1410
43. Katoh M, Matsui T, Nakajima M, Tateno C, Soeno Y, Horie T et al (2005) In vivo induction of human cytochrome P450 enzymes expressed in chimeric mice with humanized liver. Drug Metab Dispos 33:754–763
44. Katoh M, Matsui T, Okumura H, Nakajima M, Nishimura M, Naito S et al (2005) Expression of human phase II enzymes in chimeric mice with humanized liver. Drug Metab Dispos 33:1333–1340
45. Katoh M, Sawada T, Soeno Y, Nakajima M, Tateno C, Yoshizato K et al (2007) In vivo drug metabolism model for human cytochrome P450 enzyme using chimeric mice with humanized liver. J Pharm Sci 96:428–437
46. Nishimura M, Yoshitsugu H, Yokoi T, Tateno C, Kataoka M, Horie T et al (2005) Evaluation of mRNA expression of human drug-metabolizing enzymes and transporters in chimeric mouse with humanized liver. Xenobiotica 35:877–890
47. Nishimura M, Yokoi T, Tateno C, Kataoka M, Takahashi E, Horie T et al (2005) Induction of human CYP1A2 and CYP3A4 in primary culture of hepatocytes from chimeric mice with humanized liver. Drug Metab Pharmacokinet 20:121–126
48. Dandri M, Burda MR, Torok E, Pollok JM, Iwanska A, Sommer G et al (2001) Repopulation of mouse liver with human hepatocytes and in vivo infection with hepatitis B virus. Hepatology 33:981–988
49. Suemizu H, Hasegawa M, Kawai K, Taniguchi K, Monnai M, Wakui M et al (2008) Establishment of a humanized model of liver using NOD/Shi-scid IL2Rgnull mice. Biochem Biophys Res Commun 377:248–252
50. Song X, Guo Y, Duo S, Che J, Wu C, Ochiya T et al (2009) A mouse model of inducible liver injury caused by tet-on regulated urokinase for studies of hepatocyte transplantation. Am J Pathol 175:1975–1983
51. Azuma H, Paulk N, Ranade A, Dorrell C, Al-Dhalimy M, Ellis E et al (2007) Robust expansion of human hepatocytes in Fah−/−/Rag2−/−/Il2rg−/− mice. Nat Biotechnol 25:903–910
52. Bissig KD, Le TT, Woods NB, Verma IM (2007) Repopulation of adult and neonatal mice with human hepatocytes: a chimeric animal model. Proc Natl Acad Sci U S A 104:20507–20511
53. Kosaka K, Hiraga N, Imamura M, Yoshimi S, Murakami E, Nakahara T et al (2013) A novel TK-NOG based humanized mouse model for the study of HBV and HCV infections. Biochem Biophys Res Commun 441:230–235
54. Foquet L, Hermsen CC, Verhoye L, van Gemert GJ, Cortese R, Nicosia A et al (2015)

Anti-CD81 but not anti-SR-BI blocks Plasmodium falciparum liver infection in a humanized mouse model. J Antimicrob Chemother 70:1784–1787

55. Sacci JB Jr, Alam U, Douglas D, Lewis J, Tyrrell DL, Azad AF et al (2006) Plasmodium falciparum infection and exoerythrocytic development in mice with chimeric human livers. Int J Parasitol 36:353–360
56. Meuleman P, Catanese MT, Verhoye L, Desombere I, Farhoudi A, Jones CT et al (2012) A human monoclonal antibody targeting scavenger receptor class B type I precludes hepatitis C virus infection and viral spread in vitro and in vivo. Hepatology 55:364–372
57. Kneteman NM, Weiner AJ, O'Connell J, Collett M, Gao T, Aukerman L et al (2006) Anti-HCV therapies in chimeric scid-Alb/uPA mice parallel outcomes in human clinical application. Hepatology 43:1346–1353
58. Takebe Y, Saucedo CJ, Lund G, Uenishi R, Hase S, Tsuchiura T et al (2013) Antiviral lectins from red and blue-green algae show potent in vitro and in vivo activity against hepatitis C virus. PLoS One 8:e64449
59. Lin C, Gates CA, Rao BG, Brennan DL, Fulghum JR, Luong YP et al (2005) In vitro studies of cross-resistance mutations against two hepatitis C virus serine protease inhibitors, VX-950 and BILN 2061. J Biol Chem 280:36784–36791
60. Lin C, Lin K, Luong YP, Rao BG, Wei YY, Brennan DL et al (2004) In vitro resistance studies of hepatitis C virus serine protease inhibitors, VX-950 and BILN 2061: structural analysis indicates different resistance mechanisms. J Biol Chem 279:17508–17514
61. Reiser M, Hinrichsen H, Benhamou Y, Reesink HW, Wedemeyer H, Avendano C et al (2005) Antiviral efficacy of NS3-serine protease inhibitor BILN-2061 in patients with chronic genotype 2 and 3 hepatitis C. Hepatology 41:832–835
62. Thibeault D, Bousquet C, Gingras R, Lagace L, Maurice R, White PW et al (2004) Sensitivity of NS3 serine proteases from hepatitis C virus genotypes 2 and 3 to the inhibitor BILN 2061. J Virol 78:7352–7359
63. Vanwolleghem T, Meuleman P, Libbrecht L, Roskams T, De Vos R, Leroux-Roels G (2007) Ultra-rapid cardiotoxicity of the hepatitis C virus protease inhibitor BILN 2061 in the urokinase-type plasminogen activator mouse. Gastroenterology 133:1144–1155
64. Viropharma (2006) Viropharma announces presentation of HCV 796 phase Ib data at digestive disease week. Viropharma, Harrisburg, PA
65. Ohara E, Hiraga N, Imamura M, Iwao E, Kamiya N, Yamada I et al (2011) Elimination of hepatitis C virus by short term NS3-4A and NS5B inhibitor combination therapy in human hepatocyte chimeric mice. J Hepatol 54:872–878
66. Gopalsamy A, Aplasca A, Ciszewski G, Park K, Ellingboe JW, Orlowski M et al (2006) Design and synthesis of 3,4-dihydro-1H-[1]-benzothieno[2,3-c]pyran and 3,4-dihydro-1H-pyrano[3,4-b]benzofuran derivatives as non-nucleoside inhibitors of HCV NS5B RNA dependent RNA polymerase. Bioorg Med Chem Lett 16:457–460
67. Viropharma (2003) Viropharma announces results of hepatitis C Phase Ib study. Viropharma, Harrisburg, PA
68. Neumann-Haefelin C, Thimme R (2013) Adaptive immune responses in hepatitis C virus infection. Curr Top Microbiol Immunol 369:243–262
69. Dahari H, Feinstone SM, Major ME (2010) Meta-analysis of hepatitis C virus vaccine efficacy in chimpanzees indicates an importance for structural proteins. Gastroenterology 139:965–974
70. Meunier JC, Gottwein JM, Houghton M, Russell RS, Emerson SU, Bukh J et al (2011) Vaccine-induced cross-genotype reactive neutralizing antibodies against hepatitis C virus. J Infect Dis 204:1186–1190
71. Meuleman P, Bukh J, Verhoye L, Farhoudi A, Vanwolleghem T, Wang RY et al (2011) In vivo evaluation of the cross-genotype neutralizing activity of polyclonal antibodies against hepatitis C virus. Hepatology 53:755–762
72. Vanwolleghem T, Bukh J, Meuleman P, Desombere I, Meunier JC, Alter H et al (2008) Polyclonal immunoglobulins from a chronic hepatitis C virus patient protect human liver-chimeric mice from infection with a homologous hepatitis C virus strain. Hepatology 47:1846–1855
73. Meuleman P, Hesselgesser J, Paulson M, Vanwolleghem T, Desombere I, Reiser H et al (2008) Anti-CD81 antibodies can prevent a hepatitis C virus infection in vivo. Hepatology 48:1761–1768
74. WHO (2015) Hepatitis B. WHO, Geneva http://www.who.int/mediacentre/factsheets/fs204/en/
75. Lok AS (2004) Prevention of hepatitis B virus-related hepatocellular carcinoma. Gastroenterology 127:S303–S309

76. Douglas DN, Kneteman NM (2015) Generation of improved mouse models for the study of hepatitis C virus. Eur J Pharmacol 759:313–325
77. Keng CT, Sze CW, Zheng D, Zheng Z, Yong KS, Tan SQ et al (2016) Characterisation of liver pathogenesis, human immune responses and drug testing in a humanised mouse model of HCV infection. Gut 65:1744–1753
78. Wilson EM, Bial J, Tarlow B, Bial G, Jensen B, Greiner DL et al (2014) Extensive double humanization of both liver and hematopoiesis in FRGN mice. Stem Cell Res 13:404–412
79. Gutti TL, Knibbe JS, Makarov E, Zhang J, Yannam GR, Gorantla S et al (2014) Human hepatocytes and hematolymphoid dual reconstitution in treosulfan-conditioned uPA-NOG mice. Am J Pathol 184:101–109
80. Strick-Marchand H, Dusseaux M, Darche S, Huntington ND, Legrand N, Masse-Ranson G et al (2015) A novel mouse model for stable engraftment of a human immune system and human hepatocytes. PLoS One 10:e0119820
81. Bility MT, Zhang L, Washburn ML, Curtis TA, Kovalev GI, Su L (2012) Generation of a humanized mouse model with both human immune system and liver cells to model hepatitis C virus infection and liver immunopathogenesis. Nat Protoc 7:1608–1617
82. Washburn ML, Bility MT, Zhang L, Kovalev GI, Buntzman A, Frelinger JA et al (2011) A humanized mouse model to study hepatitis C virus infection, immune response, and liver disease. Gastroenterology 140:1334–1344

Chapter 33

Mouse Models for Studying HCV Vaccines and Therapeutic Antibodies

Jenna M. Gaska, Qiang Ding, and Alexander Ploss

Abstract

In spite of the immense progress in hepatitis C virus (HCV) research, efforts to prevent infection, such as generating a vaccine, have not yet been successful. The high price tag associated with current treatment options for chronic infection and the spike in new infections concurrent with growing opioid abuse are strong motivators for developing effective immunization and understanding neutralizing antibodies' role in preventing infection. Humanized mice—both human liver chimeras as well as genetically humanized models—are important platforms for testing both possible vaccine candidates as well as antibody-based therapies. This chapter details the variety of ways humanized mouse technology can be employed in pursuit of learning how HCV infection can be prevented.

Key words Hepatitis C virus, Humanized mice, Immunization, Neutralizing antibodies

1 Introduction

1.1 A Brief History of Hepatitis C Virus (HCV)

Just over three decades ago, hepatitis C virus (HCV) was still an unnamed entity, presumably responsible for the growing number of viral hepatitis cases not caused by the two hepatitis viruses known at the time, hepatitis A (HAV) and hepatitis B (HBV). Even with screening the blood supply for HAV and HBV, cases of transfusion-associated hepatitis not caused by these two viruses were still observed in the mid-1970s [1, 2]. The cause of this disease, nebulously described as non-A-non-B hepatitis (NANBH), remained unknown until the early 1980s when experiments in chimpanzees at least determined that a transmissible agent was in fact at play [3–6]. However, testing for this agent in patients, let alone isolating it, remained impossible until 1989 when ongoing efforts finally resulted in the isolation and characterization of the NANBH virus, now termed HCV [7].

Once identified, assays were rapidly developed to test patients and the blood supply for HCV [8], making blood transfusions safer and allowing for improved epidemiological data regarding the

Mansun Law (ed.), *Hepatitis C Virus Protocols*, Methods in Molecular Biology, vol. 1911,
https://doi.org/10.1007/978-1-4939-8976-8_33, © Springer Science+Business Media, LLC, part of Springer Nature 2019

virus' prevalence. By 1992, the incidence of HCV in the US blood supply was brought down to effectively zero. Since then, treatment options for HCV have undergone numerous evolutions, from the early days of interferon-α2b to pegylated interferon, the advent of ribavirin, and, since 2011, an array of direct-acting antivirals (DAAs). The arsenal of treatments now available for HCV patients have a lower incidence of side effects and cure rates of 90%+ for most genotypes.

1.2 The Continued Need for an HCV Vaccine

While highly encouraging, these achievements in understanding and treating HCV can distract from the fact that HCV continues to remain a potent global health problem, even in the developed nations where so many of these milestone findings occurred. An estimated 71.1 million people worldwide are infected with HCV and are thus at heightened risk for severe liver disease, including cirrhosis, fibrosis, and hepatocellular carcinoma (HCC) [9]. While approximately 15–45% of individuals clear their initial HCV infection, the remaining 55–85% become chronically infected [10–12]. Since the resultant liver disease from chronic infection progresses over many years, patients can go decades after the initial infection event before experiencing symptoms. Thus, these individuals can unknowingly be transmitters during this prolonged period. The Centers for Disease Control and Prevention (CDC) estimates that in the USA alone, 3.5 million people are infected with HCV and anywhere from 40% to 85% of these individuals are unaware [13]. These numbers may even be underestimated due to bias in the populations included in the surveys [14]. Due to the presence of HCV in the blood supply prior to 1992 and the lack of knowledge concerning the virus in the 1970s and 1980s, the CDC has recommended that all individuals born between 1945 and 1965 be tested for HCV as the prevalence of HCV antibodies in these individuals is four times higher than the general population [13].

Besides these increased incidences of chronic HCV infection in the aging baby boomer population, there has been a rise in acute HCV infections amongst a much younger population. Data collected from 2006 to 2012 found a growing number of injection drug users (IDUs)—predominantly white and 30 years or younger living in nonurban areas of central Appalachia—presenting with acute HCV infection upon admission to substance abuse treatment facilities [15]. Amongst IDUs around the world, including in other developed nations, HCV is considered hyperendemic, with midpoint prevalence estimates of HCV antibodies present in 60–80% of IDUs in 25 different countries, including Japan (65%), China (67%), the UK (51%), and the USA (73%) [16]. In fact, mathematical estimates of the number of transmissions stemming from an individual infected with genotypes 1a, 3a, or 4a (i.e., those seen more among IDUs and typical of trends in HCV infection in the Western world) range from 12 to 24.5 [17].

The well-lauded cure now available for HCV is unfortunately not available for *all* with HCV, underscoring the need for preventive measures such as a vaccine [18]. The extreme cost of these drug regimens—upward of $80,000—makes a worldwide rollout to all HCV patients unfeasible. In the USA, a large percentage of individuals with chronic HCV, such as IDUs and aging individuals infected pre-1992 via blood transfusion, rely on government-funded programs like Medicare and Medicaid. Unable to handle widespread coverage of the exorbitant drug costs, these services have set restrictions, such as sobriety requirements and the need for advanced liver disease progression, to limit patient eligibility [19]. These barriers only add to future healthcare costs as untreated individuals can transmit HCV over a longer period. Furthermore, a focus on curing patients only with severe liver disease is poorly misplaced, as treatment cannot undo the extreme liver damage and probability of developing liver cancer associated with these later stages.

This widespread prevalence of HCV, compounded by the inaccessibility of the most effective treatments, has made a vaccine as necessary as ever before. The holy grail of the HCV field, vaccine development, has faced many challenges, especially since besides humans, HCV only robustly infects chimpanzees. With the increased restrictions on chimpanzee use in biomedical research and the official inclusion of captive chimpanzees on the US Fish and Wildlife Services Endangered Species list, researchers now face additional funding challenges and ethical issues. Thus, to continue HCV vaccine development and related research on neutralizing antibodies (NAbs), immunocompetent small animal models are greatly needed.

1.3 Surrogate Viruses for HCV

Although this chapter will focus on altering mice to support HCV infection and provide a platform for developing vaccines and therapeutics, the inverse approach, whereby HCV or related hepaciviruses are altered to use in small animal models, is also worth noting. Early efforts to find an HCV surrogate utilized the hepacivirus George Barker virus B (GBV-B), which can infect small nonhuman primates such as tamarins [20–23], marmosets [24], and owl monkeys [25]. However, in these animals, GBV-B does not always replicate to robust levels and often causes acute infection instead of the more common persistent infection associated with HCV [24–26]. Chimeric GBV-B viruses containing, for example, parts of the HCV genome encoding the structural [27] or nonstructural [28] proteins have had greater success in achieving persistent infection in marmosets, but the utility of such models for vaccine development and drug testing remains unclear.

Over the past few years, enhanced sequencing efforts have identified viruses like HCV that also belong to the *Hepacivirus* genus of the *Flaviviridae* family. These related hepaciviruses are naturally occurring in a diverse range of species [29], including horses [30], bats [31], cattle [32, 33], black-and-white colobus monkeys [34], sharks [35], and a variety of rodents [36–38] (but none as of yet in *Mus musculus*). The genome organization of HCV and these other recently discovered hepaciviruses is similar, but the sequence divergence is substantial. For example, at the amino acid level, rodent hepaciviruses (RHVs) have 70.4% sequence divergence from HCV nonstructural proteins, with even greater divergence when comparing structural proteins [37]. Very recently, a RHV isolated from wild Norway rats in New York City [38], RHV-nr-1, was successfully used to infect both immunocompromised and immunocompetent laboratory mice, resulting in persistent or acute hepatotropic infection, respectively [39]. Thus, conventional laboratory mice do hold promise for studying the immunological response to hepaciviruses, especially if viral mutants are identified that can establish chronic infection in immunocompetent mice. However, their utility for vaccine development may be limited given the substantial antigenic divergence of these viruses from HCV. Chimeric RHVs could be constructed in which (parts of) the structural and/or nonstructural proteins are replaced, but such approaches have proven difficult as observed for HCV/GBV-B chimeras that are substantially impaired in their replicative fitness. Nonprimate hepaciviruses (NPHVs), which infect horses, are most closely related to HCV, even more so than GBV-B or primate hepaciviruses [34, 37]. However, the obvious difficulties of working in such a nontraditional research model have made NPHVs less than ideal as an HCV surrogate. With the continued efforts to sequence the viromes of diverse species, including other nonhuman primates, it is possible that more hepaciviruses will be uncovered. Until then, other approaches have focused on viral adaptation to facilitate study of HCV infection in mice. For example, the HCV strain Jc1 (genotype 2a) was successfully adapted to use murine CD81 via mutations in E1 and E2, permitting viral uptake in murine hepatocytes both in vitro and in vivo [40, 41].

1.4 Humanizing Mice to Study HCV

Mice have a long history of use in biomedical research, but they naturally do not support HCV infection. However, the plethora of tools and reagents available to study these animals and the relative ease in performing genetic manipulations have made mice suitable candidates for a variety of humanization procedures. This has allowed researchers to not only infect humanized mice with HCV but has also created new opportunities for studying vaccines and broadly neutralizing antibodies (bNAbs) in vivo. In this chapter, we will broadly describe the approaches available for generating genetically humanized mice or human liver chimeric mice and more specifically the ways to perform and assess HCV infection in such models. As vaccination regimens and antibody treatments are highly individualized to specific experiments, we will present an overview of how these methods have been successfully utilized to test HCV immunization approaches in these models.

2 Materials

Before proceeding with any work involving humanized mice and/or human pathogens, please *see* **Note 1**.

2.1 Mouse Strains

Note that all mice expressing HCV entry factors, specifically human CD81, OCLN, SCARB1 and CLDN1 should be on a reporter background enabling activation of a cellularly encoded luciferase gene that can be activated by Cre recombinase (i.e., FVB.129S6 (B6)-*Gt(ROSA)26Sor*$^{tm1(Luc)Kael}$/J [42]).

Expression of HCV entry factors in these mice can be accomplished in various ways:

1. Vector-mediated (e.g., adenoviral) expression of HCV entry factors (described in [43]).
2. HCV entry factor transgenic mice (described in [44, 45]).
3. HCV entry factor knockin mice (described in [46]).

2.2 BiCre-Jc1 Generation

1. pBi-nlsCre-Jc1 construct (described in [44]).
2. RNeasy mini kit (Qiagen).
3. T7 RiboMAX™ Express Large-Scale RNA production system (Promega).
4. Huh-7.5.1 cells (Chisari laboratory, The Scripps Research Institute [47]).
5. BTX ECM 830 electroporator (Harvard Apparatus).
6. Dulbecco's Modified Eagle Medium (DMEM, Thermo Fisher Scientific).
7. Stirred cell (Millipore).
8. 0.45 μm bottle-top filter (Millipore).

2.3 IVIS

1. IVIS Lumina II platform (Caliper Lifesciences).
2. D-Luciferin salt (Gold Biotechnology).
3. 0.2 μm filter (Thermo Fisher Scientific).
4. Isoflurane (TW Medical).
5. 28G × ½ insulin syringes (BD Biosciences).

2.4 Quantifying HCV RNA

1. ZR Viral RNA kit (Zymo Research).
2. Nuclease-Free Water (Luminex).
3. HCV primers for PCR (ordered from Integrated DNA Technologies).

4. Reverse Transcriptase (Luminex).
5. MultiCode ISOlution (Luminex).
6. Titanium Taq (Clontech).
7. HCV standard (homemade; RNA in vitro-transcribed from the plasmid, pIDTBlue, containing the Jc1-RNA sequence "GGCUCCAUCUUAGCCCUAGUCACGGCUAGCU-GUGAAAGGUCCGUGAGC" as cloned by Integrated DNA Technologies).
8. LightCycler 480 Real-Time PCR System (Roche Applied Sciences).
9. MultiCode Analysis Software v1.6.5 (Luminex).

3 Methods

3.1 Genetically Humanized Mice Can Support HCV Entry

Humanized mice for HCV research are generally created in one of two ways: by expression of human factors known to be necessary for part(s) of the HCV life cycle (Fig. 1) or by xenotransplantation. In murine cells, HCV replicons are able to replicate, but uptake of cell-culture produced virus or HCV pseudoparticles is not possible. Human CD81 [48], scavenger receptor class B type I (SCARB1) [49], and claudin-1 (CLDN1) [50] were all known human entry factors for HCV, but it was the additional identification of human occludin (OCLN) [51] as an essential entry factor. The combined expression of these four factors, or at a minimum human CD81 and human OCLN, finally led to observed uptake of HCV in murine cells in vitro [51]. This work was expanded in vivo to genetically humanize mice by adenoviral delivery of human OCLN and human CD81 [43, 44], allowing for HCV uptake in the liver of a fully immunocompetent animal. However, viral replication and infectious particle production were not observed in either these mice or those subsequently engineered to transgenically express human CD81 and human OCLN in the murine liver under the control of an albumin promoter [52]. Only upon blunting the innate immune response of these entry factor transgenic mice was the complete HCV life cycle observed as evidenced by viral RNA in both the serum and liver [52].

To express HCV entry factors by adenovirus, high particle numbers are delivered to mice (in this case, a Cre reporter strain of choice as described below) with the goal of widespread expression of these factors in the murine hepatocytes. cDNA encoding human CD81, SCARB1, CLDN1 or OCLN can been cloned into individual pAdEasy AV5 adenoviral vectors (Agilent) (for details *see* [43]). Alternatively, instead of expressing each factor in individual vectors, the cDNA of CD81 and SCARB1 or CLDN1 and OCLN can instead be linked by a sequence encoding the 2A self-cleavage

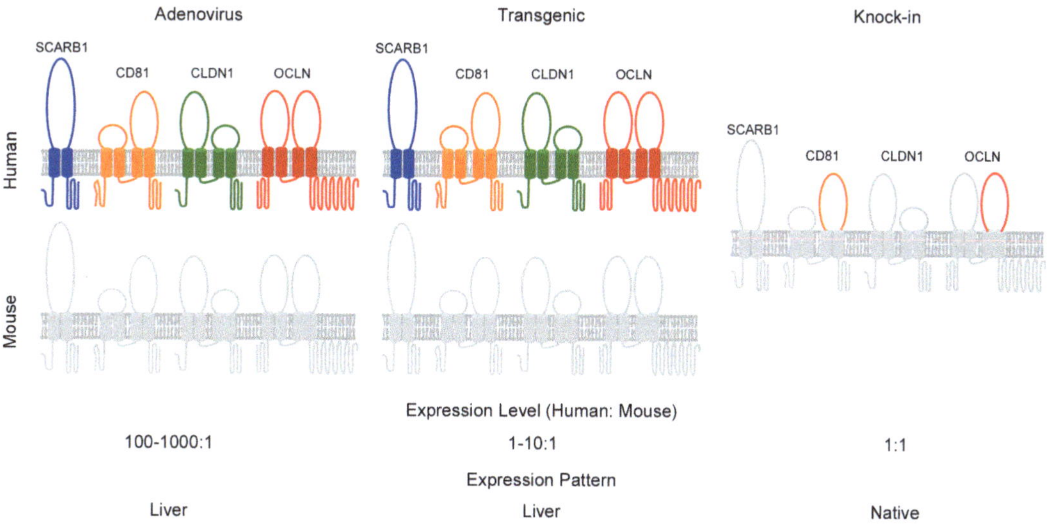

Fig. 1 Summary of genetically humanized mice supporting HCV uptake. The minimal factors required for HCV uptake in human hepatocytes are CD81, SCARB1, CLDN1, and OCLN, with human CD81 and OCLN constituting the minimal set of human factors for entry into murine hepatocytes. Thus, fully immunocompetent mice can support viral uptake by expressing these factors by adenoviral delivery or by genetic manipulation. The latter approach has been further utilized in two ways, transgenic expression of hCD81 and hOCLN under the control of an albumin promoter or knockin of humanized CD81 and OCLN where the second extracellular loops of the murine orthologs of these two proteins have been altered to facilitate HCV entry. While all these methods have proven successful, the knockin model best matches the natural tissue expression and physiological levels of CD81 and OCLN, which are expressed only in the liver and at supraphysiological levels in adenovirally transduced or transgenic mice

peptide to ensure the four factors are separate proteins following translation [53].

Although these models have been extremely useful for studying HCV infection in vivo, CD81 and OCLN are expressed only in hepatocytes and at supraphysiological levels. To overcome these drawbacks, knockin mice were recently generated that harbor CD81 and OCLN alleles minimally humanized so that the second extracellular loops of these two proteins match the relevant human sequence important for HCV entry [46]. These mice develop normally without apparent defects, express CD81 and OCLN at physiological levels across a variety of tissues, and support uptake of HCV to a similar extent as in the two models described above.

In stark contrast to the above studies in genetically humanized mice, where completion of the HCV life cycle required a dampened innate immune response, one group has successfully demonstrated persistent HCV infection in genetically humanized mice with an intact immune response. Here, human CD81 and OCLN were once more expressed under an albumin promoter but in an ICR background [45]. As genetically humanized immunocompetent C57BL/6 mice have not previously become viremic in similar experiments [44, 52], the authors point to differences in the expression of interferon-stimulated genes (ISGs) and host factors such as ApoE and miR-122 upon HCV infection as possibly responsible. While these findings are striking, genetic manipulations of ICR mice, an outbred strain, present additional technical challenges that further research will need to overcome.

3.2 Human Liver Chimeric Mice for Studying the Complete HCV Life Cycle

The second broad category of mouse humanization relevant to HCV research involves xenotransplantation to generate human liver chimeric mice (Fig. 2) with or without human immune system reconstitution. In order to accept engrafted tissue, these mice must be immunocompromised and must also have a hepatic environment conducive to the proliferation of engrafted cells over endogenous murine cells. The latter condition can be met by treating the mice with chemical agents such as tetrachloride or retrorsine, or, more typically, by changes at the genetic level. Genetic manipulation was first successfully used to induce liver failure in a mouse strain by murine urokinase plasminogen activator overexpression driven by the albumin promoter (Alb-uPA) [54]. When backcrossed with a severe combined immunodeficiency strain (SCID), the resultant SCID/Alb-uPA mice could be stably engrafted with human hepatocytes that supported HCV infection [55]. To overcome some of the disadvantages of this model, including the limited period when engraftment can be performed, a strain where the major urinary protein (MUP) promoter drives uPA expression is backcrossed with a SCID strain to form SCID/MUP-uPA mice that can undergo successful engraftment and HCV infection [56].

Knockout of the liver enzyme fumaryl acetoacetate hydrolase (FAH) has also been pursued as a method for causing selective death of endogenous murine hepatocytes. To permit actual engraftment, $FAH^{-/-}$ mice were crossed with $Rag2^{-/-}$ $IL\text{-}2R\gamma^{NULL}$ mice to create FNRG mice [57]. The absence of FAH specifically in the murine hepatocytes results in a buildup of toxic metabolites during tyrosine catabolism leaving the engrafted human hepatocytes, which still express FAH, at a growth advantage. Importantly, the

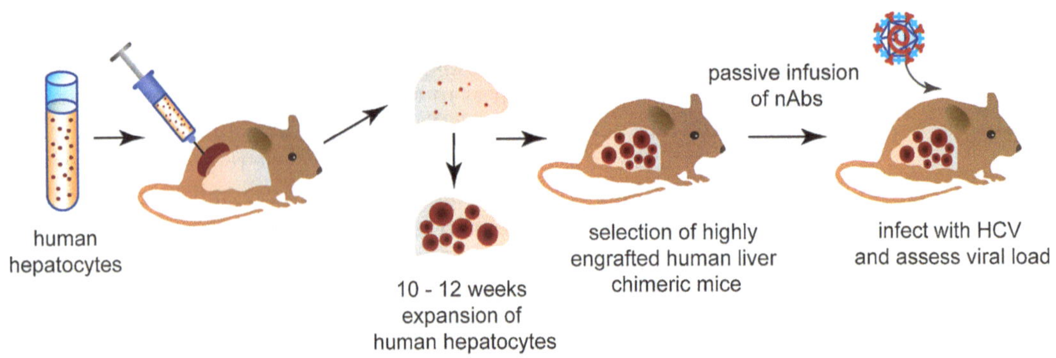

Fig. 2 The generation of human liver chimeric mice for passive immunization against HCV

FAH deficiency in murine hepatocytes can be overcome by treatment with the compound 2(2-nitro-4-trifluoromethylbenzoyl)-1,3 cyclohexanedione (NTBC). NTBC cycling thus permits "tunable" liver injury throughout the mouse's lifespan (for detailed information about generating and maintaining these mice *see* [58]). The level of engraftment corresponds to human albumin levels in the mouse's serum as determined by ELISA. Similarly, immunodeficient NOG mice expressing the transgene herpes simplex virus type 1 thymidine kinase (HSV-tk) can be treated with ganciclovir for targeted destruction of murine hepatocytes, thus allowing engrafted human hepatocytes to proliferate instead [59]. This model has also been successfully infected with HCV [60].

These numerous models permit both the stable engraftment of adult human hepatocytes in mice as well as sustained HCV infection. However, the immunocompromised nature of these mice has limited their utility for studying immune responses to HCV infection and their subsequent impact on pathogenesis. Toward this aim, dually reconstituted mice have also been generated with engraftment of both human hepatocytes and components of a human immune system [61–67]. However, widespread use of these dually engrafted mice has been limited due to variation across donors, difficulty in acquiring syngeneic human HSCs and hepatocytes, the required technical expertise, and high cost.

3.3 Assessing HCV Entry in Genetically Humanized Mice

Although HCV does not replicate in genetically humanized mice unless the innate immune response is blunted, viral entry is still supported. The ability to detect uptake of HCV in these mice is necessary to assess the efficacy of NAbs or vaccines on this step of the viral life cycle. Mouse lines containing a loxP-flanked stop cassette upstream of a marker (i.e., firefly luciferase) are readily available from commercial vendors such as the The Jackson Laboratory (Bar Harbor, see below). In the presence of Cre recombinase, the transcriptional stop cassette undergoes recombination and is effectively removed, allowing marker expression to now occur. Recombinant HCV expressing Cre recombinase can be generated (*see* Subheading 3.3.1) and injected into humanized Cre reporter mice. While below we describe the use of a genotype 2a chimera expressing Cre recombinase, this model can also be used to test HCV Cre expressing the envelope proteins of other genotypes. Upon entry of the virus, the translation of the HCV polyprotein will release the Cre recombinase, which can now move to the nucleus and facilitate recombination of the transcriptional stop cassette. In this way, hepatocytes that uptake virus will now express the marker of interest, which, in cases such as firefly luciferase, can be detected in vivo.

Examples of commonly used reporter strains available from The Jackson Laboratory (Bar Harbor ME) are FVB.129S6(B6)-*Gt(ROSA)26Sor*$^{tm1(Luc)Kael}$/J (also known as R26-LSL-FLuc) [42] and B6;129-*Gt(ROSA)26Sor*^{tm1Joe}/J (also known as R26-LSL-GNZ) [68] mice which, following activation by Cre recombinase, drive expression of firefly luciferase or a nuclear EGFP and β-galactosidase fusion protein, respectively.

3.3.1 Production of Bicistronic HCV Expressing Cre Recombinase

The sequence of the genotype 2a HCV chimera J6/JFH1 (also known as Jc1) is deposited (GenBank accession number: JF343782) and the pBi-nlsCre-Jc1 construct is available upon request upon completion of appropriate material transfer agreements.

1. Digest 10 μg of the pBi-nlsCre-Jc1 construct with XbaI to linearize the construct for in vitro transcription to produce pBi-nlsCre-Jc1 RNA. To ensure complete linearization, assess 5% volume of the digested plasmid by agarose gel electrophoresis.
2. Purify the linearized BiCre-Jc1 plasmid and use 1 μg to produce pBi-nlsCre-Jc1 RNA by in vitro transcription (T7 RiboMAX Express Large Scale RNA Production System, Promega).
3. Purify the pBi-nlsCre-Jc1 RNA using method of choice and quantify the concentration of viral RNA by NanoDrop.
4. As a quality control test, let RNA sit for 30 min on wet ice before running a few microliter on an agarose gel to check for degradation (*see* **Note 2**). If RNA is of sufficient integrity, prepare 5 μg aliquots and store at −80 °C for future use, only thawing on wet ice immediately before electroporation.
5. As all subsequent steps involve virus production, necessary biosafety precautions must be taken (*see* **Note 3**). Electroporate the pBi-nlsCre-Jc1 RNA into Huh-7.5.1 cells (DMEM containing 5% FBS) as previously described in [43]. As an additional control for electroporation efficiency, seed in a 6-well plate some cells electroporated with or without viral RNA. Collect these cells 72 h post-electroporation and assess the percentage of cells positive for NS5A by flow cytometry.
6. On the plates of electroporated cells, change the medium to DMEM without FBS 24 h post-electroporation. Collect supernatants every 6 h starting from 72 h until 144 h, always replacing the media with serum-free DMEM. Supernatants can be pooled and stored at 4 °C in a foil-covered container.
7. Filter the pooled supernatants through a 0.45 μm bottle top filter (Millipore). Store at least 1 mL of this unconcentrated virus at −80 °C to have as a control for concentration efficiency in **step 8**. Concentrate the remaining filtered supernatant using a stirred cell (Millipore). Aliquot the concentrated virus and store at −80 °C for further use.

8. Thaw an aliquot of both concentrated and unconcentrated virus to determine the viral titer ($TCID_{50}$) on Huh7.5 cells as previously described in [69].

3.3.2 Bioluminescence Imaging to Assess HCV Entry In Vivo

In vivo bioluminescence imaging allows for rapid, minimally invasive, longitudinal quantification of reporter gene activation in mice expressing at least the minimal HCV entry factors and subsequently infected with Bi-nlsCre-Jc1. Luminescence measurements generally ensure greater sensitivity through deep tissues and the skin. Here, we outline bioluminescence measurements specifically using the IVIS Lumina II platform (Caliper Life Sciences) of mice following adenoviral delivery of the four canonical human entry factors and subsequent infection with HCV Cre (equivalent imaging platforms from other manufacturers could conceivably be used). The IVIS Spectrum imager expresses the bioluminescent signal in photons per second and displays it as an intensity map. The luminescence, which is the consequence of the photon flux emitted by the luciferase-expressing cells, can be measured using a region of interest (ROI) tool.

1. Prepare stock solution of D-Luciferin at 15 mg/mL in DPBS (no calcium, no magnesium). Filter-sterilize through a 0.2 μm filter, aliquot, and store at −20 °C for future use.
2. Inject intravenously Rosa26-LSL-Fluc mice with an equal mixture of adenovirus encoding all four HCV human entry factors (10^{11} adenoviral particles per entry factor).
3. 24 h following adenoviral entry factor delivery, inject mice intravenously with 2×10^7 $TCID_{50}$ Bi-nlsCre-Jc1.
4. 72 h following injection of Bi-nlsCre-Jc1, prepare the animal for in vivo bioluminescence imaging. Anaesthetize the animal using an isoflurane vaporizer. Inject intraperitoneally appropriate volume (typically ~200 μL) of sterile 15 mg/mL D-luciferin stock solution (*see* **step 1**) so that each animal receives 150 mg D-Luciferin/kg body weight (equivalent to 10 μL of D-luciferin stock solution per gram of body weight). Place mouse in dorsal recumbency (abdomen facing up) inside the camera box of the IVIS Spectrum imager.
5. Measure the bioluminescence signal using the IVIS Lumina II in vivo bioimaging system. Due to in vivo uptake of D-Luciferin over time, capture sequential images of the mice every 5 min until luminescence saturation is reached.

3.4 Quantifying HCV RNA in Human Liver Chimeric Mouse Serum

In the case of human liver chimeric mice, reporter genomes are not necessary since HCV can robustly replicate to levels detectable in both the liver and serum of a well-engrafted animal. Thus full-length genomes such as JFH1 (genotype 2a) can be used for infections. HCV replication over time is easily tracked from relatively

Table 1
Reaction conditions for HCV RNA RT-qPCR

Reagents	Volume (μL)
Nuclease-free H_2O	Up to 25
5× MultiCode ISOlution	5
MMLV reverse transcriptase	0.5
Primer mix (5 μM forward + 5 μM reverse in nuclease-free-H_2O)	1
RNA	x
Titanium Taq	0.5

Table 2
Cycling conditions for quantifying HCV RNA in serum

Procedure	Temperature, °C	Time
Reverse transcription	50	5 min
Hot start	95	3 min
50 cycles	95 58 72	5 s 10 s[a] 20 s
Melt Curve	95 60 95	15 s 1 min 15 s[a]

[a]Data collection must be turned on at these steps for proper analysis in **step 8**

small blood volumes obtained by submandibular bleedings. Additionally, viral load can be assessed in the mouse's liver but requires perfusion of the organ as described in [58] and subsequent RNA isolation from a kit compatible for use with animal tissues. Once the RNA has been isolated, the same procedure detailed below for RT-qPCR of HCV RNA isolated from animal serum can be used. The method given below is one approach for viral RNA quantification specifically using EraGen MultiCode-RTx Technology (EraGen Biosciences).

1. Draw approximately 200 μL of whole blood by submandibular bleeding into a 1.5 mL microcentrifuge tube.
2. Allow the blood to sit upright untouched for 10–20 min at room temperature for clotting.

3. Centrifuge the blood for 10 min, 3000 rpm (956 × *g*), 4 °C. Following centrifugation, there should clearly be two layers in the tube. Transfer the upper layer (serum) to a fresh 1.5 mL microcentrifuge tube (*see* **Note 4**).
4. Purify RNA from the plasma using the ZR Viral RNA kit (Zymo Research) or an equivalent kit.
5. Quantify the viral RNA copy number by one step RT-qPCR using EraGen MultiCode-RTx Technology (EraGen Biosciences). The following primers were used for the detection of HCV RNA: 5′-GCTCACGGACCTTTCA-3′ (sense) and 5′-GGCTCCATCTTAGCCC-3′ (antisense). For use with EraGen MultiCode-RTx Technology, the sense primer requires the following modifications, which can be specified upon ordering from oligonucleotide providers such as Integrated DNA Technologies at the 5′ end: 6-FAM and int 5-Methyl dC (iMe-dC).
6. Prepare the reaction for *each* sample as in Table 1, making sure to also prepare serial dilutions of an HCV standard to generate a standard curve.
7. Run the reactions on a LightCycler 480 Real-Time PCR System (or equivalent device) using the thermal profile in Table 2.
8. Import the multicomponent data from the run into the MultiCode Analysis Software v1.6.5 for analysis.

3.5 Humanized Mice for Designing and Testing HCV Vaccines and Therapeutic Antibodies

As genetically humanized mice are still immunocompetent and do not support the full HCV life cycle, vaccine and therapeutic antibody studies utilizing these models have largely focused on whether viral uptake can be reduced. In contrast, immunodeficient human liver chimeric mice have primarily been used for passive immunization studies in which animals are infused with antibodies directed against viral antigen(s). Both prophylactic and therapeutic studies have been performed in human liver chimeric mice challenged with HCVcc or patient-derived sera. The antibodies used for passive immunization, as described in several examples below (summarized in Table 3), have been isolated from patient sera or from immunocompetent mice injected with HCV or viral components like the envelope proteins E1 and/or E2.

3.5.1 Active Vaccination

As described above, genetically humanized mice remain immunocompetent and can thus be utilized for active immunization studies. One such approach for active immunization utilizes recombinant vaccinia virus (rVV) as a vector to deliver viral genes. Previous work showed that large pieces of pathogen-derived DNA could be easily inserted into the VV genome without disrupting viral replication [70, 71]. rVVs expressing different HCV genes have been tested in

Table 3
Summary of HCV immunization studies performed in humanized mice

	Expression approach	Virus	Immunization
Genetically humanized mice	Adenovirus	HCV-Cre expressing genotypes 1b, 2a, or 4a structural proteins [43]	Active
		In vitro neutralization assay with chimeric HCVcc panel covering all seven HCV genotypes; BiCre-Jc1 (genotype 2a) in vivo [52]	Active
		HCV-Cre recombinase expressing structural proteins of Con1 (genotype 1b) or Jc1 (genotype 2a) [80]	Passive
		BiCre-Jc1 (genotype 2a) [102]	Passive
	Mouse	**Virus**	**Immunization**
Human liver chimeric mice	uPA/SCID	Heterologous HCV inoculum from an HCV genotype 1a infected patient [94]	Passive
		Patient-derived genotype 1b [96]	Passive
		H77C (genotype 1a), ED43 (genotype 4) [104])	Passive
		Prevention experiments: genotype 1b or genotype 4 HCV-infected serum; treatment experiments: Jc1 (genotype 2a), VL-JFH1 (genotype 1b/2a chimera), genotype 2a serum, genotype 4 serum [106]	Passive
		J6/JFH1 (genotype 2a) [107]	Passive
		Mouse-passaged H77C (genotype 1a) [108]	Passive
	FNRG	H77 (genotype 1a), J6/JFH1 (genotype 2a) [102]	Passive
	PXB-mouse	H77S (genotype 1a) [103]	Passive
	(cDNA-uPA/SCID)	Mouse-passaged HCR6 (genotype 1b) [105]	Passive

chimpanzees [72–74] and mice [75, 76]. In genetically humanized mice expressing human CD81 and human OCLN, injection with a rVV expressing HCV genotype 1a C-E1-E2-p7-NS2 resulted in robust antibody production against HCV E2 and reduced viral entry following subsequent challenge with different HCV genotypes [44]. An alternative to injecting a viral vector to express viral proteins is to inject the viral protein itself. Previous efforts to cost-effectively produce large-scale amounts of viral protein for clinical applications and the poor results of recombinant vaccine trials in humans had made this less appealing. However, a soluble form of E2 with a truncated transmembrane domain (sE2) was recently generated that induced NAbs in immunocompetent mice with broad, pan-genotypic activity in vitro and could be produced to scale in insect cells [53]. Furthermore, genetically humanized mice were protected against viral challenge following sE2 immunization.

Active immunization studies in mice, as in other species, can determine the type and magnitude of antibody response generated as well as the potential protection provided against subsequent viral challenge. In vitro neutralization assays can be performed using the sera of mice, which do not have to be genetically humanized, collected multiple weeks after immunization to insure sufficient time for antibody production. Sera can then be incubated with JFH1 chimeras expressing the structural proteins from multiple HCV genotypes to test for their range of neutralization. Additionally, immunized mice can be subsequently challenged with HCV in vivo. Here immunization must be performed in one of the reporter mouse strains described above and an appropriate HCV genome used (i.e., BiCre-Jc1) in order to have a readout of viral uptake since detectable replication levels in the liver and/or serum are negligible in these immunocompetent models.

3.5.2 Neutralizing Antibodies

Many of the efforts to identify NAbs capable of neutralizing different HCV genotypes predate the use of humanized mouse models. In patients, there has been widespread debate about how NAbs may affect HCV replication and the nature of their contribution to viral clearance in a given individual (reviewed in [77]). Thus, better understanding NAbs has been of great interest in designing both prophylactic and therapeutic HCV treatments. With as much as 35% difference between the sequences of HCV genotypes [78], the so-called broadly NAbs, able to act against multiple genotypes or even the variety of quasispecies present in a single infected individual, are of particular interest. This sequence diversity also underscores the need for HCV treatments with broad, pangenotypic activity to lower the risk of viral escape mutants. Prior reliance on in vitro systems has only added to the controversy concerning NAbs in patients. Thus, the advent of humanized mouse research has provided an additional platform for exploration of NAbs in the context of HCV infection.

Repeated work has shown that HCV NAbs primarily target the mediators of viral entry, the HCV envelope proteins E1 and E2 [79–87]. Thus, many efforts to induce a strong NAb response have focused on using recombinant HCV protein(s), primarily E1 and/or E2, to generate vaccines. Prior to the availability of humanized mice in the HCV field, such vaccines were tested in both nonhumanized rodent models [88–90] and humans [91]. The failure of the human recombinant vaccine trials to induce a robust NAb response [91–94] and the rise in humanized mouse models for HCV research has heightened interest in both generating better recombinant vaccines and understanding NAbs.

Many promising NAbs previously identified in other platforms have been re-examined in human liver chimeric mice. One such study found candidates able to bind HCV E2 following an extensive search of a phage-display library of antibody antigen-binding fragments (Fabs) derived from a chronically infected HCV patient. Select Fabs were used to generate recombinant monoclonal antibodies (MAbs) that were further tested in vitro. The most promising candidates from these assays specifically bound antigenic region 3 (AR3A and AR3B) of the HCV envelope glycoprotein E2, and injection of these candidates into human liver chimeric (Alb-uPA/SCID) mice was able to protect some from infection with a heterologous HCV inoculum from an HCV genotype 1a infected patient [95]. Prophylactic administration of high concentrations (200 mg/kg) of antibody AR3A or AR4A was also found to protect genetically humanized mice against challenge with genotype 1b or genotype 2a cell-culture produced HCV (HCVcc) [81]. AP33 [87, 96] and 3/11 [80], which were identified prior to the advent of humanized mice in rodents immunized with E1 and/or E2, were also tested for their ability to protect against challenge with patient-derived HCV (genotype 1b) in Alb-uPA/SCID mice [97]. Although AP33 and 3/11 recognize overlapping epitopes in E2, only AP33 (at 100 mg/kg) demonstrated the ability to protect the humanized mice from infection. In line with this work and the growing use of humanized mice for such studies, efforts have also been made to better characterize the antigenic regions of full-length and soluble E1E2 and the antibody responses they elicit specifically in mice [98]. Furthermore, vectored immunoprophylaxis, whereby adenoviruses are used to deliver NAbs, has been an approach for mouse studies of pathogens such as HIV, *Plasmodium falciparum*, and influenza [99–102]. This method has also been performed in genetically humanized mice expressing the human HCV entry factors as well as in liver chimeric FNRG mice [103]. Whether supplied individually or in combination, the NAbs tested were able to block HCV entry in the genetically humanized mice and to prevent viremia in liver chimeric FNRG mice following viral challenge. Furthermore, in this same study, liver chimeric mice already infected with J6/JFH (genotype 2a) were therapeutically treated with NAbs, leading to a drop in HCV RNA levels below the limit of detection in the sera.

3.5.3 Host Factors as Antibody Targets

While the majority of work has focused on antibodies directed against viral targets, some studies have tested the potential for antibodies against host factors as a short-term treatment for patients undergoing liver transplantation. Directed against a necessary HCV entry factor, anti-CD81 mAbs (K04 [104] and JS81 [105]) have been tested for both their prophylactic and therapeutic abilities in human liver chimeric mice. With frequent dosing, K04 treatment started post-HCV inoculation could substantially abrogate infection. Prophylactic use of either antibody did effectively protect from infection, but a high dose of Ab was required and effects on mouse health were observed.

Anti-CLDN1 mAbs have also proven successful as prophylactic treatment against HCV infection (mouse-passaged genotype 1b, 10^4 RNA copies) without any significant side effects in human liver chimeric mice [106]. A similar study demonstrated prophylactic treatment of human liver chimeric mice with a different anti-CLDN1 mAb followed by a "booster" 1 and 5 days post-infection resulted in undetectable HCV RNA in the serum for six weeks following infection [107]. Even as the concentration of antibody diminished in the sera of these mice, infection with either genotype 1b or genotype 4 sera were prevented. Therapeutic administration of anti-CLDN1 mAb was also successful in cohorts of human liver chimeric mice chronically infected with one of a variety of HCV genotypes (HCVcc Jc1; genotype 2a serum; neutralization escape variant HCVcc VL-JFH1; genotype 4). Importantly, this CLDN1-specific mAb displayed no major toxicity or adverse effects in the mice.

3.5.4 IgG as a Prophylactic

UV-inactivated HCVcc (J6/JFH1) has also been used immunize immunocompetent BALB/c mice (four separate injections at two week intervals), resulting in production of anti-E1 and anti-E2 antibodies [108]. Purified IgG from the sera of these mice was subsequently used as a prophylactic vaccine in Alb-uPA/SCID mice 1 week before infection with HCVcc (J6/JFH1). Infection was effectively inhibited at low doses (10^3 HCV RNA copies), but not at higher challenges of 10^4 or 10^5 RNA copies. Broader approaches have also been taken using anti-HCV immunoglobulin G (IgG) obtained from a chronically infected patient in Alb-uPA-SCID mice to assess its ability to protect against challenge with a genotype 1a clone derived from the acute stage of infection in the same patient [109]. Passive immunization with the patient IgG was able to confer protection against viral challenge, albeit with an extremely homogeneous inoculum.

3.6 Conclusions and Future Outlook

In this chapter, we described the numerous applications of humanized mice for assessing potential HCV vaccines and therapeutic antibodies. While this work has continued to push the field forward, several limitations of these models remain. Studies in immunocompetent, genetically humanized mice focus on humoral immunity in the context of an HCV infection that is restricted to viral uptake. Genetic humanization allows for a relatively limited number of human factors to be expressed. The ever-lengthening list of HCV entry factors thus present in an actual patient may provide additional challenges that the mouse model cannot fully recapitulate. For human liver chimeric mice, which cannot amount any kind of adaptive immune response due to their immunocompromised status, conclusions are also limited. However, efforts to combine the benefits of human liver chimeric mice—i.e., studying HCV infection past entry into hepatocytes—with the presence of a humanized immune system are ongoing. This work will aid in filling the gaps of current models and create a more accurate picture of the complex relationship between HCV and the host immune response. Furthermore, improved dually reconstituted mice will greatly strengthen our knowledge of immunopathogenesis and the downstream liver disease observed in chronically infected patients

4 Notes

1. Researchers must receive approval from their Institutional Animal Care and Use Committee (IACUC), Institutional Review Board (IRB), Institutional Biosafety Committee (IBC), or the equivalent institutions before proceeding with work involving humanized mice and/or human pathogens.
2. When running the in vitro-transcribed RNA on a gel, it is highly recommended to also run a 1 kb ladder. If the RNA is of good quality, a bright, single band should appear at around 3 kb. Beware of using RNA with excessive smearing as this is likely due to degradation and will result in decreased efficiency of subsequent steps.
3. All virus work should be performed in a BSL safety level 2 hood. Following electroporation, all cuvettes and tips that came into contact with electroporated cells need to soak in Vesphene for at least 20 min before disposal.
4. Around 200 μL of whole blood should yield at least 50 μL of plasma, which is a sufficient volume for the protocol described here.

Acknowledgments

This study is supported by grants from the National Institutes of Health (R01 AI079031, R01 AI107301, R21AI117213 to A.P.), a Research Scholar Award from the American Cancer Society (RSG-15-048-01-MPC to A.P.), a Burroughs Wellcome Fund Award for Investigators in Pathogenesis (to A.P.) and funds from Princeton University. J.M.G. was in part supported by cofunding from NIAID on iNRSA 5T32GM00738 and Q.D. by a postdoctoral fellowship from the New Jersey Commission on Cancer Research.

References

1. Alter HJ, Holland PV, Morrow AG, Purcell RH, Feinstone SM, Moritsugu Y (1975) Clinical and serological analysis of transfusion-associated hepatitis. Lancet 2:838–841
2. Feinstone SM, Kapikian AZ, Purcell RH, Alter HJ, Holland PV (1975) Transfusion-associated hepatitis not due to viral hepatitis type A or B. N Engl J Med 292:767–770
3. Alter HJ, Purcell RH, Holland PV, Popper H (1978) Transmissible agent in non-A, non-B hepatitis. Lancet 1:459–463

4. Bradley DW, Cook EH, Maynard JE, McCaustland KA, Ebert JW, Dolana GH et al (1979) Experimental infection of chimpanzees with antihemophilic (factor VIII) materials: recovery of virus-like particles associated with non-A, non-B hepatitis. J Med Virol 3:253–269
5. Hollinger FB, Gitnick GL, Aach RD, Szmuness W, Mosley JW, Stevens CE et al (1978) Non-A, non-B hepatitis transmission in chimpanzees: a project of the transfusion-transmitted viruses study group. Intervirology 10:60–68
6. Tabor E, Gerety RJ, Drucker JA, Seeff LB, Hoofnagle JH, Jackson DR et al (1978) Transmission of non-A, non-B hepatitis from man to chimpanzee. Lancet 1:463–466
7. Choo QL, Kuo G, Weiner AJ, Overby LR, Bradley DW, Houghton M (1989) Isolation of a cDNA clone derived from a blood-borne non-A, non-B viral hepatitis genome. Science 244:359–362
8. Kuo G, Choo QL, Alter HJ, Gitnick GL, Redeker AG, Purcell RH et al (1989) An assay for circulating antibodies to a major etiologic virus of human non-A, non-B hepatitis. Science 244:362–364
9. Polaris Observatory HCV Collaborators (2017) Global prevalence and genotype distribution of hepatitis C virus infection in 2015: a modelling study. Lancet Gastroenterol Hepatol 2(3):161–176
10. Thomson EC, Fleming VM, Main J, Klenerman P, Weber J, Eliahoo J et al (2011) Predicting spontaneous clearance of acute hepatitis C virus in a large cohort of HIV-1-infected men. Gut 60:837–845
11. Gerlach JT, Diepolder HM, Zachoval R, Gruener NH, Jung MC, Ulsenheimer A et al (2003) Acute hepatitis C: high rate of both spontaneous and treatment-induced viral clearance. Gastroenterology 125:80–88
12. WHO (2016) Guidelines for the screening care and treatment of persons with chronic hepatitis C infection: updated version. WHO guidelines approved by the guidelines review committee. WHO, Geneva
13. Smith BD, Morgan RL, Beckett GA, Falck-Ytter Y, Holtzman D, Teo CG et al (2012) Recommendations for the identification of chronic hepatitis C virus infection among persons born during 1945–1965. MMWR Recomm Rep 61:1–32
14. Edlin BR (2011) Perspective: test and treat this silent killer. Nature 474:S18–S19
15. Zibbell JE, Iqbal K, Patel RC, Suryaprasad A, Sanders KJ, Moore-Moravian L et al (2015) Increases in hepatitis C virus infection related to injection drug use among persons aged </=30 years – Kentucky, Tennessee, Virginia, and West Virginia, 2006–2012. MMWR Morb Mortal Wkly Rep 64:453–458
16. Nelson PK, Mathers BM, Cowie B, Hagan H, Des Jarlais D, Horyniak D et al (2011) Global epidemiology of hepatitis B and hepatitis C in people who inject drugs: results of systematic reviews. Lancet 378:571–583
17. Magiorkinis G, Sypsa V, Magiorkinis E, Paraskevis D, Katsoulidou A, Belshaw R et al (2013) Integrating phylodynamics and epidemiology to estimate transmission diversity in viral epidemics. PLoS Comput Biol 9: e1002876
18. Callaway E (2014) Hepatitis C drugs not reaching poor. Nature 508:295–296
19. Trooskin SB, Reynolds H, Kostman JR (2015) Access to costly new hepatitis C drugs: medicine, money, and advocacy. Clin Infect Dis 61:1825–1830
20. Beames B, Chavez D, Lanford RE (2001) GB virus B as a model for hepatitis C virus. ILAR J 42:152–160
21. Schaluder GG, Dawson GJ, Simons JN, Pilot-Matias TJ, Gutierrez RA, Heynen CA et al (1995) Molecular and serologic analysis in the transmission of the GB hepatitis agents. J Med Virol 46:81–90
22. Simons JN, Pilot-Matias TJ, Leary TP, Dawson GJ, Desai SM, Schlauder GG et al (1995) Identification of two flavivirus-like genomes in the GB hepatitis agent. Proc Natl Acad Sci U S A 92:3401–3405
23. Bukh J, Apgar CL, Yanagi M (1999) Toward a surrogate model for hepatitis C virus: an infectious molecular clone of the GB virus-B hepatitis agent. Virology 262:470–478
24. Lanford RE, Chavez D, Notvall L, Brasky KM (2003) Comparison of tamarins and marmosets as hosts for GBV-B infections and the effect of immunosuppression on duration of viremia. Virology 311:72–80
25. Bukh J, Apgar CL, Govindarajan S, Purcell RH (2001) Host range studies of GB virus-B hepatitis agent, the closest relative of hepatitis C virus, in New World monkeys and chimpanzees. J Med Virol 65:694–697
26. Takikawa S, Engle RE, Faulk KN, Emerson SU, Purcell RH, Bukh J (2010) Molecular evolution of GB virus B hepatitis virus during acute resolving and persistent infections in experimentally infected tamarins. J Gen Virol 91:727–733
27. Li T, Zhu S, Shuai L, Xu Y, Yin S, Bian Y et al (2014) Infection of common marmosets with

hepatitis C virus/GB virus-B chimeras. Hepatology 59:789–802

28. Zhu S, Li T, Liu B, Xu Y, Sun Y, Wang Y et al (2016) Infection of common marmosets with GB virus B chimeric virus encoding the major nonstructural proteins NS2 to NS4A of hepatitis C virus. J Virol 90:8198–8211
29. Burbelo PD, Dubovi EJ, Simmonds P, Medina JL, Henriquez JA, Mishra N et al (2012) Serology-enabled discovery of genetically diverse hepaciviruses in a new host. J Virol 86:6171–6178
30. Pfaender S, Cavalleri JM, Walter S, Doerrbecker J, Campana B, Brown RJ et al (2015) Clinical course of infection and viral tissue tropism of hepatitis C virus-like nonprimate hepaciviruses in horses. Hepatology 61:447–459
31. Quan PL, Firth C, Conte JM, Williams SH, Zambrana-Torrelio CM, Anthony SJ et al (2013) Bats are a major natural reservoir for hepaciviruses and pegiviruses. Proc Natl Acad Sci U S A 110:8194–8199
32. Corman VM, Grundhoff A, Baechlein C, Fischer N, Gmyl A, Wollny R et al (2015) Highly divergent hepaciviruses from African cattle. J Virol 89:5876–5882
33. Baechlein C, Fischer N, Grundhoff A, Alawi M, Indenbirken D, Postel A et al (2015) Identification of a novel hepacivirus in domestic cattle from Germany. J Virol 89:7007–7015
34. Lauck M, Sibley SD, Lara J, Purdy MA, Khudyakov Y, Hyeroba D et al (2013) A novel hepacivirus with an unusually long and intrinsically disordered NS5A protein in a wild Old World primate. J Virol 87:8971–8981
35. Li CX, Shi M, Tian JH, Lin XD, Kang YJ, Chen LJ et al (2015) Unprecedented genomic diversity of RNA viruses in arthropods reveals the ancestry of negative-sense RNA viruses. Elife 4:PMID:25633976
36. Drexler JF, Corman VM, Muller MA, Lukashev AN, Gmyl A, Coutard B et al (2013) Evidence for novel hepaciviruses in rodents. PLoS Pathog 9:e1003438
37. Kapoor A, Simmonds P, Scheel TK, Hjelle B, Cullen JM, Burbelo PD et al (2013) Identification of rodent homologs of hepatitis C virus and pegiviruses. MBio 4:e00216–e00213
38. Firth C, Bhat M, Firth MA, Williams SH, Frye MJ, Simmonds P et al (2014) Detection of zoonotic pathogens and characterization of novel viruses carried by commensal Rattus norvegicus in New York City. MBio 5: e01933–e01914
39. Billerbeck E, Wolfisberg R, Fahnoe U, Xiao JW, Quirk C, Luna JM et al (2017) Mouse models of acute and chronic hepacivirus infection. Science 357:204–208
40. Bitzegeio J, Bankwitz D, Hueging K, Haid S, Brohm C, Zeisel MB et al (2010) Adaptation of hepatitis C virus to mouse CD81 permits infection of mouse cells in the absence of human entry factors. PLoS Pathog 6: e1000978
41. von Schaewen M, Dorner M, Hueging K, Foquet L, Gerges S, Hrebikova G et al (2016) Expanding the host range of hepatitis C virus through viral adaptation. MBio 7: e01915-16
42. Safran M, Kim WY, Kung AL, Horner JW, DePinho RA, Kaelin WG Jr (2003) Mouse reporter strain for noninvasive bioluminescent imaging of cells that have undergone Cre-mediated recombination. Mol Imaging 2:297–302
43. Dorner M, Rice CM, Ploss A (2013) Study of hepatitis C virus entry in genetically humanized mice. Methods 59:249–257
44. Dorner M, Horwitz JA, Robbins JB, Barry WT, Feng Q, Mu K et al (2011) A genetically humanized mouse model for hepatitis C virus infection. Nature 474:208–211
45. Chen J, Zhao Y, Zhang C, Chen H, Feng J, Chi X et al (2014) Persistent hepatitis C virus infections and hepatopathological manifestations in immune-competent humanized mice. Cell Res 24:1050–1066
46. Ding Q, von Schaewen M, Hrebikova G, Heller B, Sandmann L, Plaas M et al (2017) Mice expressing minimally humanized CD81 and occludin genes support hepatitis C virus uptake in vivo. J Virol 91(4):e01799-16
47. Zhong J, Gastaminza P, Cheng G, Kapadia S, Kato T, Burton DR et al (2005) Robust hepatitis C virus infection in vitro. Proc Natl Acad Sci U S A 102:9294–9299
48. Pileri P, Uematsu Y, Campagnoli S, Galli G, Falugi F, Petracca R et al (1998) Binding of hepatitis C virus to CD81. Science 282:938–941
49. Scarselli E, Ansuini H, Cerino R, Roccasecca RM, Acali S, Filocamo G et al (2002) The human scavenger receptor class B type I is a novel candidate receptor for the hepatitis C virus. EMBO J 21:5017–5025
50. Evans MJ, von Hahn T, Tscherne DM, Syder AJ, Panis M, Wolk B et al (2007) Claudin-1 is a hepatitis C virus co-receptor required for a late step in entry. Nature 446:801–805
51. Ploss A, Evans MJ, Gaysinskaya VA, Panis M, You H, de Jong YP et al (2009) Human occludin is a hepatitis C virus entry factor

required for infection of mouse cells. Nature 457:882–886
52. Dorner M, Horwitz JA, Donovan BM, Labitt RN, Budell WC, Friling T et al (2013) Completion of the entire hepatitis C virus life cycle in genetically humanized mice. Nature 501:237–241
53. Li D, von Schaewen M, Wang X, Tao W, Zhang Y, Li L et al (2016) Altered glycosylation patterns increase immunogenicity of a subunit hepatitis C virus vaccine, inducing neutralizing antibodies which confer protection in mice. J Virol 90:10486–10498
54. Sandgren EP, Palmiter RD, Heckel JL, Daugherty CC, Brinster RL, Degen JL (1991) Complete hepatic regeneration after somatic deletion of an albumin-plasminogen activator transgene. Cell 66:245–256
55. Mercer DF, Schiller DE, Elliott JF, Douglas DN, Hao C, Rinfret A et al (2001) Hepatitis C virus replication in mice with chimeric human livers. Nat Med 7:927–933
56. Tesfaye A, Stift J, Maric D, Cui Q, Dienes HP, Feinstone SM (2013) Chimeric mouse model for the infection of hepatitis B and C viruses. PLoS One 8:e77298
57. Azuma H, Paulk N, Ranade A, Dorrell C, Al-Dhalimy M, Ellis E et al (2007) Robust expansion of human hepatocytes in Fah−/−/Rag2−/−/Il2rg−/− mice. Nat Biotechnol 25:903–910
58. von Schaewen M, Hrebikova G, Ploss A (2016) Generation of human liver chimeric mice for the study of human hepatotropic pathogens. Methods Mol Biol 1438:79–101
59. Hasegawa M, Kawai K, Mitsui T, Taniguchi K, Monnai M, Wakui M et al (2011) The reconstituted 'humanized liver' in TK-NOG mice is mature and functional. Biochem Biophys Res Commun 405:405–410
60. Kosaka K, Hiraga N, Imamura M, Yoshimi S, Murakami E, Nakahara T et al (2013) A novel TK-NOG based humanized mouse model for the study of HBV and HCV infections. Biochem Biophys Res Commun 441:230–235
61. Bility MT, Zhang L, Washburn ML, Curtis TA, Kovalev GI, Su L (2012) Generation of a humanized mouse model with both human immune system and liver cells to model hepatitis C virus infection and liver immunopathogenesis. Nat Protoc 7:1608–1617
62. Bility MT, Curtis A, Su L (2014) A chimeric mouse model to study immunopathogenesis of HCV infection. Methods Mol Biol 1213:379–388
63. Wilson EM, Bial J, Tarlow B, Bial G, Jensen B, Greiner DL et al (2014) Extensive double humanization of both liver and hematopoiesis in FRGN mice. Stem Cell Res 13:404–412
64. Gutti TL, Knibbe JS, Makarov E, Zhang J, Yannam GR, Gorantla S et al (2014) Human hepatocytes and hematolymphoid dual reconstitution in treosulfan-conditioned uPA-NOG mice. Am J Pathol 184:101–109
65. Strick-Marchand H, Dusseaux M, Darche S, Huntington ND, Legrand N, Masse-Ranson G et al (2015) A novel mouse model for stable engraftment of a human immune system and human hepatocytes. PLoS One 10:e0119820
66. Billerbeck E, Mommersteeg MC, Shlomai A, Xiao JW, Andrus L, Bhatta A et al (2016) Humanized mice efficiently engrafted with fetal hepatoblasts and syngeneic immune cells develop human monocytes and NK cells. J Hepatol 65:334–343
67. Washburn ML, Bility MT, Zhang L, Kovalev GI, Buntzman A, Frelinger JA et al (2011) A humanized mouse model to study hepatitis C virus infection, immune response, and liver disease. Gastroenterology 140:1334–1344
68. Stoller JZ, Degenhardt KR, Huang L, Zhou DD, Lu MM, Epstein JA (2008) Cre reporter mouse expressing a nuclear localized fusion of GFP and beta-galactosidase reveals new derivatives of Pax3-expressing precursors. Genesis 46:200–204
69. Lindenbach BD, Evans MJ, Syder AJ, Wolk B, Tellinghuisen TL, Liu CC et al (2005) Complete replication of hepatitis C virus in cell culture. Science 309:623–626
70. Perkus ME, Limbach K, Paoletti E (1989) Cloning and expression of foreign genes in vaccinia virus, using a host range selection system. J Virol 63:3829–3836
71. Perkus ME, Piccini A, Lipinskas BR, Paoletti E (1985) Recombinant vaccinia virus: immunization against multiple pathogens. Science 229:981–984
72. Youn JW, Hu YW, Tricoche N, Pfahler W, Shata MT, Dreux M et al (2008) Evidence for protection against chronic hepatitis C virus infection in chimpanzees by immunization with replicating recombinant vaccinia virus. J Virol 82:10896–10905
73. Ralston R, Thudium K, Berger K, Kuo C, Gervase B, Hall J et al (1993) Characterization of hepatitis C virus envelope glycoprotein complexes expressed by recombinant vaccinia viruses. J Virol 67:6753–6761
74. Rollier CS, Paranhos-Baccala G, Verschoor EJ, Verstrepen BE, Drexhage JA, Fagrouch Z et al (2007) Vaccine-induced early control of hepatitis C virus infection in chimpanzees

fails to impact on hepatic PD-1 and chronicity. Hepatology 45:602–613

75. Gomez CE, Perdiguero B, Cepeda MV, Mingorance L, Garcia-Arriaza J, Vandermeeren A et al (2013) High, broad, polyfunctional, and durable T cell immune responses induced in mice by a novel hepatitis C virus (HCV) vaccine candidate (MVA-HCV) based on modified vaccinia virus Ankara expressing the nearly full-length HCV genome. J Virol 87:7282–7300
76. Abraham JD, Himoudi N, Kien F, Berland JL, Codran A, Bartosch B et al (2004) Comparative immunogenicity analysis of modified vaccinia Ankara vectors expressing native or modified forms of hepatitis C virus E1 and E2 glycoproteins. Vaccine 22:3917–3928
77. Ball JK, Tarr AW, McKeating JA (2014) The past, present and future of neutralizing antibodies for hepatitis C virus. Antivir Res 105:100–111
78. Kuiken C, Simmonds P (2009) Nomenclature and numbering of the hepatitis C virus. Methods Mol Biol 510:33–53
79. Broering TJ, Garrity KA, Boatright NK, Sloan SE, Sandor F, Thomas WD Jr et al (2009) Identification and characterization of broadly neutralizing human monoclonal antibodies directed against the E2 envelope glycoprotein of hepatitis C virus. J Virol 83:12473–12482
80. Flint M, Maidens C, Loomis-Price LD, Shotton C, Dubuisson J, Monk P et al (1999) Characterization of hepatitis C virus E2 glycoprotein interaction with a putative cellular receptor, CD81. J Virol 73:6235–6244
81. Giang E, Dorner M, Prentoe JC, Dreux M, Evans MJ, Bukh J et al (2012) Human broadly neutralizing antibodies to the envelope glycoprotein complex of hepatitis C virus. Proc Natl Acad Sci U S A 109:6205–6210
82. Johansson DX, Voisset C, Tarr AW, Aung M, Ball JK, Dubuisson J et al (2007) Human combinatorial libraries yield rare antibodies that broadly neutralize hepatitis C virus. Proc Natl Acad Sci U S A 104:16269–16274
83. Keck ZY, Li TK, Xia J, Gal-Tanamy M, Olson O, Li SH et al (2008) Definition of a conserved immunodominant domain on hepatitis C virus E2 glycoprotein by neutralizing human monoclonal antibodies. J Virol 82:6061–6066
84. Keck ZY, Xia J, Wang Y, Wang W, Krey T, Prentoe J et al (2012) Human monoclonal antibodies to a novel cluster of conformational epitopes on HCV E2 with resistance to neutralization escape in a genotype 2a isolate. PLoS Pathog 8:e1002653
85. Schofield DJ, Bartosch B, Shimizu YK, Allander T, Alter HJ, Emerson SU et al (2005) Human monoclonal antibodies that react with the E2 glycoprotein of hepatitis C virus and possess neutralizing activity. Hepatology 42:1055–1062
86. Meunier JC, Russell RS, Goossens V, Priem S, Walter H, Depla E et al (2008) Isolation and characterization of broadly neutralizing human monoclonal antibodies to the e1 glycoprotein of hepatitis C virus. J Virol 82:966–973
87. Owsianka A, Tarr AW, Juttla VS, Lavillette D, Bartosch B, Cosset FL et al (2005) Monoclonal antibody AP33 defines a broadly neutralizing epitope on the hepatitis C virus E2 envelope glycoprotein. J Virol 79:11095–11104
88. Stamataki Z, Coates S, Evans MJ, Wininger M, Crawford K, Dong C et al (2007) Hepatitis C virus envelope glycoprotein immunization of rodents elicits cross-reactive neutralizing antibodies. Vaccine 25:7773–7784
89. Naarding MA, Falkowska E, Xiao H, Dragic T (2011) Hepatitis C virus soluble E2 in combination with QuilA and CpG ODN induces neutralizing antibodies in mice. Vaccine 29:2910–2917
90. Whidby J, Mateu G, Scarborough H, Demeler B, Grakoui A, Marcotrigiano J (2009) Blocking hepatitis C virus infection with recombinant form of envelope protein 2 ectodomain. J Virol 83:11078–11089
91. Frey SE, Houghton M, Coates S, Abrignani S, Chien D, Rosa D et al (2010) Safety and immunogenicity of HCV E1E2 vaccine adjuvanted with MF59 administered to healthy adults. Vaccine 28:6367–6373
92. Ray R, Meyer K, Banerjee A, Basu A, Coates S, Abrignani S et al (2010) Characterization of antibodies induced by vaccination with hepatitis C virus envelope glycoproteins. J Infect Dis 202:862–866
93. Law JL, Chen C, Wong J, Hockman D, Santer DM, Frey SE et al (2013) A hepatitis C virus (HCV) vaccine comprising envelope glycoproteins gpE1/gpE2 derived from a single isolate elicits broad cross-genotype neutralizing antibodies in humans. PLoS One 8: e59776
94. Meyer K, Banerjee A, Frey SE, Belshe RB, Ray R (2011) A weak neutralizing antibody response to hepatitis C virus envelope

glycoprotein enhances virus infection. PLoS One 6:e23699

95. Law M, Maruyama T, Lewis J, Giang E, Tarr AW, Stamataki Z et al (2008) Broadly neutralizing antibodies protect against hepatitis C virus quasispecies challenge. Nat Med 14:25–27
96. Clayton RF, Owsianka A, Aitken J, Graham S, Bhella D, Patel AH (2002) Analysis of antigenicity and topology of E2 glycoprotein present on recombinant hepatitis C virus-like particles. J Virol 76:7672–7682
97. Desombere I, Fafi-Kremer S, Van Houtte F, Pessaux P, Farhoudi A, Heydmann L et al (2016) Monoclonal anti-envelope antibody AP33 protects humanized mice against a patient-derived hepatitis C virus challenge. Hepatology 63:1120–1134
98. Ruwona TB, Giang E, Nieusma T, Law M (2014) Fine mapping of murine antibody responses to immunization with a novel soluble form of hepatitis C virus envelope glycoprotein complex. J Virol 88:10459–10471
99. Balazs AB, Chen J, Hong CM, Rao DS, Yang L, Baltimore D (2011) Antibody-based protection against HIV infection by vectored immunoprophylaxis. Nature 481:81–84
100. Balazs AB, Bloom JD, Hong CM, Rao DS, Baltimore D (2013) Broad protection against influenza infection by vectored immunoprophylaxis in mice. Nat Biotechnol 31:647–652
101. Balazs AB, Ouyang Y, Hong CM, Chen J, Nguyen SM, Rao DS et al (2014) Vectored immunoprophylaxis protects humanized mice from mucosal HIV transmission. Nat Med 20:296–300
102. Deal C, Balazs AB, Espinosa DA, Zavala F, Baltimore D, Ketner G (2014) Vectored antibody gene delivery protects against Plasmodium falciparum sporozoite challenge in mice. Proc Natl Acad Sci U S A 111:12528–12532
103. de Jong YP, Dorner M, Mommersteeg MC, Xiao JW, Balazs AB, Robbins JB et al (2014) Broadly neutralizing antibodies abrogate established hepatitis C virus infection. Sci Transl Med 6:254ra129
104. Ji C, Liu Y, Pamulapati C, Bohini S, Fertig G, Schraeml M et al (2015) Prevention of hepatitis C virus infection and spread in human liver chimeric mice by an anti-CD81 monoclonal antibody. Hepatology 61:1136–1144
105. Meuleman P, Hesselgesser J, Paulson M, Vanwolleghem T, Desombere I, Reiser H et al (2008) Anti-CD81 antibodies can prevent a hepatitis C virus infection in vivo. Hepatology 48:1761–1768
106. Fukasawa M, Nagase S, Shirasago Y, Iida M, Yamashita M, Endo K et al (2015) Monoclonal antibodies against extracellular domains of claudin-1 block hepatitis C virus infection in a mouse model. J Virol 89:4866–4879
107. Mailly L, Xiao F, Lupberger J, Wilson GK, Aubert P, Duong FH et al (2015) Clearance of persistent hepatitis C virus infection in humanized mice using a claudin-1-targeting monoclonal antibody. Nat Biotechnol 33:549–554
108. Akazawa D, Moriyama M, Yokokawa H, Omi N, Watanabe N, Date T et al (2013) Neutralizing antibodies induced by cell culture-derived hepatitis C virus protect against infection in mice. Gastroenterology 145 (447–455):e441–e444
109. Vanwolleghem T, Bukh J, Meuleman P, Desombere I, Meunier JC, Alter H et al (2008) Polyclonal immunoglobulins from a chronic hepatitis C virus patient protect human liver-chimeric mice from infection with a homologous hepatitis C virus strain. Hepatology 47:1846–1855

INDEX

A

Adaptive mutations 36, 73, 74, 78–80, 86, 105–107
Antibody
  avidity 365, 428, 429
  detection 165, 351, 353, 358, 359, 411, 421–430
  neutralizing 34, 39, 41, 143, 145, 231, 236, 238, 290, 291, 295, 382, 395–418, 421, 422, 433, 441–449, 472, 473, 483, 494, 496
  titration 424, 425
Antigenic domains 396
Antigenic regions (AR) 386, 392, 496
Antiviral drug data 48
Antiviral T cells 454
Apolipoproteins
  ApoB 36, 166, 235, 237
  ApoC1 .... 36, 235, 237
  ApoE 36, 38, 39, 169, 211, 235–238, 240–242, 488

B

Bi-cistronic 35
Bioinformatics analysis 48, 67, 203, 204
Bioreactor 306–308, 311, 313, 314
Biosafety 40, 41, 126, 149, 154, 167, 267, 351–353, 368, 369, 490, 498
BLAST search 48, 56
Boceprevir (BOC) 6, 10, 12

C

CD81 .... 3, 37, 169, 170, 219, 220, 224, 264, 332, 473, 484–487, 494
Cell-cell communication 324
Cell-cell contact 319–333
Cell-cell fusion 219–226, 231, 232
Cell culture model 93, 122, 157–167, 194, 239
Chimeric mouse 468–473, 491, 493
Chromatin immunoprecipitation followed by sequencing (ChIP seq) 193
Claudin-1 (CLDN1) 3, 37, 170, 219, 485–487, 497
Clinical metadata 48, 54, 56
Complement
  cascade 337
  C3/C4 promoter-luciferase plasmid 339, 340
  mediated killing 340
  regulatory protein 338, 341
Core 4, 36, 76, 93, 106, 210, 227, 264, 280, 320, 338, 433, 453
Curated sequence features 48, 65
Cyclic packaging rescue (CPR) 170, 177, 180

D

Daclatasvir (DCV) 11, 15, 16, 21, 163, 189
Dasabuvir (DSV) 7, 14, 18, 22, 23, 195
Deep mutational scanning 183–189
Direct acting antiviral drugs (DAAs) 3–25, 47, 58, 74, 121, 163, 192, 193, 195, 198, 203, 482
Drug resistance 3–25, 51, 63, 74, 184, 189, 306, 471

E

E1 .... 3, 34, 51, 105, 145, 171, 209, 220, 276, 305, 320, 382, 421, 433, 441, 484
E2
  production 94, 494, 497
  purification 107, 422
E1E2
  protein standard 296, 297, 299, 302
  quantification 295–303
  single E1/E2 cDNA template 282, 283, 285
Elbasvir (ELB) 11, 15, 23, 24
Endpoint titration 427, 428
Enzyme-linked immunosorbent assay (ELISA) 211, 323, 327, 339, 342, 358, 384–386, 390, 392, 421–430, 469, 489
Enzyme-linked immunospot (ELISpot) 350–354, 358 360
Epigenetic signature 191–205
Expression cloning 169–181

F

Flow cytometry 113, 114, 128, 177, 323, 324, 329, 331, 332, 350, 352, 355, 357, 358, 364, 365, 368, 375, 379, 400, 414, 417, 418, 454, 456, 490
Foci forming unit (FFU) 82, 160, 162, 167, 187, 195, 244

Mansun Law (ed.), *Hepatitis C Virus Protocols*, Methods in Molecular Biology, vol. 1911,
https://doi.org/10.1007/978-1-4939-8976-8, © Springer Science+Business Media, LLC, part of Springer Nature 2019

Full-length open reading frame (ORF) ...85–90
Fusion inhibitor ...220, 223, 224, 229, 231

G

GBV virus ...483, 484
GenBank ...49, 53, 280, 287, 290, 292, 490
Genotypes (GTs) ...4, 35, 47, 73, 85, 93, 105, 131, 145, 158, 183, 195, 221, 237, 276, 381, 421, 433, 441, 462, 482
Genotyping ...465
Glecaprevir (GLE) ...10, 12, 24, 25
Grazoprevir (GZR) ...10–12, 23, 24
Growth factors ...124, 125, 134, 462

H

HCV cell culture (HCVcc)
 engineering ...236, 395, 472
 neutralization ...39, 396, 494, 497
 production ...35, 36, 40, 118, 158
 purification ...118, 222
HCV core transgenic mouse ...453–457
HCV pseudo-particle (HCVpp)
 HIV Gag-packaging ...442, 443, 449
 MLV Gag-packaging ...443, 444, 449
 neutralization ...39, 298, 441–449
 production ...34, 37, 38, 40, 41, 181, 222, 287, 298, 442, 443
Hepatitis C virus (HCV)
 affinity purification ...107, 109, 115–117
 assembly ...11, 14, 34–38, 41, 74, 107, 158, 170, 209, 210, 235, 236, 264, 295, 344
 chimeric clone ...93–104, 106
 concentration ...78, 113, 115, 212, 229, 231, 265, 291, 338, 350, 351, 357, 445–447, 449, 496, 497
 entry ...3, 33–35, 37–39, 41, 52, 170, 180, 209, 219–233, 236, 263, 264, 273, 276, 290, 291, 295, 298, 306, 441, 469, 485–491, 494, 496, 497
 envelopment ...3, 34, 35, 37–39, 145, 169, 209–216, 236, 264, 295–303, 305–316, 381–392, 396, 421, 422, 469, 489, 492, 494, 496
 fractionation ...109, 115–117
 genome ...14, 33, 35, 36, 40, 41, 49, 58, 59, 74, 78–80, 90, 93, 96, 97, 108, 110, 112, 114, 139–154, 169, 281, 320, 434, 483, 491, 495
 genomic RNA ...35, 110–114, 128, 158, 275
 lineage analysis ...144
 production ...3, 34, 35, 37, 106, 107, 235, 236, 305–316, 344, 350
 purification ...106, 107, 265
 replication ...8, 9, 11, 14, 33, 35, 41, 73–82, 93, 105, 139, 142, 143, 158, 164–166, 169–181, 220, 231, 239, 263, 275, 421, 472, 491, 494
 RNA isolation ...244, 416, 492
 sequencing ...14, 76, 139–154, 471, 483
 titration ...165, 166, 239
 trafficking ...263–273
 treatment ...5–8, 10, 20, 24, 25, 47, 58, 193, 198, 203, 338, 460, 470, 475, 482, 483, 494
Hepatocellular carcinoma (HCC) ...121, 192, 239, 337–339, 345, 453, 459, 475, 482
Hepatocyte differentiation ...125, 127–129
Hepatocyte-like cells ...127–129, 195, 237
Histone-modifying enzymes ...192
Host factors ...11, 38, 123, 169–181, 474, 488, 496, 497
Human embryonic stem cells (hESC) ...122, 237
Human hepatocytes ...40, 195, 237, 461–469, 473, 487, 489
Human hepatoma cell lines
 Hep3B ...442, 445, 446, 448
 Huh7 ...35, 220, 225, 232, 277, 287, 288, 290, 291, 445, 446
 Huh7.5 ...99, 194, 195, 198, 211, 220, 222, 239, 242, 340, 423, 435–437, 491
 Huh7.5.1 ...322, 326, 331
 Huh7-Lunet ...105, 220, 222
Human induced pluripotent stem cell ...122, 127, 132
Human liver chimeric mouse ...459–475, 487–489, 491–494, 496, 497

I

Immunoblot ...128, 211, 214–216, 295–303
Immuno-fluorescent staining ...323
Immunostaining ...198, 248, 251, 252, 257, 322, 328, 329, 332, 434, 436, 437
Inducible Cas9 expression (iCas9) ...128
InFusion cloning ...93–104, 309
Innate immunity ...134, 321
Interferon ...5, 10, 105, 121, 319–333, 340, 470, 482
Internal ribosome entry site (IRES) ...4, 35, 99, 309

J

J6CF ...74–76, 79–81
JFH-1 ...73–76, 93–95, 99, 101, 104, 131, 158, 160, 162, 167, 195, 322, 328
JFH-1 ΔCore-NS2 ...95–97, 99, 103

L

Lectin
Galanthus nivalis 386, 422–424, 429
Ledipasvir (LDV) 5, 6, 8, 11, 14, 15, 17, 18, 20
Lentiviral system 308
Lipophilic dyes 263, 265, 266, 327
Lipo-viro-particles (LVPs) 36, 107, 235
Live cell imaging 256, 263–273
Long RT-PCR 90
Low-density lipoproteins (LDL) 36, 38, 107, 468
Luciferase 34, 106, 128, 221–223, 226, 230, 232, 233, 277, 284, 287, 288, 339, 434, 442, 444–446, 485, 489, 491

M

Membrane fusion 219–233
MHC class I tetramer 454
Mouse models 320, 460–464, 466, 468–475, 481–498
Multiple sequence alignment 48, 51, 56, 61, 63
Mutagenesis 8, 38, 79, 95, 97, 183, 185, 237, 241

N

NS2 .... 3, 36, 51, 93–95, 99, 103, 106, 264, 279, 280, 340, 433–435
NS3 .... 3, 4, 10–14, 16, 22–25, 35, 36, 51, 58, 74, 75, 99, 210, 238, 239, 242–244, 472
NS4A 3, 4, 10, 51, 58, 79, 472
NS5A 3, 4, 11, 13–17, 20, 21, 23–25, 36, 51, 53, 58, 61, 63, 65, 74, 113, 114, 184, 210, 236, 238, 243, 263, 264, 340, 434, 436, 490
NS5A inhibitors 6, 7, 11, 14, 16, 23–25, 58, 61, 67, 189, 195, 263
NS3/4A protease inhibitor 6, 10, 11, 23, 24, 195, 470
NS5B 3, 4, 6, 8, 14, 17, 18, 20–23, 25, 35, 36, 51, 74, 75, 80, 114, 195, 471
NS5B inhibitors 14, 17, 19

O

Occludin (OCLN) 3, 37, 170, 485–487, 494
Ombitasvir (OMV) 7, 11, 15, 16, 22, 23, 195

P

p7 .... 3, 51, 97, 99, 210, 279, 280, 293, 433
Paritaprevir (PTV) 7, 10–12, 16, 22, 23, 195
Persistent infection 157, 158, 166, 349, 453, 483
Persistently infected cell cultures 159, 162, 163
Phage display
epitope masking 382
exhaustive panning 382, 385, 386
Phylogenetic analysis 56, 59, 144, 153, 288
Pibrentasvir (PIB) 11, 15, 23–25
Plasmacytoid dendritic cell (pDC) 319–329, 331–333
pMHC class II tetramer 364, 366
pMHC multimers 352, 364, 365, 367, 374, 454
class I tetramer/pentamer 352, 358–360, 364, 367, 374
class II tetramer 364, 366, 367, 374
Posttranslational histone modifications 191
Proteinase K protection assay 209–216

Q

Quasispecies 8, 9, 13, 140, 143, 275, 494

R

Replication-competent HCV 73–82
Resistance-associated substitutions (RASs) 8, 9, 11–15, 17–20, 22–25, 58, 61, 65, 66
Ritonavir 7, 16, 22, 23, 195
RNA-dependent RNA polymerase (RdRp) 62, 139, 143, 275
RNA-seq 90, 193, 195, 197, 203–205
RNA transfection 76, 78, 215

S

scFv library 395, 410
SCID-uPA mouse 460–466, 468–475
Sequence variations 58, 61, 62, 64, 65, 67
Simeprevir (SMV) 6, 10, 12, 22
Single genome sequencing (SGS) 140, 142, 145, 151–153
Single nucleotide polymorphisms (SNPs) 62, 63, 121, 133
Sofosbuvir (SOF) 5–7, 14, 17, 18, 20–22, 24, 25, 163
SR-B1 .... 3, 173
Subgenomic replicon 35, 36, 73–75, 105, 106, 322, 471
Subtypes 5, 9, 47, 48, 50, 53, 54, 56, 58, 59, 62, 63, 67, 85, 90
Super-resolution microscopy (SRM) 247–259
Surrogate viruses for HCV 483, 484
Sustained virologic response (SVR) 5, 6, 8, 11, 17, 19–25, 192

T

T cell
  CD4 .... 349, 350, 355, 358, 363–379
  CD8 .... 320, 349, 350, 355, 356, 358, 360, 363–379, 453–457
  cell surface staining 355
  FACS analysis 352, 357, 358
  IFN-γ ELISpot 352–354, 358
  intracellular staining 351, 355, 357, 359, 365, 366, 370, 371, 376
  multi-color flow cytometry 350, 364, 365, 375
  proliferation assay 350, 352, 357, 358, 363
293T cells 34, 35, 171, 175–179, 181, 220, 221, 225, 238, 240, 277–287, 289, 296, 298, 308–313, 390, 423, 442, 443, 445
Telaprevir (TVR) 6, 10–12, 471
Tetracysteine tag 263, 271
Transmitted/founder (T/F) virus 141–145

U

uPA-SCID mouse 40, 460–464, 466, 468, 470

V

Velpatasvir (VEL) 5, 7, 11, 15, 20, 21, 24, 25
Very-low density lipoproteins (VLDL) 36, 37, 41, 107, 235, 236, 468
ViPR-HCV portal 48, 50–52, 56, 68
Virion 3, 4, 9, 11, 34, 36–38, 40, 41, 93, 116–118, 139, 140, 152, 158, 160, 166, 169, 209–211, 236, 248, 263, 264, 272, 276, 295, 381, 396, 469
Virus-liposome fusion 222, 225, 227, 228
Virus mutant library 187, 188
Virus neutralization 41, 276, 395–418
Virus Pathogen Resource (ViPR) 49, 52, 54, 61, 67
Virus-plasma membrane fusion 222, 223, 228–230
Virus titration 76, 242, 244, 386
Voxilaprevir (VOX) 10, 12, 24, 25

Y

Yeast display 395–418

FSC
www.fsc.org
MIX
Papier aus verantwortungsvollen Quellen
Paper from responsible sources
FSC® C105338

If you have any concerns about our products,
you can contact us on
ProductSafety@springernature.com

In case Publisher is established outside the EU,
the EU authorized representative is:
Springer Nature Customer Service Center GmbH
Europaplatz 3, 69115 Heidelberg, Germany

Printed by Libri Plureos GmbH
in Hamburg, Germany